*Solutions to Even-Numbered Exercises for*

# Intermediate Algebra

## Third Edition

Larson/Hostetler

# Gerry C. Fitch

Louisiana State University

Houghton Mifflin Company   Boston   New York

# Preface

This *Solutions to Even-Numbered Exercises* is a supplement to *Intermediate Algebra*, Third Edition, by Ron Larson and Robert P. Hostetler. This guide includes solutions for the even-numbered exercises in the text including chapter reviews.

These solutions give step-by-step details of each exercise. The algebraic steps are clearly shown and explanatory comments are included where appropriate. There are usually several "correct" ways to arrive at a solution to a problem in mathematics. Therefore, you should not be concerned if you have approached problems differently than we have.

We have made every effort to see that the solutions are correct. However, we would appreciate hearing about any errors or other suggestions for improvement.

We hope you find the *Solutions to Even-Numbered Exercises* to be a helpful supplement as you use the textbook, and we wish you well in your study of algebra.

Gerry C. Fitch
Louisiana State University
Baton Rouge, Louisiana 70803

# Contents

# CHAPTER P
# Prerequisites: Fundamentals of Algebra

# CHAPTER P
# Prerequisites: Fundamentals of Algebra

## Section P.1    The Real Number System

**Solutions to Even-Numbered Exercises**

2. $\left\{-\dfrac{7}{2}, -\sqrt{6}, -\dfrac{\pi}{2}, -\dfrac{3}{8}, 0, \sqrt{15}, \dfrac{10}{3}, 8, 245\right\}$

    (a) natural numbers: $\{8, 245\}$

    (b) integers: $\{0, 8, 245\}$

    (c) rational numbers: $\left\{-\dfrac{7}{2}, -\dfrac{3}{8}, 0, \dfrac{10}{3}, 8, 245\right\}$

    (d) irrational numbers: $\left\{-\sqrt{6}, -\dfrac{\pi}{2}, \sqrt{15}\right\}$

4. $\left\{-\sqrt{25}, -\sqrt{6}, -0.\overline{1}, -\dfrac{5}{3}, 0, 0.85, 3, 110\right\}$

    (a) natural numbers: $\{3, 110\}$

    (b) integers: $\{-\sqrt{25}, 0, 3, 110\}$

    (c) rational numbers: $\left\{-\sqrt{25}, -0.\overline{1}, \dfrac{-5}{3}, 0, 0.85, 3, 110\right\}$

    (d) irrational numbers: $\left\{-\sqrt{6}\right\}$

6. $\{-2, 0, 2, 4, 6, 8, 10\}$

8. $\{2, 3, 5, 7, 11, 13, 17, 19, 23\}$

10. (a) The point representing the real number 8 lies between 7 and 9.

    (b) The point representing the real number $\frac{4}{3}$ lies between 1 and 2.

    (c) The point representing the real number $-6.75$ lies between $-7$ and $-6$.

    (d) The point representing the real number $-\frac{9}{2}$ lies between $-5$ and $-4$.

12. $a = -\frac{3}{2}, b = \frac{7}{2}$

    $-\frac{3}{2} < \frac{7}{2}$

14. $a = 61.2, b = 65$

    $61.2 < 65$

16. $8 > 3$ because 8 is to the right of 3 on the number line.

18. $3.5 < 8.5$ because 3.5 is to the left of 8.5 on the number line.

20. $-2 > -5$ because $-2$ is to the right of $-5$ on the number line.

22. $-8 < 3$ because $-8$ is to the left of 3 on the number line.

24. $\frac{4}{5} < 1$ because $\frac{4}{5}$ is to the left of 1 on the number line.

26. $-\frac{3}{2} > -\frac{5}{2}$ because $-\frac{3}{2}$ is to the right of $-\frac{5}{2}$ on the number line.

28. $-\frac{5}{3} < -\frac{3}{2}$ because $-\frac{5}{3}$ lies to the left of $-\frac{3}{2}$ on the number line.

30. $-\pi < -3.1$ because $-\pi$ is to the left of $-3.1$ on the number line.

**32.** Distance $= 75 - 20 = 55$

**34.** Distance $= 32 - (-54) = 86$

**36.** Distance $= 14 - (-6) = 14 + 6 = 20$

**38.** Distance $= 125 - 0 = 125$

**40.** Distance $= 0 - (-35) = 0 + 35 = 35$

**42.** Distance $= -7 - (-12) = -7 + 12 = 5$

**44.** $|62| = 62$

**46.** $|-14| = 14$

**48.** $-|-36.5| = -36.5$

**50.** $-|-25| = -25$

**52.** $-\left|\frac{3}{8}\right| = -\frac{3}{8}$

**54.** $|-1.4| = 1.4$

**56.** $-|\pi| = -\pi$

**58.** $|-2| = |2|$ since $2 = 2$

**60.** $|150| < |-310|$ since $|-310| = 310$

**62.** $|12.5| > -|-25|$ since $|12.5| = 12.5$ and $-|-25| = -25$

**64.** $-\left|-\frac{7}{3}\right| < -\left|\frac{1}{3}\right|$ since $-\left|-\frac{7}{3}\right| = -\frac{7}{3}$ and $-\left|\frac{1}{3}\right| = -\frac{1}{3}$

**66.** Opposite: $-225$

   Absolute value: $225$

**68.** Opposite: $52$

   Absolute value: $52$

**70.** Opposite: $-\frac{7}{32}$

   Absolute value:: $\frac{7}{32}$

**72.** Opposite: $-\frac{4}{3}$

   Absolute value: $\frac{4}{3}$

**74.** Opposite: $0.4$

   Absolute value: $0.4$

**76.** Opposite of $-3$ is $3$.

Distance from $0 = 3$

**78.** Opposite of $6$ is $-6$.

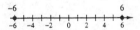

Distance from $0 = 6$

**80.** Opposite of $\frac{7}{4}$ is $-\frac{7}{4}$.

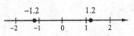

Distance from $0$ is $\frac{7}{4}$.

**82.** Opposite of $-\frac{3}{4}$ is $\frac{3}{4}$.

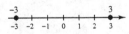

Distance from $0 = \frac{3}{4}$

**84.** Opposite of $3.5$ is $-3.5$.

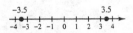

Distance from $0$ is $3.5$.

**86.** Opposite of $-1.2$ is $1.2$.

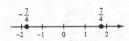

Distance from $0 = 1.2$

**88.** $y > 25$

**90.** $u \geq 16$

**92.** $35 \geq p \geq 30$ or $30 \leq p \leq 35$

**94.** $A > 10,000$

**96.** True.

**98.** False. Let $a = -3$. Then
   $|a| = |-3| = -(-3) = -a = 3$

**100.** False. $\frac{1}{6} = 0.16 \neq 0.17$

**102.** $0.15 = \frac{15}{100}$ where as $0.\overline{15} = 0.151515\ldots = \frac{15}{99}$

**104.** The real number $6$ lies farther from $-4$ than $-8$
   because the distance between $6$ and $-4$ is
   $6 - (-4) = 10$ and the distance between $-8$ and $-4$ is
   $-4 - (-8) = 4$.

## Section P.2    Operations with Real Numbers

**2.** $16 + 84 = 100$

**4.** $-16 + 84 = +(84 - 16)$
$= 68$

**6.** $16 + (-84) = -(84 - 16) = -68$

**8.** $-22 - 6 = -22 + (-6)$
$= -(22 + 6)$
$= -28$

**10.** $-5 + (-52) = -(5 + 52) = -57$

**12.** $-15 + (-6) + 32 = -21 + 32$
$= +(32 - 21)$
$= 11$

**14.** $46.08 - 35.1 - 16.25 = 46.08 + (-35.1) + (-16.25)$
$= +(46.08 - 35.1) + (-16.25)$
$= 10.98 + (-16.25)$
$= -(16.25 - 10.98)$
$= -5.27$

**16.** $\dfrac{5}{6} + \dfrac{7}{6} = \dfrac{5 + 7}{6}$
$= \dfrac{12}{6}$
$= 2$

**18.** $\dfrac{5}{9} - \dfrac{1}{9} = \dfrac{5 - 1}{9}$
$= \dfrac{4}{9}$

**20.** $\dfrac{5}{6} - \dfrac{3}{4} = \dfrac{5(2)}{6(2)} - \dfrac{3(3)}{4(3)}$
$= \dfrac{10}{12} - \dfrac{9}{12}$
$= \dfrac{10 - 9}{12}$
$= \dfrac{1}{12}$

**22.** $\dfrac{3}{11} + -\dfrac{5}{2} = -\left(\dfrac{5}{2} - \dfrac{3}{11}\right)$
$= -\left(\dfrac{5(11)}{2(11)} - \dfrac{3(2)}{11(2)}\right)$
$= \left(\dfrac{55}{22} - \dfrac{6}{22}\right)$
$= -\left(\dfrac{55 - 6}{22}\right)$
$= -\dfrac{49}{22}$

**24.** $8\dfrac{1}{2} - 4\dfrac{2}{3} = \dfrac{17}{2} - \dfrac{14}{3}$
$= \dfrac{17(3)}{2(3)} - \dfrac{14(2)}{3(2)}$
$= \dfrac{51}{6} - \dfrac{28}{6}$
$= \dfrac{51 - 28}{6}$
$= \dfrac{23}{6}$

**26.** $-36 + |-8| = -36 + 8$
$= -(36 - 8)$
$= -28$

**28.** $|-16.25| - 54.78 = 16.25 + (-54.78)$
$= -(54.78 - 16.25)$
$= -38.53$

**30.** $-\left|15\dfrac{2}{3}\right| - 12\dfrac{1}{3} = -15\dfrac{2}{3} + \left(-12\dfrac{1}{3}\right)$
$= -\left(15\dfrac{2}{3} + 12\dfrac{1}{3}\right)$
$= -28$

**32.** $5(2) = 2 + 2 + 2 + 2 + 2$

**34.** $6(-2) = (-2) + (-2) + (-2) + (-2) + (-2) + (-2)$

**36.** $9 + 9 + 9 + 9 = 4(9)$

**38.** $\left(-\dfrac{5}{22}\right) + \left(-\dfrac{5}{22}\right) + \left(-\dfrac{5}{22}\right) = 3\left(-\dfrac{5}{22}\right)$

**40.** $-7(3) = -21$

**42.** $(-4)(-7) = 28$

**44.** $7(10) = 70$

**46.** $\left(\dfrac{10}{13}\right)\left(-\dfrac{3}{5}\right) = -\dfrac{6}{13}$

**48.** $\left(\dfrac{-4}{7}\right)\left(\dfrac{-4}{5}\right) = \dfrac{16}{35}$

**50.** $\dfrac{1}{3}\left(\dfrac{2}{3}\right) = \dfrac{2}{9}$

**52.** $\dfrac{2}{3}\left(-\dfrac{18}{5}\right)\left(-\dfrac{5}{6}\right) = \dfrac{2}{3}(3) = 2$

**54.** $-\dfrac{30}{-15} = 2$

**56.** $-\dfrac{27}{-9} = 3$

**58.** $-72 \div 12 = \dfrac{-72}{12} = -6$

**60.** $\dfrac{8}{15} \div \dfrac{32}{5} = \dfrac{8}{15} \cdot \dfrac{5}{32} = \dfrac{1}{12}$

**62.** $\dfrac{-11}{12} \div \dfrac{5}{24} = \dfrac{-11}{12} \cdot \dfrac{24}{5} = \dfrac{-22}{5}$

**64.** $-3\dfrac{5}{6} \div -2\dfrac{2}{3} = -\dfrac{23}{6} \div \dfrac{-8}{3} = \dfrac{-23}{6} \cdot \dfrac{3}{-8}$
$$= \dfrac{23}{16}$$

**66.** $26\dfrac{2}{3} \div 10\dfrac{5}{6} = \dfrac{80}{3} \div \dfrac{65}{6}$
$$= \dfrac{80}{3} \cdot \dfrac{6}{65}$$
$$= \dfrac{32}{13}$$

**68.** $(-6)^5 = (-6)(-6)(-6)(-6)(-6)$

**70.** $\left(\dfrac{2}{3}\right)^3 = \dfrac{2}{3} \cdot \dfrac{2}{3} \cdot \dfrac{2}{3}$

**72.** $(0.67)^4 = (0.67)(0.67)(0.67)(0.67)$

**74.** $(-4)(-4)(-4)(-4)(-4)(-4) = (-4)^6$

**76.** $\left(\dfrac{5}{8}\right) \times \left(\dfrac{5}{8}\right) \times \left(\dfrac{5}{8}\right) \times \left(\dfrac{5}{8}\right) = \left(\dfrac{5}{8}\right)^4$

**78.** $-(5 \times 5 \times 5 \times 5 \times 5 \times 5) = -5^6$

**80.** $(-3)^2 = (-3)(-3)$
$$= 9$$

**82.** $(-3)^3 = (-3)(-3)(-3)$
$$= -27$$

**84.** $-3^4 = -(3)(3)(3)(3)$
$$= -81$$

**86.** $-\left(\dfrac{2}{3}\right)^4 = -\left(\dfrac{2}{3}\right)\left(\dfrac{2}{3}\right)\left(\dfrac{2}{3}\right)\left(\dfrac{2}{3}\right) = -\dfrac{16}{81}$

**88.** $\left(\dfrac{3}{4}\right)^3 = \left(\dfrac{3}{4}\right)\left(\dfrac{3}{4}\right)\left(\dfrac{3}{4}\right)$
$$= \dfrac{27}{64}$$

**90.** $(0.2)^4 = (0.2)(0.2)(0.2)(0.2)$
$$= 0.0016$$

**92.** $-3(0.8)^2 = -3(0.8)(0.8)$
$$= -3(0.64)$$
$$= -1.92$$

**94.** $18 - 12 + 4 = (18 - 12) + 4$
$$= 6 + 4$$
$$= 10$$

**96.** $18 + 3^2 - 12 = 18 + 9 - 12$
$$= (18 + 9) - 12$$
$$= 27 - 12$$
$$= 15$$

**98.** $6 \cdot 7 - 6^2 \div 4 = 6 \cdot 7 - 36 \div 4$
$$= (6 \cdot 7) - (36 \div 4)$$
$$= 42 - 9$$
$$= 33$$

**100.** $21 - 5(7 - 5) = 21 - 5(2)$
$$= 21 - 10$$
$$= 11$$

**102.** $72 - 8(6^2 \div 9) = 72 - 8(36 \div 9)$
$$= 72 - 8(4)$$
$$= 72 - 32$$
$$= 40$$

**104.** $18 - [4 + (17 - 12)] = 18 - [4 + 5]$
$$= 18 - 9$$
$$= 9$$

**106.** $8 \cdot 3^2 - 4(12 + 3) = 8 \cdot 9 - 4(15)$
$$= 72 - 60$$
$$= 12$$

**108.** $|(-2)^5| - (25 + 7) = |-32| - 32$
$$= 32 - 32$$
$$= 0$$

**110.** $\dfrac{9 + 6(2)}{3 + 4} = \dfrac{9 + 12}{7}$

$\qquad\qquad = \dfrac{21}{7}$

$\qquad\qquad = 3$

**112.** $\dfrac{5^3 - 50}{-15} + 27 = \dfrac{125 - 50}{-15} + 27$

$\qquad\qquad\quad = \dfrac{75}{-15} + 27$

$\qquad\qquad\quad = -5 + 27$

$\qquad\qquad\quad = 22$

**114.** $\dfrac{7^2 - 2(11)}{5^2 + 8(-2)} = \dfrac{49 - 22}{25 - 16}$

$\qquad\qquad\quad = \dfrac{27}{9}$

$\qquad\qquad\quad = 3$

**116.** $35(1032 - 4650) = 35(-3618) = -126{,}630$

**118.** $300(1.09)^{10} \approx 710.2091024 \approx 710.21$

**120.** $5(100 - 3.6^4) \div 4.1 = 5(100 - 167.9616) \div 4.1$

$\qquad\qquad\qquad\qquad = 5(-67.9616) \div 4.1$

$\qquad\qquad\qquad\qquad = -339.808 \div 4.1$

$\qquad\qquad\qquad\qquad = -82.88$

**122.** $\dfrac{1}{7} + \dfrac{1}{6} + \dfrac{1}{5} + x + \dfrac{1}{3} = 1$

Thus:

$x = 1 - \left(\dfrac{1}{7} + \dfrac{1}{6} + \dfrac{1}{5} + \dfrac{1}{3}\right)$

$\quad = 1 - \left(\dfrac{30}{210} + \dfrac{35}{210} + \dfrac{42}{210} + \dfrac{70}{210}\right)$

$\quad = 1 - \left(\dfrac{30 + 35 + 42 + 70}{210}\right)$

$\quad = 1 - \dfrac{177}{210}$

$\quad = \dfrac{210}{210} - \dfrac{177}{210}$

$\quad = \dfrac{33}{210}$

$\quad = \dfrac{11}{70}$    The sum of the parts of the circle equals 1.

**124.** $\$916{,}489.26 - \$1{,}415{,}322.62 = -\$498{,}833.36$

The company had a loss of $498,833.36

**126.**

| Year | Yearly Gain or Loss |
|------|---------------------|
| 1995 | + 0.5 |
| 1996 | − 0.2 |
| 1997 | + 1.3 |
| 1998 | − 0.2 |
| 1999 | + 0.9 |

**128.** (a) $\$75(12)(30) = \$27{,}000$

(b) $75\left[\left(1 + \dfrac{0.08}{12}\right)^{360} - 1\right]\left(1 + \dfrac{12}{0.08}\right) \approx \$112{,}522.1384$

$\qquad\qquad\qquad\qquad\qquad\qquad \approx \$112{,}522.14$

(c) $\$112{,}522.14 - 27{,}000 = \$85{,}522.14$

**130.** $l = 14$cm  $w = 8$cm

$A = lw$

$A = 14 \cdot 8$

$\quad = 112$ square centimeters

**132.** $b = 10$ ft,  $h = 7$ ft

$A = \frac{1}{2}bh$

$A = \frac{1}{2} \cdot 10 \cdot 7$

$\quad = 35$ square feet

**134.** 1 ton $\div$ 50 pounds $=$ 2000 pounds $\div$ 50 pounds $=$ 40.

There are approximately 40 bales in a ton of hay $(14)(18)(42)(40)(12) = 5,080,320$ or $5,080,520 \div 1728 = 2940$. The volume of a stack of baled hay that weighs 12 tons is approximately 5,080,320 cubic inches or 2940 cubic feet.

**136.** The reciprocal of every nonzero integer is an integer.

False. If $n$ is a nonzero integer, then its reciprocal is $\frac{1}{n}$ which is a fraction not an integer.

**138.** If a negative real number is raised to the 12th power, the result will be positive.

True. Any negative real number raised to an even numbered power will be a positive real number. Any negative real number raised to an odd-numbered power will be a negative real number.

**140.** $a \div b = b \div a$

False. Division is not commutative.

**142.** To subtract the real number $b$ from the real number $a$, add the opposite of $b$ to $a$.

**144.** If $a > 0$, then $(-a)^n = -a^n$ when $n$ is odd.

**146.** To add fractions with unlike denominators, first find the lowest common denominator.

**148.** Only common factors (not terms) of the numerator and denominator can be canceled.

# Section P.3    Properties of Real Numbers

**2.** $-5(7) = 7(-5)$

Commutative Property of Multiplication

**4.** $5 + 10 = 5$

Additive Identity Property

**6.** $2(6 \cdot 3) = (2 \cdot 6)3$

Associative Property of Multiplication

**8.** $4 \cdot \frac{1}{4} = 1$

Multiplicative Inverse Property

**10.** $(-4 \cdot 10) \cdot 8 = -4(10 \cdot 8)$

Associative Property of Multiplication

**12.** $(16 + 8) - 5 = 16 + (8 - 5)$

Associative Property of Addition

**14.** $7(9 + 15) = 7 \cdot 9 + 7 \cdot 15$

Distributive Property

**16.** $(5 + 10)(8) = 8(5 + 10)$

Commutative Property of Multiplication

**18.** $10(2x) = (10 \cdot 2)x$

Associative Property of Multiplication

**20.** $10x \cdot \frac{1}{10x} = 1$

Multiplicative Inverse Property

**22.** $2x - 2x = 0$

Additive Inverse Property

**24.** $3(2 + x) = 3 \cdot 2 + 3x$

Distributive Property

**26.** $(x + 1) - (x + 1) = 0$

Additive Inverse Property

**28.** $(6 + x) - m = 6 + (x - m)$

Associative Property of Addition

**30.** $10 + (-6) = -6 + 10$

**32.** $6 + (5 + y) = (6 + 5) + y$

**34.** $(8 - y)(4) = 8(4) - y(4)$

**36.** $13x + (-13x) = 0$

**38.** $(8x) + 0 = 8x$

**40.** (a)   Additive Inverse: $-18$

　　(b)   Multiplicative Inverse: $\frac{1}{18}$

**42.** (a)   Additive Inverse: $52$

　　(b)   Multiplicative Inverse: $-\frac{1}{52}$

**44.** (a)   Additive Inverse: $-2y$

　　(b)   Multiplicative Inverse: $\frac{1}{2y}$

**46.** (a)   Additive Inverse: $-(y - 4)$ or $-y + 4$

　　(b)   Multiplicative Inverse: $\frac{1}{y - 4}$

**48.** $(z + 6) + 10 = z + (6 + 10)$

**50.** $15 + (3 - x) = (15 + 3) - x$

**52.** $(10 \cdot 8) \cdot 5 = 10 \cdot (8 \cdot 5)$

**54.** $8(3x) = (8 \cdot 3)x$

**56.** $-3(4 - 8) = -3(4) - (-3)(8)$

**58.** $6(2x + 5) = 6 \cdot (2x) + 6 \cdot 5$

**60.** $(z - 10)(12) = z \cdot 12 - 10 \cdot 12$

**62.** $-4(10 - b) = -4(10) - (-4)(b)$

**64.** $4(x + 2) \ne 4x + 2$

　　$4(x + 2) = 4x + 8$

**66.** $-9(x + 4) \ne -9x + 36$

　　$-9(x + 4) = -9x - 36$

**68.** $(-1)(-a) + (-a) = (-1)(-a) + (1)(-a)$　　　Multiplicative Identify Property

$\qquad\qquad\qquad = (-1 + 1)(-a)$　　　Distributive Property

$\qquad\qquad\qquad = (0)(-a)$　　　Additive Inverse Property

$\qquad\qquad\qquad = 0$　　　Multiplication Property of Zero

Because $(-1)(-a) + (-a) = 0$ and $a + (-a) = 0$, it follows that $(-1)(-a) + (-a) = a + (-a)$. Using the Cancellation Property of Addition, it follows that $(-1)(-a) = a$.

**70.** 

| | |
|---|---|
| $x - 8 = 20$ | Original Equation |
| $(x - 8) + 8 = 20 + 8$ | Addition Property of Equality |
| $x + (-8 + 8) = 28$ | Associative Property of Addition |
| $x + 0 = 28$ | Additive Inverse Property |
| $x = 28$ | Additive Identity Property |

**72.** 

| | |
|---|---|
| $3x + 4 = 10$ | Original Equation |
| $(3x + 4) + (-4) = 10 + (-4)$ | Addition Property of Equality |
| $3x + [4 + (-4)] = 6$ | Associative Property of Addition |
| $3x + 0 = 6$ | Additive Inverse Property |
| $3x = 6$ | Additive Identity Property |
| $\frac{1}{3}3x = \frac{1}{3}(6)$ | Multiplication Property of Equality |
| $\left(\frac{1}{3} \cdot 3\right)x = 2$ | Associative Property of Multiplication |
| $1 \cdot x = 2$ | Multiplicative Inverse Property |
| $x = 2$ | Multiplicative Identity Property |

**74.** $15\left(1\frac{2}{3}\right) = 15\left(2 - \frac{1}{3}\right)$

$\qquad = 15(2) - 15\left(\frac{1}{3}\right)$

$\qquad = 30 - 5$

$\qquad = 25$

**76.** $5(51) = 5(50 + 1)$

$\qquad = 5(50) + 5(1)$

$\qquad = 250 + 5$

$\qquad = 255$

**78.** $12(19.95) = 12(20 - 0.05)$

$\qquad = 12(20) - 12(0.05)$

$\qquad = 240 - 0.6$

$\qquad = 239.4$

**80.** $a(b - c) = ab - ac$    Distributive Property

**82.** According to the model, the approximate annual increase in the dividend paid per share is $0.08.

**84.** 1996 Dividend per Share $= 0.88(6) + 0.21$

$\qquad\qquad\qquad\qquad = \$0.69$

Approximation $0.72 is a difference of $0.03.

**86.** The additive inverse of a real number $a$ is the number $-a$. The sum of a number and its additive inverse is the additive identity 0. For example, $8 + (-8) = 0$.

**88.** No. Zero does not have a multiplicative inverse.

**90.** To subtract the number $a$ from both sides of an equation, use the Addition Property of Equality to add $(-a)$ to both sides.

## Section P.4    Algebraic Expressions

Solutions to Even Numbered Exercises

**2.** Terms: $-16t^2, 48$

**4.** Terms: $25z^3, -4.8z^2$

**6.** Terms $14u^2, 25uv, -3v^2$

**8.** Terms $\dfrac{3}{t^2}, -\dfrac{4}{t}, 6$

**10.** The coefficient of $4x^6$ is 4.

**12.** The coefficient of $-8.4x$ is $-8.4$.

**14.** $(10 + x) - y = 10 + (x - y)$ illustrates the Associtive Property of Addition.

**16.** $(x - 2)(3) = 3(x - 2)$ illustrates the Commutative Property of Multiplication.

**18.** $5(x + 6) = (x + 6) \cdot 5$

**20.** $6x + 6 = 6 + 6x$

**22.** $-2x^4 = -2 \cdot x \cdot x \cdot x \cdot x$

**24.** $(-2x)^3 = (-2x)(-2x)(-2x)$

**26.** $(-9t)(-9t)(-9t)(-9t)(-9t)(-9t) = (-9t)^6$

**28.** $(y \cdot y \cdot y)(y \cdot y \cdot y \cdot y) = (y^3)(y^4) = y^{3+4} = y^7$

**30.** $-4^2 \cdot 4^5 = -4^{2+5} = -4^7$

**32.** $u^3 \cdot u^5 \cdot u = u^{3+5+1} = u^9$

**34.** $6^2x^3 \cdot x^5 = 36x^{3+5} = 36x^8$

**36.** $(-4x)^3 = (-4)^3(x)^3 = -64x^3$

**38.** $-2(-4x)^3 = -2(-4x)(-4x)(-4x)$

$\qquad\qquad\quad = -2(-64x^3)$

$\qquad\qquad\quad = 128x^3$

**40.** $(-5z^3)^2 = (5z^3)(-5z^3)$

$\qquad\qquad = 25z^{3+3}$

$\qquad\qquad = 25z^6$

**42.** $(-5a^2b^3)(2ab^4 = (-5 \cdot 2) \cdot (a^2 \cdot a) \cdot (b^3 \cdot b^4)$

$\qquad\qquad = -10 \cdot (a^{2+1}) \cdot (b^{3+4})$

$\qquad\qquad = -10a^3b^7$

**44.** $(3y)(2y^2) = 3 \cdot 2 \cdot y^{1+2}$

$\qquad\qquad = 6y^3$

**46.** $(-6n)(-3n^2)^2 = (-6n)(-3n^2)(-3n^2)$

$\qquad = (-6 \cdot -3 \cdot -3) \cdot n^{1+2+2}$

$\qquad = -54n^5$

**48.** $(-2a)^2(-2a)^2 = (-2a)^{2+2}$

$\qquad = (-2a)^4$

$\qquad = 16a^4$

**50.** $(10x^2y)^3(2x^4y) = 10^3 \cdot x^{2 \cdot 3} \cdot y^3(2x^4y)$

$\qquad = 1000x^6y^3(2x^4y)$

$\qquad = 2000x^{6+4}y^{3+1}$

$\qquad = 2000x^{10}y^4$

**52.** $(a^3)^k = a^{3 \cdot k}$

$\qquad = a^{3k}$

**54.** $y^{m-2} \cdot y^2 = y^{(m-2)+2}$

$\qquad = y^m$

**56.** $-2x^2 + 4x^2 = (-2 + 4)x^2 = 2x^2$

**58.** $8y + 7y - y = (8 + 7 - 1)y = 14y$

**60.** $-2a + 4b - 7a - b = (-2a - 7a) + (4b - b)$

$\qquad = -9a + 3b$

**62.** $9y + y^2 - 6y = y^2 + (9y - 6y)$

$\qquad = y^2 + 3y$

**64.** $-5y^3 + 3y - 6y^2 + 8y^3 + y - 4 = (-5y^3 + 8y^3) - 6y^2 + (3y + y) - 4$

$\qquad = 3y^3 - 6y^2 + 4y - 4$

**66.** $7x^2y + 8x^2y^2 + 2xy^2 - (xy)^2 = 7x^2y + (8x^2y^2 - x^2y^2) + 2xy^2$

$\qquad = 7x^2y + 7x^2y^2 + 2xy^2$

**68.** $8(z^3 - 4z^2 + 2) = 8z^3 - 32z^2 + 16$

**70.** $-5(-x^2 + 2y + 1) = (-5)(-x^2) + (-5)(2y) + (-5)(1)$

$\qquad = 5x^2 - 10y - 5$

**72.** $3(x + 1) + x - 6 = 3x + 3 + x - 6$

$\qquad = (3x + x) + (3 - 6)$

$\qquad = 4x - 3$

**74.** $5(a + 6) - 4(2a - 1) = 5a + 30 - 8a + 4$

$\qquad = (5a - 8a) + (30 + 4)$

$\qquad = -3a + 34$

**76.** $x(x^2 - 5) - 4(4 - x) = x^3 - 5x - 16 + 4x$

$\qquad = x^3 + (-5x + 4x) - 16$

$\qquad = x^3 - x - 16$

**78.** $x(x^2 + 3) - 3(x + 4) = x^3 + 3x - 3x - 12$

$\qquad = x^3 + (3x - 3x) - 12$

$\qquad = x^3 + 0 - 12$

$\qquad = x^3 - 12$

**80.** $4[5 - 3(x^2 + 10)] = 4[5 - 3x^2 - 30]$

$\qquad = 4[-3x^2 + (5 - 30)]$

$\qquad = 4[-3x^2 - 25]$

$\qquad = -12x^2 - 100$

**82.** $5y - y[9 + 6(y - 2)] = 5y - y[9 + 6y - 12]$

$\qquad = 5y - y[-3 + 6y]$

$\qquad = 5y + 3y - 6y^2$

$\qquad = 8y - 6y^2$

**84.** $5[3(z + 2) - (z^2 + z - 2)] = 5[3z + 6 - z^2 - z + 2]$

$\qquad = 5[-z^2 + 2z + 8]$

$\qquad = -5z^2 + 10z + 40$

**86.** $5y^3(-3y)^2 - 4(x^2)^2(x - 1) = 5y^3(9y^2) - 4x^4(x - 1)$

$\qquad = 45y^5 - 4x^5 + 4x^4$

**88.** (a)   Substitution: $\frac{3}{2}(6) - 2 = 9 - 2$

$= 7$

Value of expression: 7

(b)   Substitution: $\frac{3}{2}(-3) - 2 = -\frac{9}{2} - 2$

$= -\frac{9}{2} - \frac{4}{2}$

$= -\frac{13}{2}$

Value of expression: $-\frac{13}{2}$

**90.** (a)   Substitution: $2(2)^2 + 5(2) - 3 = 2(4) + 10 - 3$

$= 8 + 10 - 3$

$= 18 - 3$

$= 15$

Value of expression: 15

(b)   Substitution: $2(-3)^2 + 5(-3) - 3 = 2(9) - 15 - 3$

$= 18 - 15 - 3$

$= 0$

Value of expression: 0

**92.** (a)   Evaluation is not possible because division by 0 is undefined.

(b)   Substitution: $5 - \dfrac{3}{-6} = 5 - \left(-\dfrac{1}{2}\right)$

$= 5 + \dfrac{1}{2}$

$= \dfrac{10}{2} + \dfrac{1}{2}$

$= \dfrac{11}{2}$

Value of expression: $\dfrac{11}{2}$

**94.** (a)   Substitution: $6(-2) - 5(-3) = -12 + 15 = 3$

Value of expression: 3

(b)   Substitution: $6(1) - 5(1) = 6 - 5 = 1$

Value of expression: 1

**96.** (a)   Substitution: $\dfrac{0}{0 - 10} = \dfrac{0}{-10} = 0$

Value of expression: 0

(b)   Substitution: $\dfrac{4}{4 - 4} = \dfrac{4}{0} =$ undefined

Value of expression: Undefined

**98.** (a)   Substitution: $|0^2 - (-2)| = |0 + 2| = 2$

Value of expression: 2

(b)   Substitution: $|3^2 - 15| = |9 - 15| = |-6| = 6$

Value of expression: 6

**100.** (a)   Substitution: $(\$5000)(0.085)(10) = \$4250$

Value of expression: $4250

(b)   Substitution: $(\$750)(0.07)(3) = \$157.50$

Value of expression: $157.50

**102.** $A = h\left(\frac{5}{4}h + 10\right)$

$A = 12\left[\frac{5}{4}(12) + 10\right]$

$= 12[15 + 10]$

$= 12[25]$

$= 300$

**104.** $A = \frac{1}{2}(b)\left(\frac{1}{2}b + 1\right)$

$A = \frac{1}{4}b^2 + \frac{1}{2}b$

**106.**

| Year | 1990 | 1991 | 1992 | 1993 | 1994 | 1995 | 1996 |
|---|---|---|---|---|---|---|---|
| Forecast | 1830.89 | 2024.78 | 2218.67 | 2412.56 | 2606.45 | 2800.34 | 2994.23 |

**108.**

| Year | 1990 | 1991 | 1992 | 1993 | 1994 | 1995 | 1996 | 1997 |
|---|---|---|---|---|---|---|---|---|
| Price | 106.0 | 111.9 | 117.8 | 123.7 | 129.6 | 135.5 | 141.4 | 147.3 |

**110.** (a)  Square $n = 4$: $\dfrac{4(4-3)}{2} = \dfrac{4(1)}{2} = 2$ diagonals

Pentagon $n = 5$: $\dfrac{5(5-3)}{2} = \dfrac{5(2)}{2} = 5$ diagonals

Hexagon $n = 6$: $\dfrac{6(6-3)}{2} = \dfrac{6(3)}{2} = 9$ diagonals

(b)  For any natural number $n$, $n(n-3)$ is a product of an even and an odd natural number. Therefore, the product is even and $\dfrac{n(n-3)}{2}$ is a natural number.

**112.** Terms are those parts separated by addition. Factors are separated by multiplication.

**114.** To remove nested symbols of grouping remove the innermost symbols first and combine like terms. A symbol of grouping proceed by a minus sign can be removed by changing the sign of each term within the symbol.

**116.** $[x - (3 \cdot 4)] \div 5$ is equivalent to $\dfrac{x - 3 \cdot 4}{5}$.

# Section P.5    Constructing Algebraic Expressions

Solutions to Even Numbered Exercises

**2.** Five more than a number $n$ is translated into the algebraic expression $5 + n$.

**4.** The total of 25 and three times a number $n$ is translated into the algebraic expression $25 + 3n$.

**6.** Fifteen decreased by three times a number $n$ is translated into the algebraic expression $15 - 3n$.

**8.** The product of a number $y$ and 10 is decreased by 35 is translated into the algebraic expression $10y - 35$.

**10.** Seven-fifths of a number $n$ is translated into the algebraic expression $\dfrac{7}{5}n$.

**12.** The ratio of $y$ to 3 is translated into the algebraic expression $\dfrac{y}{3}$.

**14.** Twenty times the ratio of $x$ and 9 is translated into the algebraic expression $20 \cdot \dfrac{x}{9}$.

**16.** The number $u$ is tripled and the product is increased by 250 is translated into the algebraic expression $3u + 250$.

**18.** Forty percent of the cost $C$ is translated into the algebraic expression $0.40C$.

**20.** The sum of 3 and four times a number $x$, divided by 8 is translated into the algebraic expression $\dfrac{3 + 4x}{8}$.

22. The absolute value of the quotient of a number and 4 is translated into the algebraic expression $\left| \dfrac{y}{4} \right|$.

24. The sum of 10 and one-fourth the square of a number is translated into the algebraic expression $10 + \frac{1}{4}x^2$.

26. A verbal description of $x - 5$ is a number decreased by 5.

28. A verbal description of $2y + 3$ is the sum of twice a number and 3.

30. A verbal description of $4x - 5$ is four times a number decreased by 5.

32. A verbal description of $\dfrac{y}{8}$ is the quotient of a number and 8.

34. A verbal description of $\frac{2}{3}t$ is two-thirds of a number.

36. A verbal description of $-3(x + 2)$ is the sum of a number and 2 is multiplied by $-3$.

38. A verbal description of $\dfrac{x - 2}{3}$ is the difference of a number and 2 divided by 3.

40. A verbal description of $x^2 + 2$ is the square of a number increased by 2.

42. Verbal Description:    The amount of money (in dollars) represented by $x$ nickels.

Label:    $x$ = number of nickels

Algebraic Description:    $0.05x$

44. Verbal Description:    The amount of money (in cents) represented by $x$ nickels and $y$ quarters.

Labels:    $x$ = number of nickels
$y$ = number of quarters

Algebraic Description:    $5x + 25y$

46. Verbal Description:    The amount of money (in cents) represented by $m$ dimes and $n$ quarters.

Labels:    $m$ = number of dimes
$n$ = number of quarters

Algebraic Description:    $10m + 25n$

48.    $r$ = average speed

$5r$ = distance traveled

50. $t$ = number of hours traveled

$\dfrac{360}{t}$ = average speed

52.    $q$ = number of quarts of a food product

$0.65q$ = amount of food

54.    $L$ = value of the purchase

$0.06L$ = sales tax

56.    $n$ = number of children

$18 + 3n$ = total cost for the family

58.    $q$ = number of units produced per hour

$11.65 + 0.80q$ = total hourly wage

60. Verbal Description:    The sum of three consecutive integers, the first of which is $n$.

Labels:    $n$ = first integer
$n + 1$ = second integer
$n + 2$ = third integer

Algebraic Description:    $n + (n + 1) + (n + 2) = 3n + 3$

**62.** Verbal Description:    The sum of two consecutive even integers, the first of which is $2n$.

Labels:    $2n =$ first even integer
$2n + 2 =$ second even integer

Algebraic Description:    $2n + (2n + 2) = 2n + 2n + 2 = 4n + 2$

**64.** Verbal Description:    The difference of two consecutive integers, divided by 2.

Labels:    $n =$ integer
$n + 1 =$ second integer

Algebraic Description:    $\dfrac{(n + 1) - n}{2} = \dfrac{1}{2}$

**66.** Area = length $\cdot$ width

$= 2x(5x - 3) = 10x^2 - 6x$

**68.** Area $= \frac{1}{2} \cdot$ base $\cdot$ height

$= \frac{1}{2}\left(\frac{4}{5}h + 12\right) \cdot h$

$= \frac{2}{5}h^2 + 6h$

**70.** Perimeter $= 2(0.62l) + 2l$

$= 3.24l$

Area $= (0.62l)(l)$

$= 0.62l^2$

**72.** Perimeter $= (x + 2) + 5 + (x + 2) + 1 + x + 3 + x + 1$

$= (x + x + x + x) + (2 + 5 + 2 + 1 + 3 + 1)$

$= 4x + 14$

Area $= 1 \cdot (x + 2) + 3(2) + 1 \cdot (x + 2)$

$= x + 2 + 6 + x + 2$

$= (x + x) + (2 + 6 + 2)$

$= 2x + 10$

**74.** Area $= 6w \cdot w = 6w^2$

The unit of measure is square feet.

**76.**

| $n$ | 0 | 1 | 2 | 3 | 4 | 5 |
|---|---|---|---|---|---|---|
| $3n+1$ | 1 | 4 | 7 | 10 | 13 | 16 |
| Differences | | 3 | 3 | 3 | 3 | 3 |

The differences are always 3.

**78.** The differences in entries in the second row are all 4.

Thus, $a = 4$.    $4(0) + 3 = 3$    Thus, $b = 3$.

**80.** The word ratio indicates the operation of division.

**82.** (a)    No. Multiplication is commutative. $5y = y(5)$

(b)    Yes. Subtraction is not commutative. $5 - y \neq y - 5$

(c)    Yes. Division is not commutative. $\dfrac{y}{5} \neq \dfrac{5}{y}$

(d)    No. Addition is commutative. $5 + y = y + 5$

**84.** If $n$ is an integer, then $2n - 1$ and $2n + 1$ are consecutive odd integers.

Example: if $n = 4$

$2n - 1 = 7$

$2n + 1 = 9$

# Review Exercises for Chapter P

**2.** $-2 > -8$

**4.** $8.4 > -3.2$

**6.** $d = |-7 - 4|$
$\phantom{d} = |-11|$
$\phantom{d} = 11$

**8.** $d = |-8.4 - (-0.3)|$
$\phantom{d} = |-8.4 + 0.3|$
$\phantom{d} = |-8.1|$
$\phantom{d} = 8.1$

**10.** $|6| = 6$

**12.** $|-3.6| = 3.6$

**14.** $-12 + 3 = -(12 - 3) = -9$

**16.** $-154 + 86 - 240 = -(154 - 86) - 240$
$\phantom{-154 + 86 - 240} = -68 - 240$
$\phantom{-154 + 86 - 240} = -(68 + 240)$
$\phantom{-154 + 86 - 240} = -308$

**18.** $14.35 - 10.3 = 4.05$

**20.** $\dfrac{21}{16} - \dfrac{13}{16} = \dfrac{21 - 13}{16} = \dfrac{8}{16} = \dfrac{8}{8(2)} = \dfrac{1}{2}$

**22.** $\dfrac{21}{32} + \dfrac{11}{24} = \dfrac{21(3)}{32(3)} + \dfrac{11(4)}{24(4)}$
$\phantom{\dfrac{21}{32} + \dfrac{11}{24}} = \dfrac{63}{96} + \dfrac{44}{96}$
$\phantom{\dfrac{21}{32} + \dfrac{11}{24}} = \dfrac{107}{96}$

**24.** $-2\dfrac{9}{10} + 5\dfrac{3}{20} = -\dfrac{29}{10} + \dfrac{103}{20}$
$\phantom{-2\dfrac{9}{10} + 5\dfrac{3}{20}} = -\dfrac{29(2)}{10(2)} + \dfrac{103}{20}$
$\phantom{-2\dfrac{9}{10} + 5\dfrac{3}{20}} = -\dfrac{58}{20} + \dfrac{103}{20}$
$\phantom{-2\dfrac{9}{10} + 5\dfrac{3}{20}} = \dfrac{-58 + 103}{20}$
$\phantom{-2\dfrac{9}{10} + 5\dfrac{3}{20}} = \dfrac{45}{20}$
$\phantom{-2\dfrac{9}{10} + 5\dfrac{3}{20}} = \dfrac{5(9)}{5(4)}$
$\phantom{-2\dfrac{9}{10} + 5\dfrac{3}{20}} = \dfrac{9}{4}$

**26.** $(-8)(-3) = 24$

**28.** $(-16)(-15)(-4) = -960$

**30.** $\dfrac{5}{21} \cdot \dfrac{21}{5} = \dfrac{5(21)}{21(5)} = 1$

**32.** $\dfrac{85}{0} =$ undefined
Division by zero is undefined.

**34.** $-\dfrac{2}{3} \div \dfrac{4}{15} = -\dfrac{2}{3} \cdot \dfrac{15}{4}$
$\phantom{-\dfrac{2}{3} \div \dfrac{4}{15}} = -\dfrac{2(3)(5)}{3(2)(2)}$
$\phantom{-\dfrac{2}{3} \div \dfrac{4}{15}} = -\dfrac{5}{2}$

**36.** $-(-3)^4 = -81$

**38.** $2^5 = 32$

**40.** $\left(-\frac{1}{3}\right)^3 = -\frac{1}{27}$

**42.** $45 - 45 \div 3^2 = 45 - 45 \div 9$

$\qquad\qquad = 45 - 5$

$\qquad\qquad = 40$

**44.** $2 - [10 + 6(1 - 3)^2] = 16 - [10 + 6(-2)^2]$

$\qquad\qquad\qquad\qquad\quad = 16 - [10 + 6(4)]$

$\qquad\qquad\qquad\qquad\quad = 16 - [10 + 24]$

$\qquad\qquad\qquad\qquad\quad = 16 - 34$

$\qquad\qquad\qquad\qquad\quad = -18$

**46.** Multiplicative Inverse Property

**48.** Commutative Property of Multiplication

**50.** Associative Property of Multiplication

**52.** Additive Identity Property

**54.** Associative Property of Multiplication

**56.** $-5(2x - 4y) = -10x + 20y$

**58.** $x(3x + 4y) = 3x^2 + 4xy$

**60.** $6x^2 \cdot x^5 = 6x^{2+5}$

$\qquad\qquad = 6x^7$

**62.** $(12x^2y)(3x^2y^4) = 12 \cdot 3 \cdot x^{2+2} \cdot y^{1+4}$

$\qquad\qquad\qquad\quad = 36x^4y^5$

**64.** $3uv(-2uv^2)^2 = 3uv(-2)^2(u)^2(v^2)^2$

$\qquad\qquad\qquad = 3uv(4u^2v^{2\cdot2})$

$\qquad\qquad\qquad = 3uv(4u^2v^4)$

$\qquad\qquad\qquad = 12u^{1+2}v^{1+4}$

$\qquad\qquad\qquad = 12u^3v^5$

**66.** $25y + 32y = (25 + 32)y$

$\qquad\qquad = 57y$

**68.** $7r - 4 - 9 + 3r = (7 + 3)r + (-4 - 9)$

$\qquad\qquad\qquad\quad = 10r - 13$

**70.** $15 - 7(z + 2) = 15 - 7z - 14 = -7z + 1$

**72.** $30x - (10x + 80) = 30x - 10x - 80 = 20x - 80$

**74.** $-2t[8 - (6 - t)] + 5t = -2t[8 - 6 + t] + 5t$

$\qquad\qquad\qquad\qquad = -2t[2 + t] + 5t$

$\qquad\qquad\qquad\qquad = -4t - 2t^2 + 5t$

$\qquad\qquad\qquad\qquad = -2t^2 + t$

**76.** (a) $x = 0, y = 3$

Substitute: $\dfrac{0}{3 + 2} = \dfrac{0}{5} = 0$

Value of expression: 0

(b) $x = 5, y = -2$

Substitute: $\dfrac{5}{2 + (-2)} = \dfrac{5}{0} =$ undefined

Value of expression: undefined

**78.** $100 + 15n$

**80.** $\dfrac{|n + 10|}{2}$

**82.** Three less than five times a number

**84.** Negative three times the difference of a number and 10

**86.** *Verbal Description:*    The distance traveled when you travel 8 hours at an average speed of $r$ miles per hour

   *Label:*                        $r$ = average speed

   *Algebraic Description:*    $8r$

**88.** *Verbal Description:*    The sum of three consecutive odd integers the first of which is $2n + 1$

   *Labels:*                      $2n + 1$ = first integer

   $2n + 3$ = second integer

   $2n + 5$ = third integer

   *Algebraic Description:*    $(2n + 1) + (2n + 3) + (2n + 5) = 6n + 9$

**90.** $40.3 - 12.1 = 28.2$

   Thus, the difference in expenditures between television and radio is 28.2 billion dollars.

**92.** Total number of passengers $= 30.8 + 30.5 + 26.6 + 15.2 + 22.7 + 15.2$

   $= 141$ million

**94.** Average number of passengers per plane $= \dfrac{\text{Total number of passengers}}{\text{Number of airplane departures}}$

   $= \dfrac{30.5 \text{ million}}{381,000} = \dfrac{30,500,000}{381,000}$

   $\approx 80$ passengers per plane

**96.** $387 + 12(68) = 387 + 816 = 1203$

   Thus, the total amount you pay is $1203.

# CHAPTER 1
## Linear Equations and Inequalities

# CHAPTER 1
## Linear Equations and Inequalities

### Section 1.1    Linear Equations

**Solutions to Even-Numbered Exercises**

**2.** (a) $\qquad x = -1$

$5(-1) + 9 \overset{?}{=} 4$

$-5 + 9 \overset{?}{=} 4$

$4 = 4$

Yes

(b) $\qquad x = 2$

$5(2) + 9 \overset{?}{=} 4$

$10 + 9 \overset{?}{=} 4$

$19 \ne 4$

No

**4.** (a) $\qquad x = 0$

$10(0) - 3 \overset{?}{=} 7(0)$

$0 - 3 \overset{?}{=} 0$

$-3 \ne 0$

No

(b) $\qquad x = -1$

$10(-1) - 3 \overset{?}{=} 7(-1)$

$-10 - 3 \overset{?}{=} -7$

$-13 \ne -7$

No

**6.** (a) $\qquad x = 2$

$7(2) - 1 \overset{?}{=} 5(2 + 5)$

$14 - 1 \overset{?}{=} 5(7)$

$13 \ne 35$

No

(b) $\qquad x = 13$

$7(13) - 1 \overset{?}{=} 5(13 + 5)$

$91 - 1 \overset{?}{=} 5(18)$

$90 = 90$

Yes

**8.** (a) $\qquad y = -\frac{3}{2}$

$3\left(-\frac{3}{2} + 2\right) \overset{?}{=} -\frac{3}{2} - 5$

$3\left(-\frac{3}{2} + \frac{4}{2}\right) \overset{?}{=} -\frac{3}{2} - \frac{10}{2}$

$3\left(\frac{1}{2}\right) \overset{?}{=} -\frac{13}{2}$

$\frac{3}{2} \ne -\frac{13}{2}$

No

(b) $\qquad y = -5.5$

$3(-5.5 + 2) \overset{?}{=} -5.5 - 5$

$3(-3.5) \overset{?}{=} -10.5$

$-10.5 = -10.5$

Yes

**10.**

| | |
|---|---|
| $2x + 8 = 6x$ | Original equation |
| $2x - 2x + 8 = 6x - 2x$ | Addition Property of Equations |
| $8 = 4x$ | Additive Inverse Property |
| $\left(\frac{1}{4}\right)8 = \left(\frac{1}{4}\right)4x$ | Multiplication Property of Equations |
| $2 = x$ | Multiplicative Inverse Property |

Conditional equation whose solution is 2.

**12.**

| | |
|---|---|
| $\frac{2}{3}x + 4 = \frac{1}{3}x + 12$ | Original equation |
| $\frac{2}{3}x - \frac{1}{3}x + 4 = \frac{1}{3}x - \frac{1}{3}x + 12$ | Addition Property of Equations |
| $\frac{1}{3}x + 4 = 12$ | Additive Inverse Property |
| $\frac{1}{3}x + 4 - 4 = 12 - 4$ | Addition Property of Equations |
| $\frac{1}{3}x = 8$ | Additive Inverse Property |
| $3\left(\frac{1}{3}x\right) = 3(8)$ | Multiplication Property of Equations |
| $x = 24$ | Multiplicative Inverse Property |

Conditional equation whose solution is 24.

**14.** $x^2 + 3 = 8$ is not linear since the variable has an exponent of 2.

**16.** $3(x + 2) = 4x$ is linear since the variable has an exponent of 1 in each term which contains it.

**18.**

| | |
|---|---|
| $7x - 21 = 0$ | Original equation |
| $7x - 21 + 21 = 0 + 21$ | Add 21 to both sides. |
| $7x = 21$ | Combine like terms. |
| $\dfrac{7x}{7} = \dfrac{21}{7}$ | Divide both sides by 7. |
| $x = 3$ | Simplify. |

**20.**

| | |
|---|---|
| $25 - 3x = 10$ | Original equation |
| $25 - 3x + 3x = 10 + 3x$ | Add $3x$ to both sides. |
| $25 = 10 + 3x$ | Combine like terms. |
| $25 - 10 = 10 + 3x - 10$ | Subtract 10 from both sides. |
| $15 = 3x$ | Combine like terms. |
| $\dfrac{15}{3} = \dfrac{3x}{3}$ | Divide both sides by 3. |
| $5 = x$ | Simplify. |

**22.**

$x + 8 = 0$

$x + 8 - 8 = 0 - 8$

$x = -8$

**Check:**

$-8 + 8 \overset{?}{=} 0$

$0 = 0$

**24.**

$-14x = 28$

$\dfrac{-14x}{-14} = \dfrac{28}{-14}$

$x = -2$

**Check:**

$-14(-2) \overset{?}{=} 28$

$28 = 28$

**26.** $0.5t = 7$

$\dfrac{0.5t}{0.5} = \dfrac{7}{0.5}$

$t = 14$

**Check:**

$0.5(14) \overset{?}{=} 7$

$7 = 7$

**28.**

$8z - 10 = 0$

$8z - 10 + 10 = 0 + 10$

$8z = 10$

$\dfrac{8z}{8} = \dfrac{10}{8}$

$z = \dfrac{5}{4}$

**Check:**

$8\left(\dfrac{5}{4}\right) - 10 \overset{?}{=} 0$

$\dfrac{40}{4} - 10 \overset{?}{=} 0$

$10 - 10 \overset{?}{=} 0$

$0 = 0$

**30.**

$3 - 2y = 5$

$3 - 2y + 2y = 5 + 2y$

$3 = 5 + 2y$

$3 - 5 = 5 + 2y - 5$

$-2 = 2y$

$\dfrac{-2}{2} = \dfrac{2y}{2}$

$-1 = y$

**Check:**

$3 - 2(-1) \overset{?}{=} 5$

$3 + 2 \overset{?}{=} 5$

$5 = 5$

**32.**

$5y + 9 = -6$

$5y + 9 - 9 = -6 - 9$

$5y = -15$

$\dfrac{5y}{5} = \dfrac{-15}{3}$

$y = -3$

**Check:**

$5(-3) + 9 \overset{?}{=} -6$

$-15 + 9 \overset{?}{=} -6$

$-6 = -6$

**34.**

$15x - 18 = 27$

$15x - 18 + 18 = 27 + 18$

$15x = 45$

$\dfrac{15x}{15} = \dfrac{45}{15}$

$x = 3$

**Check:**

$15(3) - 18 \overset{?}{=} 27$

$45 - 18 \overset{?}{=} 27$

$27 = 27$

**36.**

$10 - 6x = -5$

$10 - 10 - 6x = -5 - 10$

$-6x = -15$

$\dfrac{-6x}{-6} = \dfrac{-15}{-6}$

$x = \dfrac{5}{2}$

**Check:**

$10 - 6\left(\dfrac{5}{2}\right) \overset{?}{=} -5$

$10 - 15 \overset{?}{=} -5$

$-5 = -5$

**38.**
$$3y + 14 = y + 20$$
$$3y + 14 - y = y + 20 - y$$
$$2y + 14 = 20$$
$$2y + 14 - 14 = 20 - 14$$
$$2y = 6$$
$$\frac{2y}{2} = \frac{6}{2}$$
$$y = 3$$

**Check:**
$$3(3) + 14 \overset{?}{=} 3 + 20$$
$$9 + 14 \overset{?}{=} 3 + 20$$
$$23 = 23$$

**40.**
$$8 - 7y = 5y - 4$$
$$8 - 7y - 5y = 5y - 5y - 4$$
$$8 - 12y = -4$$
$$8 - 8 - 12y = -4 - 8$$
$$-12y = -12$$
$$\frac{-12y}{-12} = \frac{-12}{-12}$$
$$y = 1$$

**Check:**
$$8 - 7(1) \overset{?}{=} 5(1) - 4$$
$$1 = 1$$

**42.**
$$2s - 16 = 34s$$
$$2s - 2s - 16 = 34s - 2s$$
$$-16 = 32s$$
$$\frac{-16}{32} = \frac{32s}{32}$$
$$-\frac{1}{2} = s$$

**Check:**
$$2\left(-\frac{1}{2}\right) - 16 \overset{?}{=} 34\left(-\frac{1}{2}\right)$$
$$1 - 16 \overset{?}{=} -17$$
$$-17 = -17$$

**44.**
$$24 - 2x = x$$
$$24 - 2x + 2x = x + 2x$$
$$24 = 3x$$
$$\frac{24}{3} = \frac{3x}{3}$$
$$8 = x$$

**Check:**
$$24 - 2(8) \overset{?}{=} 8$$
$$24 - 16 \overset{?}{=} 8$$
$$8 = 8$$

**46.**
$$4x = -12x$$
$$4x + 12x = -12x + 12x$$
$$16x = 0$$
$$\frac{16x}{16} = \frac{0}{16}$$
$$x = 0$$

**Check:**
$$4(0) \overset{?}{=} -12(0)$$
$$0 = 0$$

**48.**
$$6a + 2 = 6a$$
$$6a + 2 - 6a = 6a - 6a$$
$$2 = 0$$
No solution, $2 \neq 0$.

**50.**
$$6(x + 2) = 30$$
$$\frac{6(x + 2)}{6} = \frac{30}{6}$$
$$x + 2 = 5$$
$$x + 2 - 2 = 5 - 2$$
$$x = 3$$

**Check:**
$$6(3 + 2) \overset{?}{=} 30$$
$$6(5) \overset{?}{=} 30$$
$$30 = 30$$

**52.**
$$8(z - 8) = 0$$
$$\frac{8(z - 8)}{8} = \frac{0}{8}$$
$$z - 8 = 0$$
$$z - 8 + 8 = 0 + 8$$
$$z = 8$$

**Check:**
$$8(8 - 8) \overset{?}{=} 0$$
$$8(0) \overset{?}{=} 0$$
$$0 = 0$$

**54.**
$$-2(t + 3) = 9 - 5t$$
$$-2t - 6 = 9 - 5t$$
$$-2t + 5t - 6 = 9 - 5t + 5t$$
$$3t - 6 = 9$$
$$3t - 6 + 6 = 9 + 6$$
$$3t = 15$$
$$\frac{3t}{3} = \frac{15}{3}$$
$$t = 5$$

**Check:**
$$-2(5 + 3) \overset{?}{=} 9 - 5(5)$$
$$-2(8) \overset{?}{=} 9 - 25$$
$$-16 = -16$$

**56.**
$$12 = 6(y + 1) - 8y$$
$$12 = 6y + 6 - 8y$$
$$12 = -2y + 6$$
$$12 - 6 = -2y + 6 - 6$$
$$6 = -2y$$
$$\frac{6}{-2} = \frac{-2y}{-2}$$
$$-3 = y$$

**Check:**
$$12 \overset{?}{=} 6(-3 + 1) - 8(-3)$$
$$12 \overset{?}{=} 6(-2) + 24$$
$$12 \overset{?}{=} -12 + 24$$
$$12 = 12$$

**58.** $26 - (3x - 10) = 6$      **Check:**

$$26 - 3x + 10 = 6$$

$$36 - 3x = 6$$

$$36 - 3x + 3x = 6 + 3x$$

$$36 = 6 + 3x$$

$$36 - 6 = 6 + 3x - 6$$

$$30 = 3x$$

$$\frac{30}{3} = \frac{3x}{3}$$

$$10 = x$$

**Check:**

$$26 - (3 \cdot 10 - 10) \overset{?}{=} 6$$

$$26 - (30 - 10) \overset{?}{=} 6$$

$$26 - 20 \overset{?}{=} 6$$

$$6 = 6$$

**60.** $-5(x - 10) = 6(x - 10)$

$$-5x + 50 = 6x - 60$$

$$-5x - 6x + 50 = 6x - 6x - 60$$

$$-11x + 50 = -60$$

$$-11x + 50 - 50 = -60 - 50$$

$$-11x = -110$$

$$x = 10$$

**Check:**

$$-5(10 - 10) \overset{?}{=} 6(10 - 10)$$

$$-5(0) \overset{?}{=} 6(0)$$

$$0 = 0$$

**62.** $4(2 - x) = -2(x + 7) - 2x$

$$8 - 4x = -2x - 14 - 2x$$

$$8 - 4x = -4x - 14$$

$$8 - 4x + 4x = -4x + 4x - 14$$

$$8 = -14$$

No solution, $8 \neq -14$.

**64.** $-\dfrac{z}{2} = 7$

$$(-2)\left(-\frac{z}{2}\right) = (-2)7$$

$$z = -14$$

**Check:**

$$-\frac{(-14)}{2} \overset{?}{=} 7$$

$$\frac{14}{2} \overset{?}{=} 7$$

$$7 = 7$$

**66.** $z + \dfrac{1}{15} = -\dfrac{3}{10}$

$$z + \frac{1}{15} - \frac{1}{15} = -\frac{3}{10} - \frac{1}{15}$$

$$z = -\frac{9}{30} - \frac{2}{20}$$

$$z = -\frac{11}{30}$$

**Check:**

$$\left(-\frac{11}{30}\right) + \frac{1}{15} \overset{?}{=} -\frac{3}{10}$$

$$-\frac{11}{30} + \frac{2}{30} \overset{?}{=} -\frac{3}{10}$$

$$-\frac{9}{30} \overset{?}{=} -\frac{3}{10}$$

$$-\frac{3}{10} = -\frac{3}{10}$$

**68.** $\dfrac{t}{6} + \dfrac{t}{8} = 1$

$$\frac{4t}{24} + \frac{3t}{24} = 1$$

$$\frac{7t}{24} = 1$$

$$\left(\frac{24}{7}\right)\frac{7t}{24} = \left(\frac{24}{7}\right)1$$

$$t = \frac{24}{7}$$

**Check:**

$$\frac{\frac{24}{7}}{6} + \frac{\frac{24}{7}}{8} \overset{?}{=} 1$$

$$\frac{24}{7} \cdot \frac{1}{6} + \frac{24}{7} \cdot \frac{1}{8} \overset{?}{=} 1$$

$$\frac{4}{7} + \frac{3}{7} \overset{?}{=} 1$$

$$\frac{7}{7} \overset{?}{=} 1$$

$$1 = 1$$

**70.** $\dfrac{11x}{6} + \dfrac{1}{3} = 2x$    **Check:**

$$\dfrac{11x}{6} + \dfrac{1}{3} - \dfrac{1}{3} = 2x - \dfrac{1}{3}$$

$$\dfrac{11(2)}{6} + \dfrac{1}{3} \overset{?}{=} 2(2)$$

$$\dfrac{11x}{6} = 2x - \dfrac{1}{3}$$

$$\dfrac{22}{6} + \dfrac{1}{3} \overset{?}{=} 4$$

$$\dfrac{11}{6}x - 2x = 2x - 2x - \dfrac{1}{3}$$

$$\dfrac{11}{3} + \dfrac{1}{3} \overset{?}{=} 4$$

$$\dfrac{11}{6}x - \dfrac{12}{6}x = -\dfrac{1}{3}$$

$$\dfrac{12}{3} \overset{?}{=} 4$$

$$-\dfrac{1}{6}x = -\dfrac{1}{3}$$

$$4 = 4$$

$$-6\left(-\dfrac{1}{6}x\right) = \left(-\dfrac{1}{3}\right)(-6)$$

$$x = 2$$

**72.** $\dfrac{1}{9}x + \dfrac{1}{3} = \dfrac{11}{18}$    **Check:**

$$\dfrac{1}{9}x + \dfrac{1}{3} - \dfrac{1}{3} = \dfrac{11}{18} - \dfrac{1}{3}$$

$$\dfrac{1}{9}\left(\dfrac{5}{2}\right) + \dfrac{1}{3} \overset{?}{=} \dfrac{11}{18}$$

$$\dfrac{1}{9}x = \dfrac{11}{18} - \dfrac{6}{18}$$

$$\dfrac{5}{18} + \dfrac{1}{3} \overset{?}{=} \dfrac{11}{18}$$

$$\dfrac{1}{9}x = \dfrac{5}{18}$$

$$\dfrac{5}{18} + \dfrac{6}{18} \overset{?}{=} \dfrac{11}{18}$$

$$9\left(\dfrac{1}{9}x\right) = 9\left(\dfrac{5}{18}\right)$$

$$\dfrac{11}{18} = \dfrac{11}{18}$$

$$x = \dfrac{5}{2}$$

**74.** $\dfrac{8 - 3x}{4} - 4 = \dfrac{x}{6}$    **Check:**

$$12\left(\dfrac{8-3x}{4} - 4\right) = \left(\dfrac{x}{6}\right)12$$

$$\dfrac{8 - 3\left(-\frac{24}{11}\right)}{4} - 4 \overset{?}{=} \dfrac{-\frac{24}{11}}{6}$$

$$3(8 - 3x) - 48 = 2x$$

$$\dfrac{8 + \frac{72}{11}}{4} - 4 \overset{?}{=} -\dfrac{24}{11} \cdot \dfrac{1}{6}$$

$$24 - 9x - 48 = 2x$$

$$\dfrac{\frac{88}{11} + \frac{72}{11}}{4} - 4 \overset{?}{=} -\dfrac{4}{11}$$

$$-9x - 24 = 2x$$

$$\dfrac{\frac{160}{11}}{4} - 4 \overset{?}{=} -\dfrac{4}{11}$$

$$-9x - 24 + 9x = 2x + 9x$$

$$\dfrac{160}{11} \cdot \dfrac{1}{4} - 4 \overset{?}{=} -\dfrac{4}{11}$$

$$-24 = 11x$$

$$\dfrac{40}{11} - 4 \overset{?}{=} -\dfrac{4}{11}$$

$$\dfrac{-24}{11} = \dfrac{11x}{11}$$

$$\dfrac{40}{11} - \dfrac{44}{11} \overset{?}{=} -\dfrac{4}{11}$$

$$-\dfrac{24}{11} = x$$

$$-\dfrac{4}{11} = -\dfrac{4}{11}$$

**76.** $16.3 - 0.2x = 7.1$    **Check:**

$$10(16.3 - 0.2x) = 10(7.1)$$

$$16.3 - 0.2(46) \overset{?}{=} 7.1$$

$$163 - 2x = 71$$

$$16.3 - 9.2 \overset{?}{=} 7.1$$

$$163 - 2x - 163 = 71 - 163$$

$$7.1 = 7.1$$

$$-2x = -92$$

$$\dfrac{-2x}{-2} = \dfrac{-92}{-2}$$

$$x = 46$$

**78.** $6.5(1 - 2x) = 13$    **Check:**

$$6.5 - 13.0x = 13$$

$$6.5[1 - 2(-0.5)] \overset{?}{=} 13$$

$$6.5 - 6.5 - 13x = 13 - 6.5$$

$$6.5(1 + 1) \overset{?}{=} 13$$

$$-13x = 6.5$$

$$6.5(2) \overset{?}{=} 13$$

$$\dfrac{-13x}{-13} = \dfrac{6.5}{-13}$$

$$13 = 13$$

$$x = -0.5$$

**80.**   $\frac{3}{4}(6 - x) = \frac{1}{3}(4x + 5) + 2$        **Check:**

$$12\left[\frac{3}{4}(6 - x)\right] = \left[\frac{1}{3}(4x + 5) + 2\right]12$$        $\frac{3}{4}\left(6 - \frac{2}{5}\right) \stackrel{?}{=} \frac{1}{3}\left[4\left(\frac{2}{5}\right) + 5\right] + 2$

$$9(6 - x) = 4(4x + 5) + 24$$        $\frac{3}{4}\left(\frac{30}{5} - \frac{2}{5}\right) \stackrel{?}{=} \frac{1}{3}\left(\frac{8}{5} + \frac{25}{5}\right) + 2$

$$54 - 9x = 16x + 20 + 24$$        $\frac{3}{4}\left(\frac{28}{5}\right) \stackrel{?}{=} \frac{1}{3}\left(\frac{33}{5}\right) + 2$

$$54 - 9x = 16x + 44$$

$$54 - 9x - 16x = 16x + 44 - 16x$$        $\frac{21}{5} \stackrel{?}{=} \frac{11}{5} + \frac{10}{5}$

$$54 - 25x = 44$$        $\frac{21}{5} = \frac{21}{5}$

$$54 - 54 - 25x = 44 - 54$$

$$-25x = -10$$

$$\frac{-25x}{-25} = \frac{-10}{-25}$$

$$x = \frac{2}{5}$$

**82.**   *Verbal Model:*   $\boxed{\text{First integer}} + \boxed{\text{Second integer}} = 137$

*Labels:*   $n$ = first integer

$n + 1$ = second integer

*Equation:*   $n + (n + 1) = 137$

$$2n + 1 = 137$$

$$2n + 1 - 1 = 137 - 1$$

$$2n = 136$$

$$\frac{2n}{2} = \frac{136}{2}$$

$$n = 68$$

$$n + 1 = 69$$

**84.**   *Verbal Model:*   $\boxed{\text{First even integer}} + \boxed{\text{Second even integer}} = 626$

*Labels:*   $n$ = first even integer

$n + 2$ = second even integer

*Equation:*   $n + (n + 2) = 626$

$$2n + 2 = 626$$

$$2n + 2 - 2 = 626 - 2$$

$$2n = 624$$

$$\frac{2n}{2} = \frac{624}{2}$$

$$n = 312$$

$$n + 2 = 314$$

**86.**   *Verbal Model:*   $\boxed{\text{Cost of parts}} + \boxed{\text{Service call}} + 16 \cdot \boxed{\text{Number of half hours}} = 172$

*Label:*   $n$ = Number of half hours

*Equation:*   $74 + 50 + 16n = 172$

$$124 + 16n = 172$$

$$124 - 124 + 16n = 172 - 124$$

$$16n = 48$$

$$\frac{16n}{16} = \frac{48}{16}$$

$$n = 3 \text{ half hours} + \text{first half hour}$$

The repair work took 2 hours.

**88.** The object reaches its maximum height when the velocity is zero. The object then returns to the ground.

$$v = 64 - 32t$$

$$0 = 64 - 32t$$

$$0 + 32t = 64 - 32t + 32t$$

$$32t = 64$$

$$\frac{32t}{32} = \frac{64}{32}$$

$$t = 2 \text{ seconds}$$

**90.**

$$\frac{t}{12} + \frac{t}{20} = 1$$

$$60\left(\frac{t}{12} + \frac{t}{20}\right) = (1)60$$

$$5t + 3t = 60$$

$$8t = 60$$

$$\frac{8t}{8} = \frac{60}{8}$$

$$t = \frac{15}{2} = 7\frac{1}{2} \text{ hours}$$

**92.** *Verbal Model:*    $2 \cdot \boxed{\text{Length}} + 2 \cdot \boxed{\text{Width}} = \boxed{\text{Perimeter}}$

*Labels:*   $n = \text{width}$

$n + 10 = \text{length}$

*Equation:*   $2(n + 10) + 2n = 64$

$$2n + 20 + 2n = 64$$

$$4n + 20 = 64$$

$$4n + 20 - 20 = 64 - 20$$

$$4n = 44$$

$$\frac{4n}{4} = \frac{44}{4}$$

$$n = 11 \text{ meters}$$

$$n + 10 = 21 \text{ meters}$$

**94.**

$$57 = 49.3 + 1.93t$$

$$57 - 49.3 = 1.93t$$

$$7.7 = 1.93t$$

$$\frac{7.7}{1.93} = \frac{1.93t}{1.93}$$

$$4 \approx t$$

From the graph, 1994 is the year in which there were 57 million subscribers.

**96.** A real number is a solution of an equation if the equation is true when the real number is substituted into the equation.

**98.** The standard form of a linear equation is $ax + b = c$, $a \neq 0$. A linear equation is called a first-degree equation because the variable has an implied degree of 1.

**100.** Steps used to transform an equation into an equivalent equation:

(a) Simplify each side by removing symbols of grouping, combining like terms, and reducing fractions on one or both sides.

(b) Add (or subtract) the same quantity to (from) both sides of the equation.

(c) Multiply (or divide) both sides of the equation by the same nonzero real number.

(d) Interchange the two sides of the equation.

**102.** True, subtracting zero from both sides of an equation yields an equivalent equation.

# Section 1.2    Linear Equations and Problem Solving

**2.** *Verbal Model:*   $\boxed{\text{Number}} - 18 = 27$

*Label:*   Number $= x$

*Equation:*
$$x - 18 = 27$$
$$x - 18 + 18 = 27 + 18$$
$$x = 45$$

**4.** *Verbal Model:*   $\boxed{\begin{array}{c}\text{Daily}\\\text{earnings}\end{array}} = \boxed{\begin{array}{c}\text{Hourly}\\\text{rate}\end{array}} \cdot \boxed{\text{Hours}} + \boxed{\begin{array}{c}\text{Rate per}\\\text{unit}\end{array}} \cdot \boxed{\begin{array}{c}\text{Number}\\\text{of units}\end{array}}$

*Label:*   Number of units $= x$

*Equation:*
$$146 = (10) \cdot 8 + 0.75x$$
$$146 = 80 + 0.75x$$
$$146 - 80 = 80 + 0.75x - 80$$
$$66 = 0.75x$$
$$88 = x$$

**6.** Percent: 75%

Parts out of 100: 75

Decimal: 0.75

Fraction: $\frac{75}{100} = \frac{3}{4}$

**8.** Percent: $66\frac{2}{3}$%

Parts out of 100: $66\frac{2}{3}$

Decimal: 0.666 . . .

Fraction: $\frac{2}{3}$

**10.** Percent: 100%

Parts out of 100: 100

Decimal: 1.0

Fraction: 1

**12.** *Verbal Model:*   $\boxed{\begin{array}{c}\text{Compared}\\\text{number}\end{array}} = \boxed{\text{Percent}} \cdot \boxed{\begin{array}{c}\text{Base}\\\text{number}\end{array}}$

*Labels:*   Compared number $= a$
Percent $= p$
Base number $= b$

*Equation:*   $a = p \cdot b$
$a = (0.68)(800)$
$a = 544$

**14.** *Verbal Model:*   $\boxed{\begin{array}{c}\text{Compared}\\\text{number}\end{array}} = \boxed{\text{Percent}} \cdot \boxed{\begin{array}{c}\text{Base}\\\text{number}\end{array}}$

*Labels:*   Compared number $= a$
Percent $= p$
Base number $= b$

*Equation:*   $a = p \cdot b$
$a = \left(\frac{1}{3}\right)(816)$
$a = 272$

**16.** *Verbal Model:*   $\boxed{\begin{array}{c}\text{Compared}\\\text{number}\end{array}} = \boxed{\text{Percent}} \cdot \boxed{\begin{array}{c}\text{Base}\\\text{number}\end{array}}$

*Labels:*   Compared number $= a$
Percent $= p$
Base number $= b$

*Equation:*   $a = p \cdot b$
$a = (3.0)(16)$
$a = 48$

**18.** *Verbal Model:*   $\boxed{\begin{array}{c}\text{Compared}\\\text{number}\end{array}} = \boxed{\text{Percent}} \cdot \boxed{\begin{array}{c}\text{Base}\\\text{number}\end{array}}$

*Labels:*   Compared number $= a$
Percent $= p$
Base number $= b$

*Equation:*   $a = p \cdot b$
$416 = (0.65)b$

$$\frac{416}{0.65} = b$$
$$640 = b$$

**20.** *Verbal Model:* $\boxed{\text{Compared number}} = \boxed{\text{Percent}} \cdot \boxed{\text{Base number}}$

*Labels:*  Compared number = $a$

Percent = $p$

Base number = $b$

*Equation:*  $a = p \cdot b$

$168 = (3.5)b$

$\dfrac{168}{3.5} = b$

$48 = b$

**22.** *Verbal Model:* $\boxed{\text{Compared number}} = \boxed{\text{Percent}} \cdot \boxed{\text{Base number}}$

*Labels:*  Compared number = $a$

Percent = $p$

Base number = $b$

*Equation:*  $a = p \cdot b$

$496 = p(800)$

$\dfrac{496}{800} = p$

$0.62 = p$

$p = 62\%$

**24.** *Verbal Model:* $\boxed{\text{Compared number}} = \boxed{\text{Percent}} \cdot \boxed{\text{Base number}}$

*Labels:*  Compared number = $a$

Percent = $p$

Base number = $b$

*Equation:*  $a = p \cdot b$

$2.4 = p(480)$

$\dfrac{2.4}{480} = p$

$0.005 = p$

$p = 0.5\%$

**26.** *Verbal Model:* $\boxed{\text{Compared number}} = \boxed{\text{Percent}} \cdot \boxed{\text{Base number}}$

*Labels:*  Compared number = $a$

Percent = $p$

Base number = $b$

*Equation:*  $a = p \cdot b$

$900 = p(500)$

$\dfrac{900}{500} = p$

$1.8 = p$

$p = 180\%$

**28.** $\dfrac{12 \text{ ounces}}{20 \text{ ounces}} = \dfrac{12}{20} = \dfrac{6}{10} = \dfrac{3}{5}$

**30.** $\dfrac{125 \text{ cm}}{200 \text{ cm}} = \dfrac{5}{8}$

**32.** $\dfrac{1 \text{ pint}}{1 \text{ gallon}} = \dfrac{1 \text{ pint}}{8 \text{ pints}} = \dfrac{1}{8}$

**34.** $\dfrac{45 \text{ minutes}}{2 \text{ hours}} = \dfrac{45 \text{ minutes}}{120 \text{ minutes}} = \dfrac{3}{8}$

**36.** $\dfrac{y}{36} = \dfrac{6}{7}$

$y = \dfrac{6}{7} \cdot 36$

$y = \dfrac{216}{7}$

**38.** $\dfrac{5}{16} = \dfrac{x}{4}$

$4\left(\dfrac{5}{16}\right) = x$

$\dfrac{5}{4} = x$

**40.** $\dfrac{7}{8} = \dfrac{x}{2}$

$2 \cdot \dfrac{7}{8} = x$

$\dfrac{7}{4} = x$

**42.** $\dfrac{a}{5} = \dfrac{a+4}{8}$

$8a = 5(a+4)$

$8a = 5a + 20$

$3a = 20$

$a = \dfrac{20}{3}$

**44.** $\dfrac{z-3}{3} = \dfrac{z+8}{12}$

$12(z-3) = 3(z+8)$

$\dfrac{12(z-3)}{3} = \dfrac{3(z+8)}{3}$

$4(z-3) = z+8$

$4z - 12 = z + 8$

$3z - 12 = 8$

$3z = 20$

$z = \dfrac{20}{3}$

**46.** *Verbal Model:* $\boxed{\text{Amount withheld}} = \boxed{\text{Percent}} \cdot \boxed{\text{Gross monthly income}}$

*Labels:* Amount withheld $= a$
Percent $= p$
Gross monthly income $= b$

*Equation:* $a = p \cdot b$
$a = (0.065)(3800)$
$a = \$247$

**48.** *Verbal Model:* $\boxed{\text{Votes cast}} = \boxed{\text{Percent}} \cdot \boxed{\text{Eligible voters}}$

*Labels:* Votes cast $= a$
Percent $= p$
Eligible voters $= b$

*Equation:* $a = p \cdot b$
$7387 = 0.63b$
$\dfrac{7387}{0.63} = b$
$11{,}725 = b$

**50.** *Verbal Model:* $\boxed{\text{Rent}} = \boxed{\text{Percent}} \cdot \boxed{\text{Monthly income}}$

*Labels:* Rent $= a$
Percent $= p$
Monthly income $= b$

*Equation:* $748 = p \cdot 3400$
$\dfrac{748}{3400} = \dfrac{p \cdot 3400}{3400}$
$0.22 = p$
$22\% = p$

**52.** *Verbal Model:* $\boxed{\text{Tip}} = \boxed{\text{Percent}} \cdot \boxed{\text{Cost of meal}}$

*Labels:* Tip $= a$
Percent $= p$
Cost of meal $= b$

*Equation:* $5.27 = p \cdot 34.73$
$\dfrac{5.27}{34.73} = \dfrac{p \cdot 34.73}{34.73}$
$0.1517 \approx p$
$15.2\% \approx p$

**54.** *Verbal Model:* $\boxed{\text{Commission}} = \boxed{\text{Percent}} \cdot \boxed{\text{Price of home}}$

*Labels:* Commission $= a$
Percent $= p$
Price of home $= b$

*Equation:* $a = p \cdot b$
$20{,}400 = p(240{,}000)$
$\dfrac{20{,}400}{240{,}000} = p$
$0.085 = p$
$p = 8.5\%$

**56.** *Verbal Model:* $\boxed{\text{Price of van}} = \boxed{\text{Percent}} \cdot \boxed{\text{Price 3 years ago}}$

*Labels:* Price of van $= a$
Percent $= p$
Price 3 years ago $= b$

*Equation:* $a = p \cdot b$
$29{,}750 = (1.15)b$
$\dfrac{29{,}750}{1.15} = b$
$\$25{,}870 = b$

**58.** *Verbal Model:* $\boxed{\text{Area covered by rug}} = \boxed{\text{Percent}} \cdot \boxed{\text{Area of floor}}$

*Labels:*    Area covered by rug $= a$

Percent $= p$

Area of floor $= b$

*Equation:*    $a = p \cdot b$

$\pi(4)^2 = p \cdot (10)(12)$

$16\pi = 120p$

$\dfrac{16\pi}{120} = p$

$0.418879 \approx p$

$p = 41.9\%$

**60.** *Verbal Model:* $\boxed{\text{Btu obtained from coal}} = \boxed{\text{Percent}} \cdot \boxed{\text{Total Btu}}$

*Labels:*    Btu obtained from coal $= a$

Percent $= p$

Total Btu $= b$

*Equation:*    $x = (0.217)(90.6)$

$x \approx 19.7$ quadrillion Btu

**62.** Approximate increase from graph is 22 pounds.

*Verbal Model:* $\boxed{\text{Amount of increase}} = \boxed{\text{Percent}} \cdot \boxed{\text{Chicken consumption in 1980}}$

*Labels:*    Amount of increase $= a$

Percent $= p$

Chicken consumption in 1980 $= b$

*Equation:*    $a = p \cdot b$

$22 = p(41)$

$\frac{22}{41} = p$

$0.537 \approx p$

$p \approx 53.7\%$

**64.** *Verbal Model:* $\boxed{\text{Amount of fish}} = \boxed{\text{Percent}} \cdot \boxed{\text{Amount of meat}}$

*Lables:*    Amount of fish $= 14$

Percent $= p$

Amount of meat $= 14 + 62 + 48 + 62$

*Equation:*    $a = p \cdot b$

$14 = p \cdot (14 + 62 + 48 + 62)$

$14 = p(186)$

$\frac{14}{186} = p$

$0.0752 \approx p$

$p \approx 7.5\%$

**66.** $\dfrac{\text{Stock price}}{\text{Stock earning}} = \dfrac{\$56.25}{\$6.25} = \dfrac{9}{1}$

**68.** Gear ratio $= \dfrac{60 \text{ teeth}}{40 \text{ teeth}} = \dfrac{3}{2}$

**70.** $\dfrac{\text{Area of triangle 1}}{\text{Area of triangle 2}} = \dfrac{\frac{1}{2} \cdot 4 \cdot 3}{\frac{1}{2} \cdot 6 \cdot 4}$

$= \dfrac{3}{6} = \dfrac{1}{2}$

**72.** $\dfrac{\text{Total price}}{\text{Total units}} = \dfrac{\$1.29}{64} \approx \$0.0202$ per ounce

**74.** $\dfrac{\text{Total price}}{\text{Total units}} = \dfrac{\$3.49}{18} \approx \$0.1939$ per ounce

**76.** (a) Unit price $= \dfrac{\$1.79}{10.5} \approx \$0.17$ per ounce

(b) Unit price $= \dfrac{\$2.39}{16} \approx \$0.15$ per ounce

The 16-ounce package is a better buy.

**78.** (a) Unit price $= \dfrac{\$3.49}{2} \approx \$1.75$ per pound

(b) Unit price $= \dfrac{\$5.29}{3} \approx \$1.76$ per pound

The 2-pound package is a better buy.

**80.** $\dfrac{2}{5} = \dfrac{3}{x}$

$2x = 15$

$x = \dfrac{15}{2} = 7\dfrac{1}{2}$

**82.** $\dfrac{x}{5} = \dfrac{2}{3}$

$x = \dfrac{2}{3}(5)$

$x = \dfrac{10}{3} = 3\dfrac{1}{3}$

**84.** Proportion:  $\dfrac{l}{6} = \dfrac{l+15}{20}$

$20l = 6l + 90$

$14l = 90$

$l = \dfrac{90}{14}$

$l = \dfrac{45}{7} \approx 6.4$ feet

**86.** *Verbal Model:*  $\boxed{\dfrac{\text{Pounds}}{\text{Inches}}} = \boxed{\dfrac{\text{Pounds}}{\text{Inches}}}$

*Proportion:*  $\dfrac{x}{9} = \dfrac{32}{6}$

$x = 9 \cdot \dfrac{32}{6}$

$x = 48$ pounds

**88.** $\dfrac{3 \text{ cups}}{1 \text{ batch}} = \dfrac{x \text{ cups}}{3\frac{1}{2} \text{ batches}}$

$x = 3 \cdot 3\dfrac{1}{2}$

$x = 3 \cdot \dfrac{7}{2}$

$x = \dfrac{21}{2} = 10\dfrac{1}{2}$ cups

**90.** *Verbal Model:*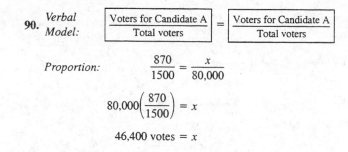

*Proportion:*  $\dfrac{870}{1500} = \dfrac{x}{80,000}$

$80,000\left(\dfrac{870}{1500}\right) = x$

$46,400$ votes $= x$

**92.** To change percents to decimals divide by 100. To change decimals to percents multiply by 100.

*Examples:* $42\% = \frac{42}{100} = 0.42$

$0.38 = (0.38)(100)\% = 38\%$

**94.** The ratio of $a$ to $b$ is $a/b$ if $a$ and $b$ have the same units.

*Examples:* Price earnings ratio, gear ratio

**96.** *Verbal Model:*  $\boxed{\begin{array}{c}\text{Original}\\\text{salary}\end{array}} - \boxed{\text{Reduction}} = \boxed{\begin{array}{c}\text{Reduced}\\\text{salary}\end{array}}$

*Equation:*  $x - 0.07x = 0.93x$

*Verbal Model:*  $\boxed{\text{Reduction}} = \boxed{\text{Percent}} \cdot \boxed{\begin{array}{c}\text{Reduced}\\\text{salary}\end{array}}$

*Equation:*  $a = p \cdot b$

$0.07x = p \cdot 0.93x$

$\dfrac{0.07x}{0.93x} = p$

$7.53\% \approx p$

Base number is smaller.

# Section 1.3    Business and Scientific Problems

**2.** *Verbal Model:* $\boxed{\text{Selling price}} = \boxed{\text{Cost}} + \boxed{\text{Markup}}$

*Labels:* Selling price = 113.67

Cost = 84.20

Markup = $x$

*Equation:* $113.67 = 84.20 + x$

$113.67 - 84.20 = x$

$\$29.47 = x$

*Verbal Model:* $\boxed{\text{Markup}} = \boxed{\text{Markup rate}} \cdot \boxed{\text{Cost}}$

*Labels:* Markup = 29.47

Markup rate = $x$

Cost = 84.20

*Equation:* $29.47 = x \cdot 84.20$

$\dfrac{29.47}{84.20} = x$

$35\% = x$

**4.** *Verbal Model:* $\boxed{\text{Selling price}} = \boxed{\text{Cost}} + \boxed{\text{Markup}}$

*Labels:* Selling price = 603.72

Cost = $x$

Markup = 184.47

*Equation:* $603.72 = x + 184.47$

$603.72 - 184.47 = x$

$\$419.25 = x$

*Verbal Model:* $\boxed{\text{Markup}} = \boxed{\text{Markup rate}} \cdot \boxed{\text{Cost}}$

*Labels:* Markup = 184.47

Markup rate = $x$

Cost = 419.25

*Equation:* $184.47 = x \cdot 419.25$

$\dfrac{184.47}{419.25} = x$

$44\% = x$

**6.** *Verbal Model:* $\boxed{\text{Selling price}} = \boxed{\text{Cost}} + \boxed{\text{Markup}}$

*Labels:* Selling price = 16,440.50

Cost = $x$

Markup = 3890.50

*Equation:* $16,440.50 = x + 3890.50$

$16,440.50 - 3890.50 = x$

$\$12,550.00 = x$

*Verbal Model:* $\boxed{\text{Markup}} = \boxed{\text{Markup rate}} \cdot \boxed{\text{Cost}}$

*Labels:* Markup = 3890.50

Markup rate = $x$

Cost = 12,550.00

*Equation:* $3890.50 = x \cdot 12,550.00$

$\dfrac{3890.50}{12,550.00} = x$

$31\% = x$

**8.** *Verbal Model:* $\boxed{\text{Markup}} = \boxed{\text{Markup rate}} \cdot \boxed{\text{Cost}}$

*Labels:* Markup = $x$

Markup rate = $33\frac{1}{3}\%$

Cost = 732.00

*Equation:* $x = 33\frac{1}{2}\% \cdot 732.00$

$x = \frac{1}{3} \cdot 732.00$

$x = \$244.00$

*Verbal Model:* $\boxed{\text{Selling price}} = \boxed{\text{Cost}} + \boxed{\text{Markup}}$

*Labels:* Selling price = $x$

Cost = 732.00

Markup = 244.00

*Equation:* $x = 732.00 + 244.00$

$x = \$976.00$

**10.** *Verbal Model:*   $\boxed{\text{Sale price}} = \boxed{\text{List price}} - \boxed{\text{Discount}}$

*Labels:*   Sale price = 79.73
List price = 119.00
Discount = $x$

*Equation:*   $79.73 = 119.00 - x$

$x = 119.00 - 79.73$

$x = \$39.27$

*Verbal Model:*   $\boxed{\text{Discount}} = \boxed{\text{Discount rate}} \cdot \boxed{\text{List price}}$

*Labels:*   Discount = 39.27
Discount rate = $x$
List price = 119.00

*Equation:*   $39.27 = x \cdot 119.00$

$\dfrac{39.27}{119.00} = x$

$33\% = x$

**14.** *Verbal Model:*   $\boxed{\text{Sale price}} = \boxed{\text{Percent}} \cdot \boxed{\text{List price}}$

*Labels:*   Sale price = 15.92
Percent = 0.80
List price = $x$

*Equation:*   $15.92 = 0.80 \cdot x$

$\dfrac{15.92}{0.80} = x$

$\$19.90 = x$

*Verbal Model:*   $\boxed{\text{Sale price}} = \boxed{\text{List price}} - \boxed{\text{Discount}}$

*Labels:*   Sale price = 15.92
List price = 19.90
Discount = $x$

*Equation:*   $15.92 = 19.90 - x$

$x = 19.90 - 15.92$

$x = \$3.98$

**12.** *Verbal Model:*   $\boxed{\text{Sale price}} = \boxed{\text{List price}} - \boxed{\text{Discount}}$

*Labels:*   Sale price = $x$
List price = 345.00
Discount = 134.55

*Equation:*   $x = 345.00 - 134.55$

$x = \$210.45$

*Verbal Model:*   $\boxed{\text{Discount}} = \boxed{\text{Discount rate}} \cdot \boxed{\text{List price}}$

*Labels:*   Discount = 134.55
Discount rate = $x$
List price = 345.00

*Equation:*   $134.55 = x \cdot 345.00$

$\dfrac{134.55}{345.00} = x$

$39\% = x$

**16.** *Verbal Model:*   $\boxed{\text{Sale price}} = \boxed{\text{List price}} - \boxed{\text{Discount}}$

*Labels:*   Sale price = 257.32
List price = $x$
Discount = 202.18

*Equation:*   $257.32 = x - 202.18$

$257.32 + 202.18 = x$

$\$459.50 = x$

*Verbal Model:*   $\boxed{\text{Discount}} = \boxed{\text{Discount rate}} \cdot \boxed{\text{List price}}$

*Labels:*   Discount = 202.18
Discount rate = $x$
List price = 459.50

*Equation:*   $202.18 = x \cdot 459.50$

$\dfrac{202.18}{459.50} = x$

$44\% = x$

**18.** *Verbal Model:*    $\boxed{\text{Selling price}} = \boxed{\text{Cost}} + \boxed{\text{Markup}}$

*Labels:*    Selling price = 35.00

Cost = 18.75

Markup = $x$

*Equation:*    $35.00 = 18.75 + x$

$35.00 - 18.75 = x$

$\$16.25 = x$

**20.** *Verbal Model:*    $\boxed{\text{Selling price}} = \boxed{\text{Cost}} + \boxed{\text{Markup}}$

*Labels:*    Selling price = 60

Cost = 35

Markup = $x$

*Equation:*    $60 = 35 + x$

$60 - 35 = x$

$25 = x$

*Verbal Model:*    $\boxed{\text{Markup}} = \boxed{\text{Markup rate}} \cdot \boxed{\text{Cost}}$

*Labels:*    Markup = 25

Markup rate = $x$

Cost = 35

*Equation:*    $25 = x \cdot 35$

$\frac{25}{35} = x$

$0.714 \approx x$

$71.4\% \approx x$

**22.** *Verbal Model:*    $\boxed{\text{Sale price}} = \boxed{\text{List price}} - \boxed{\text{Discount}}$

*Labels:*    Sale price = 0.75

List price = 1.75

Discount = $x$

*Equation:*    $0.75 = 1.75 - x$

$x = 1.75 - 0.75$

$x = \$1.00$

**24.** *Verbal Model:*    $\boxed{\text{Sale price}} = \boxed{\text{List price}} - \boxed{\text{Discount}}$

*Labels:*    Sale price = 10

List price = 14

Discount = $x$

*Equation:*    $10 = 14 - x$

$x = 14 - 10$

$x = \$4$

*Verbal Model:*    $\boxed{\text{Discount}} = \boxed{\text{Discount rate}} \cdot \boxed{\text{List price}}$

*Labels:*    Discount = 4

Discount rate = $x$

List price = 14

*Equation:*    $4 = x(14)$

$\frac{4}{14} = x$

$0.286 \approx x$

$28.6\% \approx x$

**26.** *Verbal Model:* $\boxed{\text{Surcharge}} = \boxed{\begin{array}{c}\text{Surcharge}\\\text{rate}\end{array}} \cdot \boxed{\text{Premium}}$

*Labels:*  Surcharge = $x$
Surcharge rate = 20%
Premium = 862

*Equation:*  $x = 20\% \cdot 862$

$x = \$172.40$

*Verbal Model:* $\boxed{\begin{array}{c}\text{Current}\\\text{premium}\end{array}} = \boxed{\begin{array}{c}\text{Previous}\\\text{premium}\end{array}} + \boxed{\text{Surcharge}}$

*Labels:*  Current premium = $x$
Previous premium = 862
Surcharge = 172.40

*Equation:*  $x = 862.00 + 172.40$

$x = \$1034.40$

**28.** *Verbal Model:* $\boxed{\text{Markup}} = \boxed{\begin{array}{c}\text{Markup}\\\text{rate}\end{array}} \cdot \boxed{\text{Cost}}$

*Labels:*  Markup = $x$
Markup rate = 30%
Cost = 22.60

*Equation:*  $x = 30\% \cdot 22.60$

$x = \$6.78$

*Verbal Model:* $\boxed{\begin{array}{c}\text{Selling}\\\text{price}\end{array}} = \boxed{\text{Cost}} + \boxed{\text{Markup}}$

*Labels:*  Selling price = $x$
Cost = 22.60
Markup = 6.78

*Equation:*  $x = 22.60 + 6.78$

$x = \$29.38$

*Verbal Model:* $\boxed{\begin{array}{c}\text{Number of}\\\text{spoiled bananas}\end{array}} = \boxed{\begin{array}{c}\text{Total}\\\text{bananas}\end{array}} + \boxed{\begin{array}{c}\text{Percent}\\\text{to spoil}\end{array}}$

*Labels:*  Spoiled bananas = $x$
Total bananas = 100
Percent = 10%

*Equation:*  $x = 100 \cdot 10\%$

$x = 10$

Cost per pound $= \dfrac{\text{Total cost}}{\text{Number of pounds}}$

$= \dfrac{29.38}{90} = \$0.33/\text{lb}$

**30.** *Verbal Model:* $\boxed{\text{Commission}} = \boxed{\begin{array}{c}\text{Commission}\\\text{rate}\end{array}} \cdot \boxed{\text{Sales}}$

*Labels:*  Commission = $x$
Commission rate = 6%
Sales = 5500

*Equation:*  $x = 6\% \cdot 5500$

$x = \$330$

*Verbal Model:* $\boxed{\begin{array}{c}\text{Weekly}\\\text{pay}\end{array}} = \boxed{\text{Salary}} + \boxed{\text{Commission}}$

*Labels:*  Weekly pay = $x$
Salary = 375
Commission = 330

*Equation:*  $x = 375 + 330$

$x = \$705$

**32.** *Verbal Model:* $\boxed{\begin{array}{c}\text{Total}\\\text{bill}\end{array}} = \boxed{\begin{array}{c}\text{First 1/2}\\\text{hour charge}\end{array}} + \boxed{\begin{array}{c}\text{Additional 1/2}\\\text{hour charges}\end{array}}$

*Labels:*  Total bill = 104
First 1/2 hour charge = 50
Number of 1/2 hours = $x$
Additional 1/2 hour charges = $18x$

*Equation:*  $104 = 50 + 18x$

$54 = 18x$

$3 = x$

The length of a service call is 2 hours.

**34.** *Verbal Model:*  $\boxed{\text{Total bill}} = \boxed{\text{Parts charge}} + \boxed{\text{Labor charge}}$

*Labels:*    Total bill = 648

Parts charge = 315

Charge per hour = $x$

Labor charge = $9x$

*Equation:*    $648 = 315 + 9x$

$333 = 9x$

$\$37 \text{ per hour} = x$

**36.** *Verbal Model:*  $\boxed{\text{Amount of solution 1}} + \boxed{\text{Amount of solution 2}} = \boxed{\text{Amount of final solution}}$

*Labels:*    Percent of solution 1 = 50%

Liters of solution 1 = $x$

Percent of solution 2 = 75%

Liters of solution 2 = $10 - x$

Percent of final solution = 60%

Liters of final solution = 10

*Equation:*  $0.50x + 0.75(10 - x) = 0.60(10)$

$0.50x + 7.5 - 0.75x = 6$

$-0.25x = -1.5$

$x = 6 \text{ liters at } 50\%$

$10 - x = 4 \text{ liters at } 75\%$

**38.** *Verbal Model:*  $\boxed{\text{Amount of solution 1}} + \boxed{\text{Amount of solution 2}} = \boxed{\text{Amount of final solution}}$

*Labels:*    Percent of solution 1 = 60%

Gallons of solution 1 = $x$

Percent of solution 2 = 80%

Gallons of solution 2 = $55 - x$

Percent of final solution = 75%

Gallons of final solution = 55

*Equation:*  $0.60x + 0.80(55 - x) = 0.75(55)$

$0.60x + 44 - 0.80x = 41.25$

$-0.20x = -2.75$

$x = 13.75 \text{ gallons at } 60\%$

$55 - x = 41.25 \text{ gallons at } 80\%$

**40.** *Verbal Model:*  $\boxed{\text{Total cost}} = \boxed{\text{Cost of first nut}} + \boxed{\text{Cost of second nut}}$

*Labels:*    Total number of pounds = 100

Cost per pound = 4.13

Number of pounds of first nut = $x$

Cost per pound of first nut = 3.88

Number of pounds of second nut = $100 - x$

Cost per pound of second nut = 4.88

*Equation:*  $100(4.13) = x(3.88) + (100 - x)(4.88)$

$413 = 3.88x + 488 - 4.88x$

$-75 = -1.00x$

$75 = x \text{ pounds at } \$3.88$

$25 = 100 - x \text{ pounds at } \$4.88$

**42.** *Verbal Model:*  $\boxed{\text{Total sales}} = \boxed{\text{Adult sales}} + \boxed{\text{Children sales}}$

*Labels:*    Total sales = 1350

Number of adult tickets = $4x$

Price of adult tickets = 6

Number of children tickets = $x$

Price of children tickets = 3

*Equation:*    $1350 = 6(4x) + 3x$

$1350 = 24x + 3x$

$1350 = 27x$

$50 \text{ children tickets} = x$

**44.** *Verbal Model:*  $\boxed{\begin{array}{c}\text{Original}\\\text{gas/oil}\\\text{mixture}\end{array}} + \boxed{\text{Gasoline}} = \boxed{\begin{array}{c}\text{Final}\\\text{gas/oil}\\\text{mixture}\end{array}}$

*Labels:*    Percent of gasoline in original mixture = $\frac{40}{41}$

Number of gallons in original mixture = 2.5

Number of gallons of gasoline added = $x$

Percent of gasoline in final mixture = $\frac{50}{51}$

Number of gallons in final mixture = $2.5 + x$

*Equation:*  $\frac{40}{41}(2.5) + x = \frac{50}{51}(2.5 + x)$

$\frac{100}{41} + x = \frac{125}{51} + \frac{50}{51}x$

$\frac{1}{51}x = \frac{125}{51} - \frac{100}{41}$

$x = 51\left(\frac{125}{51} - \frac{100}{41}\right)$

$x \approx 0.61 \text{ gallon}$

**46.** *Verbal Model:*   $\boxed{\text{Distance}} = \boxed{\text{Rate}} \cdot \boxed{\text{Time}}$

*Labels:*   Distance = $d$

Rate = 45

Time = 10

*Equation:*   $d = 45 \cdot 10$

$d = 450$ feet

**48.** *Verbal Model:*   $\boxed{\text{Distance}} = \boxed{\text{Rate}} \cdot \boxed{\text{Time}}$

*Labels:*   Distance = 250

Rate = 32

Time = $t$

*Equation:*   $250 = 32 \cdot t$

$\frac{250}{32} = t$

$\frac{125}{16}$ seconds = $t$

**50.** *Verbal Model:*   $\boxed{\text{Distance}} = \boxed{\text{Rate}} \cdot \boxed{\text{Time}}$

*Labels:*   Distance = 385

Rate = $r$

Time = 7

*Equation:*   $385 = r \cdot 7$

$\frac{385}{7} = r$

55 mph = $r$

**52.** *Verbal Model:*   $\boxed{\text{Distance}} = \boxed{\text{Rate}} \cdot \boxed{\text{Time}}$

*Labels:*   Distance = 12

Rate = 8

Time = $x$

*Equation:*   $12 = 8x$

$\frac{12}{8} = x$

$1\frac{1}{2}$ hours = $x$

**54.** *Verbal Model:*   $\boxed{\text{Distance}} = \boxed{\text{Rate}} \cdot \boxed{\text{Time}}$

*Labels:*   Distance of truck 1 = $x$

Rate of truck 1 = 52

Time of truck 1 = $4\frac{1}{2}$

Distance of truck 2 = $y$

Rate of truck 2 = 56

Time of truck 2 = $4\frac{1}{2}$

*Equations:*   $x = 52 \cdot 4\frac{1}{2}$

$x = 234$ miles

$y = 56 \cdot 4\frac{1}{2}$

$y = 252$ miles

The two trucks are $252 - 234 = 18$ miles apart.

**56.** *Verbal Model:*   $\boxed{\text{Distance}} = \boxed{\text{Rate}} \cdot \boxed{\text{Time}}$

*Labels:*   Distance = 93,000,000

Rate = 186,282.369

Time = $x$

*Equation:*   $93,000,000 = 186,282.369x$

$\frac{93,000,000}{186,282.369} = x$

$499.2420941$ seconds $\approx x$

$8.32$ minutes $\approx x$

**58.** *Verbal Model:*   $\boxed{\text{Distance}} = \boxed{\text{Rate}} \cdot \boxed{\text{Time}}$

*Labels:*   Distance = 5

Rates = 30 and 45

Time = $x$

*Equation:*   $5 = 45x - 30x$

$5 = 15x$

$\frac{5}{15} = x$

20 minutes = $x$

**60.** (a) Typist 1's rate = $\frac{1}{5}$ job per hour

Typist 2's rate = $\frac{1}{8}$ job per hour

(b) *Verbal Model:* $\boxed{\text{Work done}} = \boxed{\begin{array}{c}\text{Work done}\\\text{by first}\\\text{person}\end{array}} + \boxed{\begin{array}{c}\text{Work done}\\\text{by second}\\\text{person}\end{array}}$

*Labels:*    Work done = 1

Rate for first person = $\frac{1}{5}$

Time for first person = $t$

Rate for second person = $\frac{1}{8}$

Time for second person = $t$

*Equation:*    $1 = \left(\dfrac{1}{5}\right)(t) + \left(\dfrac{1}{8}\right)(t)$

$1 = \left(\dfrac{1}{5} + \dfrac{1}{8}\right)t$

$1 = \left(\dfrac{13}{40}\right)t$

$\dfrac{1}{13/40} = t$

$3\dfrac{1}{13} \text{ hours} = \dfrac{40}{13} \text{ hours} = t$

**62.** *Verbal Model:* $\boxed{\begin{array}{c}\text{Work}\\\text{done}\end{array}} = \boxed{\begin{array}{c}\text{Work done}\\\text{by smaller}\\\text{pump}\end{array}} + \boxed{\begin{array}{c}\text{Work done}\\\text{by larger}\\\text{pump}\end{array}}$

*Labels:*    Work done = 1

Rate of smaller pump = $\frac{1}{30}$

Rate of larger pump = $\frac{1}{15}$

Time for each pump = $t$

*Equation:*    $1 = \left(\dfrac{1}{30}\right)(t) + \left(\dfrac{1}{15}\right)(t)$

$1 = \left(\dfrac{1}{30} + \dfrac{1}{15}\right)t$

$1 = \dfrac{3}{30}t$

$\dfrac{1}{3/30} = t$

10 minutes = $t$

**64.**    $A = P + Prt$

$A - P = Prt$

$\dfrac{A - P}{Pt} = r$

**66.**    $S = C + rC$

$S = C(1 + r)$

$\dfrac{S}{1 + r} = C$

**68.**    $A = \dfrac{1}{2}(a + b)h$

$2A = (a + b)h$

$\dfrac{2A}{h} = a + b$

$\dfrac{2A}{h} - a = b$

$\dfrac{2A - ah}{h} = b$

**70.** *Common formula:* $SA = 2lw + 2wh + 2lh$

*Equation:* $SA = 2(3)(4) + 2(4)(2) + 2(3)(2)$

$SA = 24 + 16 + 12$

$SA = 52$ square units

**72.** *Common formula:* $V = \frac{4}{3}\pi r^3$

*Equation:* $V = \frac{4}{3}\pi(6)^3$

$V = 288\pi$

$V \approx 904.8$ cubic meters

**74.** *Verbal Model:* Perimeter = 2 [Width] + 2 [Height]

*Labels:* Perimeter = 18

Width = $x$

Height = $1.25x$

*Equation:* $18 = 2x + 2(1.25x)$

$18 = 2x + 2.5x$

$18 = 4.5x$

$\dfrac{18}{4.5} = x$

$4 \text{ feet} = x$

**76.** *Verbal Model:* Perimeter = 2 [Width] + 2 [Length]

*Labels:* Perimeter = 64

Width = $x$

Length = $3x$

*Equation:* $64 = 2x + 2(3x)$

$64 = 2x + 6x$

$64 = 8x$

$8 \text{ inches} = x$

$24 \text{ inches} = 3x$

**78.** *Verbal Model:* [Interest] = [Principal] · [Rate] · [Time]

*Labels:* Interest = 400

Principal = 2500

Rate = $r$

Time = 2

*Equation:* $400 = (2500)(r)(2)$

$400 = 5000r$

$\dfrac{400}{5000} = r$

$8\% = 0.08 = r$

**80.** *Verbal Model:* [Interest] = [Principal] · [Rate] · [Time]

*Labels:* Interest = $I$

Principal = 36,000

Rate = 13%

Time = $\frac{1}{2}$

*Equation:* $I = (36{,}000)(0.13)\left(\frac{1}{2}\right)$

$I = \$2340$

[Amount of payment] = [Principal] + [Interest]

$= 36{,}000 + 2340$

$= \$38{,}340$

**82.** *Verbal Model:* [Interest] = [Principal] · [Rate] · [Time]

*Labels:* Interest = 400

Principal at 7% = $x$

Principal at 5% = $7000 - x$

Time = 1

*Equation:* $400 = 0.07x + 0.05(7000 - x)$

$400 = 0.07x + 350 - 0.05x$

$50 = 0.02x$

$\dfrac{50}{0.02} = x$

$\$2500 = x$

**84.** (a) $y = 6.88 + 0.209t, \ 0 \le t < 7$

From the graph, 1994 was the year when the average hourly wage was $7.72.

$7.72 = 6.88 + 0.209t$

$0.84 = 0.209t$

$\dfrac{0.84}{0.209} = t$

$4.0191 \approx t$

$4 \approx t$

Yes, the result is the same, 1994.

(b) The average annual hourly raise for cafeteria workers during this 8-year period is $0.209. Determine the average hourly wage for each year using the model. The difference between each two consecutive years is $0.209.

**86.** (a)  *Verbal Model:*  Total cost = Standard service + Premium movie cost

Total cost = Standard service + 11.91 · Number of channels

(b)  $C = 31.20 + 11.91x$

| $x$ | 1 | 2 | 3 | 4 | 5 |
|---|---|---|---|---|---|
| $C$ | $43.11 | $55.02 | $66.93 | $78.84 | $90.75 |

(c)  *Verbal Model:*  Total cost = Standard service + Pay-per-view service

Total cost = Standard service + 2.99 + 3.95 · Number of movies

(d)  $C = 31.20 + 2.99 + 3.95x$

$= 34.19 + 3.95x$

| $x$ | 1 | 2 | 3 | 4 | 5 | 6 | 7 | 8 |
|---|---|---|---|---|---|---|---|---|
| $C$ | $38.14 | $42.09 | $46.04 | $49.99 | $53.94 | $57.89 | $61.84 | $65.79 |

(e)  *Verbal Model:*  Cost of two premium movie channels = Percent · Total cost

*Labels:*   Cost of two premium movie channels = 23.82

Percent = $x$

Total cost = 55.02

*Equation:*   $23.82 = x \cdot 55.02$

$\dfrac{23.82}{55.02} = x$

$43.3\% \approx x$

**88.** To find the sale price of an item, subtract the list price minus the discount rate times the list price.

**90.** Yes, if the sides of a square are doubled, the area doubles. If a square has side $s$, then the perimeter is $4s$. So if a square has side $2s$, then the perimeter is $4(2s) = 8s$ which is double $4s$.

**92.** The volume of a right circular cylinder is the area of the base times the height. The base is a circle whose area is $\pi r^2$. So, $V = \pi r^2 h$.

# Section 1.4    Linear Inequalities

**2.** (a) $3(0) + 2 < \dfrac{7(0)}{5}$

$0 + 2 < \dfrac{0}{5}$

$2 < 0$

No

(b) $3(4) + 2 < \dfrac{7(4)}{5}$

$12 + 2 < \dfrac{28}{5}$

$14 < \dfrac{28}{5}$

No

(c) $3(-4) + 2 < \dfrac{7(-4)}{5}$

$-12 + 2 < -\dfrac{28}{5}$

$-10 < -\dfrac{28}{5}$

Yes

(d) $3(-1) + 2 < \dfrac{7(-1)}{5}$

$-3 + 2 < -\dfrac{7}{5}$

$-1 < -\dfrac{7}{5}$

No

**4.** (a) $-2 < \dfrac{3-0}{2} \le 2$

$-2 < \dfrac{3}{2} \le 2$

Yes

(b) $-2 < \dfrac{3-3}{2} \le 2$

$-2 < \dfrac{0}{2} \le 2$

$-2 < 0 \le 2$

Yes

(c) $-2 < \dfrac{3-9}{2} \le 2$

$-2 < -\dfrac{6}{2} \le 2$

$-2 < -3 \le 2$

No

(d) $-2 < \dfrac{3-(-12)}{2} \le 2$

$-2 < \dfrac{15}{2} \le 2$

No

**6.** Matches graph (b).

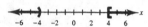

**8.** Matches graph (c).

**10.** Matches graph (c).

**12.** Matches graph (e).

**14.** Matches graph (b).

**16.** $x > -6$

**18.** $x \le -2.5$

**20.** $-1 < x \le 5$

**22.** $9 \ge x \ge 3$

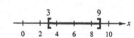

**24.** $-\dfrac{15}{4} < x < -\dfrac{5}{2}$

**26.** $x \le -4$ or $x > 0$

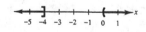

**28.** $x \le -1$ or $x \ge 1$

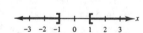

**30.**    $5 - \dfrac{1}{3}x > 8$

$5 - \dfrac{1}{3}x + \dfrac{1}{3}x > 8 + \dfrac{1}{3}x$

$5 > 8 + \dfrac{1}{3}x$

**32.**    $x + 1 < 0$

$x + 1 - 1 < 0 - 1$

$x < -1$

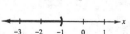

**34.**    $z - 4 > 0$

$z - 4 + 4 > 0 + 4$

$z > 4$

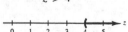

**36.** $3x \geq 12$

$\dfrac{3x}{3} \geq \dfrac{12}{3}$

$x \geq 4$

**38.** $-6x \leq 24$

$\dfrac{-6x}{-6} \geq \dfrac{24}{-6}$

$x \geq -4$

**40.** $-\dfrac{1}{5}x > -2$

$(-5)\left(-\dfrac{1}{5}x\right) < (-5)(-2)$

$x < 10$

**42.** $1 - y \geq -5$

$1 - y - 1 \geq -5 - 1$

$-y \geq -6$

$(-1)(-y) \leq (-1)(-6)$

$y \leq 6$

**44.** $3x + 4 \leq 22$

$3x + 4 - 4 \leq 22 - 4$

$3x \leq 18$

$\dfrac{3x}{3} \leq \dfrac{18}{3}$

$x \leq 6$

**46.** $12 - 5x > 5$

$12 - 5x - 12 > 5 - 12$

$-5x > -7$

$\dfrac{-5x}{-5} < \dfrac{-7}{-5}$

$x < \dfrac{7}{5}$

**48.** $21x - 11 \leq 6x + 19$

$21x - 11 - 6x \leq 6x + 19 - 6x$

$15x - 11 \leq 19$

$15x - 11 + 11 \leq 19 + 11$

$15x \leq 30$

$\dfrac{15x}{15} \leq \dfrac{30}{15}$

$x \leq 2$

**50.** $6x - 1 > 3x - 11$

$6x - 3x - 1 > 3x - 3x - 11$

$3x - 1 > -11$

$3x - 1 + 1 > -11 + 1$

$3x > -10$

$\dfrac{3x}{3} > -\dfrac{10}{3}$

$x > -\dfrac{10}{3}$

**52.** $\dfrac{x}{6} - \dfrac{x}{4} \leq 1$

$12\left(\dfrac{x}{6} - \dfrac{x}{4}\right) \leq 12(1)$

$2x - 3x \leq 12$

$-x \leq 12$

$(-1)(-x) \geq 12(-1)$

$x \geq -12$

**54.** $\dfrac{x + 3}{6} + \dfrac{x}{8} \geq 1$

$24\left(\dfrac{x + 3}{6} + \dfrac{x}{8}\right) \geq 24(1)$

$4(x + 3) + 3x \geq 24$

$4x + 12 + 3x \geq 24$

$7x + 12 \geq 24$

$7x + 12 - 12 \geq 24 - 12$

$7x \geq 12$

$\dfrac{7x}{7} \geq \dfrac{12}{7}$

$x \geq \dfrac{12}{7}$

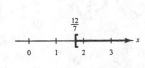

**56.** $\dfrac{4x}{7} + 1 > \dfrac{x}{2} + \dfrac{5}{7}$

$14\left(\dfrac{4x}{7} + 1\right) > 14\left(\dfrac{x}{2} + \dfrac{5}{7}\right)$

$8x + 14 > 7x + 10$

$8x - 7x + 14 > 7x - 7x + 10$

$x + 14 > 10$

$x + 14 - 14 > 10 - 14$

$x > -4$

**58.**     $-6 \leq 3x - 9 < 0$

$-6 + 9 \leq 3x - 9 + 9 < 0 + 9$

$3 \leq 3x < 9$

$\dfrac{3}{3} \leq \dfrac{3x}{3} < \dfrac{9}{3}$

$1 \leq x < 3$

**60.**     $-10 \leq 4 - 7x < 10$

$-10 - 4 \leq 4 - 4 - 7x < 10 - 4$

$-14 \leq -7x < 6$

$\dfrac{-14}{-7} \geq \dfrac{-7x}{-7} > \dfrac{6}{-7}$

$2 \geq x > -\dfrac{6}{7}$

$-\dfrac{6}{7} < x \leq 2$

**62.**     $-2 < -\dfrac{1}{2}s \leq 0$

$-2 \cdot -2 > -2 \cdot -\dfrac{1}{2}s \geq 0 \cdot -2$

$4 > s \geq 0$

$0 \leq s < 4$

**64.**     $0 \leq \dfrac{x - 5}{2} < 4$

$0 \leq x - 5 < 8$

$0 + 5 \leq x - 5 + 5 < 8 + 5$

$5 \leq x < 13$

**66.**     $-\dfrac{2}{3} < \dfrac{x - 4}{-6} \leq \dfrac{1}{3}$

$4 > x - 4 \geq -2$

$4 + 4 > x - 4 + 4 \geq -2 + 4$

$8 > x \geq 2$

$2 \leq x < 8$

**68.**     $8 - 3x > 5 \qquad \text{and} \qquad x - 5 \geq -10$

$8 - 8 - 3x > 5 - 8 \quad \text{and} \quad x - 5 + 5 \geq -10 + 5$

$-3x > -3 \qquad \text{and} \qquad x \geq -5$

$\dfrac{-3x}{-3} < \dfrac{-3}{-3}$

$x < 1 \qquad \text{and} \qquad x \geq -5$

$-5 \leq x < 1$

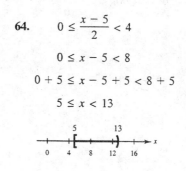

**70.**     $9 - x \leq 3 + 2x \qquad \text{and} \qquad 3x - 7 \leq -22$

$9 - x - 2x \leq 3 + 2x - 2x \quad \text{and} \quad 3x - 7 + 7 \leq -22 + 7$

$9 - 3x \leq 3 \qquad\qquad \text{and} \qquad 3x \leq -15$

$9 - 9 - 3x \leq 3 - 9 \qquad \text{and} \qquad \dfrac{3x}{3} \leq \dfrac{-15}{3}$

$-3x \leq -6 \qquad\qquad \text{and} \qquad x \leq -5$

$\dfrac{-3x}{-3} \geq \dfrac{-6}{-3}$

$x \geq 2$

No solution

**72.**    $\dfrac{x}{3} - 2 \geq 1$        or        $5 + \dfrac{3}{4}x \leq -4$

$3\left(\dfrac{x}{3} - 2\right) \geq (1)3$    or    $4\left(5 + \dfrac{3}{4}x\right) \leq (-4)4$

$\quad x - 6 \geq 3$        or        $20 + 3x \leq -16$

$x - 6 + 6 \geq 3 + 6$  or  $20 - 20 + 3x \leq -16 - 20$

$\qquad x \geq 9$        or        $3x \leq -36$

$\qquad\qquad\qquad\qquad\qquad \dfrac{3x}{3} \leq \dfrac{-36}{3}$

$\qquad\qquad\qquad\qquad\qquad\quad x \leq -12$

*(number line: shaded at $x \leq -12$ and $x \geq 9$; marks at $-14$ $-12$ $-8$ $-4$ $0$ $4$ $8$ $12$, point at 9)*

**74.**    $3x + 10 \leq -x - 6$        or        $\dfrac{x}{2} + 5 < \dfrac{5}{2}x - 4$

$3x + x + 10 \leq -x + x - 6$  or  $2\left(\dfrac{x}{2} + 5\right) < \left(\dfrac{5}{2}x - 4\right)2$

$\quad 4x + 10 \leq -6$        or        $x + 10 < 5x - 8$

$4x + 10 - 10 \leq -6 - 10$  or  $x - 5x + 10 < 5x - 5x - 8$

$\qquad 4x \leq -16$        or        $-4x + 10 < -8$

$\qquad \dfrac{4x}{4} \leq \dfrac{-16}{4}$    or    $-4x + 10 - 10 < -8 - 10$

$\qquad\quad x \leq -4$        or        $-4x < -18$

$\qquad\qquad\qquad\qquad\qquad \dfrac{-4x}{-4} > \dfrac{-18}{-4}$

$\qquad\qquad\qquad\qquad\qquad\quad x > \dfrac{9}{2}$

*(number line: marks at $-8$ $-6$ $-4$ $-2$ $0$ $2$ $4$ $6$ $8$, point at $\frac{9}{2}$)*

**76.**    $2(4 - z) \geq 8(1 + z)$

$\quad 8 - 2z \geq 8 + 8z$

$8 - 2z + 2z \geq 8 + 8z + 2z$

$\qquad 8 \geq 8 + 10z$

$\quad 8 - 8 \geq 8 + 10z - 8$

$\qquad 0 \geq 10z$

$\qquad \dfrac{0}{10} \geq \dfrac{10z}{10}$

$\qquad 0 \geq z$

$\qquad z \leq 0$

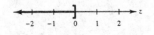

**78.**    $16 < 4(y + 2) - 5(2 - y)$

$16 < 4y + 8 - 10 + 5y$

$\quad 16 < 9y - 2$

$16 + 2 < 9y - 2 + 2$

$\quad 18 < 9y$

$\quad \dfrac{18}{9} < \dfrac{9y}{9}$

$\quad 2 < y$

$\quad y > 2$

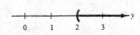

**80.**  $x < -2$ or $x > 5$

$\{x | x < -2\} \cup \{x | x > 5\}$

**82.**  $-7 < x < -1$

$\{x | x > -7\} \cap \{x | x < -1\}$

**84.**  $-4.5 \leq x \leq 2$

$\{x | x \geq -4.5\} \cap \{x | x \leq 2\}$

**86.**  $\{x | x > 2\} \cap \{x | x < 8\}$

**88.**  $\{x | x \geq -1\} \cup \{x | x < -6\}$

**90.**  $\{x | x < 0\} \cup \{x | x \geq \frac{2}{3}\}$

**92.** $y > -2$

**94.** $m \geq 4$

**96.** $450 \leq x \leq 500$

**98.** $t$ is less than 4.

**100.** $t$ is at least $-4$, but no more than 4.

**102.** $x$ is no more than 5 and is more than $-2$.

**104.** *Verbal Model:* $\boxed{\text{Rent}} + \boxed{\text{Food}} + \boxed{\begin{array}{c}\text{Other}\\\text{costs}\end{array}} \leq \boxed{\begin{array}{c}\text{Monthly}\\\text{budget}\end{array}}$

*Labels:* Rent = 600

Food = 350

Other costs = $C$

Monthly budget = 1800

*Inequality:* $600 + 350 + C \leq 1800$

$$950 + C \leq 1800$$

$$C \leq \$850$$

**106.** *Verbal Model:* $\boxed{\begin{array}{c}\text{Elevation of}\\\text{San Francisco}\end{array}} < \boxed{\begin{array}{c}\text{Elevation}\\\text{of Dallas}\end{array}} < \boxed{\begin{array}{c}\text{Elevation}\\\text{of Denver}\end{array}}$

The elevation of San Francisco is less than (<) the elevation of Denver.

**108.** *Verbal Model:* $\boxed{\text{Cost}} < \boxed{25{,}000}$

*Label:* Cost = $0.58m + 7800$

*Inequality:* $0.58m + 7800 < 25{,}000$

$$0.58x + 7800 - 7800 < 25{,}000 - 7800$$

$$0.58m < 172{,}000$$

$$\frac{0.58m}{0.58} < \frac{172{,}000}{0.58}$$

$$m < 29{,}655.17$$

$$m \leq 29{,}655$$

**110.** *Verbal Model:* $\boxed{R} > \boxed{C}$

*Labels:* $R = 105.45x$

$C = 78x + 25{,}850$

*Inequality:* $105.45x > 78x + 25{,}850$

$$105.45x - 78x > 25{,}850$$

$$27.45x > 25{,}850$$

$$\frac{27.45x}{27.45} > \frac{25{,}850}{27.45}$$

$$x > 941.71$$

$$x \geq 942 \text{ units}$$

**112.** *Verbal Model:* $\boxed{\begin{array}{c}\text{Cost of}\\\text{first minute}\end{array}} + \boxed{\begin{array}{c}\text{Cost of additional}\\\text{minutes}\end{array}} \leq \boxed{15}$

*Label:* Number of additional minutes = $t$

*Inequality:* $1.45 + 0.95t \leq 15$

$$1.45 + 0.95t - 1.45 \leq 15 - 1.45$$

$$0.95t \leq 13.55$$

$$\frac{0.95t}{0.95} \leq \frac{13.55}{0.95}$$

$$t \leq 14.26$$

Since $t$ represents the additional minutes after the first minute, the call must be less than 15.26 minutes. If a portion of a minute is billed as a full minute, then the call must be less than or equal to 15 minutes.

**114.** *Verbal Model:* $100 \leq \boxed{\text{Perimeter}} \leq 120$

*Label:* Perimeter = $2x + 28$

*Inequality:* $100 \leq 2x + 28 \leq 120$

$$100 - 28 \leq 2x + 28 - 28 \leq 120 - 28$$

$$72 \leq 2x \leq 92$$

$$\frac{72}{2} \leq \frac{2x}{2} \leq \frac{92}{2}$$

$$36 \leq x \leq 46$$

**116.**  $\dfrac{1}{3}n > 7$

$3\left(\dfrac{1}{3}n\right) > 3(7)$

$n > 21$

**118.** *Verbal Model:*  $\boxed{\text{Second plan}} > \boxed{\text{First plan}}$

*Label:*    Number of units produced $= x$

*Inequality:*  $1000 + 0.04x > 3000$

$0.04x > 2000$

$\dfrac{0.04x}{0.04} > \dfrac{2000}{0.04}$

$x > \$50{,}000$

**120.** *Verbal Model:*  $\boxed{\begin{array}{c}\text{Air pollutant}\\\text{emission}\end{array}} < 5$

*Label:*    Air pollutant emission $= 5.890 - 0.276t$

*Inequality:*    $5.890 - 0.276t < 5$

$5.890 - 5.890 - 0.276t < 5 - 5.890$

$-0.276t < -0.890$

$\dfrac{-0.276t}{-0.276} > \dfrac{-0.890}{-0.276}$

$t > 3.22$

$t = 4 \rightarrow$ year 1994

$t = 5 \rightarrow$ year 1995

**122.** Adding $-5$ and subtracting 5 from both sides of an inequality is the same.

**124.** A linear inequality that has an unbounded solution set is $4x - 3 > 0$.

$4x - 3 > 0$

$4x - 3 + 3 > 0 + 3$

$4x > 3$

$\dfrac{4x}{4} > \dfrac{3}{4}$

$x > \dfrac{3}{4}$

Solution set is all real numbers greater than $\frac{3}{4}$.

**126.** An example of "reversing an inequality symbol" is $3x - 2 \le 4$ and $-(3x - 2) \ge -4$.

**128.**  $-3 \le x \le 10$

$3 \ge -x \ge -10$

# Section 1.5    Absolute Value Equations and Inequalities

**2.**  $|2(3) - 16| \overset{?}{=} 10$

$|6 - 16| \overset{?}{=} 10$

$|-10| \overset{?}{=} 10$

$10 = 10$

Yes

**4.**  $\left|\dfrac{1}{2}(6) + 4\right| \overset{?}{=} 8$

$|3 + 4| \overset{?}{=} 8$

$|7| \overset{?}{=} 8$

$7 = 8$

No

**6.**  $7 - 2t = 5$ or $7 - 2t = -5$

**8.** $22k + 6 = 9$ or $22k + 6 = -9$

**10.** $|x| = 3$

$x = 3$ or $x = -3$

**12.** $|s| = 16$

$s = 16$ or $s = -16$

**14.** $|x| = -82$

An absolute value cannot be negative. No solution

**16.** $\left|\frac{1}{3}x\right| = 2$

$\frac{1}{3}x = 2$   or   $\frac{1}{3}x = -2$

$x = 6$          $x = -6$

**18.** $|z - 100| = 100$

$z - 100 = 100$   or   $z - 100 = -100$

$z = 200$          $z = 0$

**20.** $|7a + 6| = 8$

$7a + 6 = 8$   or   $7a + 6 = -8$

$7a = 2$          $7a = -14$

$a = \frac{2}{7}$          $a = -2$

**22.** $|3 - 5x| = 13$

$3 - 5x = 13$   or   $3 - 5x = -13$

$-5x = 10$          $-5x = -16$

$x = -2$          $x = \frac{16}{5}$

**24.** $|20 - 5t| = 50$

$20 - 5t = 50$   or   $20 - 5t = -50$

$-5t = 30$          $-5t = -70$

$t = -6$          $t = 14$

**26.** $|3x - 2| = -5$

An absolute value cannot be negative. No solution

**28.** $\left|3 - \frac{4}{5}x\right| = 1$

$3 - \frac{4}{5}x = 1$   or   $3 - \frac{4}{5}x = -1$

$-\frac{4}{5}x = -2$          $-\frac{4}{5}x = -4$

$x = \frac{5}{2}$          $x = 5$

**30.** $|2 - 1.054x| = 2$

$2 - 1.054x = 2$   or   $2 - 1.054x = -2$

$-1.054x = 0$          $-1.054x = -4$

$x = 0$          $x \approx 3.80$

**32.** $|6x - 4| - 7 = 3$

$|6x - 4| = 10$

$6x - 4 = 10$   or   $6x - 4 = -10$

$6x = 14$          $6x = -6$

$x = \frac{14}{6}$          $x = -1$

$x = \frac{7}{3}$

**34.** $|5 - 2x| + 10 = 6$

$|5 - 2x| = -4$

An absolute value cannot be negative. No solution

**36.** $4|5x + 1| = 24$

$|5x + 1| = 6$

$5x + 1 = 6$   or   $5x + 1 = -6$

$5x = 5$          $5x = -7$

$x = 1$          $x = -\frac{7}{5}$

**38.** $2|4 - 3x| - 6 = -2$

$2|4 - 3x| = 4$

$|4 - 3x| = 2$

$4 - 3x = 2$   or   $4 - 3x = -2$

$-3x = -2$          $-3x = -6$

$x = \frac{2}{3}$          $x = 2$

**40.** $|10 - 3x| = |x + 7|$

$10 - 3x = x + 7$   or   $10 - 3x = -(x + 7)$

$10 = 4x + 7$          $10 - 3x = -x - 7$

$3 = 4x$          $17 = 2x$

$\frac{3}{4} = x$          $\frac{17}{2} = x$

**42.** $|x - 2| = |2x - 15|$

$x - 2 = 2x - 15$    or    $x - 2 = -(2x - 15)$

$\qquad 13 = x \qquad\qquad\qquad x - 2 = -2x + 15$

$\qquad\qquad\qquad\qquad\qquad\qquad 3x = 17$

$\qquad\qquad\qquad\qquad\qquad\qquad x = \frac{17}{3}$

**44.** $|5x + 4| = |3x + 25|$

$5x + 4 = 3x + 25$    or    $5x + 4 = -(3x + 25)$

$\qquad 2x = 21 \qquad\qquad\qquad 5x + 4 = -3x - 25$

$\qquad x = \frac{21}{2} \qquad\qquad\qquad\qquad 8x = -29$

$\qquad\qquad\qquad\qquad\qquad\qquad\qquad x = -\frac{29}{8}$

**46.** $3|2 - 3x| = |9x + 21|$

$3(2 - 3x) = 9x + 21$    or    $3(2 - 3x) = -(9x + 21)$

$6 - 9x = 9x + 21 \qquad\qquad 6 - 9x = -9x - 21$

$6 - 18x = 21 \qquad\qquad\qquad 6 = -21$

$-18x = 15 \qquad\qquad$ No solution

$x = \dfrac{15}{-18}$

$x = -\dfrac{5}{6}$

**48.** $|t + 2| = 6$

**50.** (a) $|-7| \leq 5$

$\qquad 7 \leq 5$

$\qquad$ No

(b) $|-4| \leq 5$

$\qquad 4 \leq 5$

$\qquad$ Yes

(c) $|4| \leq 5$

$\qquad 4 \leq 5$

$\qquad$ Yes

(d) $|9| \leq 5$

$\qquad 9 \leq 5$

$\qquad$ No

**52.** (a) $|16 - 3| > 5$

$\qquad |13| > 5$

$\qquad\ \ 13 > 5$

$\qquad$ Yes

(b) $|3 - 3| > 5$

$\qquad |0| > 5$

$\qquad\ 0 > 5$

$\qquad$ No

(c) $|-2 - 3| > 5$

$\qquad |-5| > 5$

$\qquad\ \ 5 > 5$

$\qquad$ No

(d) $|-3 - 3| > 5$

$\qquad |-6| > 5$

$\qquad\ \ 6 > 5$

$\qquad$ Yes

**54.** $-5 \leq 6x + 7 \leq 5$

**56.** $8 - x > 25$ or $8 - x < -25$

**58.**

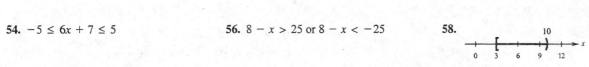

**60.**

**62.** $|x| < 6$

$\quad -6 < x < 6$

**64.** $|y| \geq 4$

$\quad y \leq -4$ or $y \geq 4$

**66.** $|4z| \leq 9$

$\quad -9 \leq 4z \leq 9$

$\quad -\frac{9}{4} \leq z \leq \frac{9}{4}$

**68.** $\left|\dfrac{t}{2}\right| < 4$

$\quad -4 < \dfrac{t}{2} < 4$

$\quad -8 < t < 8$

**70.** $|x - 3| \leq 6$

$\quad -6 \leq x - 3 \leq 6$

$\quad -3 \leq x \leq 9$

**72.** $|x - 4| \geq 3$

$\quad x - 4 \geq 3$    or    $x - 4 \leq -3$

$\qquad\ x \geq 7 \qquad\qquad\quad x \leq 1$

**74.** $|3x + 4| < 2$

$\quad -2 < 3x + 4 < 2$

$\quad -6 < 3x < -2$

$\quad -2 < x < -\frac{2}{3}$

**76.** $|3t + 1| > 5$

$\quad 3t + 1 > 5$    or    $3t + 1 < -5$

$\qquad 3t > 4 \qquad\qquad\quad 3t < -6$

$\qquad t > \frac{4}{3} \qquad\qquad\quad t < -2$

**78.** $|8 - 7x| < -6$

Absolute value is never negative.
No solution

**80.** $|4x - 5| > -3$

$-\infty < x < \infty$

Absolute value is always positive.

**82.** $\dfrac{|s - 3|}{5} > 4$

$|s - 3| > 20$

$s - 3 < -20$   or   $s - 3 > 20$

$s < -17$              $s > 23$

**84.** $\dfrac{|a + 6|}{2} \geq 16$

$|a + 6| \geq 32$

$a + 6 > 32$   or   $a + 6 \leq -32$

$a \geq 26$              $a \leq -38$

**86.** $\left|\dfrac{x}{8} + 1\right| < 0$

An absolute value cannot be less
than zero. No solution

**88.** $|1.5t - 8| \leq 16$

$-16 \leq 1.5t - 8 \leq 16$

$-8 \leq 1.5t \leq 24$

$\dfrac{-8}{1.5} \leq \dfrac{1.5t}{1.5} \leq \dfrac{2.4}{1.5}$

$-5.\overline{3} \leq t \leq 16$

**90.** $\left|3 - \dfrac{x}{4}\right| > 0.15$

$3 - \dfrac{x}{4} > 0.15$   or   $3 - \dfrac{x}{4} < -0.15$

$12 - x > 0.6$         $12 - x < -0.6$

$-x > -11.4$         $-x < -12.6$

$x < 11.4$         $x > 12.6$

$x < \dfrac{57}{5}$         $x > \dfrac{63}{5}$

**92.** $-4|2x - 7| > -12$

$|2x - 7| < 3$

$-3 < 2x - 7 < 3$

$4 < 2x < 10$

$2 < x < 5$

**94.** $\left|8 - \dfrac{2}{3}x\right| + 6 \geq 10$

$\left|8 - \dfrac{2}{3}x\right| \geq 4$

$8 - \dfrac{2}{3}x \geq 4$   or   $8 - \dfrac{2}{3}x \leq -4$

$-\dfrac{2}{3}x \geq -4$         $-\dfrac{2}{3}x \leq -12$

$x \leq 6$         $x \geq 18$

**96.** $|2x - 1| \leq 3$

Keystrokes: [Y=] [ABS] [(] 2 [X,T,θ] [−] 1 [)] [≤] 3 [GRAPH]

$-1 \leq x \leq 2$

**98.** $|7r - 3| > 11$

Keystrokes: [Y=] [ABS] [(] 7 [X,T,θ] [−] 3 [)] [>] 11 [GRAPH]

$r < -\dfrac{8}{7}$ or $r > 2$

**100.** $|a + 1| - 4 < 0$

Keystrokes: [Y=] [ABS] [(] [X,T,θ] [+] 1 [)] [−] 4 [<] 0 [GRAPH]

$-5 < a < 3$

**102.** Matches graph (c).

$|x - 4| < 1$

$-1 < x - 4 < 1$

$3 < x < 5$

**104.** Matches graph (a).

$|2(x - 4)| \geq 4$

$2(x - 4) \geq 4$   or   $2(x - 4) \leq -4$

$x - 4 \geq 2$         $x - 4 \leq -2$

$x \geq 6$         $x \leq 2$

**106.** $(-4, 4)$

$|x| < 4$

**108.** $[-13, -9]$

$$-13 \le x \le -9$$
$$-2 \le x + 11 \le 2$$
$$|x + 11| \le 2$$

**110.** $|x| > 2$

**112.** $|x - 16| < 5$

**114.** $\left|\dfrac{t - 15.6}{1.9}\right| < 1$

$$-1 < \dfrac{t - 15.6}{1.9} < 1$$
$$-1.9 < t - 15.6 < 1.9$$
$$13.7 < t < 17.5$$

**116.** Maximum error for each bag is $\frac{1}{2}$ ounce. For four bags, the maximum error is $4\left(\frac{1}{2}\right) = 2$ ounces. The greatest amount you can expect to get is $4(16) + 2 = 66$ ounces. The least amount is $4(16) - 2 = 62$ ounces.

**118.** An absolute value equation that has only one solution is $|x| = 0$.

**120.** The solution of $|x| > 3$ is all real numbers more than 3 units from 0.

**122.** The graph of $|y + 2| > 4$ can be described as all real numbers more than 4 units from $-2$.

**124.** $|2x - 6| \le 6$ since:

$$-6 \le 2x - 6 \le 6$$
$$0 \le 2x \le 12$$
$$0 \le x \le 6$$

# Review Exercises for Chapter 1

**2.** (a) $3\left(3 - \frac{9}{2}\right) = -\frac{9}{2}$

$$3\left(\frac{6}{2} - \frac{9}{2}\right) = -\frac{9}{2}$$
$$3\left(-\frac{3}{2}\right) = -\frac{9}{2}$$
$$-\frac{9}{2} = -\frac{9}{2}$$

Solution

(b) $3\left[3 - \left(-\frac{2}{3}\right)\right] = -\left(-\frac{2}{3}\right)$

$$3\left(\frac{9}{3} + \frac{2}{3}\right) = \frac{2}{3}$$
$$3\left(\frac{11}{3}\right) = \frac{2}{3}$$
$$11 = \frac{2}{3}$$

Not a solution

**4.** (a) $\dfrac{-12 + 2}{6} = \dfrac{7}{2}$

$$\dfrac{-10}{6} = \dfrac{7}{2}$$
$$-\dfrac{5}{3} = \dfrac{7}{2}$$

Not a solution

(b) $\dfrac{19 + 2}{6} = \dfrac{7}{2}$

$$\dfrac{21}{6} = \dfrac{7}{2}$$
$$\dfrac{7}{2} = \dfrac{7}{2}$$

Solution

**6.** $x - 7 = 3$

$$x - 7 + 7 = 3 + 7$$
$$x = 10$$

Check:

$$10 - 7 \overset{?}{=} 3$$
$$3 = 3$$

**8.** $11x = 44$

$$\dfrac{11x}{11} = \dfrac{44}{11}$$
$$x = 4$$

Check:

$$11(4) \overset{?}{=} 44$$
$$44 = 44$$

**10.** $\dfrac{1}{10}x = 5$

$$10\left(\dfrac{1}{10}x\right) = (5)10$$
$$x = 50$$

Check:

$$\dfrac{1}{10}(50) \overset{?}{=} 5$$
$$5 = 5$$

**12.**

$$3 - 2x = 9$$
$$3 - 3 - 2x = 9 - 3$$
$$-2x = 6$$
$$\frac{-2x}{-2} = \frac{6}{-2}$$
$$x = -3$$

**Check:**

$$3 - 2(-3) \overset{?}{=} 9$$
$$3 + 6 \overset{?}{=} 9$$
$$9 = 9$$

**14.**

$$3 + 6x = 51$$
$$3 + 6x - 3 = 51 - 3$$
$$6x = 48$$
$$\frac{6x}{6} = \frac{48}{6}$$
$$x = 8$$

**Check:**

$$3 + 6(8) \overset{?}{=} 51$$
$$3 + 48 \overset{?}{=} 51$$
$$51 = 51$$

**16.**

$$9 - 2x = 4x - 7$$
$$9 - 9 - 2x = 4x - 7 - 9$$
$$-2x = 4x - 16$$
$$-2x - 4x = 4x - 4x - 16$$
$$-6x = -16$$
$$\frac{-6x}{-6} = \frac{-16}{-6}$$
$$x = \frac{8}{3}$$

**Check:**

$$9 - 2\left(\frac{8}{3}\right) \overset{?}{=} 4\left(\frac{8}{3}\right) - 7$$
$$9 - \frac{16}{3} \overset{?}{=} \frac{32}{3} - 7$$
$$\frac{27}{3} - \frac{16}{3} \overset{?}{=} \frac{32}{3} - \frac{21}{3}$$
$$\frac{11}{3} = \frac{11}{3}$$

**18.**

$$-2(x + 4) = 2x - 7$$
$$-2x - 8 = 2x - 7$$
$$-2x - 2x - 8 = 2x - 2x - 7$$
$$-4x - 8 = -7$$
$$-4x - 8 + 8 = -7 + 8$$
$$-4x = 1$$
$$\frac{-4x}{-4} = \frac{1}{-4}$$
$$x = -\frac{1}{4}$$

**Check:**

$$-2\left(-\frac{1}{4} + 4\right) \overset{?}{=} 2\left(-\frac{1}{4}\right) - 7$$
$$\frac{1}{2} - 8 \overset{?}{=} -\frac{1}{2} - 7$$
$$\frac{1}{2} - \frac{16}{2} \overset{?}{=} -\frac{1}{2} - \frac{14}{2}$$
$$-\frac{15}{2} = -\frac{15}{2}$$

**20.**

$$7x + 2(7 - x) = 8$$
$$7x + 14 - 2x = 8$$
$$5x + 14 = 8$$
$$5x + 14 - 14 = 8 - 14$$
$$5x = -6$$
$$\frac{5x}{5} = \frac{-6}{5}$$
$$x = -\frac{6}{5}$$

**Check:**

$$7\left(-\frac{6}{5}\right) + 2\left[7 - \left(\frac{6}{5}\right)\right] \overset{?}{=} 8$$
$$-\frac{42}{5} + 2\left(\frac{35}{5} + \frac{6}{5}\right) \overset{?}{=} 8$$
$$-\frac{42}{5} + 2\left(\frac{41}{5}\right) \overset{?}{=} 8$$
$$-\frac{42}{5} + \frac{82}{5} \overset{?}{=} 8$$
$$\frac{40}{5} \overset{?}{=} 8$$
$$8 = 8$$

**22.**
$$8(x - 2) = 3(x - 2)$$
$$8x - 16 = 3x - 6$$
$$8x - 16 - 3x = 3x - 6 - 3x$$
$$5x - 16 = -6$$
$$5x - 16 + 16 = -6 + 16$$
$$5x = 10$$
$$\frac{5x}{5} = \frac{10}{5}$$
$$x = 2$$

**Check:**
$$8(2 - 2) \stackrel{?}{=} 3(2 - 2)$$
$$8(0) \stackrel{?}{=} 3(0)$$
$$0 = 0$$

**24.**
$$\frac{1}{4}s + \frac{3}{8} = \frac{5}{2}$$
$$8\left(\frac{1}{4}s + \frac{3}{8}\right) = 8\left(\frac{5}{2}\right)$$
$$2s + 3 = 20$$
$$2s + 3 - 3 = 20 - 3$$
$$2s = 17$$
$$\frac{2s}{2} = \frac{17}{2}$$
$$s = \frac{17}{2}$$

**Check:**
$$\frac{1}{4}\left(\frac{17}{2}\right) + \frac{3}{8} \stackrel{?}{=} \frac{5}{2}$$
$$\frac{17}{8} + \frac{3}{8} \stackrel{?}{=} \frac{5}{2}$$
$$\frac{20}{8} \stackrel{?}{=} \frac{5}{2}$$
$$\frac{5}{2} = \frac{5}{2}$$

**26.**
$$2.5x - 6.2 = 3.7x - 5.8$$
$$2.5x - 3.7x - 6.2 = 3.7x - 3.7x - 5.8$$
$$-1.2x - 6.2 = -5.8$$
$$-1.2x - 6.2 + 6.2 = -5.8 + 6.2$$
$$-1.2x = 0.4$$
$$\frac{-1.2x}{-1.2} = \frac{0.4}{-1.2}$$
$$x = -\frac{1}{3}$$

**Check:**
$$2.5\left(-\frac{1}{3}\right) - 6.2 \stackrel{?}{=} 3.7\left(-\frac{1}{3}\right) - 5.8$$
$$-\frac{2.5}{3} - \frac{18.6}{3} \stackrel{?}{=} -\frac{3.7}{3} - \frac{17.4}{3}$$
$$-\frac{21.1}{3} = -\frac{21.1}{3}$$

**28.**

| Percent | Parts out of 100 | Decimal | Fraction |
|---|---|---|---|
| $16\frac{2}{3}\%$ | $16\frac{2}{3}$ | $0.1\overline{6}$ | $\frac{1}{6}$ |

**30.**
*Verbal Model:* $\boxed{\text{Compared number}} = \boxed{\text{Percent}} \cdot \boxed{\text{Base number}}$

*Labels:* Compared number $= a$
Percent $= p$
Base number $= b$

*Equation:* $a = p \cdot b$
$$a = 0.004(7350)$$
$$a = 29.4$$

**32.**
*Verbal Model:* $\boxed{\text{Compared number}} = \boxed{\text{Percent}} \cdot \boxed{\text{Base number}}$

*Labels:* Compared number $= a$
Percent $= p$
Base number $= b$

*Equation:* $a = p \cdot b$
$$498 = 0.83 \cdot b$$
$$\frac{498}{0.83} = b$$
$$600 = b$$

**34.**
*Verbal Model:* $\boxed{\text{Compared number}} = \boxed{\text{Percent}} \cdot \boxed{\text{Base number}}$

*Labels:* Compared number $= a$
Percent $= p$
Base number $= b$

*Equation:* $a = p \cdot b$
$$162.5 = p(6500)$$
$$\frac{162.5}{6500} = p$$
$$0.025 = p \text{ or } 2.5\%$$

**36.** $\dfrac{3 \text{ quarts}}{5 \text{ pints}} = \dfrac{6 \text{ pints}}{5 \text{ pints}} = \dfrac{6}{5}$

**38.** $\dfrac{3 \text{ meters}}{150 \text{ centimeters}} = \dfrac{300 \text{ centimeters}}{150 \text{ centimeters}} = \dfrac{2}{1}$

**40.** $\dfrac{x}{16} = \dfrac{5}{12}$

$16\left(\dfrac{x}{16}\right) = 16\left(\dfrac{5}{12}\right)$

$x = \dfrac{20}{3}$

**42.** $\dfrac{x+1}{3} = \dfrac{x-1}{2}$

$2(x+1) = 3(x-1)$

$2x + 2 = 3x - 3$

$2x - 3x + 2 = 3x - 3x - 3$

$-x + 2 = -3$

$-x + 2 - 2 = -3 - 2$

$-x = -5$

$x = 5$

**44.** *Verbal Model:*   $\boxed{\text{Selling price}} = \boxed{\text{Cost}} + \boxed{\text{Markup}}$

*Labels:*   Selling price = 31.33
Cost = 23.50
Markup = $x$

*Equation:*   $31.33 = 23.50 + x$

$31.33 - 23.50 = x$

$\$7.83 = x$

*Verbal Model:*   $\boxed{\text{Markup}} = \boxed{\text{Markup rate}} \cdot \boxed{\text{Cost}}$

*Labels:*   Markup = 7.83
Markup rate = $x$
Cost = 23.50

*Equation:*   $7.83 = x \cdot 23.50$

$\dfrac{7.83}{23.50} = x$

$33.3\% \approx x$

**46.** *Verbal Model:*   $\boxed{\text{Selling price}} = \boxed{\text{Cost}} + \boxed{\text{Markup}}$

*Labels:*   Selling price = 895.00
Cost = $x$
Markup = 223.75

*Equation:*   $895.00 = x + 223.75$

$895.00 - 223.75 = x$

$\$671.25 = x$

*Verbal Model:*   $\boxed{\text{Markup}} = \boxed{\text{Markup rate}} \cdot \boxed{\text{Cost}}$

*Labels:*   Markup = 223.75
Markup rate = $x$
Cost = 671.25

*Equation:*   $223.75 = x \cdot 671.25$

$\dfrac{223.75}{671.25} = x$

$33.3\% \approx x$

**48.** *Verbal Model:* [Sale price] = [List price] − [Discount]

*Labels:* Sale price = 279.98

List price = 559.95

Discount = $x$

*Equation:* $279.98 = 559.95 - x$

$x = 559.95 - 279.98$

$x = \$279.97$

*Verbal Model:* [Discount] = [Discount rate] · [List price]

*Labels:* Discount = 279.97

Discount rate = $x$

List price = 559.95

*Equation:* $279.97 = x \cdot 559.95$

$\frac{279.97}{559.95} = x$

$50\% \approx x$

**50.** *Verbal Model:* [Sale price] = [List price] − [Discount]

*Labels:* Sale price = $x$

List price = 39.00

Discount = 15.60

*Equation:* $x = 39.00 - 15.60$

$x = \$23.40$

*Verbal Model:* [Discount] = [Discount rate] · [List price]

*Labels:* Discount = 15.60

Discount rate = $x$

List price = 39.00

*Equation:* $15.60 = x \cdot 39.00$

$\frac{15.60}{39.00} = x$

$40\% = x$

**52.** $\frac{2}{3}u - 4v = 2v + 3$

$\frac{2}{3}u - 3 = 6v$

$\frac{1}{9}u - \frac{1}{2} = v$

$\frac{1}{18}(2u - 9) = v$

**54.** $S = 2\pi r^2 + 2\pi rh$

$S - 2\pi r^2 = 2\pi rh$

$\frac{S - 2\pi r^2}{2\pi r} = h$

**56.** $x + 8 > 5$

$x + 8 - 8 > 5 - 8$

$x > -3$

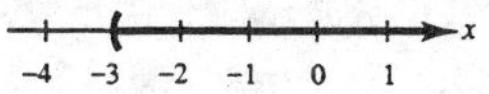


**58.** $-11x \geq 44$

$x \leq -4$

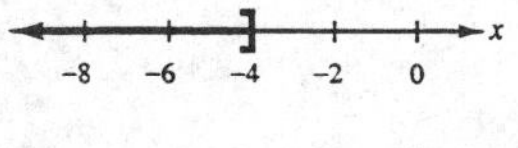


**60.** $3x - 11 \leq 7$

$3x - 11 + 11 \leq 7 + 11$

$3x \leq 18$

$\frac{3x}{3} \leq \frac{18}{3}$

$x \leq 6$

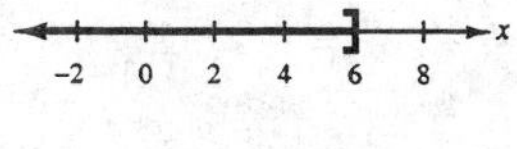


**62.** $12 - 3x < 4x - 2$

$12 - 3x - 4x < 4x - 4x - 2$

$12 - 7x < -2$

$12 - 12 - 7x < -2 - 12$

$-7x < -14$

$\frac{-7x}{-7} > \frac{-14}{-7}$

$x > 2$

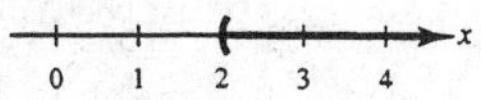


**64.** $\frac{x}{4} - 2 < \frac{3x}{8} + 5$

$8\left(\frac{x}{4} - 2\right) < \left(\frac{3x}{8} + 5\right)8$

$2x - 16 < 3x + 40$

$2x - 3x - 16 < 3x - 3x + 40$

$-x - 16 < 40$

$-x - 16 + 16 < 40 + 16$

$-x < 56$

$x > -56$

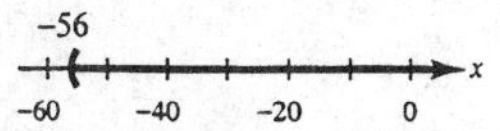

**66.** $-13 \le 3 - 4x < 13$

$-16 \le -4x < 10$

$4 \ge x > -\dfrac{10}{4}$

$4 \ge x > -\dfrac{5}{2}$

$-\dfrac{5}{2} < x \le 4$

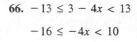

**68.** $12 \ge \dfrac{x-3}{2} > 1$

$24 \ge x - 3 > 2$

$27 \ge x > 5$

$5 < x \le 27$

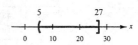

**70.**

$6 - 2x \le 1$ or $10 - 4x > -6$

$6 - 6 - 2x \le 1 - 6$ or $10 - 10 - 4x > -6 - 10$

$-2x \le -5$ or $-4x > -16$

$\dfrac{-2x}{-2} \ge \dfrac{-5}{-2}$ or $\dfrac{-4x}{-4} < \dfrac{-16}{-4}$

$x \ge \dfrac{5}{2}$ or $x < 4$

$-\infty < x < \infty$

**72.** $3(2 - y) \ge 2(1 + y)$

$6 - 3y \ge 2 + 2y$

$4 \ge 5y$

$\dfrac{4}{5} \ge y$

**74.** $x \ge 0$

**76.** $V < 27$

**78.** $|x| = -4$

No solution. Absolute value cannot be negative.

**80.** $|2x + 3| = 7$

$2x + 3 = 7$ or $2x + 3 = -7$

$2x = 4$       $2x = -10$

$x = 2$       $x = -5$

**82.** $|x - 2| - 2 = 4$

$|x - 2| = 6$

$x - 2 = 6$ or $x - 2 = -6$

$x = 8$      $x = -4$

**84.** $|5x + 6| = |2x - 1|$

$5x + 6 = 2x - 1$ or $5x + 6 = -(2x - 1)$

$3x = -7$      $5x + 6 = -2x + 1$

$x = -\dfrac{7}{3}$      $7x = -5$

$x = -\dfrac{5}{7}$

**86.** $|t + 3| > 2$

$t + 3 > 2$ or $t + 3 < -2$

$t > -1$ or $t < -5$

**88.** $\left|\dfrac{t}{3}\right| < 1$

$-1 < \dfrac{t}{3} < 1$

$-3 < t < 3$

**90.** $|5x - 1| < 9$

$-9 < 5x - 1 < 9$

$-8 < 5x < 10$

$-\dfrac{8}{5} < x < 2$

**92.** $|2y - 1| + 4 < -1$

$|2y - 1| < -5$

No solution. Absolute value cannot be negative.

**94.** $|5(1 - x)| \le 25$

Keystrokes: Y= ABS ( 5 − 5 X,T,θ ) ≤ 25 GRAPH

$-4 \le x \le 6$ or $[-4, 6]$

**96.** $[-18, -12]$

$-18 \le x \le -12$

$-3 \le x + 15 \le 3$

$|x + 15| \le 3$

**98.** *Verbal Model:* $\boxed{\text{First integer}} + \boxed{\text{Second integer}} = \boxed{\text{Sum}}$

*Labels:* First integer $= x$

Second integer $= x + 2$

Sum $= 74$

*Equation:* $x + (x + 2) = 74$

$2x + 2 = 74$

$2x = 72$

$x = 36, \quad x + 2 = 38$

**100.** *Verbal Model:* $\boxed{\text{Defective parts}} = \boxed{\text{Percent rate}} \cdot \boxed{\text{Sample}}$

*Labels:* Defective parts $= 6$

Percent rate $= 1.6\%$

Sample $= x$

*Equation:* $6 = 0.016 \cdot x$

$\dfrac{6}{0.016} = x$

$375 = x$

**102.** *Verbal Model:* $\boxed{\text{Retirement}} = \boxed{\text{Percent rate}} \cdot \boxed{\text{Gross income}}$

*Labels:* Retirement $= 216$

Percent rate $= x$

Gross income $= 3200$

*Equation:* $216 = x \cdot 3200$

$\dfrac{216}{3200} = x$

$6.75\% = x$

**104.** *Verbal Model:* $\dfrac{\text{Tax}}{\text{Assessed value}} = \dfrac{\text{Tax}}{\text{Assessed value}}$

*Proportion:* $\dfrac{1680}{105{,}000} = \dfrac{x}{125{,}000}$

$x = \dfrac{(1680)(125{,}000)}{105{,}000}$

$x = \$2000$

**106.** *Verbal Model:* $\dfrac{\text{Inches}}{\text{Miles}} = \dfrac{\text{Inches}}{\text{Miles}}$

*Proportion:* $\dfrac{\frac{1}{3}}{50} = \dfrac{3\frac{1}{4}}{x}$

$\dfrac{1}{3}x = 50\left(3\dfrac{1}{4}\right)$

$\dfrac{1}{3}x = 50\left(\dfrac{13}{4}\right)$

$\dfrac{1}{3}x = \dfrac{325}{2}$

$3\left(\dfrac{1}{3}x\right) = 3\left(\dfrac{325}{2}\right)$

$x = \dfrac{975}{2}$

$x = 487.5$ miles

**108.** *Verbal Model:* $\dfrac{\text{Leg 1}}{\text{Leg 2}} = \dfrac{\text{Leg 1}}{\text{Leg 2}}$

*Proportion:* $\dfrac{2}{6} = \dfrac{x}{9}$

$6x = 18$

$x = 3$

**110.** *Verbal Model:* $\dfrac{\text{Flagpole's height}}{\text{Length of flagpole's shadow}} = \dfrac{\text{Lamp post's height}}{\text{Length of lamp post's shadow}}$

*Proportion:* $\dfrac{h}{30} = \dfrac{5}{3}$

$h = 30 \cdot \dfrac{5}{3} = 50$ feet

**112.** *Verbal Model:* $\boxed{\text{Increase}} = \boxed{\text{Percent}} \cdot \boxed{\begin{array}{c}1999\\ \text{revenue}\end{array}}$

*Labels:*    Increase = 157.60

Percent = $x$

1999 revenue = 4521.40

*Equation:*    $157.60 = x \cdot 4521.40$

$\dfrac{157.60}{4521.40} = x$

$x \approx 3.5\%$

**114.** *Verbal Model:* $\boxed{\text{Markup}} = \boxed{\begin{array}{c}\text{Markup}\\ \text{rate}\end{array}} \cdot \boxed{\text{Cost}}$

*Labels:*    Markup = $x$

Markup rate = 35%

Cost = 259.95

*Equation:*    $x = 0.35 \cdot 259.95$

$x = \$90.98$

*Verbal Model:* $\boxed{\begin{array}{c}\text{Selling}\\ \text{price}\end{array}} = \boxed{\text{Cost}} + \boxed{\text{Markup}}$

*Labels:*    Selling price = $x$

Cost = 259.95

Markup = 90.98

*Equation:*    $x = 259.95 + 90.98$

$x = \$350.93$

**116.** *Verbal Model:* $\boxed{\text{Discount}} = \boxed{\begin{array}{c}\text{Discount}\\ \text{rate}\end{array}} \cdot \boxed{\begin{array}{c}\text{List}\\ \text{price}\end{array}}$

*Labels:*    Discount = $x$

Discount rate = 25%

List price = 259

*Equation:*    $x = 0.25 \cdot \$259$

$x = \$64.75$

*Verbal Model:* $\boxed{\begin{array}{c}\text{Sale}\\ \text{price}\end{array}} = \boxed{\begin{array}{c}\text{List}\\ \text{price}\end{array}} - \boxed{\text{Discount}}$

*Labels:*    Sale price = $x$

List price = 259

Discount = 64.75

*Equation:*    $x = 259 - 64.75$

$x = \$194.25$

**118.** *Verbal Model:* $\boxed{\begin{array}{c}\text{Total}\\ \text{salary}\end{array}} = \boxed{\begin{array}{c}\text{Weekly}\\ \text{salary}\end{array}} + \boxed{\text{Rate}} \cdot \boxed{\begin{array}{c}\text{Total}\\ \text{sales}\end{array}}$

*Labels:*    Total salary = 650

Weekly salary = 150

Rate = 6%

*Equation:*    $650 = 150 + 0.06x$

$500 = 0.06x$

$\$8333.33 = x$

**120.** *Verbal Model:* $\boxed{\begin{array}{c}\text{Amount of}\\ \text{solution 1}\end{array}} + \boxed{\begin{array}{c}\text{Amount of}\\ \text{solution 2}\end{array}} = \boxed{\begin{array}{c}\text{Amount of}\\ \text{final solution}\end{array}}$

*Labels:*    Percent of solution 1 = 25%

Gallons of solution 1 = $x$

Percent of solution 2 = 50%

Gallons of solution 2 = $8 - x$

Percent of final solution = 40%

Gallons of final solution = 8

*Equation:*    $0.25x + 0.50(8 - x) = 0.40(8)$

$0.25x + 4 - 0.50x = 3.2$

$-0.25x = -0.8$

$x = 3.2$ gallons of solution 1

$8 - x = 4.8$ gallons of solution 2

**122.** *Verbal Model:*   $\boxed{\text{Distance}} = \boxed{\text{Rate}} \cdot \boxed{\text{Time}}$

*Labels:*   Distance = 330 miles

Rate = 52 mph

Time = $t$

*Equation:*   $330 = 52 \cdot t$

$$\frac{330}{52} = t$$

6.35 hours = $t$

**124.** *Verbal Model:*   $\boxed{\text{Distance}} = \boxed{\text{Rate}} \cdot \boxed{\text{Time}}$

*Labels:*   Distance = 80 miles at 40 mph

320 miles at $x$ mph

400 miles at 50 mph

Rates = 40, $r$, 50

Time = 8 hours

*Equation:*   $\dfrac{80}{40} + \dfrac{320}{r} = \dfrac{400}{50}$

$$2 + \frac{320}{r} = 8$$

$$\frac{320}{r} = 6$$

$$320 = 6r$$

$$53\tfrac{1}{3} \text{ mph} = r$$

**126.** *Verbal Model:*   $\boxed{\begin{array}{c}\text{Work}\\\text{done}\end{array}} = \boxed{\begin{array}{c}\text{Work done}\\\text{by person 1}\end{array}} + \boxed{\begin{array}{c}\text{Work done}\\\text{by person 2}\end{array}}$

*Labels:*   Work done = $\dfrac{1}{2}$

Rate of person 1 = $\dfrac{1}{8}$

Rate of person 2 = $\dfrac{1}{10}$

Time = $t$

*Equation:*   $\dfrac{1}{2} = \dfrac{1}{8}(t) + \dfrac{1}{10}(t)$

$$\frac{1}{2} = \frac{5}{40}t + \frac{4}{20}t$$

$$\frac{1}{2} = \frac{9}{40}t$$

$$\left(\frac{40}{9}\right)\left(\frac{1}{2}\right) = \left(\frac{40}{9}\right)\left(\frac{9}{40}t\right)$$

$$\frac{20}{9} = t$$

$$t \approx 2.22 \text{ hours}$$

**128.** *Verbal Model:*   $\boxed{\text{Interest}} = \boxed{\text{Principal}} \cdot \boxed{\text{Rate}} \cdot \boxed{\text{Time}}$

*Labels:*   Interest = 37.50

Principal = 500.00

Rate = $r$

Time = 1

*Equation:*   $37.50 = 500.00 \cdot r \cdot 1$

$$\frac{37.50}{500.00} = r$$

$$7.5\% = r$$

**130.** *Verbal Model:*   $\boxed{\text{Interest}} = \boxed{\text{Principal}} \cdot \boxed{\text{Rate}} \cdot \boxed{\text{Time}}$

*Labels:*   Interest = $i$

Principal = 3,250,000

Rate = 12%

Time = 2

*Equation:*   $i = (3,250,000)(0.12)(2)$

$i = \$780,000$

*Verbal Model:*   $\boxed{\begin{array}{c}\text{Total}\\\text{repaid}\end{array}} = \boxed{\text{Principal}} + \boxed{\text{Interest}}$

*Labels:*   Total repaid = $x$

Principal = 3,250,000

Interest = 780,000

*Equation:*   $x = 3,250,000 + 780,000$

$x = \$4,030,000$

**132.**

*Verbal Model:*  $\boxed{\text{Interest}} = \boxed{\text{Principal}} \cdot \boxed{\text{Rate}} \cdot \boxed{\text{Time}}$

*Labels:*  Interest = $i$
Principal = $\$1000$
Rate = $7\%$
Time = $\frac{1}{2}$

*Equation:*  $i = 1000 \cdot 0.07 \cdot \frac{1}{2}$
$i = \$35$

*Verbal Model:*  $\boxed{\text{New principal}} = \boxed{\text{Principal}} + \boxed{\text{Interest}}$

*Labels:*  New principal = $x$
Principal = 1000
Interest = 35

*Equation:*  $x = 1000 + 35$
$x = \$1035$

*Verbal Model:*  $\boxed{\text{Interest}} = \boxed{\text{New principal}} \cdot \boxed{\text{Rate}} \cdot \boxed{\text{Time}}$

*Labels:*  Interest = $i$
New principal = 1035
Rate = $7\%$
Time = $\frac{1}{2}$

*Equation:*  $i = 1035 \cdot 0.07 \cdot \frac{1}{2}$
$i = \$36.23$

*Verbal Model:*  $\boxed{\text{Total interest}} = \boxed{\begin{array}{c}\text{Interest} \\ \text{in first} \\ \text{6 months}\end{array}} + \boxed{\begin{array}{c}\text{Interest} \\ \text{in second} \\ \text{6 months}\end{array}}$

*Labels:*  Total interest = $i$
Interest in first 6 months = 35
Interest in second 6 months = 36.23

*Equation:*  $i = 35 + 36.23$
$i = \$71.23$

**136.**

*Verbal Model:*  $\boxed{\begin{array}{c}\text{Total} \\ \text{cost}\end{array}} \geq \boxed{\begin{array}{c}\text{Cost} \\ \text{of first} \\ \text{minute}\end{array}} + 0.49 \cdot \boxed{\begin{array}{c}\text{Cost of} \\ \text{additional} \\ \text{minutes}\end{array}}$

*Labels:*  Total cost = 7.50
Cost of first minute = 0.99
Number of additional minutes = $x$

*Inequality:*  $7.50 \geq 0.99 + 0.49x > 0.99$
$6.51 \geq 0.49x > 0$
$13.3 \geq x > 0$
$0 < \text{length of call} \leq 14$

**134.**

*Verbal Model:*  $\text{Perimeter} = 2 \cdot \boxed{\text{Length}} + 2 \cdot \boxed{\text{Width}}$

*Labels:*  Perimeter = 110 feet
Length = $l$
Width = $\frac{3}{5}l$

*Equation:*
$$110 = 2l + 2\left(\tfrac{3}{5}l\right)$$
$$110 = 2l + \tfrac{6}{5}l$$
$$550 = 10l + 6l$$
$$550 = 16l$$
$$34.375 \text{ feet} = l$$
$$20.625 \text{ feet} = \tfrac{3}{5}l$$

**138.**  $|t - 77| < 27$
$-27 < t - 77 < 27$
$50 < t < 104$

# CHAPTER 2
# Graphs and Functions

# CHAPTER 2
# Graphs and Functions

## Section 2.1    The Rectangular Coordinate System

**Solutions to Even-Numbered Exercises**

**2.**

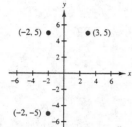

$(-2, 5)$ is 2 units to the left of the vertical axis and 5 units above the horizontal axis.

$(-2, -5)$ is 2 units to the left of the vertical axis and 5 units below the horizontal axis.

$(3, 5)$ is 3 units to the right of the vertical axis and 5 units above the horizontal axis.

**4.**

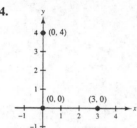

$(0, 4)$ is 0 units right or left of the vertical axis and 4 units above the horizontal axis.

$(0, 0)$ is the origin.

$(3, 0)$ is 3 units to the right of the vertical axis and 0 units above or below the horizontal axis.

**6.**

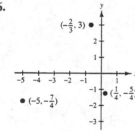

$\left(-\frac{2}{3}, 3\right)$ is $\frac{2}{3}$ unit to the left of the vertical axis and 3 units above the horizontal axis.

$\left(\frac{1}{4}, -\frac{5}{4}\right)$ is $\frac{1}{4}$ unit to the right of the vertical axis and $\frac{5}{4}$ units below the horizontal axis.

$\left(-5, -\frac{7}{4}\right)$ is 5 units to the left of the vertical axis and $\frac{7}{4}$ units below the horizontal axis.

**8.**

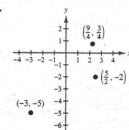

$(-3, -5)$ is 3 units to the left of the vertical axis and 5 units below the horizontal axis.

$\left(\frac{9}{4}, \frac{3}{4}\right)$ is $\frac{9}{4}$ units to the right of the vertical axis and $\frac{3}{4}$ units above the horizontal axis.

$\left(\frac{5}{2}, -2\right)$ is $\frac{5}{2}$ units to the right of the vertical axis and 2 units below the horizontal axis.

**10.**

| Point | Position | Coordinates |
|-------|----------|-------------|
| A | 3 left, 2 above | $(-3, 2)$ |
| B | 4 right, 1 below | $(4, -1)$ |
| C | $\frac{1}{2}$ left, $-2$ below | $\left(-\frac{1}{2}, -2\right)$ |

**12.**

| Point | Position | Coordinates |
|-------|----------|-------------|
| A | $\frac{7}{2}$ left, 2 below | $\left(-\frac{7}{2}, -2\right)$ |
| B | 5 right, 1 above | $(5, 1)$ |
| C | 0 right or left, 4 above | $(0, 4)$ |

**14.**

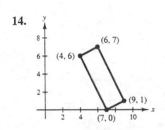

**16.**

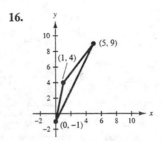

**18.**

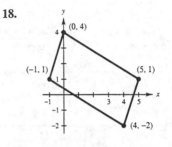

**20.**

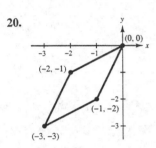

**22.** Point 10 units right of $y$-axis and 4 units below $x$-axis $= (10, -4)$.

**24.** Point 2 units right of $y$-axis and 5 units above $x$-axis $= (2, 5)$.

**26.** Coordinates of point are equal in magnitude and opposite in sign and point is 7 units right of $y$-axis $= (7, -7)$.

**28.** Point is on negative $y$-axis 5 units from the origin $= (0, -5)$.

**30.** $(4, -2)$ is in Quadrant IV.

**32.** $\left(-\frac{5}{11}, -\frac{3}{8}\right)$ is in Quadrant III.

**34.** $(-6.2, 8.05)$ is in Quadrant II.

**36.** $(x, y)$, $x > 0$, $y > 0$ is in Quadrant I.

**38.** $(-10, y)$ is in Quadrant II or III.

**40.** $(x, 5)$ is in Quadrant I or II.

**42.** $(x, y)$, $xy < 0$ is in Quadrant II or IV.

**44.**

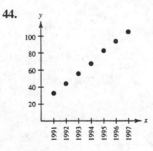

**46.**

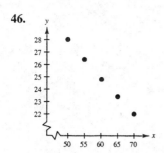

The relationship between $x$ and $y$ is as the value of $x$ increases the value of $y$ decreases.

**48.** $(-3, 5)$ shifted 6 units right and 3 units down $= (3, 2)$

$(-1, 2)$ shifted 6 units right and 3 units down $= (5, -1)$

$(-3, -1)$ shifted 6 units right and 3 units down $= (3, -4)$

$(-5, 2)$ shifted 6 units right and 3 units down $= (1, -1)$

**50.**

| $x$ | $-2$ | $0$ | $2$ | $4$ | $6$ |
|---|---|---|---|---|---|
| $y = \frac{3}{4}x + 2$ | $\frac{1}{2}$ | $2$ | $\frac{7}{2}$ | $5$ | $\frac{13}{2}$ |

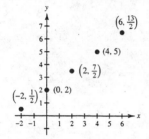

**52.**

| $x$ | $-6$ | $-3$ | $0$ | $\frac{3}{4}$ | $10$ |
|---|---|---|---|---|---|
| $y = \frac{4}{3}x - \frac{1}{3}$ | $-\frac{25}{3}$ | $-\frac{13}{3}$ | $-\frac{1}{3}$ | $\frac{2}{3}$ | $13$ |

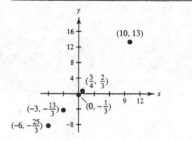

**54.**

| $x$ | $-2$ | $0$ | $2$ | $4$ | $6$ |
|---|---|---|---|---|---|
| $y = |3x - 4| + 1$ | $11$ | $5$ | $3$ | $9$ | $15$ |

*Keystrokes:* Y= ABS ( 3 X,T,θ − 4 ) + 1 GRAPH

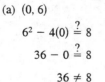
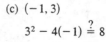

**56.** $y^2 - 4x = 8$

(a) $(0, 6)$

$6^2 - 4(0) \stackrel{?}{=} 8$

$36 - 0 \stackrel{?}{=} 8$

$36 \neq 8$

Not a solution

(b) $(-4, 2)$

$2^2 - 4(-4) \stackrel{?}{=} 8$

$4 + 16 \stackrel{?}{=} 8$

$20 \neq 8$

Not a solution

(c) $(-1, 3)$

$3^2 - 4(-1) \stackrel{?}{=} 8$

$9 + 4 \stackrel{?}{=} 8$

$13 \neq 8$

Not a solution

(d) $(7, 6)$

$6^2 - 4(7) \stackrel{?}{=} 8$

$36 - 28 \stackrel{?}{=} 8$

$8 = 8$

Solution

**58.** $5x - 2y + 50 = 0$

(a) $(-10, 0)$

$5(-10) - 2(0) + 50 \stackrel{?}{=} 0$

$-50 - 0 + 50 \stackrel{?}{=} 0$

$0 = 0$

Solution

(b) $(-5, 5)$

$5(-5) - 2(5) + 50 \stackrel{?}{=} 0$

$-25 - 10 + 50 \stackrel{?}{=} 0$

$15 \neq 0$

Not a solution

**— CONTINUED —**

**58.**    — CONTINUED —

(c)  (0, 25)

$$5(0) - 2(25) + 50 \overset{?}{=} 0$$

$$0 - 50 + 50 \overset{?}{=} 0$$

$$0 = 0$$

Solution

(d)  (20, −2)

$$5(20) - 2(-2) + 50 \overset{?}{=} 0$$

$$100 + 4 + 50 \overset{?}{=} 0$$

$$154 \neq 0$$

Not a solution

**60.**  $y = \frac{5}{8}x - 2$

(a)  (0, 0)

$$0 \overset{?}{=} \frac{5}{8}(0) - 2$$

$$0 \overset{?}{=} 0 - 2$$

$$0 \neq -2$$

Not a solution

(b)  (8, 3)

$$3 \overset{?}{=} \frac{5}{8}(8) - 2$$

$$3 \overset{?}{=} 5 - 2$$

$$3 = 3$$

Solution

(c)  (−16, −7)

$$-7 \overset{?}{=} \frac{5}{8}(-16) - 2$$

$$-7 \overset{?}{=} -10 - 2$$

$$7 \neq -12$$

Not a solution

(d)  $\left(-\frac{8}{5}, 3\right)$

$$3 \overset{?}{=} \frac{5}{8}\left(-\frac{8}{5}\right) - 2$$

$$3 \overset{?}{=} -1 - 2$$

$$3 \neq -3$$

Not a solution

**62.**  $d = |1 - 8|$

$= |-7|$

$= 7$

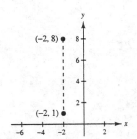

Vertical line

**64.**  $d = |130 - (-120)|$

$= |130 + 120|$

$= |250|$

$= 250$

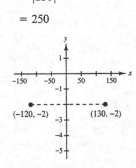

Horizontal line

**66.**  $d = |-10 - 1|$

$= |-11|$

$= 11$

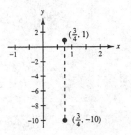

Vertical line

**68.**  $d = \left|\frac{11}{2} - \frac{1}{2}\right|$

$= |5|$

$= 5$

Horizontal line

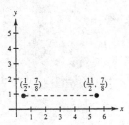

**70.** $d = \sqrt{(5-8)^2 + (2-3)^2}$

$= \sqrt{(-3)^2 + (1)^2}$

$= \sqrt{9+1}$

$= \sqrt{10}$

**72.** $d = \sqrt{(15-3)^2 + (5-10)^2}$

$= \sqrt{(12)^2 + (-5)^2}$

$= \sqrt{144 + 25}$

$= \sqrt{169}$

$= 13$

**74.** $d = \sqrt{(0-2)^2 + [-5-(-8)]^2}$

$= \sqrt{(-2)^2 + (3)^2}$

$= \sqrt{4+9}$

$= \sqrt{13}$

**76.** $d = \sqrt{[10-(-5)]^2 + (-3-4)^2}$

$= \sqrt{(10+5)^2 + (-3-4)^2}$

$= \sqrt{15^2 + (-7)^2}$

$= \sqrt{225 + 49}$

$= \sqrt{274}$

**78.** $d = \sqrt{\left(\dfrac{3}{2} - \dfrac{1}{2}\right)^2 + (2-1)^2}$

$= \sqrt{\left(\dfrac{2}{2}\right)^2 + (1)^2}$

$= \sqrt{1+1}$

$= \sqrt{2}$

**80.**

$d_1 = \sqrt{(0-2)^2 + (9-2)^2} = \sqrt{4+49} = \sqrt{53}$

$d_2 = \sqrt{(2-9)^2 + (2-4)^2} = \sqrt{49+4} = \sqrt{53}$

$d_3 = \sqrt{(0-9)^2 + (9-4)^2} = \sqrt{81+25} = \sqrt{106}$

$\left(\sqrt{53}\right)^2 + \left(\sqrt{53}\right)^2 \overset{?}{=} \left(\sqrt{106}\right)^2$

$53 + 53 \overset{?}{=} 106$

$106 = 106$

By Pythagorean Theorem, it is a right triangle.

**82.**

$d_1 = \sqrt{(-4-3)^2 + (3-5)^2} = \sqrt{49+4} = \sqrt{53}$

$d_2 = \sqrt{(3-5)^2 + [5-(-2)]^2} = \sqrt{4+49} = \sqrt{53}$

$d_3 = \sqrt{(-4-5)^2 + [3-(-2)]^2} = \sqrt{81+25} = \sqrt{106}$

$\left(\sqrt{53}\right)^2 + \left(\sqrt{53}\right)^2 \overset{?}{=} \left(\sqrt{106}\right)^2$

$53 + 53 \overset{?}{=} 106$

$106 = 106$

By Pythagorean Theorem, it is a right triangle.

**84.** $d = \sqrt{(-1-2)^2 + (6-4)^2} = \sqrt{(-3)^2 + 2^2} = \sqrt{9+4} = \sqrt{13}$

$d = \sqrt{[-3-(-1)]^2 + (1-6)^2} = \sqrt{(-2)^2 + (-5)^2} = \sqrt{4+25} = \sqrt{29}$

$d = \sqrt{(-3-2)^2 + (1-4)^2} = \sqrt{(-5)^2 + (-3)^2} = \sqrt{25+9} = \sqrt{34}$

$\sqrt{13} + \sqrt{29} \neq \sqrt{34}$    Not collinear

**86.** $d = \sqrt{(1-2)^2 + (1-4)^2} = \sqrt{(-1)^2 + (-3)^2} = \sqrt{1+9} = \sqrt{10}$

$d = \sqrt{(0-1)^2 + (-2-1)^2} = \sqrt{(-1)^2 + (-3)^2} = \sqrt{1+9} = \sqrt{10}$

$d = \sqrt{(0-2)^2 + (-2-4)^2} = \sqrt{(-2)^2 + (-6)^2} = \sqrt{4+36} = \sqrt{40} = 2\sqrt{10}$

$\sqrt{10} + \sqrt{10} = 2\sqrt{10}$    Collinear

**88.** $M = \left(\dfrac{-3+7}{2}, \dfrac{-2+2}{2}\right) = \left(\dfrac{4}{2}, \dfrac{0}{2}\right) = (2, 0)$

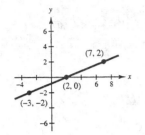

**90.** $M = \left(\dfrac{2+9}{2}, \dfrac{7+(-1)}{2}\right) = \left(\dfrac{11}{2}, \dfrac{6}{2}\right) = \left(\dfrac{11}{2}, 3\right)$

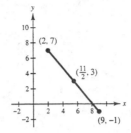

**92.**

| $x$ | 2 | 4 | 8 | 10 | 20 |
|-----|-----|-----|-----|-----|-----|
| $y = 0.75x + 8$ | 9.5 | 11 | 14 | 15.5 | 23 |

$y = 0.75(2) + 8$   $\quad$ $y = 0.75(4) + 8$   $\quad$ $y = 0.75(8) + 8$   $\quad$ $y = 0.75(10) + 8$   $\quad$ $y = 0.75(20) + 8$

$\quad = 1.5 + 8$   $\qquad\qquad = 3 + 8$   $\qquad\qquad = 6 + 8$   $\qquad\qquad = 7.5 + 8$   $\qquad\qquad = 15 + 8$

$\quad = 9.5$   $\qquad\qquad\quad = 11$   $\qquad\qquad\quad = 14$   $\qquad\qquad\quad = 15.5$   $\qquad\qquad\quad = 23$

For each additional 4 units produced, the hourly wage increases by \$3.

**94.** $x^2 = 8x^2 + 8^2$

$\quad x = \sqrt{64 + 64}$

$\quad x = \sqrt{128}$

$\quad x = 8\sqrt{2}$

Handrail $= 4\left(8\sqrt{2}\right) = 32\sqrt{2} \approx 45.25$ inches

**96.** $d = \sqrt{[-1-(-5)]^2 + [4-(-2)]^2} = \sqrt{4^2 + 6^2} = \sqrt{16 + 36} = \sqrt{52} = 2\sqrt{13}$

$\quad d = \sqrt{[3-(-1)]^2 + (-1-4)^2} = \sqrt{4^2 + (-5)^2} = \sqrt{16 + 25} = \sqrt{41}$

$\quad d = \sqrt{[3-(-5)]^2 + [-1-(-2)]^2} = \sqrt{8^2 + 1^2} = \sqrt{64 + 1} = \sqrt{65}$

$\quad P = 2\sqrt{13} + \sqrt{41} + \sqrt{65} \approx 21.68$

**98.** The $x$-coordinate measures the distance from the $y$-axis to the point. The $y$-coordinate measures the distance from the $x$-axis to the point.

**100.** $(-3, 4)$ is not a solution point of $y = 4x + 15$ because

$\quad 4 \neq 4(-3) + 15$

$\quad 4 \neq 3$

**102.** The Pythagorean Theorem states for a right triangle with hypotenuse $c$ and sides $a$ and $b$ you have $a^2 + b^2 = c^2$.

**104.**

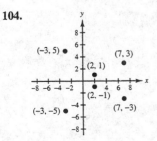

When the sign of the $y$-coordinate is changed, the point is on the opposite side of the $x$-axis as the original point.

## Section 2.2   Graphs of Equations

**2.** $y = 2 + x$

Matches graph (b)

**4.** $y = x^2$

Matches graph (a)

**6.** $y = |x|$

Matches graph (c)

**8.**

| $x$ | $-2$ | $-1$ | 0 | 1 | 2 |
|---|---|---|---|---|---|
| $y = -2x$ | 4 | 2 | 0 | $-2$ | $-4$ |
| Solution | $(-2, 4)$ | $(-1, 2)$ | $(0, 0)$ | $(1, -2)$ | $(2, -4)$ |

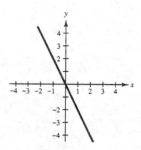

**10.**

| $x$ | $-2$ | $-1$ | 0 | 1 | 2 |
|---|---|---|---|---|---|
| $y = -x + 2$ | 4 | 3 | 2 | 1 | 0 |
| Solution | $(-2, 4)$ | $(-1, 3)$ | $(0, 2)$ | $(1, 1)$ | $(2, 0)$ |

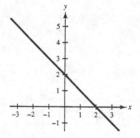

**12.**

| $x$ | $-2$ | $-1$ | 0 | 1 | 2 |
|---|---|---|---|---|---|
| $y = 3x + 2$ | $-4$ | $-1$ | 2 | 5 | 8 |
| Solution | $(-2, -4)$ | $(-1, -1)$ | $(0, 2)$ | $(1, 5)$ | $(2, 8)$ |

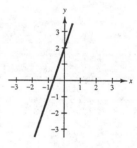

**14.**

| $x$ | $-2$ | $-1$ | 0 | 1 | 2 |
|---|---|---|---|---|---|
| $y = \frac{1}{2}x + 2$ | 1 | $\frac{3}{2}$ | 2 | $\frac{5}{2}$ | 3 |
| Solution | $(-2, 1)$ | $\left(-1, \frac{3}{2}\right)$ | $(0, 2)$ | $\left(1, \frac{5}{2}\right)$ | $(2, 3)$ |

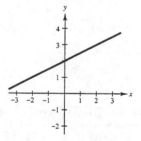

**16.**

| $x$ | $-2$ | $-1$ | 0 | 1 | 2 |
|---|---|---|---|---|---|
| $y = x^2$ | 4 | 1 | 0 | 1 | 4 |
| Solution | $(-2, 4)$ | $(-1, 1)$ | $(0, 0)$ | $(1, 1)$ | $(2, 4)$ |

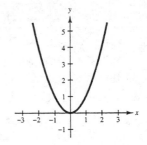

**18.**

| $x$ | $-2$ | $-1$ | 0 | 1 | 2 |
|---|---|---|---|---|---|
| $y = 1 - x^2$ | $-3$ | 0 | 1 | 0 | $-3$ |
| Solution | $(-2, -3)$ | $(-1, 0)$ | $(0, 1)$ | $(1, 0)$ | $(2, -3)$ |

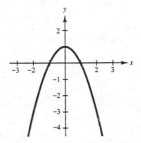

**20.**

| $x$ | $-2$ | $-1$ | 0 | 1 | 2 |
|---|---|---|---|---|---|
| $y = x^2 - x$ | 6 | 2 | 0 | 0 | 2 |
| Solution | $(-2, 6)$ | $(-1, 2)$ | $(0, 0)$ | $(1, 0)$ | $(2, 2)$ |

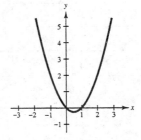

**22.**

| $x$ | $-2$ | $-1$ | 0 | 1 | 2 |
|---|---|---|---|---|---|
| $y = x^2 + 3x - 4$ | $-6$ | $-6$ | $-4$ | 0 | 6 |
| Solution | $(-2, -6)$ | $(-1, -6)$ | $(0, -4)$ | $(1, 0)$ | $(2, 6)$ |

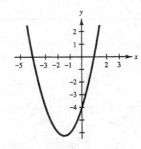

**24.**

| $x$ | $-2$ | $-1$ | 0 | 1 | 2 |
|---|---|---|---|---|---|
| $y = -|x|$ | $-2$ | $-1$ | 0 | $-1$ | $-2$ |
| Solution | $(-2, -2)$ | $(-1, -1)$ | $(0, 0)$ | $(1, -1)$ | $(2, -2)$ |

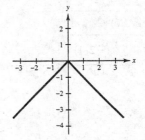

**26.**

| $x$ | $-2$ | $-1$ | 0 | 1 | 2 |
|---|---|---|---|---|---|
| $y = |x| - 1$ | 1 | 0 | $-1$ | 0 | 1 |
| Solution | $(-2, 1)$ | $(-1, 0)$ | $(0, -1)$ | $(1, 0)$ | $(2, 1)$ |

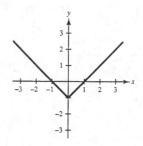

**28.**

| $x$ | $-2$ | $-1$ | 0 | 1 | 2 |
|---|---|---|---|---|---|
| $y = |x - 1|$ | 3 | 2 | 1 | 0 | 1 |
| Solution | $(-2, 3)$ | $(-1, 2)$ | $(0, 1)$ | $(1, 0)$ | $(2, 1)$ |

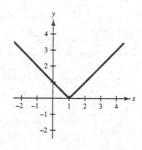

**30.**

| $x$ | $-2$ | $-1$ | 0 | 1 | 2 |
|---|---|---|---|---|---|
| $y = x^3$ | $-8$ | $-1$ | 0 | 1 | 8 |
| Solution | $(-2, -8)$ | $(-1, -1)$ | $(0, 0)$ | $(1, 1)$ | $(2, 8)$ |

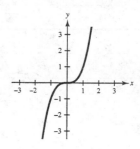

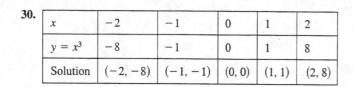

**32.** $y = x + 2$

   $y$-intercept:  $y = 0 + 2$

$$y = 2 \qquad (0, 2)$$

   $x$-intercept:  $0 = x + 2$

$$-2 = x \qquad (-2, 0)$$

**34.** $y = 12 - \frac{2}{5}x$

   $y$-intercept:  $y = 12 - \frac{2}{5}(0)$

$$y = 12 \qquad (0, 12)$$

   $x$-intercept:  $0 = 12 - \frac{2}{5}x$

$$\frac{2}{5}x = 12$$

$$x = 30 \qquad (30, 0)$$

**36.** $3x - 2y + 12 = 0$

   $y$-intercept:  $3(0) - 2y + 12 = 0$

$$0 - 2y + 12 = 0$$

$$-2y = -12$$

$$y = 6 \qquad (0, 6)$$

   $x$-intercept:  $3x - 2(0) + 12 = 0$

$$3x - 0 + 12 = 0$$

$$3x = -12$$

$$x = -4 \qquad (-4, 0)$$

**38.** $2x + 3y - 8 = 0$

   $y$-intercept:  $2(0) + 3y - 8 = 0$

$$3y = 8$$

$$y = \frac{8}{3} \qquad \left(0, \frac{8}{3}\right)$$

   $x$-intercept:  $2x + 3(0) - 8 = 0$

$$2x = 8$$

$$x = 4 \qquad (4, 0)$$

**40.** $y = |x| + 4$

$y$-intercept: $y = |0| + 4$

$\qquad y = 4 \quad (0, 4)$

$x$-intercept: $0 = |x| + 4$

$\qquad -4 = |x|$

$\qquad$ None

**42.** $y = |x - 4|$

$y$-intercept: $y = |0 - 4|$

$\qquad y = 4 \quad (0, 4)$

$x$-intercept: $0 = |x - 4|$

$\qquad 0 = x - 4$

$\qquad 4 = x \quad (4, 0)$

**44.** $y = |x + 3| - 1$

$y$-intercept: $y = |0 + 3| - 1$

$\qquad y = 3 - 1$

$\qquad y = 2 \quad (0, 2)$

$x$-intercept: $0 = |x + 3| - 1$

$\qquad 1 = |x + 3|$

$\qquad 1 = x + 3 \quad$ or $\quad x + 3 = -1$

$\qquad -2 = x \qquad\qquad x = -4 \quad (-2, 0), (-4, 0)$

**46.** $3x - y + 9 = 0$

**Estimate:** $y$-intercept $\approx 9$

$\qquad\quad x$-intercept $\approx -3$

**Check:** $3(0) - y + 9 = 0$

$\qquad\qquad\qquad y = 9 \quad (0, 9)$

$\qquad\quad 3x - 0 + 9 = 0$

$\qquad\qquad\qquad 3x = -9$

$\qquad\qquad\qquad x = -3 \quad (-3, 0)$

**48.** $x = 3$

**Estimate:** no $y$-intercept

$\qquad\quad x$-intercept $\approx 3$

**Check:** $x \neq 0$

$\qquad\quad x = 3 \quad (3, 0)$

**50.** $y = 3x + 12$

*Keystrokes:* [Y=] 3 [X,T,$\theta$] [+] 12 [GRAPH]

**Estimate:** $y$-intercept $\approx 12$, $x$-intercept $\approx -4$

**Check:** $y = 3(0) + 12$

$\qquad\quad y = 12 \quad (0, 12)$

**Check:** $0 = 3x + 12$

$\qquad\quad -12 = 3x$

$\qquad\quad -4 = x \quad (-4, 0)$

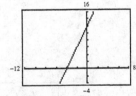

**52.** $y = (x + 2)(x - 3)$

*Keystrokes:* $\boxed{Y=}$ $\boxed{(}$ $\boxed{X,T,\theta}$ $\boxed{+}$ 2 $\boxed{)}$ $\boxed{(}$ $\boxed{X,T,\theta}$ $\boxed{-}$ 3 $\boxed{)}$ $\boxed{GRAPH}$

**Estimate:** $y$-intercept $\approx -6$, $x$-intercepts $\approx -2, 3$

**Check:** $y = (0 + 2)(0 - 3)$

$\qquad y = -6 \quad (0, -6)$

**Check:** $0 = (x + 2)(x - 3)$

$\qquad x = -2 \quad x = 3$

$\qquad (-2, 0), (3, 0)$

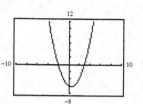

**54.** $y = |2x - 4| + 1$

*Keystrokes:* $\boxed{Y=}$ $\boxed{ABS}$ $\boxed{(}$ 2 $\boxed{X,T,\theta}$ $\boxed{-}$ 4 $\boxed{)}$ $\boxed{+}$ 1 $\boxed{GRAPH}$

Estimate: $y$-intercept $\approx 5$, no $x$-intercepts

**Check:** $y = |2(0) - 4| + 1$

$\qquad y = 4 + 1$

$\qquad y = 5 \quad (0, 5)$

**Check:** $\quad 0 = |2x - 4| + 1$

$\qquad -1 = |2x - 4|$

$\qquad$ None

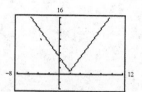

**56.** $y = x - 3$

$y = 0 - 3$

$y = -3 \quad (0, -3)$

$0 = x - 3$

$3 = x \quad (3, 0)$

$y = 1 - 3$

$y = -2 \quad (1, -2)$

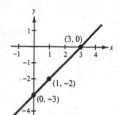

**58.** $y = -4x + 8$

$0 = -4x + 8$

$4x = 8$

$x = 2 \quad (2, 0)$

$y = -4(0) + 8$

$y = 0 + 8$

$y = 8 \quad (0, 8)$

$y = -4(1) + 8$

$y = -4 + 8$

$y = 4 \quad (1, 4)$

**60.** $y - 2x = -4$

$0 - 2x = -4$

$-2x = -4$

$\quad x = 2 \quad (2, 0)$

$y - 2(0) = -4$

$y - 0 = -4$

$\quad y = -4 \quad (0, -4)$

$y - 2(1) = -4$

$y - 2 = -4$

$\quad y = -2 \quad (1, -2)$

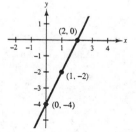

**62.** $3x + 4y = 12$

$3(0) + 4y = 12$

$0 + 4y = 12$

$4y = 12$

$\qquad y = 3 \quad (0, 3)$

$3x + 4(0) = 12$

$3x + 0 = 12$

$3x = 12$

$\qquad x = 4 \quad (4, 0)$

$3(1) + 4y = 12$

$3 + 4y = 12$

$4y = 9$

$\qquad y = \frac{9}{4} \quad \left(1, \frac{9}{4}\right)$

**64.**  $5x - y = 10$

$5(0) - y = 10$

$0 - y = 10$

$-y = 10$

$y = -10$    $(0, -10)$

$5x - (0) = 10$

$5x = 10$

$x = 2$    $(2, 0)$

$5(1) - y = 10$

$5 - y = 10$

$-y = 5$

$y = -5$    $(1, -5)$

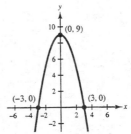

**66.**  $y = 9 - x^2$

$0 = 9 - x^2$

$x^2 = 9$

$x = 3$  or  $x = -3$    $(3, 0)$ or $(-3, 0)$

$y = 9 - 0^2$

$y = 9$    $(0, 9)$

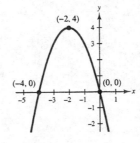

**68.**  $y = x^2 - 4$

$y = 0^2 - 4$

$= -4$    $(0, -4)$

$y = 2^2 - 4$

$= 0$    $(2, 0)$

$y = (-2)^2 - 4$

$= 0$    $(-2, 0)$

**70.**  $y = -x(x + 4)$

$y = -(0)(0 + 4)$

$y = 0$    $(0, 0)$

$0 = -x(x + 4)$

$x = 0$  or  $x = -4$    $(0, 0), (-4, 0)$

$y = -(-2)(-2 + 4)$

$y = 2(2)$

$y = 4$    $(-2, 4)$

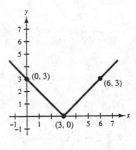

**72.**  $y = |x| + 2$

$y = |-1| + 2$

$y = 1 + 2$

$y = 3$    $(-1, 3)$

$y = |0| + 2$

$y = 0 + 2$

$y = 2$    $(0, 2)$

$y = |1| + 2$

$y = 1 + 2$

$y = 3$    $(1, 3)$

**74.**  $y = |x - 3|$

$0 = |x - 3|$

$x = 3$    $(3, 0)$

$y = |0 - 3|$

$y = |-3|$

$y = 3$    $(0, 3)$

$y = |6 - 3|$

$y = |3|$

$y = 3$    $(6, 3)$

**76.** $y = |x| + |x - 2|$

$y = |0| + |0 - 2|$

$y = 0 + |-2|$

$y = 0 + 2$

$y = 2 \quad (0, 2)$

$y = |1| + |1 - 2|$

$y = 1 + |-1|$

$y = 1 + 1$

$y = 2 \quad (1, 2)$

$y = |3| + |3 - 2|$

$y = 3 + |1|$

$y = 3 + 1$

$y = 4 \quad (3, 4)$

**78.** $y = 20{,}000 - 3000t$

$y = 20{,}000 - 3000(0)$

$y = 20{,}000 \quad (0, 20{,}000)$

$y = 20{,}000 - 3000(6)$

$y = 2000 \quad (6, 2000)$

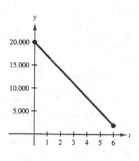

**80.** $0 \le x \le 10$

$(0, 55{,}000), (10, 10{,}000)$

$$m = \frac{55{,}000 - 10{,}000}{0 - 10} = \frac{45{,}000}{-10} = -4500$$

$y = -4500x + 55{,}000$

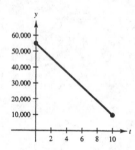

**82.** (a)

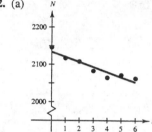

(b) The model is a good representation of the data.

(c) $N = -13.6(10) + 2134$

$N = -136 + 2134$

$N = 1998$

(d) The model becomes negative.

**84.**

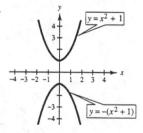

When the expression for $y$ is multiplied by $-1$, the graph is reflected in the $x$-axis.

**86.** There are an infinite number of solution points that make up the graph of $y = 2x - 1$.

**88.** It is possible to sketch the graph of an equation showing profit is decreasing at a slower rate than in the past.

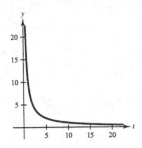

# Section 2.3    Slope and Graphs of Linear Equations

**2.** $(0, 5)$ and $(4, 3)$

$$m = \frac{3 - 5}{4 - 0} = \frac{-2}{4} = \frac{-1}{2}$$

**4.** $(2, 0)$ and $(3, 4)$

$$m = \frac{4 - 0}{3 - 2} = \frac{4}{1} = 4$$

**6.** $(0, 5)$ and $(3, 5)$

$$m = \frac{5 - 5}{3 - 0} = \frac{0}{3} = 0$$

**8.** (a) $m = -\frac{5}{2} \implies L_2$

(b) $m$ is undefined $\implies L_3$

(c) $m = 2 \implies L_1$

**10.** $m = \dfrac{-4 - 0}{-3 - 0} = \dfrac{-4}{-3} = \dfrac{4}{3}$    Line rises.

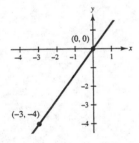

**12.** $m = \dfrac{1 - 0}{-2 - 0} = \dfrac{1}{-2}$    Line falls.

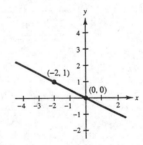

**14.** $m = \dfrac{-5 - 1}{4 - 7} = \dfrac{-6}{-3} = 2$    Line rises.

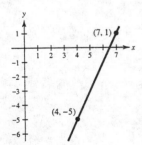

**16.** $m = \dfrac{2 - 2}{-9 - 2} = \dfrac{0}{-11} = 0$    Line is horizontal.

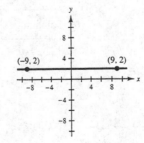

**18.** $m = \dfrac{8 - 4}{-3 - (-3)} = \dfrac{4}{0} =$ undefined   Line is vertical.

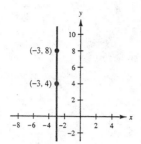

**20.** $m = \dfrac{\frac{2}{3} - (-1)}{3 - \frac{1}{2}} \cdot \dfrac{6}{6} = \dfrac{4 + 6}{18 - 3} = \dfrac{10}{15} = \dfrac{2}{3}$   Line rises.

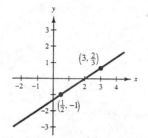

**22.** $m = \dfrac{\frac{1}{2} - \left(-\frac{1}{2}\right)}{\frac{5}{8} - \left(-\frac{3}{2}\right)} = \dfrac{\frac{1}{2} + \frac{1}{2}}{\frac{5}{8} + \frac{3}{2}} = \dfrac{1}{\frac{5}{8} + \frac{12}{8}} = \dfrac{1}{\frac{17}{8}} = \dfrac{8}{17}$   Line rises.

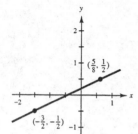

**24.** $m = \dfrac{4.5 - 4.5}{3 - 0} = \dfrac{0}{3} = 0$   Line is horizontal.

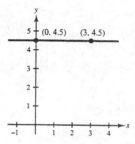

**26.**

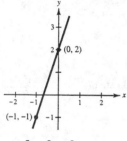

| $x$ | $-1$ | $0$ | $1$ |
|---|---|---|---|
| $y = 3x + 2$ | $-1$ | $2$ | $5$ |
| Solution | $(-1, -1)$ | $(0, 2)$ | $(1, 5)$ |

$m = \dfrac{5 - 2}{1 - 0} = \dfrac{3}{1} = 3$

**28.**

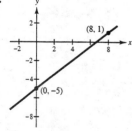

| $x$ | $-1$ | $0$ | $1$ |
|---|---|---|---|
| $y = \frac{3}{4}x - 5$ | $-5\frac{3}{4}$ | $-5$ | $-4\frac{1}{4}$ |
| Solution | $\left(-1, -5\frac{3}{4}\right)$ | $(0, -5)$ | $\left(1, -4\frac{1}{4}\right)$ |

$m = \dfrac{-4\frac{1}{4} - (-5)}{1 - 0} = -\dfrac{17}{4} + \dfrac{20}{4} = \dfrac{3}{4}$

**30.**

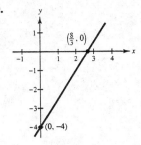

| $x$ | $-1$ | $0$ | $1$ |
|---|---|---|---|
| $y = \frac{3}{2}x - 4$ | $-5\frac{1}{2}$ | $-4$ | $-2\frac{1}{2}$ |
| Solution | $\left(-1, -5\frac{1}{2}\right)$ | $(0, -4)$ | $\left(1, -2\frac{1}{2}\right)$ |

$$3x - 2y = 8$$
$$-2y = -3x + 8$$
$$y = \frac{3}{2}x - 4$$
$$m = \frac{-2\frac{1}{2} - (-4)}{1 - 0} = -\frac{5}{2} + \frac{8}{2} = \frac{3}{2}$$

**32.**
$$\frac{3}{4} = \frac{0 - (-2)}{5 - x}$$
$$\frac{3}{4} = \frac{2}{5 - x}$$
$$3(5 - x) = 8$$
$$15 - 3x = 8$$
$$-3x = -7$$
$$x = \frac{7}{3}$$

**34.** $\quad -6 = \dfrac{y = 20}{2 - (-3)}$
$$-6 = \frac{y - 20}{5}$$
$$-30 = y - 20$$
$$-10 = y$$

**36.** $\dfrac{y - 3}{x - (-4)}$ is not possible.

$$x + 4 = 0$$

Vertical line:

$$(-4, 1), (-4, 2), (-4, 3)$$

Any point with an $x$-coordinate of $-4$.

**38.** $2 = \dfrac{y + 5}{x + 1}$

Let $x = 0$, solve for $y$:

$$2 = \frac{y + 5}{1}$$
$$2 = y + 5$$
$$-3 = y$$
$$(0, -3)$$

Let $x = 1$, solve for $y$:

$$2 = \frac{y + 5}{2}$$
$$4 = y + 5$$
$$-1 = y$$
$$(1, -1)$$

**40.** $-3 = \dfrac{y - 6}{x - (-2)}$

Let $x = -1$, solve for $y$:

$$-3 = \frac{y - 6}{1}$$
$$-3 = y - 6$$
$$3 = y$$
$$(-1, 3)$$

Let $x = 0$, solve for $y$:

$$-3 = \frac{y - 6}{2}$$
$$-6 = y - 6$$
$$0 = y$$
$$(0, 0)$$

**42.** $-\dfrac{3}{4} = \dfrac{y - 1}{x + 1}$

Let $x = 3$, solve for $y$:

$$-\frac{3}{4} = \frac{y - 1}{4}$$
$$-3 = y - 1$$
$$-2 = y$$
$$(3, -2)$$

Let $x = 7$, solve for $y$:

$$-\frac{3}{4} = \frac{y - 1}{8}$$
$$-6 = y - 1$$
$$-5 = y$$
$$(7, -5)$$

**44.** $2x + 4y = 16$
$$4y = -2x + 16$$
$$y = -\frac{2}{4}x + 4$$
$$y = -\frac{1}{2}x + 4$$

**46.** $3x - 2y = -10$

$-2y = -3x - 10$

$y = \frac{3}{2}x + 5$

**48.** $8x - 6y + 1 = 0$

$-6y = -8x - 1$

$y = \frac{4}{3}x + \frac{1}{6}$

**50.** $y = -\frac{2}{3}x + 4$

**52.** $y = 4 - 2x$

$m = -2$

$(0, 4)$

**54.** $4x + 8y = -1$

$8y = -4x - 1$

$y = -\frac{4}{8}x - \frac{1}{8}$

$y = -\frac{1}{2}x - \frac{1}{8}$

$m = -\frac{1}{2}; \left(0, -\frac{1}{8}\right)$

**56.** $6y - 5x + 18 = 0$

$6y = 5x - 18$

$y = \frac{5}{6}x - 3$

$m = \frac{5}{6}; (0, -3)$

**58.** $x - y - 5 = 0$

$-y = -x + 5$

$y = x - 5$

slope $= 1$   $y$-intercept $= -5$

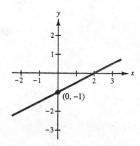

**60.** $x - y = 0$

$-y = -x$

$y = x$

slope $= 1$   $y$-intercept $= 0$

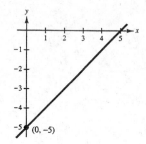

**62.** $x - 2y - 2 = 0$

$-2y = -x + 2$

$y = \frac{1}{2}x - 1$

slope $= \frac{1}{2}$   $y$-intercept $= -1$

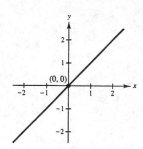

**64.** $8x + 6y - 3 = 0$

$6y = -8x + 3$

$y = -\frac{8}{6}x + \frac{3}{6}$

$y = -\frac{4}{3}x + \frac{1}{2}$

slope $= -\frac{4}{3}$   $y$-intercept $= \frac{1}{2}$

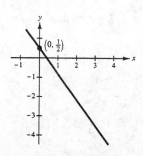

**66.**

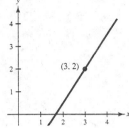

Locate a second point with the slope of $\frac{3}{2}$.

$$m = \frac{3}{2} = \frac{\text{change in } y}{\text{change in } x}$$

**70.**

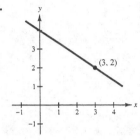

**68.**

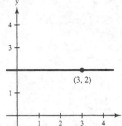

Locate a second point with the slope of 0.

Line is horizontal.

Locate a second point with the slope of $-\left(\frac{2}{3}\right)$.

$$m = -\frac{2}{3} = \frac{\text{change in } y}{\text{change in } x}$$

**72.**   $3x + 5y + 15 = 0$
$3(0) + 5y + 15 = 0$
$\qquad\qquad 5y = -15$
$\qquad\qquad\quad y = -3 \qquad (0, -3)$
$3x - 5(0) - 15 = 0$
$\qquad\qquad 3x = -15$
$\qquad\qquad\ x = -5 \qquad (-5, 0)$

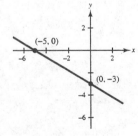

**74.**   $3x - 5y - 15 = 0$
$3(0) - 5y - 15 = 0$
$\qquad\qquad -5y = 15$
$\qquad\qquad\quad y = -3 \qquad (0, -3)$
$3x - 5(0) - 15 = 0$
$\qquad\qquad 3x = 15$
$\qquad\qquad\ x = 5 \qquad (5, 0)$

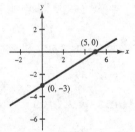

**76.** $L_1$: $y = 3x - 2$

$L_2$: $y = 3x + 1$

$m_1 = 3$ and $m_2 = 3$

$m_1 = m_2$ so the lines are parallel.

**78.** $L_1$: $y = -\frac{2}{3}x - 5$

$L_2$: $y = \frac{3}{2}x + 1$

$m_1 = -\frac{2}{3}$ and $m_2 = \frac{3}{2}$

$m_1 \cdot m_2 = -1$ so the lines are perpendicular.

**80.** $\dfrac{1}{10} = \dfrac{4}{x}$

  $x = 40$ feet

  $c = \sqrt{4^2 + 40^2} = \sqrt{1616} = 4\sqrt{101} \approx 40.2$ feet

**82.** $\dfrac{4}{5} = \dfrac{h}{15}$

  $h = \dfrac{4 \cdot 15}{5} = 12$ feet

**84.** (a) *Verbal Model:* | Total Amount | $-$ | Amount invested in 6% fund |

  *Expression:* $8000 - x$

  (b) *Verbal Model:* | Total Interest | $=$ | Principal in 6% account | $\cdot$ | Rate of 6% account | $\cdot$ | Time for 6% account |

  $+$ | Principal in $7\frac{1}{2}$% account | $\cdot$ | Rate of $7\frac{1}{2}$% account | $\cdot$ | Time for $7\frac{1}{2}$% account |

  *Equation:* $y = x(0.06)(1) + (8000 - x)(0.075)(1)$

  $y = 600 - 0.015x$

  (c) *Keystrokes:* [Y=] 600 [−] 0.15 [X,T,θ] [GRAPH]

  (d) As the amount invested at 6% increases, the total interest decreases.

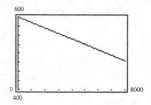

**86.** Yes, any pair of points on a line can be used to calculate the slope of the line. When different pairs of points are selected, the change in $y$ and the change in $x$ are the lengths of the sides of similar triangles. Corresponding sides of similar triangles are proportional.

**88.** The line with slope $-3$ is steeper. There is a vertical change of 3 units for each 1 unit change in $x$. The slope $3/2$ means that there is a vertical change of 3 units for every 2 unit change in $x$.

**90.** The $x$-coordinate of the $x$-intercept is the same as the solution of the equation when $y = 0$.

## Section 2.4    Equations of Lines

**2.** $y = \frac{2}{3}x - 2$

  Matches graph (d)

**4.** $y = -3x + 2$

  Matches graph (c)

**6.**    $-2 = \dfrac{y + 4}{x - 1}$

  $-2(x - 1) = y + 4$

  $-2x + 2 = y + 4$

  $-2 = 2x + y$

**8.**    $\dfrac{2}{3} = \dfrac{y - 9}{x - 6}$

  $2(x - 6) = 3(x - 9)$

  $2x - 12 = 3y - 27$

  $2x - 3y = -15$

**10.**    $-\dfrac{1}{6} = \dfrac{y - (3/2)}{x - (-2)}$

  $x + 2 = -6\left(y - \dfrac{3}{2}\right)$

  $x + 2 = -6y + 9$

  $x + 6y = 7$

**12.** $y - 0 = \frac{2}{3}(x - 0)$

$\quad\quad y = \frac{2}{3}x$

**14.** $y - 2 = 3(x - 0)$

$\quad\quad y - 2 = 3x$

**16.** $y - (-3) = \frac{1}{3}(x - 0)$

$\quad\quad y + 3 = \frac{1}{3}x$

**18.** $y - (-1) = 3(x - 4)$

$\quad\quad y + 1 = 3(x - 4)$

**20.** $y - (-8) = -\frac{2}{3}(x - 6)$

$\quad\quad y + 8 = -\frac{2}{3}(x - 6)$

**22.** $y - \left(-\frac{3}{2}\right) = 1(x - 1)$

$\quad\quad y + \frac{3}{2} = 1(x - 1)$

**24.** $y - \frac{1}{2} = -\frac{2}{5}\left[x - \left(-\frac{5}{2}\right)\right]$

$\quad\quad y - \frac{1}{2} = -\frac{2}{5}\left(x + \frac{5}{2}\right)$

**26.** $y - 5 = 0(x + 8)$   or

$\quad\quad y - 5 = 0$

**28.** $m = \dfrac{-5 - 0}{3 - 0} = \dfrac{-5}{3}$

$\quad\quad y - 0 = -\dfrac{5}{3}(x - 0)$

$\quad\quad\quad y = -\dfrac{5}{3}x$

$\quad\quad\quad 3y = -5x$

$\quad 5x + 3y = 0$

**30.** $m = \dfrac{0 - (-2)}{2 - 0} = \dfrac{2}{2} = 1$

$\quad\quad y - (-2) = 1(x - 0)$

$\quad\quad\quad y + 2 = x$

$\quad\quad x - y - 2 = 0$

**32.** $m = \dfrac{8 - (-2)}{2 - 1} = \dfrac{10}{1}$

$\quad\quad y - 8 = 10(x - 2)$

$\quad\quad y - 8 = 10x - 20$

$\quad 10x - y - 12 = 0$

**34.** $m = \dfrac{5 - 4}{3 - 5} = \dfrac{1}{-2}$

$\quad\quad y - 4 = \dfrac{1}{-2}(x - 5)$

$\quad\quad y - 4 = -\dfrac{1}{2}x + \dfrac{5}{2}$

$\quad\quad 2y - 8 = -x + 5$

$\quad x + 2y - 13 = 0$

**36.** $m = \dfrac{\frac{7}{3} - \frac{1}{3}}{4 - (-1)} = \dfrac{\frac{6}{3}}{5} = \dfrac{2}{5}$

$\quad\quad y - \dfrac{1}{3} = \dfrac{2}{5}[x - (-1)]$

$\quad\quad y - \dfrac{1}{3} = \dfrac{2}{5}x + \dfrac{2}{5}$

$\quad\quad 15y - 5 = 6x + 6$

$\quad 6x - 15y + 11 = 0$

**38.** $m = \dfrac{-\frac{2}{5} - \frac{3}{5}}{\frac{3}{4} - (-4)} = \dfrac{-\frac{5}{5}}{\frac{3}{4} + \frac{16}{4}} = \dfrac{-1}{\frac{19}{4}} = -\dfrac{4}{19}$

$\quad\quad y - \dfrac{3}{5} = -\dfrac{4}{19}[x - (-4)]$

$\quad\quad y - \dfrac{3}{5} = -\dfrac{4}{19}x - \dfrac{16}{19}$

$\quad\quad 95y - 57 = -20x - 80$

$\quad 20x + 95y + 23 = 0$

**40.** $m = \dfrac{2.3 - (-8)}{6 - 2} = \dfrac{10.3}{4} = \dfrac{103}{40}$

$$y - (-8) = \dfrac{103}{40}(x - 2)$$

$$y + 8 = \dfrac{103x}{40} - \dfrac{103}{20}$$

$$40y + 320 = 103x - 206$$

$$103x - 40y - 526 = 0$$

**42.** $m = \dfrac{-3.4 - 0.6}{3 - (-5)} = \dfrac{-4}{8} = -\dfrac{1}{2}$

$$y - 0.6 = -\dfrac{1}{2}[x - (-5)]$$

$$y - \dfrac{6}{10} = -\dfrac{1}{2}(x + 5)$$

$$y - \dfrac{3}{5} = -\dfrac{1}{2}x - \dfrac{5}{2}$$

$$10\left(y - \dfrac{3}{5}\right) = 10\left(-\dfrac{1}{2}x - \dfrac{5}{2}\right)$$

$$10y - 6 = -5x - 25$$

$$5x + 10y + 19 = 0$$

**44.** $m = \dfrac{0 - 10}{5 - 0} = \dfrac{-10}{5} = -2$

$$y = 10 = -2(x - 0)$$

$$y - 10 = -2x$$

$$y = -2x + 10$$

**46.** $m = \dfrac{3 - (-3)}{4 - (-6)} = \dfrac{3 + 3}{4 + 6} = \dfrac{6}{10} = \dfrac{3}{5}$

$$y - 3 = \dfrac{3}{5}(x - 4)$$

$$y - 3 = \dfrac{3}{5}x - \dfrac{12}{5}$$

$$y = \dfrac{3}{5}x - \dfrac{3}{5}$$

**48.** $x = 2$ because every $x$-coordinate is 2.

**50.** $y = -3$ because every $y$-coordinate is $-3$.

**52.** $y = 4$ because both points have a $y$-coordinate of 4.

**54.** $x + 6y = 12$

$$6y = -x + 12$$

$$y = -\dfrac{1}{6}x + 2 \quad \text{slope} = -\dfrac{1}{6}$$

(a) $y - 4 = -\dfrac{1}{6}(x + 3)$

$$y - 4 = -\dfrac{1}{6}x - \dfrac{1}{2}$$

$$y = -\dfrac{1}{6}x + \dfrac{7}{2}$$

(b) $y - 4 = 6(x + 3)$

$$y - 4 = 6x + 18$$

$$y = 6x + 22$$

**56.** $3x + 10y = 24 \quad \text{slope} = -\dfrac{3}{10}$

$$10y = -3x + 24$$

$$y = -\dfrac{3}{10}x + \dfrac{24}{10}$$

(a) $y - (-4) = -\dfrac{3}{10}(x - 6)$

$$y + 4 = -\dfrac{3}{10}x + \dfrac{9}{5}$$

$$y = -\dfrac{3}{10}x + \dfrac{9}{5} - \dfrac{20}{5}$$

$$y = -\dfrac{3}{10}x - \dfrac{11}{5}$$

(b) $y - (-4) = \dfrac{10}{3}(x - 6)$

$$y + 4 = \dfrac{10}{3}x - 20$$

$$y = \dfrac{10}{3}x - 24$$

**58.** $2x + 5y - 12 = 0$

$$5y = -2x + 12$$

$$y = -\frac{2}{5}x + \frac{12}{5} \quad \text{slope} = -\frac{2}{5}$$

(a) $y - (-10) = -\frac{2}{5}[x - (-5)]$

$$y + 10 = -\frac{2}{5}(x + 5)$$

$$y + 10 = -\frac{2}{5}x - 2$$

$$y = -\frac{2}{5}x - 12$$

(b) $y - (-10) = \frac{5}{2}[x - (-5)]$

$$y + 10 = \frac{5}{2}(x + 5)$$

$$y + 10 = \frac{5}{2}x + \frac{25}{2}$$

$$y = \frac{5}{2}x + \frac{25}{2} - \frac{20}{2}$$

$$y = \frac{5}{2}x + \frac{5}{2}$$

**60.** $-5x + 4y = 0$

$$\frac{5}{4}x = y \quad \text{slope} = \frac{5}{4}$$

(a) $y - \frac{9}{4} = \frac{5}{4}\left(x = \frac{5}{8}\right)$

$$y - \frac{9}{4} = \frac{5}{4}x - \frac{25}{32}$$

$$y = \frac{5}{4}x - \frac{25}{32} + \frac{72}{32}$$

$$y = \frac{5}{4}x + \frac{47}{32}$$

(b) $y - \frac{9}{4} = -\frac{4}{5}\left(x - \frac{5}{8}\right)$

$$y - \frac{9}{4} = -\frac{4}{5}x + \frac{1}{2}$$

$$y = -\frac{4}{5}x + \frac{2}{4} + \frac{9}{4}$$

$$y = -\frac{4}{5}x + \frac{11}{4}$$

**62.** $x - 10 = 0$

$x = 0$   The slope is undefined because the line is vertical.

(a)  $x = 3$

$x - 3 = 0$

(b)  $y = -4$

$y + 4 = 0$

**64.** $\frac{x}{-6} + \frac{y}{2} = 1$

$$-\frac{x}{6} + \frac{y}{2} = 1$$

**66.** $\dfrac{x}{-\frac{8}{3}} + \dfrac{y}{-4} = 1$

$$-\frac{3x}{8} = \frac{y}{4} = 1$$

**68.** $m = \dfrac{10 - 5}{50 - 41} = \dfrac{5}{9}$

$C - 5 = \dfrac{5}{9}(F - 41)$

$C - 5 = \dfrac{5}{9}F - \dfrac{205}{9}$

$C = \dfrac{5}{9}F - \dfrac{205}{9} + \dfrac{45}{9}$

$C = \dfrac{5}{9}F - \dfrac{160}{9}$

$C = \dfrac{5}{9}(72) - \dfrac{160}{9}$

$C = 40 - \dfrac{160}{9}$

$C \approx 22.2°$

**70.** $m = \dfrac{250,000 - 150,000}{5 - 3} = \dfrac{100,000}{2} = 50,000$

$S - 250,000 = 50,000(t - 5)$

$S - 250,000 = 50,000t - 250,000$

$S = 50,000t$

$S = 50,000(6)$

$S = \$300,000$

**72.** $m = \dfrac{157 - 141}{100 - 50} = \dfrac{16}{50} = \dfrac{8}{25} = 0.32$

$C - 141 = 0.32(x - 50)$

$C - 141 = 0.32x - 16$

$C = 0.32x + 125$

The sales representative is reimbursed $0.32 per mile.

**74.** (a) $C = 0.34x + 150$

(b)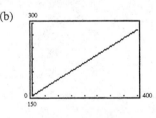

$C = 0.34(230) + 150$

$C = 78.20 + 150$

$C = \$228.20$

estimate $= \$228.20$

(c) estimate 147 miles

$200 = 0.34x + 150$

$50 = 0.34x$

$\dfrac{50}{0.34} = x$

$147.0588 = x$

$147 \approx x$ miles

**76.** (a) $(0, \$27{,}500)$   $(5, \$12{,}000)$

$$m = \frac{12{,}000 - 27{,}500}{5 - 0} = \frac{-15{,}500}{5} = -3100$$

$$V - 27{,}500 = -3100(t - 0)$$

$$V = -3100t + 27{,}500$$

(b) $V = -3100(2) + 27{,}500$

$V = -6200 + 27{,}500$

$V = \$21{,}300$

**78.** (a) $(0.80, 6000)$   $(1, 4000)$

$$m = \frac{6000 - 4000}{0.80 - 1.00} = \frac{2000}{-0.20} = -10{,}000$$

$$x - 4000 = -10{,}000(p - 1)$$

$$x - 4000 = -10{,}000p + 10{,}000$$

$$x = -10{,}000p + 14{,}000$$

(c) $x = -10{,}000(0.90) + 14{,}000$

$x = -9000 + 14{,}000$

$x = 5000$

Thus, if the price is $0.90, the demand will be 5000 cans.

(b) $x = -10{,}000(1.10) + 14{,}000$

$x = -11{,}000 + 14{,}000$

$x = 3000$

Thus, if the price is $1.10, the demand will be 3000 cans.

**80.** (a) and (b)

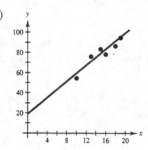

(d) $y = 4(17) + 19 = 68 + 19 = 87$

(c) Two points taken from the "best-fitting" line sketched in part (b) are $(12, 67)$ and $(20, 99)$.

$$m = \frac{99 - 67}{20 - 12} = \frac{32}{8} = 4$$

$$y - 67 = 4(x - 12)$$

$$y - 67 = 4x - 48$$

$$y = 4x + 19$$

**82.** (d)

| $x$ | 0 | 1 | 2 | 3 | 4 |
|---|---|---|---|---|---|
| $y$ | \$15,900 | \$14,000 | \$12,100 | \$10,200 | \$8300 |

(e) $y - 15{,}900 = -1900(x - 0)$

$\qquad\quad y = -1900x + 15{,}900$

(f) $y = -1900(5) + 15{,}900 = -9500 + 15{,}900 = \$6400$

No. $y = \$6400$ when $x = 5$.

**84.** Point-slope form: $y - y_1 = m(x - x_1)$

Slope-intercept form: $y = mx + b$

General form: $ax + by + c = 0$

**86.** The variable $y$ is missing in the equation of a vertical line because any point on a vertical line is independent of $y$.

# Section 2.5    Relations and Functions

**2.** Domain = {3, 4, 6, 8}

Range = {−2, 3, 5, 10}

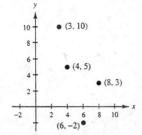

**4.** Domain = {−3}

Range = {2, 5, 6}

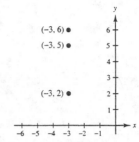

**6.** (8, 480), (10, 600), (7.5, 450), (4, 240)

**8.** (−1, −1), (0, 0), (1, 1), (2, 8), (3, 27), (4, 64)

**10.** (1977, Carter), (1981, Reagan), (1985, Reagan), (1989, Bush), (1993, Clinton), (1997, Clinton)

**12.** Yes, this relation is a function because each element in the domain is assigned exactly one element in the range.

**14.** No, this relation is not a function as 100 in the domain is paired with two numbers in the range (25 and 30).

**16.** Yes, this relation is a function as each number in the domain is paired with exactly one number in the range.

**18.** Yes, this relation is a function because each element in the domain is assigned exactly one element in the range.

**20.** No, this relation is not a function as 0 in the domain is paired with two numbers in the range (1 and 20) as is 1 (8 and 15).

**22.** Yes, this relation is a function as each number in the domain is paired with exactly one number in the range.

**24.** (a) No   (b) Yes   (c) No   (d) Yes

**26.**
$$x^2 + 4y^2 = 16$$
$$0^2 + 4(2)^2 \overset{?}{=} 16$$
$$16 = 16$$
$$0^2 + 4(-2)^2 \overset{?}{=} 16$$
$$16 = 16$$

Both (0, 2) and (0, −2) are solutions of $x^2 + 4y^2 = 16$ which implies $y$ is not a function of $x$.

**28.**
$$|y - 2| = x$$
$$|4 - 2| \overset{?}{=} 2$$
$$2 = 2$$
$$|0 - 2| \overset{?}{=} 2$$
$$2 = 2$$

Both (2, 4) and (2, 0) are solutions of $|y - 2| = x$ which implies $y$ is not a function of $x$.

**30.** $y = 3 - 8x$ represents $y$ as a function of $x$ because there is one value of $y$ associated with one value of $x$.

**32.** $x - 9y + 3 = 0$ represents $y$ as a function of $x$ because there is one value of $y$ associated with one value of $x$.

**34.** $y = (x + 2)^2 + 3$ represents $y$ as a function of $x$ because there is one value of $y$ associated with one value of $x$.

**36.** $f(x) = 6 - 2x$

    (a) $f(3) = 6 - 2(3) = 6 - 6 = 0$

    (b) $f(-4) = 6 - 2(-4) = 6 + 8 = 14$

    (c) $f(n) = 6 - 2n$

    (d) $f(n - 2) = 6 - 2(n - 2) = 6 - 2n + 4 = 10 - 2n$

**38.** $f(x) = \sqrt{x + 8}$

    (a) $f(1) = \sqrt{1 + 8} = \sqrt{9} = 3$

    (b) $f(-4) = \sqrt{-4 + 8} = \sqrt{4} = 2$

    (c) $f(h) = \sqrt{h + 8}$

    (d) $f(h - 8) = \sqrt{h - 8 + 8} = \sqrt{h}$

**40.** $f(x) = \dfrac{2x}{x - 7}$

    (a) $f(2) = \dfrac{2(2)}{2 - 7} = \dfrac{4}{-5}$

    (b) $f(-3) = \dfrac{2(-3)}{-3 - 7} = \dfrac{-6}{-10} = \dfrac{3}{5}$

    (c) $f(t) = \dfrac{2t}{t - 7}$

    (d) $f(t + 5) = \dfrac{2(t + 5)}{t + 5 - 7} = \dfrac{2t + 10}{t - 2}$

**42.** $f(x) = 3 - 7x$

    (a) $f(-1) = 3 - 7(-1) = 3 + 7 = 10$

    (b) $f\left(\dfrac{1}{2}\right) = 3 - 7\left(\dfrac{1}{2}\right) = 3 - \dfrac{7}{2} = \dfrac{6}{2} - \dfrac{7}{2} = -\dfrac{1}{2}$

    (c) $f(t) + f(-2) = 3 - 7(t) + 3 - 7(-2) = 3 - 7t + 3 + 14 = -7t + 20$

    (d) $f(2t - 3) = 3 - 7(2t - 3) = 3 - 14t + 21 = 24 - 14t$

**44.** $h(x) = x^2 - 2x$

    (a) $h(2) = 2^2 - 2(2) = 4 - 4 = 0$

    (b) $h(0) = 0^2 - 2(0) = 0$

    (c) $h(1) - h(-4) = [1^2 - 2(1)] - [(-4)^2 - 2(-4)] = 1 - 2 - 16 - 8 = -25$

    (d) $h(4t) = (4t)^2 - 2(4t) = 16t^2 - 8t$

**46.** $h(x) = \sqrt{2x - 3}$

    (a) $h(4) = \sqrt{2(4) - 3} = \sqrt{5}$

    (b) $h(2) = \sqrt{2(2) - 3} = \sqrt{1} = 1$

    (c) $h(4n) = \sqrt{2(4n) - 3} = \sqrt{8n - 3}$

    (d) $h(n + 2) = \sqrt{2(n + 2) - 3} = \sqrt{2n + 4 - 3} = \sqrt{2n + 1}$

**48.** $g(x) = \dfrac{|x + 1|}{x + 1}$

(a) $g(2) = \dfrac{|2 + 1|}{2 + 1} = \dfrac{|3|}{3} = \dfrac{3}{3} = 1$

(b) $g\left(-\dfrac{1}{3}\right) = \dfrac{\left|-\dfrac{1}{3} + 1\right|}{-\dfrac{1}{3} + 1} = \dfrac{\left|\dfrac{2}{3}\right|}{\dfrac{2}{3}} = \dfrac{\dfrac{2}{3}}{\dfrac{2}{3}} = 1$

(c) $g(-4) = \dfrac{|-4 + 1|}{-4 + 1} = \dfrac{|-3|}{-3} = \dfrac{3}{-3} = -1$

(d) $g(3) + g(-5) = \dfrac{|3 + 1|}{3 + 1} + \dfrac{|-5 + 1|}{-5 + 1} = \dfrac{|4|}{4} + \dfrac{|-4|}{-4} = \dfrac{4}{4} + \dfrac{4}{-4} = 1 + (-1) = 0$

**50.** $f(x) = \dfrac{x + 2}{x - 3}$

(a) $f(-3) = \dfrac{-3 + 2}{-3 - 3} = \dfrac{-1}{-6} = \dfrac{1}{6}$

(b) $f\left(-\dfrac{3}{2}\right) = \dfrac{-\dfrac{3}{2} + 2}{-\dfrac{3}{2} - 3} \cdot \dfrac{2}{2} = \dfrac{-3 + 4}{-3 - 6} = \dfrac{1}{-9}$

(c) $f(4) + f(8) = \dfrac{4 + 2}{4 - 3} + \dfrac{8 + 2}{8 - 3} = \dfrac{6}{1} + \dfrac{10}{5} = 6 + 2 = 8$

(d) $f(x - 5) = \dfrac{x - 5 + 2}{x - 5 - 3} = \dfrac{x - 3}{x - 8}$

**52.** $f(x) = \begin{cases} -x, \text{ if } x \le 0 \\ 6 - 3x, \text{ if } x > 0 \end{cases}$

(a) $f(0) = -0 = 0$

(b) $f\left(-\dfrac{3}{2}\right) = -\left(-\dfrac{3}{2}\right) = \dfrac{3}{2}$

(c) $f(4) = 6 - 3(4) = 6 - 12 = -6$

(d) $f(-2) + f(25) = -(-2) + 6 - 3(25) = 2 + 6 - 75 = -67$

**54.** $f(x) = \begin{cases} x^2, \text{ if } x < 1 \\ x^2 - 3x + 2, \text{ if } x \ge 1 \end{cases}$

(a) $f(1) = (1)^2 - 3(1) + 2 = 1 - 3 + 2 = 0$

(b) $f(-1) = (-1)^2 = 1$

(c) $f(2) = (2)^2 - 3(2) + 2 = 4 - 6 + 2 = 0$

(d) $f(-3) + f(3) = (-3)^2 + (3)^2 - 3(3) + 2 = 9 + 9 - 9 + 2 = 11$

**56.** $f(x) = 3x + 4$

(a) $\dfrac{f(x + 1) - f(1)}{x} = \dfrac{3(x + 1) + 4 - [3(1) + 4]}{x} = \dfrac{3x + 3 + 4 - 3 - 4}{x} = \dfrac{3x}{x} = 3$

(b) $\dfrac{f(x - 5) - f(5)}{x} = \dfrac{3(x - 5) + 4 - [3(5) + 4]}{x} = \dfrac{3x - 15 + 4 - 19}{x} = \dfrac{3x - 30}{x} = \dfrac{3(x - 10)}{x}$

**58.** The domain of $f(x) = 4x - 3$ is all real numbers $x$.

**60.** The domain of $g(x) = \dfrac{x + 5}{x + 4}$ is all real numbers $x$ such that $x \neq -4$ because $x + 4 \neq 0$ means $x \neq -4$.

**62.** The domain if $g(s) = \dfrac{s - 2}{(s - 6)(s - 10)}$ is all real numbers $s$ such that $s \neq 6, 10$ because $(s - 6)(s - 10) \neq 0$ means $s - 6 \neq 0$ and $s \neq 6$ and $s - 10 \neq 0$ and $s \neq 10$.

**64.** The domain of $f(x) = \sqrt{2 - x}$ is all real numbers $x$ such that $x \leq 2$ because $2 - x \geq 0$ means $-x \geq -2$ and $x \leq 2$.

**66.** The domain of $G(x) = \sqrt{8 - 3x}$ is all real numbers $x$ such that $x \leq \dfrac{8}{3}$ because $8 - 3x \geq 0$ means $-3x \geq -8$ and $x \leq \dfrac{8}{3}$.

**68.** The domain of $f(x) = |x + 3|$ is all real numbers $x$.

**70.** Domain: $\{-3, -1, 2, 5\}$

   Range: $\{-2, 0, 3, 4\}$

**72.** Domain: $\left\{\frac{1}{2}, \frac{3}{4}, 1, \frac{5}{4}\right\}$

   Range: $\{4, 5, 6, 7\}$

**74.** Domain: $s > 0$

   Range: $A > 0$

**76.** Domain: $r > 0$

   Range: $V > 0$

**78.** *Verbal Model:* $\boxed{\text{Surface}} = 6 \cdot \boxed{\text{Length of edge}}^2$

   *Labels:* Surface $= S(x)$

   Length of edge $= x$

   *Function:* $S(x) = 6x^2$

**80.** *Verbal Model:* $\boxed{\begin{array}{c}\text{Length of}\\\text{diagonal}\end{array}} = \sqrt{2} \cdot \boxed{\begin{array}{c}\text{Length of}\\\text{side}\end{array}}$

   *Labels:* Length of diagonal $= L(x)$

   Length of side $= x$

   *Function:* $L = \sqrt{2}\,x$

**82.** *Verbal Model:* $\boxed{\text{Total Cost}} = \boxed{\begin{array}{c}\text{Fixed}\\\text{Costs}\end{array}} + \boxed{\begin{array}{c}\text{Variable}\\\text{Costs}\end{array}}$

   *Labels:* Total cost $= C(x)$

   Fixed costs $= 8000$

   Variable costs $= 1.95x$

   Number of games $= x$

*Function:* $C(x) = 8000 + 1.95x, x > 0$

**84.** *Verbal Model:* $\boxed{\text{Area}} = \boxed{\text{Side}} \cdot \boxed{\text{Side}}$

   *Label:* Area $= A(x)$

   *Function:* $A(x) = (32 - 2x)(32 - 2x)$

   $\phantom{Function: A(x)} = 2(16 - x) \cdot 2(16 - x)$

   $\phantom{Function: A(x)} = 4(16 - x)^2$

**86.** (a) $P(1600) = 50\sqrt{1600} - 0.5(1600) - 500$

   $\phantom{P(1600)} = 50(40) - 800 - 500$

   $\phantom{P(1600)} = 2000 - 800 - 500$

   $\phantom{P(1600)} = \$700$

   (b) $P(2500) = 50\sqrt{2500} - 0.5(2500) - 500$

   $\phantom{P(2500)} = 50(50) - 1250 - 500$

   $\phantom{P(2500)} = 2500 - 1250 - 500$

   $\phantom{P(2500)} = \$750$

**88.** $W(a) = \begin{cases} 12h, & 0 < h \leq 40 \\ 18(h - 40) + 480, & h > 40 \end{cases}$

   (a) $W(30) = 12(30) = \$360$

   $W(40) = 12(40) = \$480$

   $W(45) = 18(45 - 40) + 480 = 90 + 480 = \$570$

   $W(50) = 18(50 - 40) + 480 = 180 + 480 = \$660$

   (b) No. $h < 0$ is not in the domain of $W$.

**90.** $f(1993) = 55,000,000$

  55,000,000 students

**92.** (a) This is a correct mathematical use of the
     word function.

  (b) This is not a correct mathematical use of the
     word function.

**94.** A relation is any set of ordered pairs. A function is a relation in which no two ordered pairs have the same first component and different second components.

**96.** The domain is the set of inputs of the function and the range is the set of outputs of the function.

## Section 2.6    Graphs of Functions

**2.**

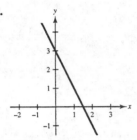

Domain: $-\infty < x < \infty$

Range: $-\infty < y < \infty$

**4.**

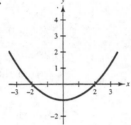

Domain: $-\infty < x < x$

Range: $[-1, \infty)$ or $y \geq -1$

**6.**

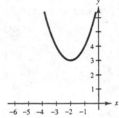

Domain: $-\infty < x < \infty$

Range: $[3, \infty)$ or $y \geq 3$

**8.**

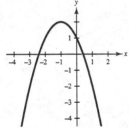

Domain: $-\infty < x < \infty$

Range: $[-\infty, 2)$ or $-\infty < y \leq 2$

**10.**

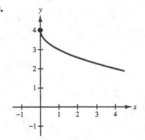

Domain: $0 \leq x < \infty$

Range: $-\infty < y \leq 4$

**12.**

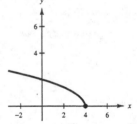

Domain: $(-\infty, 4]$ or $-\infty < x \leq 4$

Range: $[0, \infty)$ or $0 \leq y < \infty$

**14.**

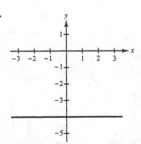

Domain: $-\infty < x < \infty$

Range: $y = -4$

**16.**

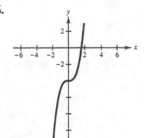

Domain: $-\infty < x < x$

Range: $-\infty < y < \infty$

**18.**

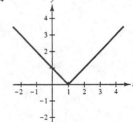

Domain: $-\infty < x < \infty$

Range: $0 \leq y < \infty$

**20.**

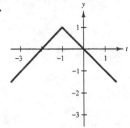

Domain: $-\infty < t < \infty$

Range: $(-\infty, 1]$ or $-\infty < y \le 1$

**22.**

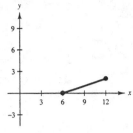

Domain: $6 \le x \le 12$ or $[6, 12]$

Range: $0 \le y \le 2$ or $[0, 2]$

**24.**

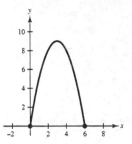

Domain: $0 \le x \le 6$ or $[0, 6]$

Range: $0 \le y \le 9$ or $[0. 9]$

**26.**

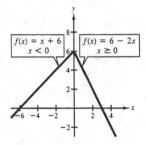

Domain: $-\infty < x < \infty$

Range: $(-\infty, 6]$ or $-\infty < y \le 6$

**28.**

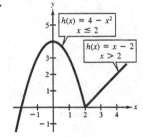

Domain: $-\infty < x < \infty$

Range: $-\infty < y < \infty$

**30.** *Keystrokes:*

Domain: $-\infty < x < \infty$

Range: $[0, \infty)$ or $0 \le y < \infty$

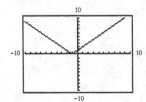

**32.** *Keystrokes:*

Domain: $[-2, 2]$ or $-2 \le t \le 2$

Range: $[0, 2]$ or $0 \le t \le 2$

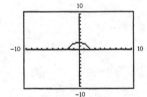

**34.** Yes, $y = x^2 - 2x$ passes the Vertical Line Test and is a function of $x$.

**36.** No, $y$ is not a function of $x$ by the Vertical Line Test.

**38.** Yes, $y$ is a function by the Vertical Line Test.

**40.** No, $|y| = x$ does not pass the Vertical Line Test so $y$ is not a function of $x$.

**42.**

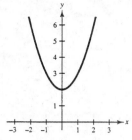

*y* is a function of *x*.

**44.**

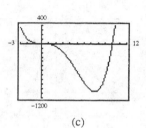

*y* is not a function of *x*.

**46.** (d) graph matches $f(x) = (x - 2)^2$

**48.** (c) graph matches $f(x) = |x + 2|$

**50.** (c) shows the most complete graph.

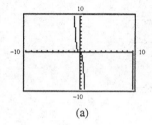

(a)

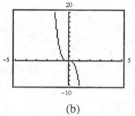

(b)

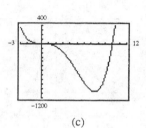

(c)

**52.** (a)  Vertical shift 3 units upward

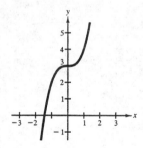

(b)  Vertical shift 5 units downward

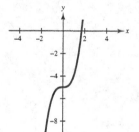

(c)  Horizontal shift 3 units to the right

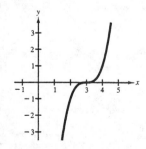

(d)  Horizontal shift 2 units to the left

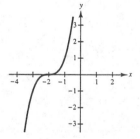

—CONTINUED—

**52.** —CONTINUED—

(e) Reflection in the y-axis

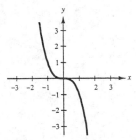

(f) Reflection in the x-axis

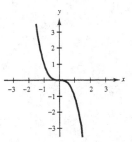

(g) Vertical shift 2 units upward
Reflection in the x-axis
Horizontal shift 1 unit to the right

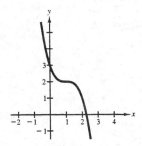

(h) Horizontal shift 2 units to the left
Vertical shift 3 units downward

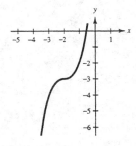

**54.** *Keystrokes:*

$$\boxed{Y=}\ \boxed{ABS}\ \boxed{(}\ \boxed{X,T,\theta}\ \boxed{+}\ 3\ \boxed{)}\ \boxed{GRAPH}$$

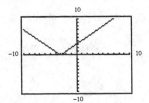

Horizontal shift 3 units to the left

**56.** *Keystrokes:*

$$\boxed{Y=}\ \boxed{ABS}\ \boxed{(}\ \boxed{(-)}\ \boxed{X,T,\theta}\ \boxed{)}\ \boxed{GRAPH}$$

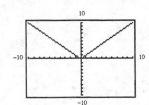

Reflection in the y-axis

**58.** *Keystrokes:*

$$\boxed{Y=}\ 5\ \boxed{-}\ \boxed{ABS}\ \boxed{X,T,\theta}\ \boxed{GRAPH}$$

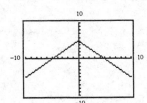

Reflection in the x-axis and vertical shift 5 units upward

**60.** Graph is shifted 3 units upward

$$h(x) = x^2 + 3$$

**62.** Graph is reflected in the x-axis and shifted 1 unit upward

$$h(x) = -x^2 + 1$$

**64.** Graph is reflected in the x-axis and shifted 3 units downward

$$h(x) = -x^2 - 3$$

**66.** Graph is shifted 1 unit downward

$h(x) = x^2 - 1$

**68.** $f(x) = \sqrt{x} + 1$

**70.** $f(x) = \sqrt{x - 3}$

**72.** $f(x) = \sqrt{1 - x}$

**74.** (a) $y = f(x) - 1$

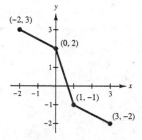

(b) $y = f(x + 1)$

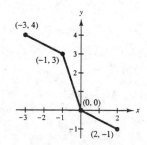

(c) $y = f(x - 1)$

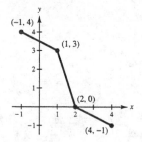

(d) $y = -f(x - 2)$

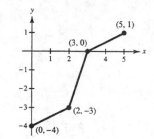

(e) $y = f(-x)$

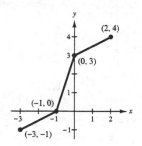

(f) $y = f(x) + 2$

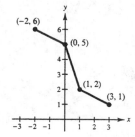

**76.** *Keystrokes:*

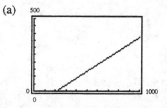

(a)

(b)   $0 = 0.47x - 100$

$100 = 0.47x$

$213 \approx x$

(c) $300 = 0.47x - 100$

$400 = 0.47x$

$851 \approx x$

**78.** (a) *Verbal Model:*  | Perimeter | = 2 | Length | + 2 | Width | + 8 | Width of walkway |

*Labels:*  Perimeter $= P(x)$

Length $= 40$ ft

Width $= 30$ ft

Width of walkway $= x$

*Function:* $P(x) = 2(40) + 2(30) + 8x$

$$= 80 + 60 + 8x$$

$$P(x) = 140 + 8x$$

(b) *Keystrokes:* Y= 140 + 8 X,T,θ GRAPH

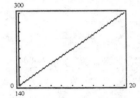

(c) Slope $= 8$ so for each 1-foot increase in the width of the walkway the perimeter increases by 8 feet.

**80.** (a) $T$ is a function of $t$ because to each $t$ there corresponds one and only one temperature $T$.

(b) $T(4) = 60°$, $T(15) = 72°$

(c) If the thermostat were reprogrammed to produce a temperature $H$ where $H(t) = T(t - 1)$, all the temperature changes would occur 1 hour later.

(d) If the thermostat were reprogrammed to produce a temperature $H(t) = T(t) - 1$, the temperature would be decreased by 1 degree.

**82.** Use the Vertical Line Test to determine if an equation represents $y$ as a function of $x$. If the graph of an equation has the property that no vertical line intersects the graph at two (or more) points, the equation represents $y$ as a function of $x$.

**84.** $g(x) = -f(x)$ is a reflection in the $x$-axis of the graph of $f(x)$.

**86.** $g(x) = f(x - 2)$ is a horizontal shift 2 units to the right of the graph of $f(x)$.

# Review Exercises for Chapter 2

**2.**

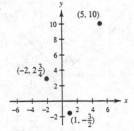

**4.**

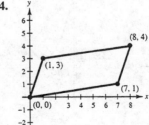

**6.** Quadrant III

**8.** Quadrants I, III

**10.** (a) $(3, 10)$    $3(3) - 2(10) + 18 \overset{?}{=} 0$

$$9 - 20 + 18 \overset{?}{=} 0$$

$$7 \neq 0 \text{ no}$$

(c) $(-4, 3)$    $3(-4) - 2(3) + 18 \overset{?}{=} 0$

$$-12 - 6 + 18 \overset{?}{=} 0$$

$$0 = 0 \text{ yes}$$

(b) $(0, 9)$    $3(0) - 2(9) + 18 \overset{?}{=} 0$

$$0 - 18 + 18 \overset{?}{=} 0$$

$$0 = 0 \text{ yes}$$

(d) $(-8, 0)$    $3(-8) - 2(0) + 18 \overset{?}{=} 0$

$$-24 - 0 + 18 \overset{?}{=} 0$$

$$-6 \neq 0 \text{ no}$$

**12.** $d = \sqrt{(6 - 2)^2 + [-5 - (-5)]^2} = \sqrt{4^2 + 0^2} = \sqrt{16} = 4$

**14.** $d = \sqrt{[3 - (-2)]^2 + (-2 - 10)^2} = \sqrt{25 + 144} = \sqrt{169} = 13$

**16.** $y = x^2 + 4$

Matches graph (b)

**18.** $y = \sqrt{x = 4}$

Matches graph (d)

**20.** $y = \frac{3}{4}x - 2$

$$0 = \frac{3}{4}x - 2$$

$$2 = \frac{3}{4}x$$

$$\frac{8}{3} = x \qquad \left(\frac{8}{3}, 0\right)$$

$$y = \frac{3}{4}(0) - 2$$

$$y = 0 - 2$$

$$y = -2 \qquad (0, -2)$$

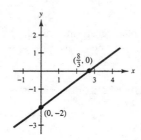

**22.** $3x + 4y + 12 = 0$

$$3x + 4(0) + 12 = 0$$

$$3x = -12$$

$$x = -4 \qquad (-4, 0)$$

$$3(0) + 4y + 12 = 0$$

$$4y = -12$$

$$y = -3 \qquad (0, -3)$$

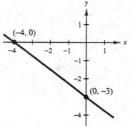

**24.** $y = (x - 2)^2$

$$y = (0 - 2)^2 = 4 \qquad (0, 4)$$

$$0 = (x - 2)^2$$

$$0 = x - 2$$

$$2 = x \qquad (2, 0)$$

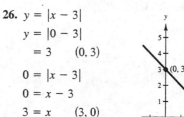

**26.** $y = |x - 3|$

$$y = |0 - 3|$$

$$= 3 \qquad (0, 3)$$

$$0 = |x - 3|$$

$$0 = x - 3$$

$$3 = x \qquad (3, 0)$$

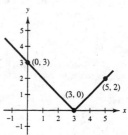

**28.** $y + 3x + 9$

  $y$-intercept: $y + 3(0) + 9$

  $+ 9$    $(0, 9)$

  $x$-intercept: $0 + 3x + 9$

  $-9 + 3x$

  $-3 + x$    $(-3, 0)$

**30.** $5x + 4y + 10$

  $y$-intercept: $5(0) + 4y + 10$

  $4y + 10$

  $y + \frac{10}{4}$

  $y + \frac{5}{2}$    $\left(0, \frac{5}{2}\right)$

  $x$-intercept: $5x + 4(0) + 10$

  $5x + 10$

  $x + 2$    $(2, 0)$

**32.** $y + |x| + 4$

  $y$-intercept: $y + |0| + 4$

  $+ 4$    $(0, 4)$

  $x$-intercept: $0 + |x| + 4$

  $-4 + |x|$

  None

**34.** $y + |3 - 6x| - 15$

  $y$-intercept: $y + |3 - 6(0)| - 15$

  $+ 3 - 15$

  $+ -12$    $(0, -12)$

  $x$-intercepts: $0 + |3 - 6x| - 15$

  $15 + |3 - 6x|$

$15 + 3 - 6x$ or $-15 + 3 - 6x$

$12 + -6x$    $-18 + -6x$

$-2 + x$    $3 + x$

$(-2, 0),\ (3, 0)$

**36.** *Keystrokes:* $\boxed{Y=}\ \boxed{(}\ 1\ \boxed{\div}\ 4\ \boxed{)}\ \boxed{(}\ \boxed{X,T,\theta}\ \boxed{-}\ 2\ \boxed{)}\ \boxed{\wedge}\ 3\ \boxed{GRAPH}$

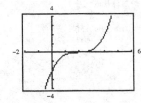    $(2, 0),\ (0, -2)$

**38.** *Keystrokes:* $\boxed{Y=}\ 3\ \boxed{-}\ \boxed{ABS}\ \boxed{(}\ \boxed{X,T,\theta}\ \boxed{-}\ 3\ \boxed{)}\ \boxed{GRAPH}$

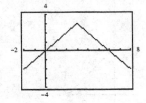    $(0, 0),\ (6, 0)$

**40.** *Keystrokes:* [Y=] [X,T,θ] [−] 2 [√] [X,T,θ] [GRAPH]

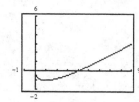

(0, 0), (4, 0)

**42.** $m = \dfrac{-8 - 5}{3 - (-2)} = \dfrac{-13}{5}$

**44.** $m = \dfrac{8 - 2}{7 - 7} = \dfrac{6}{0}$ is undefined.

**46.** $m = \dfrac{6 - 0}{\frac{7}{2} - 0} = \dfrac{6}{\frac{7}{2}} = \dfrac{12}{7}$

**48.**  $m = \dfrac{3 - 1}{8 - 2} = \dfrac{2}{6} = \dfrac{1}{3}$

$\dfrac{1}{3} = \dfrac{t - 1}{1 - 2}$

$\dfrac{1}{3} = \dfrac{t - 1}{-1}$

$-1 = 3(t - 1)$

$-1 = 3t - 3$

$2 = 3t$

$\dfrac{2}{3} = t$

**50.** $2 = \dfrac{y - \left(\frac{1}{2}\right)}{x - (-4)}$

$2 = \dfrac{y - \frac{1}{2}}{x + 4}$

$\left(-3, \dfrac{5}{2}\right), \left(-2, \dfrac{9}{2}\right)$

**52.** $-\dfrac{1}{3} = \dfrac{y - \left(-\frac{3}{2}\right)}{x - (-3)}$

$-\dfrac{1}{3} = \dfrac{y + \frac{3}{2}}{x + 3}$

$\left(0, -\dfrac{5}{2}\right), \left(3, -\dfrac{7}{2}\right)$

There are many solutions to this problem.

**54.** $0 = \dfrac{y - (-2)}{x - 7}$

$0 = \dfrac{y + 2}{x - 7}$

$(0, -2), (-3, -2)$

There are many solutions to this problem.

**56.** $x - 3y - 6 = 0$

$-3y = -x + 6$

$y = \tfrac{1}{3}x - 2$

**58.** $y - 6 = 0$

$y = 6$

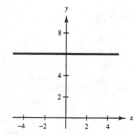

**60.** $L_1$: $y + 2x - 5$

$L_2$: $y + 2x + 3$

$m_1 + 2, m_2 + 2$

$m_1 + m_2$ so the lines are parallel

**62.** $L_1$: $y + -0.3x - 2$

$L_2$: $y + 0.3x + 1$

$m_1 + -0.3, m_2 + 0.3$

$m_1 \neq m_2, m_1 \cdot m_2 \neq -1$ so the lines are neither

**64.** $L_1$: $4x + 3y - 6 + 0$　　　　$L_2$: $3x - 4y - 8 + 0$　　　　$m_1 \cdot m_2 + -1$

$3y + -4x + 6$　　　　$-4y + -3x + 8$　　　　so the lines are perpendicular

$y + -\frac{4}{3}x + 2$　　　　$y + \frac{3}{4}x - 2$

$m_1 + -\frac{4}{3}$　　　　$m_2 + \frac{3}{4}$

**66.** $y - (-5) + 3[x - (-5)]$

$y + 5 + 3(x + 5)$

$y + 5 + 3x + 15$

$3x - y + 10 + 0$

**68.** $y - (-2) + -2(x - 5)$

$y + 2 + -2x + 10$

$2x + y - 8 + 0$

**70.** $y - \left(-\frac{4}{3}\right) + \frac{3}{2}[x - (-2)]$

$y + \frac{4}{3} + \frac{3}{2}(x + 2)$

$y + \frac{4}{3} + \frac{3}{2}x + 3$

$6\left(y + \frac{4}{3}\right) + 6\left(\frac{3}{2}x + 3\right)$

$6y + 8 + 9x + 18$

$9x - 6y + 10 + 0$

**72.** $x + 7$

$x - 7 + 0$

**74.** $m + \dfrac{10 - 10}{6 - 0} + \dfrac{0}{6} + 0$

$y + 10$

$y - 10 + 0$

**76.** $m + \dfrac{-7 - 2}{4 - (-10)} + \dfrac{-9}{14}$

$y - (-7) + -\dfrac{9}{14}(x - 4)$

$y + 7 + -\dfrac{9}{14}x + \dfrac{18}{7}$

$14(x + 7) + 14\left(-\dfrac{9}{14}x + \dfrac{18}{7}\right)$

$14y + 98 + -9x + 36$

$9x + 14y + 62 + 0$

**78.** $m + \dfrac{5 - 0}{\frac{5}{2} - \frac{5}{2}} + \dfrac{5}{0}$　　　$m$ is undefined.

$x + \dfrac{5}{2}$

$x - \dfrac{5}{2} + 0$　or　$2x - 5 + 0$

**80.** $2x + 4y = 1$

$\quad 4y = -2x + 1$

$\quad y = -\frac{1}{2}x + \frac{1}{4} \quad m = -\frac{1}{2}$

(a) $\quad y - 5 = -\frac{1}{2}[x - (-1)]$

$\quad y - 5 = -\frac{1}{2}(x + 1)$

$\quad y - 5 = -\frac{1}{2}x - \frac{1}{2}$

$\quad 2(y - 5) = 2\left(-\frac{1}{2}x - \frac{1}{2}\right)$

$\quad 2y - 10 = -x - 1$

$\quad x + 2y - 9 = 0$

(b) $y - 5 = 2[x - (-1)]$

$\quad y - 5 = 2(x + 1)$

$\quad y - 5 = 2x + 2$

$\quad 0 = 2x - y + 7$

**82.** $4x - 3y = 12$

$\quad -3y = -4x + 12$

$\quad y = \frac{4}{3}x - 4 \quad m = \frac{4}{3}$

(a) $\quad y - 3 = \frac{4}{3}\left(x - \frac{3}{8}\right)$

$\quad y - 3 = \frac{4}{3}x - \frac{1}{2}$

$\quad 6(y - 3) = 6\left(\frac{4}{3}x - \frac{1}{2}\right)$

$\quad 6y - 18 = 8x - 3$

$\quad 0 = 8x - 6y + 15$

(b) $\quad y - 3 = -\frac{3}{4}\left(x - \frac{3}{8}\right)$

$\quad y - 3 = -\frac{3}{4}x + \frac{9}{32}$

$\quad 32(y - 3) = 32\left(-\frac{3}{4}x + \frac{9}{32}\right)$

$\quad 32y - 96 = -24x + 9$

$\quad 24x + 32y - 105 = 0$

**84.** Yes, this relation is a function because each number in the domain is paired to only one number in the range.

**86.** No, this relation is not a function because the 35 in the domain is paired to two numbers (5 and 7) in the range.

**88.** $h(x) = x(x - 8)$

(a) $h(8) = 8(8 - 8) = 8(0) = 0$

(b) $h(10) = 10(10 - 8) = 10(2) = 20$

(c) $h(-3) + h(4) = [-3(-3 - 8)] + [4(4 - 8)] = (-3)(-11) + (4)(-4) = 33 - 16 = 17$

(d) $h(4t) = (4t)(4t - 8) = 16t^2 - 32t$

**90.** $g(x) = \dfrac{|x + 4|}{4}$

(a) $g(0) = \dfrac{|0 + 4|}{4} = \dfrac{|4|}{4} = \dfrac{4}{4} = 1$

(b) $g(-8) = \dfrac{|-8 + 4|}{4} = \dfrac{|-4|}{4} = \dfrac{4}{4} = 1$

(c) $g(2) - g(-5) = \dfrac{|2 + 4|}{4} - \dfrac{|-5 + 4|}{4} = \dfrac{|6|}{4} - \dfrac{|-1|}{4} = \dfrac{6}{4} - \dfrac{1}{4} = \dfrac{5}{4}$

(d) $g(x - 2) = \dfrac{|x - 2 + 4|}{4} = \dfrac{|x + 2|}{4}$

**92.** $h(x) + \begin{cases} x^3, & \text{if } x \le 1 \\ (x - 1)^2, & \text{if } x > 1 \end{cases}$

(a) $h(2) + (2 - 1)^2 + 1 + 1^2 + 1 + 1 + 1 + 2$

(b) $h\left(-\dfrac{1}{2}\right) + \left(-\dfrac{1}{2}\right)^3 + -\dfrac{1}{8}$

(c) $h(0) + 0^3 + 0$

(d) $h(4) - h(3) + (4 - 1)^2 + 1 - [(3 - 1)^2 + 1] + 3^2 + 1 - [2^2 + 1] + 9 + 1 - [4 + 1] + 9 + 1 - 5 + 5$

**94.** $f(x) + 7x + 10$

(a) $\dfrac{f(x + 1) - f(1)}{x} + \dfrac{7(x + 1) + 10 - [7(1) + 10]}{x} + \dfrac{7x + 7 + 10 - 7 - 10}{x} + \dfrac{7x}{x} + 7$

(b) $\dfrac{f(x - 5) - f(5)}{x} + \dfrac{7(x - 5) + 10 - [7(5) + 10]}{x} + \dfrac{7x - 35 + 10 - 35 - 10}{x} + \dfrac{7x - 70}{x} + \dfrac{7(x - 10)}{x}$

**96.** $s - 1 \ne 0$    $s + 5 \ne 0$

$s \ne 1$    $s \ne -5$

Domain: $(-\infty, -5) \cup (-5, 1) \cup (1, \infty)$ or all real values of $s$ such that $s \ne 1$ and $s \ne -5$

**98.** Domain: all real values of $x$.

$-\infty < x < \infty$

**100.**
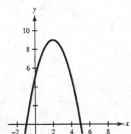

**102.**
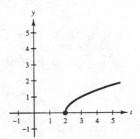

**104.**
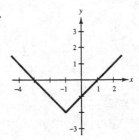

**106.**
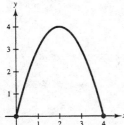

**108.**
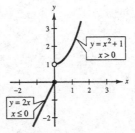

**110.** Yes, $y$ is a function of $x$, because $y$ passes the Vertical Line Test.

**112.** No, $y$ is not a function of $x$, because $y$ does not pass the Vertical Line Test.

**114.** $h(x) + \sqrt{x} + 3$ is a vertical shift 3 units upward.

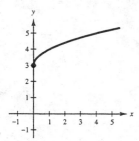

**116.** $h(x) + 1 - \sqrt{x + 4}$ is a reflection in the *x*-axis, horizontal shift 4 units to the left, and a vertical shift 1 unit upward.

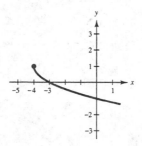

**118.** $y + (x - 1)^2$

Horizontal shift 1 unit to the right

**120.** $y + -x^2 + 1$

Reflection in the *x*-axis and a vertical shift 1 unit upward

**122.** *Verbal Model:* $\boxed{\dfrac{\text{Rise}}{\text{Run}}} + \boxed{\dfrac{\text{Rise}}{\text{Run}}}$

*Proportion:* $-\dfrac{12}{100} + \dfrac{-1500}{x}$

$$x + \dfrac{-1500(100)}{-12}$$

$$x + 12{,}500 \text{ feet}$$

**124.** (0, $8,000), (10, $3,000)

$$m + \dfrac{3{,}000 - 8{,}000}{10 - 0} + \dfrac{-5{,}000}{10} + -500$$

$$V - 8{,}000 + -500(t - 0)$$

$$V - 8{,}000 + -500t$$

$$V + -500t + 8000$$

Domain: $0 \le t \le 10$

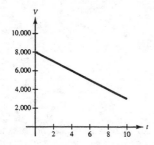

**126.** $d^2 + x^2 + y^2$

$$d + (4.75 + 0.88)^2 + (4.75 + 0.88)^2$$

$$d^2 + (5.63)^2 + (5.63)^2$$

$$d^2 + 31.6969 + 31.6969$$

$$d^2 + 63.3938$$

$$d \approx 7.96$$

**128.** *Verbal Model:*  $\boxed{\text{Perimeter}} + 2\boxed{\text{Length}} + 2\boxed{\text{Width}}$

$$100 + 2\,\text{Length} + 2x$$

$$\frac{100 - 2x}{2} + \text{Length}$$

$$50 - x + \text{Length}$$

*Verbal Model:*  $\boxed{\text{Area}} + \boxed{\text{Length}} \cdot \boxed{\text{Width}}$

*Labels:*  Area $+ A(x)$

Length $+ 50 - x$

Width $+ x$

*Function:*  $A(x) + x[50 - x]$

$$0 < x \le 25$$

# CHAPTER 3
## Polynomials and Factoring

# CHAPTER 3
# Polynomials and Factoring

## Section 3.1    Adding and Subtracting Polynomials

**Solutions to Even-Numbered Exercises**

**2.** Standard form: $5t + 3$

Degree: 1

Leading coefficient: 5

**4.** Standard form: $3x^2 - 4x + 8$

Degree: 2

Leading coefficient: 3

**6.** Standard form: $6z^4 - 16z^2 + 8z$

Degree: 4

Leading coefficient: 6

**8.** Standard form: $4t^5 - t^2 + 6t + 3$

Degree: 5

Leading coefficient: 4

**10.** Standard form: 28

Degree: 0

Leading coefficient: 28

**12.** Standard form: $-\frac{1}{2}at^2 + 48$

Degree: 2

Leading coefficient: $-\frac{1}{2}a$

**14.** $-6y + 3 + y^3$ is a trinomial

**16.** $t^3$ is a monomial

**18.** $25 - 2y^2$ is a binomial

**20.** A trinomial of degree 4 and leading coefficient of $-2$ is any trinomial beginning $-2x^4$ and containing two other terms of degree less than 4 such as $-2x^4 + x - 4$.

**22.** A monomial of degree 0 is any constant such as 16 or 8 or $-4$.

**24.** $x^3 - 4x^{1/3}$ is not a polynomial because the second term is not of the form $ax^k$ ($k$ must be a nonnegative integer).

**26.** $\dfrac{2}{x - 4}$ is not a polynomial because the expression is not of the form $ax^k$.

**28.** $(6 - 2x) + 4x = 6 + (-2x + 4x) = 6 + 2x$

**30.** $(3x + 1) + (6x - 1) = (3x + 6x) + (1-1) = 9x$

**32.** $(3x^3 - 2x + 8) + (3x - 5) = 3x^3 + (-2x + 3x) + (8 - 5) = 3x^3 + x + 3$

**34.** $(z^3 + 6z - 2) + (3z^2 - 6z) = z^3 + 3z^2 + (6z - 6z) + (-2)$

$$= z^3 + 3z^2 - 2$$

**36.** $(y^5 - 4y) + (3y - y^5) + (y^5 - 5) = (y^5 - y^5 + y^5) + (-4y + 3y) - 5$

$$= y^5 - y - 5$$

**38.** $(3a^2 + 5a) + (7 - a^2 - 5a) + (2a^2 + 8) = (3a^2 - a^2 + 2a^2) + (5a - 5a) + (7 + 8)$

$$= 4a^2 + 15$$

**40.** $(2 - \frac{1}{4}y^2 + y^4) + (\frac{1}{3}y^4 - \frac{3}{2}y^2 - 3) = (y^4 + \frac{1}{3}y^4) + (-\frac{1}{4}y^2 - \frac{3}{2}y^2) + (2 - 3)$

$$= (\tfrac{3}{3}y^4 + \tfrac{1}{3}y^4) + (-\tfrac{1}{4}y^2 - \tfrac{6}{4}y^2) - 1$$

$$= \tfrac{4}{3}y^4 - \tfrac{7}{4}y^2 - 1$$

**42.** $(0.13x^4 - 2.25x - 1.63) + (5.3x^4 + 1.76x^2 + 1.29x) =$

$(0.13x^4 + 5.3x^4) + 1.76x^2 + (-2.25x + 1.29x) - 1.63 =$

$5.43x^4 + 1.76x^2 - 0.96x - 1.63$

**44.**
$$
\begin{array}{r}
3x^4 - 2x^2 - 9 \\
-5x^4 + x^2 \phantom{ - 9} \\
\hline
-2x^4 - x^2 - 9
\end{array}
$$

**46.**
$$
\begin{array}{r}
4x^3 + 8x^2 - 5x + 3 \\
x^3 - 3x^2 \phantom{-5x} - 7 \\
\hline
5x^3 + 5x^2 - 5x - 4
\end{array}
$$

**48.**
$$
\begin{array}{r}
-16t^2 + 48t + 64 \\
-32t + 16 \\
\hline
-16t^2 + 16t + 80
\end{array}
$$

**50.**
$$
\begin{array}{r}
1.7y^3 - 6.2y^2 \phantom{+ 2.2y} + 5.9 \\
-3.5y^3 + 6.7y^2 + 2.2y \phantom{+ 5.9} \\
\hline
-1.8y^3 + 0.5y^2 + 2.2y + 5.9
\end{array}
$$

**52.** $(5y^4 - 2) - (3y^4 + 2) = (5y^4 - 2) + (-3y^4 - 2)$

$$= (5y^4 - 3y^4) + (-2 + -2)$$

$$= 2y^4 - 4$$

**54.** $(5q^2 - 3q + 5) - (4q^2 - 3q - 10) = (5q^2 - 3q + 5) + (-4q^2 + 3q + 10)$

$$= (5q^2 - 4q^2) + (-3q + 3q) + (5 + 10)$$

$$= q^2 + 15$$

**56.** $(-10s^2 - 5) - (2s^2 + 6s) = (-10s^2 - 5) + (-2s^2 - 6s)$

$$= (-10s^2 - 2s^2) + (-6s) + (-5)$$

$$= -12s^2 - 6s - 5$$

**58.** $\left(12 - \frac{2}{3}x + \frac{1}{2}x^2\right) - \left(x^3 + 3x^2 - \frac{1}{6}x\right) = \left(12 - \frac{2}{3}x + \frac{1}{2}x^2\right) + \left(-x^3 - 3x^2 + \frac{1}{6}x\right)$

$$= 12 + \left(\frac{-2}{3}x + \frac{1}{6}x\right) + \left(\frac{+1}{2}x^2 - 3x^2\right) - x^3$$

$$= 12 + \left(\frac{-4}{6}x + \frac{1}{6}x\right) + \left(\frac{1}{2}x^2 - \frac{6}{2}x^2\right) - x^3$$

$$= 12 - \frac{3}{6}x - \frac{5}{2}x^2 - x^3$$

$$= 12 - \frac{1}{2}x - \frac{5}{2}x^2 - x^3$$

**60.** $(u^3 - 9.75u^2 + 0.12u - 3) - (0.7u^3 - 6.9u^2 - 4.83) = (u^3 - 9.75u^2 + 0.12u - 3) + (-0.7u^3 + 6.9u^2 + 4.83)$

$$= (u^3 - 0.7u^3) + (-9.75u^2 + 6.9u^2) + 0.12u + (-3 + 4.83)$$

$$= 0.3u^3 - 2.85u^2 + 0.12u + 1.83$$

**62.** $(y^2 + 3y^4) - (y^4 - (y^2 - 8y)) = (y^2 + 3y^4) + (-y^4 + (y^2 - 8y))$

$$= (y^2 + y^2) + (3y^4 - y^4) - 8y$$

$$= 2y^4 + 2y^2 - 8y$$

**64.**
$$3t^4 - 5t^2 \Rightarrow 3t^4 - 5t^2$$
$$\underline{-(-t^4 + 2t^2 - 14) \Rightarrow t^4 - 2t^2 + 14}$$
$$4t^4 - 7t^2 + 14$$

**66.**
$$4x^2 + 5x - 6 \Rightarrow 4x^2 + 5x - 6$$
$$\underline{-(2x^2 - 4x + 5) \Rightarrow -2x^2 + 4x - 5}$$
$$2x^2 + 9x - 11$$

**68.** $(13x^3 - 9x^2 + 4x - 5) - (5x^3 + 7x + 3)$

$$13x^3 - 9x^2 + 4x - 5) \Rightarrow 13x^3 - 9x^2 + 4x - 5$$
$$\underline{-(5x^3 \qquad + 7x + 3) \Rightarrow -5x^3 \qquad - 7x - 3}$$
$$8x^3 - 9x^2 - 3x - 8$$

**70.** $(2x^2 + 1) - (x^2 - 2x + 1) = (2x^2 + 1) + (-x^2 + 2x - 1)$

$$= (2x^2 - x^2) + 2x + (1-1)$$

$$= x^2 + 2x$$

**72.** $(15 - 2y + y^2) + (3y^2 - 6y + 1) - (4y^2 - 8y + 16) = (y^2 + 3y^2 - 4y^2) + (-2y - 6y + 8y) + (15 + 1 - 16)$

$$= 0$$

**74.** $(p^3 + 4) - [(p^2 + 4) + (3p - 9)] = (p^3 + 4) - [p^2 + 3p + (4 - 9)]$

$$= (p^3 + 4) - [p^2 + 3p - 5]$$

$$= (p^3 + 4) + (-p^2 - 3p + 5)$$

$$= p^3 - p^2 - 3p + (4 + 5)$$

$$= p^3 - p^2 - 3p + 9$$

**76.** $(5x^4 - 3x^2 + 9) - [(2x^4 + x^3 - 7x^2) - (x^2 + 6)] = (5x^4 - 3x^2 + 9) + (-2x^4 - x^3 + 7x^2 + x^2 + 6)$

$$= (5x^4 - 2x^4) - x^3 + (-3x^2 + 7x^2 + x^2) + (9 + 6)$$

$$= 3x^4 - x^3 + 5x^2 + 15$$

**78.** $(x^3 - 2x^2 - x) - 5(2x^3 + x^2 - 4x) = (x^3 - 2x^2 - x) - 10x^3 - 5x^2 + 20x$

$$= (x^3 - 10x^3) + (-2x^2 - 5x^2) + (-x + 20x)$$

$$= -9x^3 - 7x^2 + 19x$$

**80.** $-10(v + 2) + 8(v - 1) - 3(v - 9) = -10v - 20 + 8v - 8 - 3v + 27$

$$= (-10v + 8v - 3v) + (-20 - 8 + 27)$$

$$= -5v - 1$$

**82.** $9(7x^2 - 3x + 3) - 4(15x + 2) - (3x^2 - 7x) = 63x^2 - 27x + 27 - 60x - 8 - 3x^2 + 7x$

$$= (63x^2 - 3x^2) + (-27x - 60x + 7x) + (27 - 8)$$

$$= 60x^2 - 80x + 19$$

**84.** $3x^2 - 2[3x + (9 - x^2)] = 3x^2 - 2[3x + 9 - x^2]$

$$= 3x^2 - 6x - 18 + 2x^2$$

$$= (3x^2 + 2x^2) - 6x - 18$$

$$= 5x^2 - 6x - 18$$

**86.** $y_1$   [Y=] [(] [(] 1 [÷] 2 [)] [X,T,θ] [^] 3 [+] 2 [X,T,θ] [)] [+] [(] [X,T,θ] [^] 3 [−] [X,T,θ] [x²]

     [−] [X,T,θ] [+] 1 [)] [ENTER]

   $y_2$   [(] 3 [÷] 2 [)] [X,T,θ] [^] 3 [−] [X,T,θ] [x²] [+] [X,T,θ] [+] 1 [GRAPH]

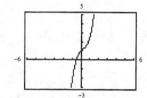

$y_1$ and $y_2$ represent equivalent expressions since the graphs of $y_1$ and $y_2$ are identical.

**88.** $h(x) = f(x) - g(x)$

$$= (4x^3 - 3x^2 + 7) - (9 - x - x^2 - 5x^3)$$

$$= (4x^3 - 3x^2 + 7) + (-9 + x + x^2 + 5x^3)$$

$$= (4x^3 + 5x^3) + (-3x^2 + x^2) + x + (7 - 9)$$

$$= 9x^3 - 2x^2 + x - 2$$

**90.**

| Polynomial | Value | Substitute | Simplify |
|---|---|---|---|
| $h(t) = -16t^2 + 256$ | (a) $t = 0$ | $-16(0)^2 + 256$ | 256 feet |
| | (b) $t = 1$ | $-16(1)^2 + 256$ | 240 feet |
| | (c) $t = \frac{5}{2}$ | $-16\left(\frac{5}{2}\right)^2 + 256$ | 156 feet |
| | (d) $t = 4$ | $-16(4)^2 + 256$ | 0 feet |

At time $t = 0$, the object is at 256 feet and drops, reaching the ground at time $t = 4$.

**92.**

| Polynomial | Value | Substitute | Simplify |
|---|---|---|---|
| $h(t) = -16t^2 + 96t$ | (a) $t = 0$ | $-16(0)^2 + 96(0)$ | 0 feet |
| | (b) $t = 2$ | $-16(2)^2 + 96(2)$ | 128 feet |
| | (c) $t = 3$ | $-16(3)^2 + 96(3)$ | 144 feet |
| | (d) $t = 6$ | $-16(6)^2 + 96(6)$ | 0 feet |

At $t = 0$, the object is at a height of 0 feet (on the ground). It moves upward, reaches a maximum height and returns downward. At time $t = 6$, the object is again on the ground.

**94.** The free-falling object was thrown upward.

$h(t) = -16(0)^2 + 50(0)$

$h(0) = 0 + 0$

$h(0) = 0$ feet

**96.** The free-falling object was thrown upward.

$h(0) = -16(0)^2 + 32(0) + 300$

$h(0) = 0 + 0 + 300$

$h(0) = 300$ feet

**98.** $h(t) = -16t^2 + 200$

$h(1) = -16(1)^2 + 200 = -16 + 200 = 184$ feet

$h(2) = -16(2)^2 + 200 = -16(4) + 200 = -64 + 200 = 136$ feet

$h(3) = -16(3)^2 + 200 = -16(9) + 200 = -144 + 200 = 56$ feet

**100.** *Verbal Model:* $\boxed{\text{Profit}} = \boxed{\text{Revenue}} - \boxed{\text{Cost}}$

$\quad$ *Equation:* $\quad P = R - C$

$\qquad\qquad\quad P = 17x - (12x + 8000)$

$\qquad\qquad\quad P = 5x - 8000$

$\qquad\qquad\quad P = 5(10,000) - 8000$

$\qquad\qquad\quad P = \$42,000$

**102.** Perimeter $= (4x + 2) + (2x + 10) + (2x - 5) + (x + 3) + 2x + 4x$

$\qquad\qquad\quad = 15x + 10$

**104.** Area of Region $= 5 \cdot x + 5 \cdot x + 5 \cdot 3x$ or $5 \cdot [x + x + 3x]$

$\qquad\qquad\qquad\quad = 5x + 5x + 15x \qquad$ or $5[5x]$

$\qquad\qquad\qquad\quad = 25x \qquad\qquad\quad$ or $25x$

**106.** Area $= (4x + 7)5 - 2x(4)$

$\qquad\quad = 20x + 35 - 8x$

$\qquad\quad = 12x + 35$

**108.** (a) $T = R + B$

$\qquad\quad T = (1.1x) + (0.14x^2 - 4.43x + 58.40)$

$\qquad\qquad = 0.14x^2 + (1.1x - 4.43x) + 58.40$

$\qquad\qquad = 0.14x^2 - 3.33x + 58.4$

$\quad$ (b) Keystrokes

$\quad y_1$ $\boxed{Y=}$ 1.1 $\boxed{X,T,\theta}$ $\boxed{\text{ENTER}}$

$\quad y_2$ .14 $\boxed{X,T,\theta}$ $\boxed{x^2}$ $\boxed{-}$ 4.43 $\boxed{X,T,\theta}$ $\boxed{+}$ 58.4 $\boxed{\text{ENTER}}$

$\quad y_3$ .14 $\boxed{X,T,\theta}$ $\boxed{x^2}$ $\boxed{-}$ 3.33 $\boxed{X,T,\theta}$ $\boxed{+}$ 58.4 $\boxed{\text{GRAPH}}$

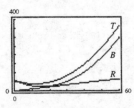

$\quad$ (c) $T = 0.14(30)^2 - 3.33(30) + 58.4 = 84.5$ feet

$\qquad\quad T = 0.14(60)^2 - 3.33(60) + 58.4 = 362.6$ feet

**110.** Addition (or subtraction) separates terms.
$\quad$ Multiplication separates factors.

**112.** Yes, two third-degree polynomials can be added to produce a second-degree polynomial. For example,
$(x^3 - Rx^2 + 4) + (-x^3 + x^2 + 3x) = -x^2 + 3x + 4$

**114.** To subtract one polynomial from another, add the opposite. You can do this by changing the sign of each of the terms of the polynomial that is being subtracted and then adding the resulting like terms.

## Section 3.2    Multiplying Polynomials

**2.** $z^2 \cdot z^3 = (z \cdot z) \cdot (z \cdot z \cdot z)$

$\qquad = z^{2+3}$

$\qquad = z^5$

**4.** $(2y)^3 = (2y) \cdot (2y) \cdot (2y) = (2 \cdot 2 \cdot 2)(y \cdot y \cdot y) = 2^3 y^3 = 8y^3$

**6.** $(x^2)^3 = x^2 \cdot x^2 \cdot x^2 = x \cdot x \cdot x \cdot x \cdot x \cdot x = x^6$

**8.** $\dfrac{y^5}{y^3} = \dfrac{y \cdot y \cdot y \cdot y \cdot y}{y \cdot y \cdot y}$

$\qquad = y^{5-3}$

$\qquad = y^2$

**10.** $\left(\dfrac{3}{t}\right)^5 = \left(\dfrac{3}{t}\right)\left(\dfrac{3}{t}\right)\left(\dfrac{3}{t}\right)\left(\dfrac{3}{t}\right)\left(\dfrac{3}{t}\right) = \dfrac{3 \cdot 3 \cdot 3 \cdot 3 \cdot 3}{t \cdot t \cdot t \cdot t \cdot t} = \dfrac{3^5}{t^5} = \dfrac{243}{t^5}$

**12.** (a) $5^2 y^4 \cdot y^2 = 25(y^4 \cdot y^2) = 25y^{4+2} = 25y^6$

(b) $(5y)^2 \cdot y^4 = 25y^2 \cdot y^4 = 25(y^2 \cdot y^4 = 25y^{2+4} = 25y^6$

**14.** (a) $(-5z^3)^2 = (-5)^2(z^3)^2 = 25z^{3 \cdot 2} = 25z^6$

(b) $(-5z)^4 = (-5)^4 z^4 = 625z^4$

**16.** (a) $(6xy^7)(-x) = -6 \cdot x \cdot x \cdot y^7 = -6x^{1+1}y^7 = -6x^2 y^7$

(b) $(x^5 y^3)(2y^3) = 2 \cdot x^5 \cdot y^{3+3} = 2x^5 y^6$

**18.** (a) $(3y)^3(2y^2) = (3)^3 y^3(2y^2)$

$\qquad = 27 \cdot 2 \cdot y^3 \cdot y^2$

$\qquad = 54y^{3+2}$

$\qquad = 54y^5$

(b) $3y^3 \cdot 2y^2 = 3 \cdot 2 \cdot y^3 \cdot y^2$

$\qquad = 6y^{3+2}$

$\qquad = 6y^5$

**20.** (a) $-(m^3 n^2)(mn^3) = -(m^3 n^2)(mn^3) = -m^{3+1} \cdot n^{2+3} = -m^4 n^5$

(b) $-(m^3 n^2)^2(-mn^3) = -(m^3)^2(n^2)^2(-mn^3)$

$\qquad = -m^6 n^4(-1)mn^3$

$\qquad = m^6 mn^4 n^3$

$\qquad = m^{6+1}n^{4+3}$

$\qquad = m^7 n^7$

**22.** (a) $\dfrac{28x^2y^3}{2xy^2} = \dfrac{28}{2} \cdot \dfrac{x^2}{x} \cdot \dfrac{y^3}{y^2}$

$\qquad\qquad = \dfrac{28}{2} \cdot x^{2-1} \cdot y^{3-2}$

$\qquad\qquad = 14xy$

   (b) $\dfrac{24xy^2}{8y} = \dfrac{24}{8} \cdot x \cdot \dfrac{y^2}{y}$

$\qquad\qquad = \dfrac{24}{8} \cdot x \cdot y^{2-1}$

$\qquad\qquad = 3xy$

**24.** (a) $\left(\dfrac{2a}{3y}\right)^5 = \dfrac{(2)^5 a^5}{(3)^5 y^5} = \dfrac{32a^5}{243y^5}$

   (b) $-\left(\dfrac{2a}{3y}\right)^2 = -\dfrac{(2)^2 a^2}{(3)^2 y^2} = -\dfrac{4a^2}{9y^2}$

**26.** (a) $\dfrac{(-3xy)^3}{9xy^2} = \dfrac{(-3)^3 x^3 y^3}{9xy^2} = \dfrac{-27x^3 y^3}{9xy^2} = -3x^{3-1}y^{3-2} = -3x^2 y$

   (b) $\dfrac{(-3xy)^4}{-3(xy)^2} = \dfrac{(-3)^4 x^4 y^4}{-3x^2 y^2} = \dfrac{81x^4 y^4}{-3x^2 y^2} = -27x^{4-2}y^{4-2} = -27x^2 y^2$

**28.** (a) $\left[\dfrac{(3x^2)(2x)^2}{(-2x)(6x)}\right]^2 = \left[\dfrac{(3x^2)(4x^2)}{-12x^2}\right]^2$

$\qquad\qquad = \left[\dfrac{12x^4}{-12x^2}\right]^2$

$\qquad\qquad = [-x^2]^2$

$\qquad\qquad = x^4$

   (b) $\left[\dfrac{(3x^2)(2x)^4}{(-2x)(6x)}\right]^2 = \left[\dfrac{(3x^2)(16x^4)}{(4x^2)(6x)}\right]^2$

$\qquad\qquad = \left[\dfrac{48x^6}{24x^3}\right]^2$

$\qquad\qquad = [2x^3]^2$

$\qquad\qquad = (2)^2(x^3)^2$

$\qquad\qquad = 4x^6$

**30.** (a) $\dfrac{x^{3n}y^{2n-1}}{x^n y^{n+3}} = x^{3n-n}y^{(2n-1)-(n+3)} = x^{2n}y^{2n-1-n-3} = x^{2n}y^{n-4}$

   (b) $\dfrac{x^{4n-6}y^{n+10}}{x^{2n-5}y^{n-2}} = x^{4n-6-(2n-5)}y^{(n+10)-(n-2)}$

$\qquad\qquad = x^{4n-6-2n+5}y^{n+10-n+2} = x^{2n-1}y^{12}$

**32.** $(-6n)(3n^2) = (-6 \cdot 3)(n \cdot n^2) = -18n^{1+2} = -18n^3$

**34.** $5z(2z - 7) = 5z(2z) + 5z(-7) = 10z^2 - 35z$

**36.** $3y(-3y^2 + 7y - 3) = 3y(-3y^2) + 3y(7y) + 3y(-3)$

$\qquad\qquad = -9y^3 + 21y^2 - 9y$

**38.** $-3a^2(8 - 2a - a^2) = -3a^2(8) - 3a^2(-2a) - 3a^2(-a^2)$

$\qquad\qquad = -24a^2 + 6a^3 + 3a^4$

**40.** $-y^4(7y^3 - 4y^2 + y - 4) = -7y^7 + 4y^6 - y^5 + 4y^4$

**42.** $4t(-3t)(t^2 - 1) = (-12t^2)(t^2 - 1)$

$\qquad\qquad = (-12t^2)(t^2) - (-12t^2)(1)$

$\qquad\qquad = -12t^4 + 12t^2$

**44.** $ab^3(2a - 9a^2b + 3b) = ab^3(2a) + ab^3(-9a^2b) + ab^3(3b)$

$\qquad\qquad = 2a^2b^3 - 9a^3b^4 + 3ab^4$

**46.** $(x - 5)(x - 3) = x^2 - 3x - 5x + 15 = x^2 - 8x + 15$

**48.** $(x + 7)(x - 1) = x^2 - x + 7x - 7 = x^2 + 6x - 7$

**50.** $(x - 6)(x + 6) = x^2 + 6x - 6x - 36 = x^2 - 36$

**52.** $(3x + 1)(x - 4) = 3x^2 - 12x + x - 4 = 3x^2 - 11x - 4$

**54.** $(4x + 7)(3x + 7) = 12x^2 + 28x + 21x + 49 = 12x^2 + 49x + 49$

**56.** $(6x^2 + 2)(9 - 2x) = 54x^2 - 12x^3 + 18 - 4x = -12x^3 + 54x^2 - 4x + 18$

**58.** $\left(5t - \frac{3}{4}\right)(2t - 16) = 10t^2 - 80t - \frac{3}{2}t + 12$

$$= 10t^2 - \frac{160}{2}t - \frac{3}{2}t + 12$$

$$= 10t^2 - \frac{163}{2}t + 12$$

**60.** $(2x - y)(3x - 2y) = 6x^2 - 4xy - 3xy + 2y^2$

$$= 6x^2 - 7xy + 2y^2$$

**62.** $(s - 3t)(s + t) - (s - 3t)(s - t) = s^2 + st - 3ts - 3t^2 - s^2 + st + 3ts - 3t^2$

$$= 2st - 6t^2$$

**64.** $(z + 2)(z^2 - 4z + 4) = (z + 2)z^2 + (z + 2)(-4z) + (z + 2)4$

$$= z^3 + 2z^2 - 4z^2 - 8z + 4z + 8$$

$$= z^3 - 2z^2 - 4z + 8$$

**66.** $(2t + 3)(t^2 - 5t + 1) = (2t + 3)t^2 + (2t + 3)(-5t) + (2t + 3)(1)$

$$= 2t^3 + 3t^2 - 10t^2 - 15t + 2t + 3$$

$$= 2t^3 - 7t^2 - 13t + 3$$

**68.** $(2x^2 - 5x + 1)(3x - 4) = (3x - 4)2x^2 + (3x - 4)(-5x) + (3x - 4)1$

$$= 6x^3 - 8x^2 - 15x^2 + 20x + 3x - 4$$

$$= 6x^3 - 23x^2 + 23x - 4$$

**70.** $(x^2 + 4)(x^2 - 2x - 4) = (x^2 + 4)x^2 + (x^2 + 4)(-2x) + (x^2 + 4)(-4)$

$$= x^2 \cdot x^2 + 4x^2 - x^2(2x) - 8x - 4x^2 - 16$$

$$= x^4 + 4x^2 - 2x^3 - 8x - 4x^2 - 16$$

$$= x^4 - 2x^3 - 8x - 16$$

**72.** $(2x^2 - 3)(2x^2 - 2x + 3) = (2x^2 - 3)(2x^2) + (2x^2 - 3)(-2x) + (2x^2 - 3)(3)$

$$= 4x^4 - 6x^2 - 4x^3 + 6x + 6x^2 - 9$$

$$= 4x^4 - 4x^3 + 6x - 9$$

**74.** $(y^2 + 3y + 5)(2y^2 - 3y - 1) = y^2(2y^2 - 3y - 1) + 3y(2y^2 - 3y - 1) + 5(2y^2 - 3y - 1)$

$$= 2y^4 - 3y^3 - y^2 + 6y^3 - 9y^2 - 3y + 10y^2 - 15y - 5$$

$$= 2y^4 + 3y^3 - 18y - 5$$

**76.**
$$
\begin{array}{r}
4x^4 \;-\; 6x^2 + 9 \\
\times \qquad\quad 2x^2 + 3 \\
\hline
+\;12x^4 - 18x^2 + 27 \\
8x^6 - 12x^4 + 18x^2 \\
\hline
8x^6 \qquad\qquad\qquad + 27
\end{array}
$$

**78.**
$$
\begin{array}{r}
z^2 \;+\; z + 1 \\
\times \qquad\quad z - 2 \\
\hline
-\;2z^2 - 2z - 2 \\
z^3 + z^2 \;+\; z \\
\hline
z^3 - z^2 \;-\; z - 2
\end{array}
$$

**80.**
$$
\begin{array}{r}
2s^2 \;-\; 5s + 6 \\
\times \qquad\quad 3s - 4 \\
\hline
-\;8s^2 + 20s - 24 \\
6s^3 - 15s^2 + 18s \\
\hline
6s^3 - 23s^2 + 38s - 24
\end{array}
$$

**82.**
$$
\begin{array}{r}
y^2 \;+\; 3y + 5 \\
2y^2 \;-\; 3y - 1 \\
\hline
-\;y^2 \;-\; 3y - 5 \\
-\;3y^3 - 9y^2 - 15y \\
2y^4 + 6y^3 + 10y^2 \\
\hline
2y^4 + 3y^3 \qquad\quad - 18y - 5
\end{array}
$$

**84.** $(x - 5)(x + 5) = (x)^2 - (5)^2 = x^2 - 25$

**86.** $(x + 1)(x - 1) = (x)^2 - (1)^2 = x^2 - 1$

**88.** $(4 + 3z)(4 - 3z) = (4)^2 - (3z)^2 = 16 - 9z^2$

**90.** $(8 - 3x)(8 + 3x) = (8)^2 - (3x)^2 = 64 - 9x^2$

**92.** $(5u + 12v)(5u - 12v) = (5u)^2 - (12v)^2$
$$= 25u^2 - 144v^2$$

**94.** $(8x - 5y)(8x + 5y) = (8x)^2 - (5y)^2$
$$= 64x^2 - 25y^2$$

**96.** $\left(\frac{2}{3}x + 7\right)\left(\frac{2}{3}x - 7\right) = \left(\frac{2}{3}x\right)^2 - (7)^2 = \frac{4}{9}x^2 - 49$

**98.** $(4a - 0.1b)(4a + 0.1b) = (4a)^2 - (0.1b)^2 = 16a^2 - 0.01b^2$

**100.** $(x + 9)^2 = (x)^2 + 2(x)(9) + (9)^2 = x^2 + 18x + 81$

**102.** $(u - 7)^2 = (u)^2 - 2(u)(7) + (7)^2 = u^2 - 14u + 49$

**104.** $(3x + 8)^2 = (3x)^2 + 2(3x)(8) + (8)^2 = 9x^2 + 48x + 64$

**106.** $(5 - 3z)^2 = (5)^2 - 2(5)(3z) + (3z)^2 = 25 - 30z + 9z^2$

**108.** $(2x + 5y)^2 = (2x)^2 + 2(2x)(5y) + (5y)^2$
$$= 4x^2 + 20xy + 25y^2$$

**110.** $[(x - 4) - y]^2 = (x - 4)^2 - 2(x - 4)(y) + (y)^2$
$$= (x)^2 - 2(x)(4) + (4)^2 - 2xy + 8y + y^2$$
$$= x^2 - 8x + 16 - 2xy + 8y + y^2$$

**112.** $[z + (y + 1)][z - (y + 1)] = (z)^2 - (y + 1)^2$
$$= z^2 - ((y)^2 + 2(y)(1) + 1^2)$$
$$= z^2 - (y^2 + 2y + 1)$$
$$= z^2 - y^2 - 2y - 1$$

**114.** $(y - 2)^3 = (y - 2)(y - 2)(y - 2)$
$$= (y^2 - 4y + 4)(y - 2)$$
$$= y^2(y - 2) - 4y(y - 2) + 4(y - 2)$$
$$= y^3 - 2y^2 - 4y^2 + 8y + 4y - 8$$
$$= y^3 - 6y^2 + 12y - 8$$

**116.** $(u - v)^3 = (u - v)(u - v)(u - v)$
$$= (u^2 - 2uv + v^2)(u - v)$$
$$= u^2(u - v) - 2uv(u - v) + v^2(u - v)$$
$$= u^3 - u^2v - 2u^2v + 2uv^2 + uv^2 - v^3$$
$$= u^3 - 3u^2v + 3uv^2 - v^3$$

**118.** *Keystrokes:*

$y_1$  Y=  (  X,T,θ  −  3  )  $x^2$  ENTER

$y_2$  X,T,θ  $x^2$  −  6  X,T,θ  +  9  GRAPH

$y_1 = y_2$ because $(x - 3)^2 = (x)^2 - 2(x)(-3) - (3)^2 = x^2 + 6x - 9$

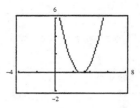

**120.** *Keystrokes:*

$y_1$  Y=  (  X,T,θ  +  (  1  ÷  2  )  )  (  X,T,θ  −  (  1  ÷  2  )  )  ENTER

$y_2$  X,T,θ  $x^2$  −  (  1  ÷  4  )  GRAPH

$y_1 = y_2$ because $\left(x + \frac{1}{2}\right)\left(x - \frac{1}{2}\right) = (x)^2 - \left(\frac{1}{2}\right)^2 = x^2 - \frac{1}{4}$

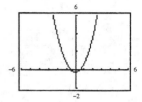

**122.** (a) $f(y + 2) = 2(y + 2)^2 - 5(y + 2) + 4$

$\qquad\qquad = 2(y^2 + 4y + 4) - 5y - 10 + 4$

$\qquad\qquad = 2y^2 + 8y + 8 - 5y - 10 + 4$

$\qquad\qquad = 2y^2 + 3y + 2$

(b) $f(1 + h) - f(1) = 2(1 + h)^2 - 5(1 + h) + 4 - [2(1)^2 - 5(1) + 4]$

$\qquad\qquad = 2(1 + 2h + h^2) - 5 - 5h + 4 - [2 - 5 + 4]$

$\qquad\qquad = 2 + 4h + 2h^2 - 5 - 5h + 4 - (1)$

$\qquad\qquad = 2h^2 - h$

**124.** (a) *Verbal Model:* $\boxed{\text{Volume}} = \boxed{\text{Length}} \cdot \boxed{\text{Width}} \cdot \boxed{\text{Height}}$

$\qquad$ *Function:*  $V(n) = (2n - 2) \cdot (2n + 2) \cdot (2n)$

$\qquad\qquad\qquad = (4n^2 - 4)(2n)$

$\qquad\qquad\qquad = 8n^3 - 8n$

(b) $2n - 2 = 6, n = 4, V(4) = 8(4)^3 - 8(4)$

$\qquad\qquad\qquad = 8(64) - 32$

$\qquad\qquad\qquad = 512 - 32$

$\qquad\qquad\qquad = 480$ cubic inches

(c) *Verbal Model:* $\boxed{\text{Area}} = \boxed{\text{Length}} \cdot \boxed{\text{Width}}$

$\qquad$ *Function:*  $A(n) = (2n + 2) \cdot (2n - 2)$

$\qquad\qquad\qquad = 4n^2 - 4$

(d) $\qquad$ $Area = (2n + 2 + 2)(2n - 2 + 2)$

$\qquad\qquad = (2n + 4)(2n)$

$\qquad\qquad = 4n^2 + 8n$

$\quad A(n + 4) = (2(n + 4) + 2) \cdot (2(n + 4) - 2)$

$\qquad\qquad = (2n + 8 + 2)(2n + 8 - 2)$

$\qquad\qquad = (2n + 10)(2n + 6)$

$\qquad\qquad = 4n^2 + 12n + 20n + 60$

$\qquad\qquad = 4n^2 + 32n + 60$

**126.** *Verbal Model:*

$$\boxed{\begin{array}{c}\text{Area of}\\\text{Shaded}\\\text{Region}\end{array}} = \boxed{\begin{array}{c}\text{Area of}\\\text{Outside}\\\text{Rectangle}\end{array}} - \boxed{\begin{array}{c}\text{Area of}\\\text{Inside}\\\text{Rectangle}\end{array}}$$

*Equation:*    $A = l \cdot w - l \cdot w$

$$A = 3x \cdot x - (x - 3)2$$

$$A = 3x^2 - 2x + 6$$

**128.** *Verbal Model:*

$$\boxed{\begin{array}{c}\text{Area of}\\\text{shaded region}\end{array}} = \boxed{\begin{array}{c}\text{Area of}\\\text{rectangle}\end{array}} - \boxed{\begin{array}{c}\text{Area of}\\\text{triangle}\end{array}}$$

*Equation:*    $A = (x + 5)(x + 2x) - \frac{1}{2}(x)(x + 5)$

$$= (x + 5)(3x) - \frac{1}{2}x(x + 5)$$

$$= 3x^2 + 15x - \frac{1}{2}x^2 - \frac{5}{2}x$$

$$= \frac{5}{2}x^2 + \frac{25}{2}x$$

**130.** $A = \frac{1}{2}b \cdot h$

$$= \frac{1}{2}(3x)(x + 5)$$

$$= \frac{3}{2}x(x + 5)$$

$$= \frac{3}{2}x^2 + \frac{15}{2}x$$

**132.** Interest $= 1000(1 + 0.095)^2$

$$= 1000(1.095)^2$$

$$= 1000(1.199025)$$

$$= 1199.025$$

$$\approx \$1199.03$$

**134.** Area $= l \cdot w$

$$= (x + a)(x + a)$$

$$= x^2 + 2ax + a^2$$

$$Area = (x \cdot x) + (a \cdot x) + (a \cdot x) + (a \cdot a)$$

$$= x^2 + ax + ax + a^2$$

$$= x^2 + 2ax + a^2$$

Formula $= (x + a)(x + a) = x^2 + 2ax + a^2$

Distributive Property

**136.** (a)    $(x + y)^2 = (x + y)(x + y)$

$$= x^2 + xy + yx + y^2$$

$$= x^2 + 2xy + y^2$$

(b)    $(x - y)^2 = (x - y)(x - y)$

$$= x^2 - xy - xy + y^2$$

$$= x^2 - 2xy + y^2$$

(c)   $(x - y)(x + y) = x^2 + xy - xy - y^2$

$$= x^2 - y^2$$

**138.** Rules for exponents:

$$a^m \cdot a^n = a^{m+n}$$

$$(ab)^m = a^m b^m$$

$$(a^m)^n = a^{mn}$$

$$\frac{a^m}{a^n} = a^{m-n}, m > n, a \neq 0$$

$$\left(\frac{a}{b}\right)^m = \frac{a^m}{b^m}, b \neq 0$$

**140.** Example: $(x + 2)(x - 3) = x(x - 3) + 2(x - 3)$

$$= x^2 - 3x + 2x - 6$$

$$= x^2 - x - 6$$

**142.** The degree of the product of two polynomials of degrees $m$ and $n$ is $m + n$.

## Section 3.3    Factoring Polynomials

**2.**    $36 = 2^2 \cdot 3^2$

$$150 = 2 \cdot 3 \cdot 5^2$$

$$100 = 2^2 \cdot 5^2$$

GCF $= 2$

**4.**   $27x^4 = 3^3 x^4$

$$18x^3 = 2 \cdot 3^2 x^3$$

GCF $= 9x^3$

**6.** $-45y = (-1) \cdot 3^2 \cdot 5 \cdot y$

$$150y^3 = 2 \cdot 3 \cdot 5^2 \cdot y^3$$

GCF $= 3 \cdot 5 \cdot y = 15y$

**8.** $16x^2y = 2^4x^2y$

$84xy^2 = 2^2 \cdot 3 \cdot 7 \cdot x \cdot y^2$

$36x^2y^2 = 2^2 \cdot 3^2 \cdot x^2 \cdot y^2$

$\text{GCF} = 2^2 \cdot x \cdot y = 4xy$

**10.** $66(3 - y) = 2 \cdot 3 \cdot 11 \cdot (3 - y)$

$44(3 - y)^2 = 2^2 \cdot 11 \cdot (3 - y)^2$

$\text{GCF} = 2 \cdot 11 \cdot (3 - y) = 22(3 - y)$

**12.** $5x + 5 = 5(x + 1)$

**14.** $-15t - 10 = -5(3t + 2)$

**16.** $14z^3 + 21 = 7(2z^3 + 3)$

**18.** $-a^3 - 4a = -a(a^2 + 4)$

**20.** $36y^4 + 24y^2 = 12y^2(3y^2 + 2)$

**22.** $16 - 3y^3$ is prime.
No common factor other than 1.

**24.** $9 - 27y - 15y^2 = 3(3 - 9y - 5y^2)$

**26.** $4uv + 6u^2v^2 = 2uv(2 + 3uv)$

**28.** $4x^2 - 2xy + 3y^2$ Prime
No common factor other than 1.

**30.** $17x^5y^3 - xy^2 + 34y^2 = y^2(17x^5y - x + 34)$

**32.** $32 - x^4 = -1(-32 + x^4)$

$= -(x^4 - 32)$

**34.** $15 - 5x = -5(-3 + x) = -5(x - 3)$

**36.** $12x - 6x^2 - 18 = -6(-2x + x^2 + 3) = -6(x^2 - 2x + 3)$

**38.** $-2t^3 + 4t^2 + 7 = -1(2t^3 - 4t^2 - 7)$

$= -(2t^3 - 4t^2 - 7)$

**40.** $3z + \frac{3}{8} = \frac{1}{8}(24z + 3)$

**42.** $\frac{1}{3}x - \frac{5}{6} = \frac{1}{6}(2x - 5)$

**44.** $7t(s + 9) - 6(s + 9) = (s + 9)(7t - 6)$

**46.** $6(4t - 3) - 5t(4t - 3) = (4t - 3)(6 - 5t)$

**48.** $4(5y - 12) + 3y^2(5y - 12) = (5y - 12)(4 + 3y^2)$

**50.** $2y^2(y^2 + 6)^3 + 7(y^2 + 6)^3 = (y^2 + 6)^3(2y^2 + 7)$

**52.** $(3x + 7)(2x - 1) + (x - 6)(2x - 1) = (2x - 1)(3x + 7 + x - 6)$

$= (2x - 1)(4x + 1)$

**54.** $x^2 - 7x + x - 7 = (x^2 - 7x) + (x - 7)$

$= x(x - 7) + (x - 7)$

$= (x - 7)(x + 1)$

**56.** $y^2 + 3y + 4y + 12 = (y^2 + 3y) + (4y + 12)$

$= y(y + 3) + 4(y + 3)$

$= (y + 3)(y + 4)$

**58.** $t^3 - 11t^2 + t - 11 = (t^3 - 11t^2) + (t - 11)$

$= t^2(t - 11) + (t - 11)$

$= (t - 11)(t^2 + 1)$

**60.** $3s^3 + 6s^2 + 5s + 10 = (3s^3 + 6s^2) + (5s + 10)$

$= 3s^2(s + 2) + 5(s + 2)$

$= (s + 2)(3s^2 + 5)$

**62.** $4u^4 - 6u - 2u^3 + 3 = (4u^4 - 6u) + (-2u^3 + 3)$

$= 2u(2u^3 - 3) - 1(2u^3 - 3)$

$= (2u^3 - 3)(2u - 1)$

**64.** $10u^4 - 8u^2v^3 - 12v^4 + 15u^2v = (10u^4 - 8u^2v^3) + (-12v^4 + 15u^2v)$

$$= 2u^2(5u^2 - 4v^3) + 3v(-4v^3 + 5u^2)$$

$$= (5u^2 - 4v^3)(2u^2 + 3v)$$

**66.** $y^2 - 144 = y^2 - (12)^2$

$\qquad = (y - 12)(y + 12)$

**68.** $16 - b^2 = 4^2 - b^2$

$\qquad = (4 - b)(4 + b)$

**70.** $9z^2 - 25 = (3z)^2 - (5)^2$

$\qquad = (3z - 5)(3z + 5)$

**72.** $49 - 64x^2 = 7^2 - (8x)^2$

$\qquad = (7 - 8x)(7 + 8x)$

**74.** $9u^2 - v^2 = (3u)^2 - (v)^2$

$\qquad = (3u - v)(3u + v)$

**76.** $100a^2 - 49b^2 = (10a)^2 - (7b)^2$

$\qquad = (10a - 7b)(10a + 7b)$

**78.** $v^2 - \frac{9}{25} = v^2 - \left(\frac{3}{5}\right)^2$

$\qquad = \left(v - \frac{3}{5}\right)\left(v + \frac{3}{5}\right)$

**80.** $\frac{1}{4}x^2 - \frac{36}{49}y^2 = \left(\frac{1}{2}x\right)^2 - \left(\frac{6}{7}y\right)^2$

$\qquad = \left(\frac{1}{2}x - \frac{6}{7}y\right)\left(\frac{1}{2}x + \frac{6}{7}y\right)$

**82.** $(x - 3)^2 - 4 = (x - 3)^2 - (2)^2$

$\qquad = [(x - 3) - 2][(x - 3) + 2]$

$\qquad = (x - 5)(x - 1)$

**84.** $36 - (y - 6)^2 = (6)^2 - (y - 6)^2$

$\qquad = [6 - (y - 6)][6 + (y - 6)]$

$\qquad = [6 - y + 6][6 + y - 6]$

$\qquad = [-y + 12][y]$

$\qquad = y(12 - y)$

**86.** $(3y - 1)^2 - (x + 6)^2 = [(3y - 1) - (x + 6)][(3y - 1) + (x + 6)]$

$$= [3y - 1 - x - 6][3y - 1 + x + 6]$$

$$= (3y - x - 7)(3y + x + 5)$$

$$\text{or}$$

$$= -(x - 3y + 7)(x + 3y + 5)$$

**88.** $t^3 - 27 = t^3 - (3)^3$

$\qquad = (t - 3)(t^2 + 3t + 9)$

**90.** $z^3 + 125 = z^3 + 5^3$

$\qquad = (z + 5)(z^2 - 5z + 25)$

**92.** $27s^3 + 64 = (3s)^3 + (4)^3$

$\qquad = (3s + 4)(9s^2 - 12s + 16)$

**94.** $64v^3 - 125 = (4v)^3 - (5)^3$

$\qquad = (4v - 5)(16v^2 + 20v + 25)$

**96.** $m^3 - 8n^3 = m^3 - (2n)^3$

$\qquad = (m - 2n)(m^2 + 2mn + 4n^2)$

**98.** $u^3 + 125v^3 = u^3 + (5v)^3$

$\qquad = (u + 5v)(u^2 - 5uv + 25v^2)$

**100.** $8y^2 - 18 = 2(4y^2 - 9)$

$\qquad = 2((2y)^2 - 3^2)$

$\qquad = 2(2y - 3)(2y + 3)$

**102.** $a^3 - 16a = a(a^2 - 16)$

$\qquad = a[a^2 - (4)^2]$

$\qquad = a(a - 4)(a + 4)$

**104.** $u^4 - 16 = (u^2)^2 - (4)^2$

$\qquad = (u^2 - 4)(u^2 + 4)$

$\qquad = [(u^2 - (2)^2][u^2 + 4]$

$\qquad = (u - 2)(u + 2)(u^2 + 4)$

**106.** $6x^5 + 30x^3 = 6x^3(x^2 + 5)$

**108.** $2u^6 + 54v^6 = 2(u^6 + 27v^6)$

$\qquad = 2[(u^2)^3 + (3v^2)^3]$

$\qquad = 2(u^2 + 3v^2)(u^4 - 3u^2v^2 + 9v^4)$

**110.** $81 - 16y^{4n} = 9^2 - (4y^{2n})^2$

$\qquad = (9 - 4y^{2n})(9 + 4y^{2n})$

$\qquad = [3^2 - (2y^n)^2](9 + 4y^{2n})$

$\qquad = (3 - 2y^n)(3 + 2y^n)(9 + 4y^{2n})$

**112.** *Keystrokes:*

$y_1$

$y_2$ [X,T,θ] [x²] [(] [X,T,θ] [−] 2 [)] [GRAPH]

$y_1 = y_2$

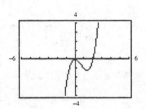

**114.** *Keystrokes:*

$y_1$ [Y=] [X,T,θ] [(] [X,T,θ] [+] 1 [)] [−] 4 [(] [X,T,θ] [+] 1 [)] [ENTER]

$y_2$ [(] [X,T,θ] [+] 1 [)] [(] [X,T,θ] [−] 4 [)] [GRAPH]

$y_1 = y_2$

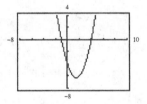

**116.** $6x^3 - 8x^2 + 9x - 12 = (6x^3 - 8x^2) + (9x - 12)$

$\qquad = 2x^2(3x - 4) + 3(3x - 4)$

$\qquad = (3x - 4)(2x^2 + 3)$

or $6x^3 - 8x^2 + 9x - 12 = (6x^3 + 9x) + (-8x^2 - 12)$

$\qquad = 3x(2x^2 + 3) - 4(2x^2 + 3)$

$\qquad = (2x^2 + 3)(3x - 4)$

**118.** $R = 1000x - 0.4x^2$

$\quad = x(1000 - 0.4x)$

$R = xp$

$p = 1000 - 0.4x$

**120.** $kQx - kx^2 = kx(Q - x)$

**122.** $A = 32w - w^2$

$\quad = w(32 - w)$

$\quad = w \cdot l$

Thus, $l = 32 - w$

**124.** $S = \pi r^2 + 2\pi rh$

$\quad = \pi r(r + 2h)$

**126.** (a)

| Solid | Length | Width | Height | Volume |
|-------|--------|-------|--------|--------|
| Entire cube | $a$ | $a$ | $a$ | $a^3$ |
| Solid I | $a$ | $a$ | $a - b$ | $a^2(a - b)$ |
| Solid II | $a$ | $a - b$ | $b$ | $ab(a - b)$ |
| Solid III | $a - b$ | $b$ | $b$ | $b^2(a - b)$ |
| Solid IV | $b$ | $b$ | $b$ | $b^3$ |

(b) Solid I + Solid II + Solid III

$$a^2(a - b) + ab(a - b) + b^2(a - b) = (a - b)(a^2 + ab + b^2)$$

(c) If the smaller cube is removed from the larger, the remaining solid has a volume of $a^3 - b^3$ and is composed of the three rectangular boxes labeled Solid I, Solid II, and Solid III. From part (b) we have $a^3 - b^3 = (a - b)(a^2 + ab + b^2)$.

**128.** Check a result after factoring by multiplying the factors to see if the product is the original polynomial.

**130.** Factor used as a noun expresses any one of the expressions that when multiplied together yields the product.

Factor used as a verb is the process of finding the expressions that, when multiplied together, yield the given product.

**132.** An example of a polynomial that is prime with respect to the integers is $x^2 + 1$.

# Section 3.4    Factoring Trinomials

**2.** $z^2 + 6z + 9 = z^2 + 2(3z) + (3)^2$
$$= (z + 3)(z + 3)$$
$$= (z + 3)^2$$

**4.** $y^2 - 14y + 49 = y^2 - 2(7)y + (7)^2$
$$= (y - 7)^2$$

**6.** $4z^2 + 28z + 49 = (2z)^2 + 2(2z)(7) + (7)^2$
$$= (2z + 7)(2z + 7)$$
$$= (2z + 7)^2$$

**8.** $4x^2 - 4x + 1 = (2x)^2 - 2(2x)(1) + (1)^2$
$$= (2x - 1)^2$$

**10.** $x^2 - 14xy + 49y^2 = x^2 - 2(x)(7y) + (7y)^2$
$$= (x - 7y)^2$$

**12.** $4y^2 + 20yz + 25z^2 = (2y)^2 + 2(2y)(5z) + (5z)^2$
$$= (2y + 5z)^2$$

**14.** $4x^2 - 32x + 64 = 4(x^2 - 8x + 16)$
$$= 4(x^2 - 2(x)(4) + 4^2)$$
$$= 4(x - 4)^2$$

**16.** $3u^3 - 48u^2 + 192u = 3u(u^2 - 16u + 64)$
$$= 3u(u^2 - 2(u)(8) + 8^2)$$
$$= 3u(u - 8)^2$$

**18.** $-18y^3 - 12y^2 - 2y = -2y(9y^2 + 6y + 1)$
$$= -2y((3y)^2 + 2(3y)(1) + 1^2)$$
$$= -2y(3y + 1)^2$$

**20.** $\frac{1}{9}x^2 + \frac{8}{15}x + \frac{16}{25} = \left(\frac{1}{3}x\right)^2 + 2\left(\frac{1}{3}x\right)\left(\frac{4}{5}\right) + \left(\frac{4}{5}\right)^2 = \left(\frac{1}{3}x + \frac{4}{5}\right)^2$

or

$$= \frac{25}{225}x^2 + \frac{120}{225}x + \frac{144}{225}$$

$$= \frac{1}{225}(25x^2 + 120x + 144)$$

$$= \frac{1}{225}((5x)^2 + 2(5x)(12) + 12^2) = \frac{1}{225}(5x + 12)^2$$

**22.** $x^2 + bx + \frac{9}{16} = x^2 + bx + \left(\frac{3}{4}\right)^2$

(a) $\qquad b = \frac{3}{2}$

$x^2 + \frac{3}{2} \cdot x + \frac{9}{16} = x^2 + 2\left(\frac{3}{4}\right)x + \left(\frac{3}{4}\right)^2$

$\qquad\qquad\qquad = x^2 + \frac{3}{2}x + \frac{9}{16}$

or

(b) $\qquad b = -\frac{3}{2}$

$x^2 - \frac{3}{2}x + \frac{9}{16} = x^2 + 2\left(-\frac{3}{4}\right)x + \left(-\frac{3}{4}\right)^2$

$\qquad\qquad\qquad = x^2 - \frac{3}{2}x + \frac{9}{16}$

**24.** $16x^2 + bxy + 25y^2$

(a) $\qquad b = 40$

$16x^2 + 40xy + 25y^2 = (4x)^2 + 2(4x)(5y) + (5y)^2$

$\qquad\qquad\qquad = (4x + 5y)^2$

or

(b) $\qquad b = -40$

$16x^2 - 40xy + 25y^2 = (4x)^2 - 2(4x)(5y) + (5y)^2$

$\qquad\qquad\qquad = (4x - 5y)^2$

**26.** $x^2 + 12x + c$

$c = 36$

$x^2 + 12x + 36 = (x)^2 + 2(6)x + (6)^2$

$\qquad\qquad\quad = (x + 6)^2$

**28.** $z^2 - 20z + c$

$c = 100$

$z^2 - 20z + 100 = (z)^2 - 2(10)z + (10)^2$

$\qquad\qquad\qquad = (z - 10)^2$

**30.** $a^2 + 2a - 8 = (a + 4)(a - 2)$

**32.** $y^2 + 6y + 8 = (y + 4)(y + 2)$

**34.** $x^2 + 7x + 12 = (x + 4)(x + 3)$

**36.** $z^2 + 2z - 24 = (z - 4)(z + 6)$

**38.** $x^2 + 7x + 10 = (x + 5)(x + 2)$

**40.** $x^2 - 10x + 24 = (x - 6)(x - 4)$

**42.** $m^2 - 3m - 10 = (m - 5)(m + 2)$

**44.** $x^2 + 4x - 12 = (x + 6)(x - 2)$

**46.** $y^2 - 35y + 300 = (y - 15)(y - 20)$

**48.** $u^2 + 5uv + 6v^2 = (u + 3v)(u + 2v)$

**50.** $a^2 - 21ab + 110b^2 = (a - 10b)(a - 11b)$

**52.** $b = 15 \qquad x^2 + 15x + 14 = (x + 14)(x + 1)$

$\quad b = -15 \qquad x^2 - 15x + 14 = (x - 14)(x - 1)$

$\quad x = 9 \qquad\quad x^2 + 9x + 14 = (x + 7)(x + 2)$

$\quad x = -9 \qquad x^2 - 9x + 14 = (x - 7)(x - 2)$

**54.** $b = 6 \qquad x^2 + 6x - 7 = (x + 7)(x - 1)$

$\quad b = -6 \qquad x^2 - 6x - 7 = (x - 7)(x + 1)$

**56.** $b = 17 \qquad x^2 + 17x - 38 = (x + 19)(x - 2)$

$\quad b = -17 \qquad x^2 - 17x - 38 = (x - 19)(x + 2)$

$\quad b = 37 \qquad x^2 + 37x - 38 = (x + 38)(x - 1)$

$\quad b = -37 \qquad x^2 - 37x - 38 = (x - 38)(x + 1)$

**58.** $c = 8 \qquad x^2 + 9x + 8 = (x + 8)(x + 1)$

$c = 14 \qquad x^2 + 9x + 14 = (x + 7)(x + 2)$

$c = -10 \qquad x^2 + 9x - 10 = (x + 10)(x - 1)$

$c = -36 \qquad x^2 + 9x - 36 = (x + 12)(x - 3)$

There are more possibilities.

**60.** There are many possibilities such as:

$c = 11 \qquad x^2 - 12x + 11 = (x - 11)(x - 1)$

$c = 20 \qquad x^2 - 12x + 20 = (x - 2)(x - 10)$

$c = 27 \qquad x^2 - 12x + 27 = (x - 9)(x - 3)$

Also note that if $c$ is a negative number, there are many possibilities for $c$ such as the following:

$c = -13 \qquad x^2 - 12x - 13 = (x - 13)(x + 1)$

$c = -28 \qquad x^2 - 12x - 28 = (x - 14)(x + 2)$

$c = -45 \qquad x^2 - 12x - 45 = (x - 15)(x + 3)$

**62.** $5x^2 + 19x + 12 = (x + 3)(5x + 4)$

**64.** $5c^2 + 11c - 12 = (c + 3)(5c - 4)$

**66.** $3y^2 - y - 30 = (y + 3)(3y - 10)$

**68.** $5x^2 + 7x + 2 = (5x + 2)(x + 1)$

**70.** $3x^2 + 8x + 5 = (3x + 5)(x + 1)$

**72.** $2t^2 - 13t + 20 = (2t - 5)(t - 4)$

**74.** $4y^2 - 5y - 9 = (y + 1)(4y - 9)$

**76.** $2z^2 + 3z + 8$ is Prime

**78.** $10x^2 - 24x - 18 = 2(5x^2 - 12x - 9)$

$= 2(5x + 3)(x - 3)$

**80.** $20x^2 + x - 12 = (5x + 4)(4x - 3)$

**82.** $-6x^2 + 5x - 6$ is Prime

**84.** $2 + 5x - 12x^2 = (2 - 3x)(1 + 4x)$

**86.** $12x^2 + 32x - 12 = 4(3x^2 + 8x - 3)$

$= 4(3x - 1)(x + 3)$

**88.** $12x^2 + 42x^3 - 54x^4 = 6x^2(2 + 7x - 9x^2)$

$= 6x^2(2 + 9x)(1 - x)$

**90.** $6u^2 - 5uv - 4v^2 = (3u - 4v)(2u + v)$

**92.** $10x^2 + 9xy - 9y^2 = (5x - 3y)(2x + 3y)$

**94.** $2x^2 + 9x + 9 = (2x^2 + 6x) + (3x + 9)$

$= 2x(x + 3) + 3(x + 3)$

$= (x + 3)(2x + 3)$

**96.** $6x^2 - x - 15 = (6x^2 - 10x) + (9x - 15)$

$= 2x(3x - 5) + 3(3x - 5)$

$= (3x - 5)(2x + 3)$

**98.** $12x^2 - 28x + 15 = (12x^2 - 18x) + (-10x + 15)$

$= 6x(2x - 3) - 5(2x - 3)$

$= (6x - 5)(2x - 3)$

**100.** $20y^2 - 45 = 5(4y^2 - 9)$

$= 5[(2y)^2 - (3)^2]$

$= 5(2y - 3)(2y + 3)$

**102.** $16z^3 - 56z^2 + 49z =$

$(16z^2 - 56z + 49) = z[(4z)^2 - 2(4z)(7) + (7)^2]$

$= z(4z - 7)^2$

**104.** $3t^3 - 24 = 3(t^3 - 8)$

$= 3[t^3 - (2)^3]$

$= 3(t - 2)(t^2 + 2t + 4)$

**106.** $8m^3n + 20m^2n^2 - 48mn^3 = 4mn(2m^2 + 5mn - 12n^2)$

$= 4mn(2m - 3n)(m + 4n)$

**108.** $x^3 - 7x^2 - 4x + 28 = (x^3 - 7x^2) + (-4x + 28)$

$= x^2(x - 7) - 4(x - 7)$

$= (x - 7)(x^2 - 4)$

$= (x - 7)(x - 2)(x + 2)$

**110.** $(x + 7y)^2 - 4a^2 = [(x + 7y) - 2a][(x + 7y) + 2a]$
$$= (x + 7y - 2a)(x + 7y + 2a)$$

**112.** $a^2 - 2ab + b^2 - 16 = (a^2 - 2ab + b^2) - 16$
$$= (a - b)^2 - 4^2$$
$$= (a - b - 4)(a - b + 4)$$

**114.** $x^4 - 16y^4 = (x^2)^2 - (4y^2)^2$
$$= (x^2 - 4y^2)(x^2 + 4y^2)$$
$$= (x - 2y)(x + 2y)(x^2 + 4y^2)$$

**116.** *Keystrokes:*

$y_1$ [Y=] 4 [X,T,θ] [x²] [−] 4 [X,T,θ] [+] 1 [ENTER]

$y_2$ [(] 2 [X,T,θ] [−] 1 [)] [x²] [GRAPH]

$y_1 = y_2$

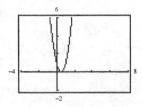

**118.** *Keystrokes:*

$y_1$ [Y=] 3 [X,T,θ] [x²] [−] 8 [X,T,θ] [−] 16 [ENTER]

$y_2$ [(] 3 [X,T,θ] [+] 4 [)] [(] [X,T,θ] [−] 4 [)] [GRAPH]

$y_1 = y_2$

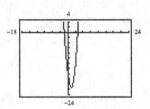

**120.** $a^2 + 2a + 1 = (a + 1)^2$ matches graph (a).

**122.** $ab + a + b - 1 = (a + 1)(b + 1)$ matches graph (d).

**124.** *Verbal Model:*

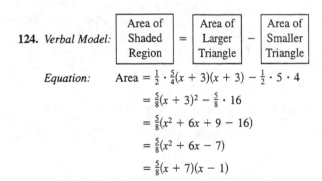

| Area of Shaded Region | | Area of Larger Triangle | | Area of Smaller Triangle |
|---|---|---|---|---|
| | = | | − | |

*Equation:*    Area $= \frac{1}{2} \cdot \frac{5}{4}(x + 3)(x + 3) - \frac{1}{2} \cdot 5 \cdot 4$
$$= \frac{5}{8}(x + 3)^2 - \frac{5}{8} \cdot 16$$
$$= \frac{5}{8}(x^2 + 6x + 9 - 16)$$
$$= \frac{5}{8}(x^2 + 6x - 7)$$
$$= \frac{5}{8}(x + 7)(x - 1)$$

**126.** (a) $8n^3 + 12n^2 - 2n - 3 = (8n^3 + 12n^2) + (-2n - 3)$
$$= 4n^2(2n + 3) - 1(2n + 3)$$
$$= (4n^2 - 1)(2n + 3)$$
$$= (2n - 1)(2n + 1)(2n + 3)$$

(b) If n = 15,

$2n - 1 = 2(15) - 1 = 29$

$2n + 1 = 2(15) + 1 = 31$

$2n + 3 = 2(15) + 3 = 33$

**128.** An example of a prime trinomial is $x^2 + x + 1$.

**130.** $3x + 6 = 3(x + 2)$ and $3x - 9 = 3(x - 3)$

    *so* $(3x + 6)(3x - 9) = 3(x + 2) \cdot 3(x - 3)$
$$= 9(x + 2)(x - 3) \; not \; 3(x + 2)(x - 3)$$

**132.** $(2x - 4)(x + 1)$ is not in completely factored form because $2x - 4$ has the common factor 2, $2x - 4 = 2(x - 2)$

# Section 3.5    Solving Polynomial Equations

**2.** $z(z + 6) = 0$

$z = 0 \qquad z + 6 = 0$

$\qquad\qquad z = -6$

**4.** $(s - 16)(s + 15) = 0$

$s - 16 = 0 \qquad s + 15 = 0$

$s = 16 \qquad s = -15$

**6.** $17(t - 3)(t + 8) = 0$

$t - 3 = 0 \qquad t + 8 = 0$

$t = 3 \qquad t = -8$

**8.** $(5x - 3)(x - 8) = 0$

$5x - 3 = 0 \qquad x - 8 = 0$

$5x = 3 \qquad x = 8$

$x = \frac{3}{5}$

**10.** $\frac{1}{5}x(x - 2)(3x + 4) = 0$

$\frac{1}{5}x = 0 \qquad x - 2 = 0 \qquad 3x + 4 = 0$

$x = 0 \qquad x = 2 \qquad 3x = -4$

$x = -\frac{4}{3}$

**12.** $(y - 39)(2y + 7)(y + 12) = 0$

$y - 39 = 0 \qquad 2y + 7 = 0 \qquad y + 12 = 0$

$y = 39 \qquad 2y = -7 \qquad y = -12$

$y = -\frac{7}{2}$

**14.** $3x^2 + 9x = 0$

$3x(x + 3) = 0$

$x = 0 \qquad x + 3 = 0$

$\qquad x = -3$

**16.** $4x^2 - 6x = 0$

$2x(2x - 3) = 0$

$2x = 0 \qquad 2x - 3 = 0$

$x = 0 \qquad 2x = 3$

$x = \frac{3}{2}$

**18.** $x(x - 15) + 3(x - 15) = 0$

$(x - 15)(x + 3) = 0$

$x - 15 = 0 \qquad x + 3 = 0$

$x = 15 \qquad x = -3$

**20.** $x(x + 10) - 2(x + 10) = 0$

$(x + 10)(x - 2) = 0$

$x + 10 = 0 \qquad x - 2 = 0$

$x = -10 \qquad x = 2$

**22.** $x^2 - 121 = 0$

$(x - 11)(x + 11) = 0$

$x - 11 = 0 \qquad x + 11 = 0$

$x = 11 \qquad x = -11$

**24.** $25z^2 - 100 = 0$

$25(z^2 - 4) = 0$

$25(z - 2)(z + 2) = 0$

$z - 2 = 0 \qquad z + 2 = 0$

$z = 2 \qquad z = -2$

**26.** $x^2 - x - 12 = 0$

$(x - 4)(x + 3) = 0$

$x - 4 = 0 \qquad x + 3 = 0$

$x = 4 \qquad x = -3$

**28.** $20 - 9x + x^2 = 0$

$(5 - x)(4 - x) = 0$

$5 - x = 0 \qquad 4 - x = 0$

$5 = x \qquad 4 = x$

**30.** $14x^2 + 9x = -1$

$14x^2 + 9x + 1 = 0$

$(7x + 1)(2x + 1) = 0$

$7x + 1 = 0 \qquad 2x + 1 = 0$

$x = -\frac{1}{7} \qquad x = -\frac{1}{2}$

**32.** $11 + 32y - 3y^2 = 0$

$(11 - y)(1 + 3y) = 0$

$11 - y = 0 \qquad 1 + 3y = 0$

$y = 11 \qquad 3y = -1$

$y = -\frac{1}{3}$

**34.** $a^2 + 4a + 10 = 6$

  $a^2 + 4a + 4 = 0$

  $(a + 2)(a + 2) = 0$

  $a + 2 = 0$

  $\quad a = -2$

**36.** $x^2 - 12x + 21 = -15$

  $x^2 - 12x + 36 = 0$

  $(x - 6)(x - 6) = 0$

  $x - 6 = 0$

  $\quad x = 6$

**38.** $16t^2 + 48t + 40 = 4$

  $16t^2 + 48t + 36 = 0$

  $4(4t^2 + 12t + 9) = 0$

  $4(2t + 3)(2t + 3) = 0$

  $2t + 3 = 0$

  $2t = -3$

  $t = -\frac{3}{2}$

**40.** $s(s + 4) = 96$

  $s^2 + 4s = 96$

  $s^2 + 4s - 96 = 0$

  $(s + 12)(s - 8) = 0$

  $s + 12 = 0 \quad s - 8 = 0$

  $\quad s = -12 \quad s = 8$

**42.** $x(x - 4) = 12$

  $x^2 - 4x = 12$

  $x^2 - 4x - 12 = 0$

  $(x - 6)(x + 2) = 0$

  $x - 6 = 0 \quad x + 2 = 0$

  $\quad x = 6 \quad x = -2$

**44.** $3u(3u + 1) = 20$

  $9u^2 + 3u = 20$

  $9u^2 + 3u - 20 = 0$

  $(3u - 4)(3u + 5) = 0$

  $3u - 4 = 0 \quad 3u + 5 = 0$

  $3u = 4 \quad 3u = -5$

  $u = \frac{4}{3} \quad u = -\frac{5}{3}$

**46.** $(x - 8)(x - 7) = 20$

  $x^2 - 7x - 8x + 56 = 20$

  $x^2 - 15x + 36 = 0$

  $(x - 12)(x - 3) = 0$

  $x - 12 = 0 \quad x - 3 = 0$

  $\quad x = 12 \quad x = 3$

**48.** $(u - 6)(u + 4) = -21$

  $u^2 - 2u - 24 = -21$

  $u^2 - 2u - 3 = 0$

  $(u - 3)(u + 1) = 0$

  $u - 3 = 0 \quad u + 1 = 0$

  $\quad u = 3 \quad u = -1$

**50.** $(s + 4)^2 - 49 = 0$

  $[(s + 4) - 7][(s + 4) + 7] = 0$

  $(s - 3)(s + 11) = 0$

  $s - 3 = 0 \quad s + 11 = 0$

  $\quad s = 3 \quad s = -11$

**52.** $1 - (y + 3)^2 = 0$

  $[1 - (y + 3)][1 + (y + 3)] = 0$

  $(-y - 2)(y + 4) = 0$

  $-y - 2 = 0 \quad y + 4 = 0$

  $\quad -y = 2 \quad y = -4$

  $\quad y = -2$

**54.** $x^3 + 18x^2 + 45x = 0$

  $x(x^2 + 18x + 45) = 0$

  $x(x + 15)(x + 3) = 0$

  $x = 0 \quad x + 15 = 0 \quad x + 3 = 0$

  $\quad x = -15 \quad x = -3$

**56.** $3u^3 - 5u^2 - 2u = 0$

  $u(3u^2 - 5u - 2) = 0$

  $u(3u + 1)(u - 2) = 0$

  $u = 0 \quad 3u + 1 = 0 \quad u - 2 = 0$

  $\quad 3u = -1 \quad u = 2$

  $\quad u = -\frac{1}{3}$

**58.** $16(3 - u) - u^2(3 - u) = 0$

  $(3 - u)(16 - u^2) = 0$

  $(3 - u)(4 - u)(4 + u) = 0$

  $3 - u = 0 \quad 4 - u = 0 \quad 4 + u = 0$

  $\quad 3 = u \quad 4 = u \quad u = -4$

**60.** $x^3 - 2x^2 - 4x + 8 = 0$

$x^2(x - 2) - 4(x - 2) = 0$

$(x - 2)(x^2 - 4) = 0$

$(x - 2)(x - 2)(x + 2) = 0$

$x - 2 = 0 \quad x - 2 = 0 \quad x + 2 = 0$

$x = 2 \qquad x = 2 \qquad x = -2$

**62.** $v^3 + 4v^2 - 4v - 16 = 0$

$v^2(v + 4) - 4(v + 4) = 0$

$(v + 4)(v^2 - 4) = 0$

$(v + 4)(v - 2)(v + 2) = 0$

$v + 4 = 0 \quad v - 2 = 0 \quad v + 2 = 0$

$v = -4 \qquad v = 2 \qquad v = -2$

**64.** $x^4 + 2x^3 - 9x^2 - 18x = 0$

$x^3(x + 2) - 9x(x + 2) = 0$

$(x + 2)(x^3 - 9x) = 0$

$(x + 2)x(x^2 - 9) = 0$

$(x + 2)x(x - 3)(x + 3) = 0$

$x + 2 = 0 \quad x = 0 \quad x - 3 = 0 \quad x + 3 = 0$

$x = -2 \qquad\qquad x = 3 \qquad x = -3$

**66.** $9x^4 - 15x^3 - 9x^2 + 15x = 0$

$3x^3(3x - 5) - 3x(3x - 5) = 0$

$(3x - 5)(3x^3 - 3x) = 0$

$(3x - 5)3x(x^2 - 1) = 0$

$(3x - 5)(3x)(x - 1)(x + 1) = 0$

$3x - 5 = 0 \quad 3x = 0 \quad x - 1 = 0 \quad x + 1 = 0$

$x = \frac{5}{3} \qquad x = 0 \qquad x = 1 \qquad x = -1$

**68.** From the graph, the $x$-intercept is $(2, 0)$.
The solution of the equation $0 = x^2 - 4x + 4$ is 2.

$0 = x^2 - 4x + 4$

$0 = (x - 2)(x - 2)$

$x - 2 = 0 \quad x - 2 = 0$

$x = 2 \qquad x = 2$

**70.** From the graph, the $x$-intercepts are $(-1, 0)$, $(1, 0)$, and $(3, 0)$. The solutions to the equation are $-1$, $1$, and 3.

$0 = x^3 - 3x^2 - x + 3$

$0 = x^2(x - 3) - 1(x - 3)$

$0 = (x - 3)(x^2 - 1)$

$0 = (x - 3)(x - 1)(x + 1)$

$x - 3 = 0 \quad x - 1 = 0 \quad x + 1 = 0$

$x = 3 \qquad x = 1 \qquad x = -1$

**72.** *Keystrokes:*

Y= X,T,θ x² − 11 X,T,θ + 28 GRAPH

The $x$-intercepts are 4 and 7, so the solutions are 4 and 7.

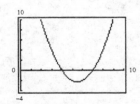

**74.** *Keystrokes:*

Y= ( X,T,θ − 2 ) x² − 9 GRAPH

The $x$-intercepts are $-1$ and 5, so the solutions are $-1$ and 5.

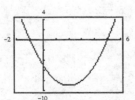

**76.** *Keystrokes:*

Y= X,T,θ ^ 3 − 4 X,T,θ GRAPH

The $x$-intercepts are $-2$, 0, and 2, so the solutions are $-2$, 0, and 2.

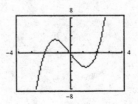

**78.** *Keystrokes:*

$\boxed{\text{Y=}}$ 2 $\boxed{+}$ $\boxed{\text{X,T,}\theta}$ $\boxed{-}$ 2 $\boxed{\text{X,T,}\theta}$ $\boxed{x^2}$ $\boxed{-}$ $\boxed{\text{X,T,}\theta}$ $\boxed{\wedge}$ 3 $\boxed{\text{GRAPH}}$

The *x*-intercepts are $-2$, $-1$, and 1, so the solutions are $-2$, $-1$, and 1.

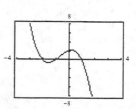

**80.** $ax^2 - ax = 0$

$ax(x - 1) = 0$

$ax = 0 \quad x - 1 = 0$

$x = \dfrac{0}{a} \qquad x = 1$

$x = 0 \qquad x = 1$

**82.** $x = 1$ and $x = 6$

$(x - 1)(x - 6) = 0$

$x^2 - 7x + 6 = 0$

**84.** *Verbal Model:* $\boxed{\text{Number}} + \boxed{\text{Its Square}} = \boxed{72}$

*Labels:*     Number $= x$

Its Square $= x^2$

*Equation:*     $x + x^2 = 72$

$x^2 + x - 72 = 0$

$(x + 9)(x - 8) = 0$

$x + 9 = 0 \quad x - 8 = 0$

$x = -9 \qquad x = 8$

reject

**86.** *Verbal Model:* $\boxed{\text{Area}} = \boxed{\text{Length}} \cdot \boxed{\text{Width}}$

*Labels:*     Length $= 2\frac{1}{4}w$

Width $= w$

*Equation:*     $900 = 2\frac{1}{4}w \cdot w$

$900 = \frac{9}{4}w^2$

$0 = \frac{9}{4}w^2 - 900$

$0 = \frac{9}{4}(w^2 - 400)$

$0 = \frac{9}{4}(w - 20)(w + 20)$

$w - 20 = 0 \quad w + 20 = 0$

$w = 20 \qquad w = -20$

Thus, the width of the rectangle is 20 inches.

$2\frac{1}{4}(20) = \frac{9}{4}(20) = 45$

Thus, the length of the rectangle is 45 inches.

**88.** *Verbal Model:* $\boxed{\substack{\text{Area of} \\ \text{exposed picture}}} = \boxed{\text{Length}} \cdot \boxed{\text{Width}}$

*Labels:*     Length $= (28 - 2w)$

Width $= (20 - 2w)$

*Equation:*     $468 = (28 - 2w)(20 - 2w)$

$468 = 560 - 56w - 40w + 4w^2$

$0 = 92 - 96w + 4w^2$

$0 = 4(23 - 24w + w^2)$

$0 = 2(1 - w)(23 - w)$

$1 - w = 0 \quad 23 - w = 0$

$23 = w$

$1 = w \qquad$ reject

The width of the frame is 1 cm.

**90.** *Verbal Model:* $\boxed{\text{Area}} = \frac{1}{2} \boxed{\text{Base}} \cdot \boxed{\text{Height}}$

*Labels:*     Base $= x$

Height $= x - 4$

*Equation:*     $70 = \frac{1}{2}x(x - 4)$

$70 = \frac{1}{2}x^2 - 2x$

$0 = \frac{1}{2}x^2 - 2x - 70$

$0 = x^2 - 4x - 140$

$0 = (x - 14)(x + 10)$

$x - 14 = 0 \quad x + 10 = 0$

$x = 14 \qquad x = -10$

reject

The base of the triangle is 14 inches.

The height is $14 - 4 = 10$ inches.

**92.** $S = x^2 + 4xh$

$880 = x^2 + 4x(6)$

$0 = x^2 + 24x - 880$

$0 = (x - 20)(x + 44)$

$0 = x - 20 \quad 0 = x + 44$

$20 = x \qquad -44 = x$

$20'' \times 20'' \qquad$ reject

**94.** $-16t^2 + 48t + 64 = 0$

$-16(t^2 - 3t - 4) = 0$

$-16(t - 4)(t + 1) = 0$

$t - 4 = 0 \quad t + 1 = 0$

$t = 4 \qquad t = -1$

reject

Thus, the object reaches the ground after 4 seconds.

**96.** *Verbal Model:* $\boxed{\text{Revenue}} = \boxed{\text{Cost}}$

*Equation:* $\quad 60x - x^2 = 75 + 40x$

$0 = x^2 - 20x + 75$

$0 = (x - 5)(x - 15)$

$x - 5 = 0 \qquad x - 15 = 0$

$x = 5 \text{ units} \qquad x = 15 \text{ units}$

**98.** (a) $3(x + 6)^2 - 10(x + 6) - 8 = 0$

let $u = (x + 6)$

$3u^2 - 10u - 8 = 0$

$(3u + 2)(u - 4) = 0$

$3u + 2 = 0 \qquad u - 4 = 0$

$u = -\dfrac{2}{3} \qquad u = 4$

$x + 6 = -\dfrac{2}{3} \quad x + 6 = 4$

$x = -\dfrac{20}{3} \qquad x = -2$

or

$3(x^2 + 12x + 36) - 10x - 60 - 8 = 0$

$3x^2 + 36x + 108 - 10x - 68 = 0$

$3x^2 + 26x + 40 = 0$

$(3x + 20)(x + 2) = 0$

$3x + 20 = 0 \qquad x + 2 = 0$

$3x = -20 \qquad x = -2$

$x = \dfrac{-20}{3}$

(b) $8(x + 2)^2 - 18(x + 2) + 9 = 0$

let $u = (x + 2)$

$8u^2 - 18u + 9 = 0$

$(4u - 3)(2u - 3) = 0$

$4u - 3 = 0 \qquad 2u - 3 = 0$

$u = \dfrac{3}{4} \qquad u = \dfrac{3}{2}$

$x + 2 = \dfrac{3}{4} \qquad x + 2 = \dfrac{3}{2}$

$x = -\dfrac{5}{4} \qquad x = -\dfrac{1}{2}$

or

$8(x^2 + 4x + 4) - 18x - 36 + 9 = 0$

$8x^2 + 32x + 32 - 18x - 36 + 9 = 0$

$8x^2 + 14x + 5 = 0$

$(4x + 5)(2x + 1) = 0$

$4x + 5 = 0 \qquad 2x + 1 = 0$

$4x = -5 \qquad 2x = -1$

$x = \dfrac{-5}{4} \qquad x = \dfrac{-1}{2}$

**100.** An example of how the Zero-Factor Property can be used to solve a quadratic equation: If $x(x - 2) = 0$, then $x = 0$ or $x - 2 = 0$. The solutions are $x = 0$ and $x = 2$.

**102.** Yes, it is possible for a quadratic equation to have only one solution. For example:

$x^2 + 2x + 1 = 0$

$(x + 1)^2 = 0$

$x = -1$

# Review Exercises for Chapter 3

**2.** $z^2 - 2 + 4z^{-2}$ is not a polynomial because the exponents of a variable must be a natural number.

**4.** Standard form: $2x^6 + x^5 - 5x^3 - 7$

Leading coefficient: 2

Degree: 6

**6.** Standard form: $-8x^7 + x^5 - 2x^3 + 9x$

Leading coefficient: $-8$

Degree: 7

**8.** Trinomial of degree 5 and leading coefficient $-6$:

$-6x^5 + 2x - 4$

**10.** Binomial of degree 2 and leading coefficient 7:

$7x^2 - 6x$

**12.** $(6x + 1) + (x^2 - 4x) =$

$x^2 + (6x - 4x) + 1 =$

$x^2 + 2x + 1$

**14.** $(7 - 12x^2 + 8x^3) + (x^4 - 6x^3 + 7x^2 - 5) =$

$x^4 + (8x^3 - 6x^3) + (-12x^2 + 7x^2) + (7 - 5) = x^4 + 2x^3 - 5x^2 + 2$

**16.** $(10y^2 + 3) - (y^2 + 4y - 9) = (10y^2 + 3) + (-y^2 - 4y + 9)$

$= (10y^2 - y^2) - 4y + (3 + 9)$

$= 9y^2 - 4y + 12$

**18.** $(7x^4 - 10x^2 + 4x) + (x^3 - 3x) - (3x^4 - 5x^2 + 1) =$

$(7x^4 - 3x^4) + x^3 + (-10x^2 + 5x^2) + (4x - 3x) - 1 = 4x^4 + x^3 - 5x^2 + x - 1$

**20.** $(7z^2 + 6z) - 3(5z^2 + 2z) = 7z^2 + 6z - 15z^2 - 6z$

$= (7z^2 - 15z^2) + (6z - 6z)$

$= -8z^2$

**22.** $(16a^3 + 5a) - 5[a + (2a^3 - 1)] = 16a^3 + 5a - 5a - 10a^3 + 5$

$= (16a^3 - 10a^3) + (5a - 5a) + 5$

$= 6a^3 + 5$

**24.** $-3y^2 \cdot y^4 = -3y^{2+4} = -3y^6$

**26.** $(v^4)^2 = v^{4 \cdot 2} = v^8$

**28.** $(-3y)^2(2) = (-3)^2 y^2(2)$

$= 9y^2(2)$

$= 18y^2$

**30.** $(12x^2y)(3x^2y^4)^2 = (12x^2y)(9x^4y^8)$

$= (12 \cdot 9)x^{2+4}y^{1+8}$

$= 108x^6y^9$

**32.** $\dfrac{15m^3}{25m} = \left(\dfrac{15}{25}\right)m^{3-1} = \dfrac{3}{5}m^2$

**34.** $-\dfrac{(-2x^2y^3)^2}{-3xy^2} = -\dfrac{4(x^2)^2(y^3)^2}{-3xy^2}$

$= -\dfrac{4x^4y^6}{-3xy^2}$

$= \dfrac{4x^{4-1}y^{6-2}}{3}$

$= \dfrac{4x^3y^4}{3}$

**36.** $\left(-\dfrac{1}{2}y^2\right)^3 = \left(-\dfrac{1}{2}\right)^3(y^2)^3$

$= -\dfrac{1}{8}y^6$

**38.** $(-4y)^2(y-2) = 16y^2(y-2)$

$= 16y^3 - 32y^2$

**40.** $-2y(5y^2 - y - 4) = -10y^3 + 2y^2 + 8y$

**42.** $(x+6)(x-9) = x^2 - 9x + 6x - 54$

$= x^2 - 3x - 54$

**44.** $(4x-1)(2x-5) = 8x^2 - 20x - 2x + 5$

$= 8x^2 - 22x + 5$

**46.** $(3y^2 + 2)(4y^2 - 5) = 3y^2(4y^2 - 5) + 2(4y^2 - 5)$

$= 12y^4 - 15y^2 + 8y^2 - 10$

$= 12y^4 - 7y^2 - 10$

**48.** $(5s^3 + 4s - 3)(4s - 5) = 5s^3(4s - 5) + 4s(4s - 5) - 3(4s - 5)$

$= 20s^4 - 25s^3 + 16s^2 - 20s - 12s + 15$

$= 20s^4 - 25s^3 + 16s^2 - 32s + 15$

**50.** $(3v+2)(-5v) + 5v(3v+2) = -15v^2 - 10v + 15v^2 + 10v$

$= 0$

**52.** $(8-3x)^2 = (8)^2 - 2(8)(3x) + (3x)^2$

$= 64 - 48x + 9x^2$

**54.** $(u+4v)^2 = u^2 + 2(u)(4v) + (4v)^2$

$= u^2 + 8uv + 16v^2$

**56.** $(7a+4)(7a-4) = (7a)^2 - (4)^2$

$= 49a^2 - 16$

**58.** $(5x-2y)(5x+2y) = (5x)^2 - (2y)^2 = 25x^2 - 4y^2$

**60.** $[(m-5) + n]^2 = (m-5)^2 + 2(m-5)n + n^2$

$= m^2 - 2m(5) + 5^2 + 2nm - 10n + n^2$

$= m^2 - 10m + 25 + 2mn - 10n + n^2$

**62.** $8y - 12y^4 = 4y(2 - 3y^3)$

**64.** $(u-9v)(u-v) + v(u-9v) = (u-9v)(u-v+v)$

$= u(u-9v)$

**66.** $y^3 + 4y^2 - y - 4 = (y^3 + 4y^2) + (-y - 4)$

$= y^2(y+4) - (y+4)$

$= (y+4)(y^2 - 1)$

$= (y+4)(y-1)(y+1)$

**68.** $x^3 + 7x^2 + 3x + 21 = (x^3 + 7x^2) + (3x + 21)$

$= x^2(x+7) + 3(x+7)$

$= (x+7)(x^2 + 3)$

**70.** $b^2 - 900 = b^2 - (30)^2$

$= (b-30)(b+30)$

**72.** $16y^2 - 49 = (4y)^2 - (7)^2$

$\qquad = (4y - 7)(4y + 7)$

**74.** $(y - 3)^2 - 16 = (y - 3)^2 - (4)^2$

$\qquad\qquad = (y - 3 - 4)(y - 3 + 4)$

$\qquad\qquad = (y - 7)(y + 1)$

**76.** $t^3 - 125 = t^3 - 5^3$

$\qquad = (t - 5)(t^2 + 5t + 25)$

**78.** $27x^3 + 64 = (3x)^3 + (4)^3$

$\qquad = (3x + 4)(9x^2 - 12x + 16)$

**80.** $y^2 + 16y + 64 = y^2 + 2(8)y + (8)^2$

$\qquad\qquad = (y + 8)^2$

**82.** $u^2 - 10uv + 25v^2 = u^2 - 2(u)(5v) + (5v)^2$

$\qquad\qquad = (u - 5v)^2$

**84.** $x^2 - 12x + 32 = (x - 8)(x - 4)$

**86.** $5x^2 + 11x - 12 = (5x - 4)(x + 3)$

**88.** $12x^2 - 13x - 14 = (4x - 7)(3x + 2)$

**90.** $3b + 27b^3 = 3b(1 + 9b^2)$

**92.** $x^3 + 3x^2 - 4x - 12 = x^2(x + 3) - 4(x + 3)$

$\qquad\qquad = (x + 3)(x^2 - 4)$

$\qquad\qquad = (x + 3)(x - 2)(x + 2)$

**94.** $x^2 - \frac{2}{3}x + \frac{1}{9} = x^2 - 2\left(\frac{1}{3}\right)x + \left(\frac{1}{3}\right)^2$

$\qquad\qquad = \left(x - \frac{1}{3}\right)^2$

**96.** $3x^3 + 23x^2 - 8x = x(3x - 1)(x + 8)$

**98.** $u^6 - 8v^6 = (u^2)^3 - (2v^2)^3$

$\qquad = (u^2 - 2v^2)(u^4 + 2u^2v^2 + 4v^4)$

**100.** $4t^2 - 12t = -9$

$4t^2 - 12t + 9 = 0$

$(2t - 3)(2t - 3) = 0$

$2t - 3 = 0$

$2t = 3$

$t = \frac{3}{2}$

**102.** $x^2 - 25x = -150$

$x^2 - 25x + 150 = 0$

$(x - 15)(x - 10) = 0$

$x - 15 = 0 \quad x - 10 = 0$

$\qquad x = 15 \qquad x = 10$

**104.** $3x(4x + 7) = 0$

$3x = 0 \quad 4x + 7 = 0$

$\quad x = 0 \qquad 4x = -7$

$\qquad\qquad x = -\frac{7}{4}$

**106.** $(x + 3)^2 - 25 = 0$

$(x + 3 - 5)(x + 3 + 5) = 0$

$(x - 2)(x + 8) = 0$

$x - 2 = 0 \quad x + 8 = 0$

$\quad x = 2 \qquad x = -8$

**108.** $b^3 - 6b^2 - b + 6 = 0$

$b^2(b - 6) - (b - 6) = 0$

$(b - 6)(b^2 - 1) = 0$

$(b - 6)(b - 1)(b + 1) = 0$

$b - 6 = 0 \quad b - 1 = 0 \quad b + 1 = 0$

$\quad b = 6 \qquad b = 1 \qquad b = -1$

**110.** *Keystrokes:*

Y=  X,T,θ  ^  3  −  6  X,T,θ  $x^2$  GRAPH

The $x$-intercepts are 0 and 6, so the solutions are 0 and 6.

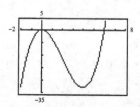

**112.** $x = \frac{5}{2}$   $x = 3$

$(x - \frac{5}{2})(x - 3) = 0$

$(2x - 5)(x - 3) = 0$

$2x^2 - 6x - 5x + 15 = 0$

$2x^2 - 11x + 15 = 0$

**114.** $x = -3$   $x = 3$   $x = 5$

$(x + 3)(x - 3)(x - 5) = 0$

$(x^2 - 9)(x - 5) = 0$

$x^3 - 5x^2 - 9x + 45 = 0$

**116.** *Verbal Model:*   $\boxed{\text{Profit}}$ = $\boxed{\text{Revenue}}$ − $\boxed{\text{Cost}}$

*Equation:*   $P(x) = 20x - (12x + 3000)$

$= 8x - 3000$

$P(1200) = 8(1200) - 3000$

$= 9600 - 3000$

$= \$6600$

**118.** $10p^3(1 - p)^2 = 10p^3(1 - p)(1 - p)$

$= 10p^3(1 - 2p + p^2)$

$= 10p^3 - 20p^4 + 10p^5$

**120.** *Perimeter* $= (13a - 26) + [12(a - 2)] + [5(a - 2)]$

$= 13a - 26 + 12a - 24 + 5a - 10$

$= (13a + 12a + 5a) + (-26 - 24 - 10)$

$= 30a - 60$

*Area* $= \frac{1}{2} \cdot 5(a - 2) \cdot 12(a - 2)$

$= 6(a - 2)^2$

**122.** *Verbal Model:*   $\boxed{\begin{array}{c}\text{Area of}\\\text{Shaded}\\\text{Region}\end{array}}$ = $\boxed{\begin{array}{c}\text{Area of}\\\text{Larger}\\\text{Triangle}\end{array}}$ − $\boxed{\begin{array}{c}\text{Area of}\\\text{Smaller}\\\text{Triangle}\end{array}}$

*Equation:*   Area $= \frac{1}{2} \cdot (3x + 10)(3x) - \frac{1}{2}(3x)(2x)$

$= \frac{3}{2}x(3x + 10) - (3x)(x)$

$= \frac{9}{2}x^2 + 15x - 3x^2$

$= \frac{9}{2}x^2 - \frac{6}{2}x^2 + 15x$

$= \frac{3}{2}x^2 + 15x$ *or* $\frac{3}{2}x(x + 10)$

**124.** *Verbal Model:* $\boxed{\text{Area}} = \boxed{\text{Length}} \cdot \boxed{\text{Width}}$

*Equation:* 
$$432 = x\left(\tfrac{3}{4}x\right)$$
$$432 = \tfrac{3}{4}x^2$$
$$0 = \tfrac{3}{4}x^2 - 432$$
$$0 = 3x^2 - 1728$$
$$0 = 3(x^2 - 576)$$
$$0 = 3(x - 24)(x + 24)$$
$$x - 24 = 0 \quad x + 24 = 0$$
$$x = 24 \qquad x = -24$$
$$\text{reject}$$

The length of the rectangel is 24 inches.

The width is $\tfrac{3}{4}(24) = 18$ inches.

**128.** *Verbal Model:* $\boxed{\begin{array}{c}\text{First odd}\\\text{integer}\end{array}} \cdot \boxed{\begin{array}{c}\text{Second odd}\\\text{integer}\end{array}} = \boxed{195}$

*Equation:*
$$(2n + 1)(2n + 3) = 195$$
$$4n^2 + 8n + 3 = 195$$
$$4n^2 + 8n - 192 = 0$$
$$4(n^2 + 2n - 48) = 0$$
$$(n + 8)(n - 6) = 0$$
$$n + 8 = 0 \qquad n - 6 = 0$$
$$n = -8 \qquad n = 6$$
$$\text{reject} \qquad 2n + 1 = 13$$
$$\qquad\qquad 2n + 3 = 15$$

**126.**
$$-16t^2 + 32t + 48 = 0$$
$$-16(t^2 - 2t - 3) = 0$$
$$-16(t - 3)(t + 1) = 0$$
$$t - 3 = 0 \quad t + 1 = 0$$
$$t = 3 \qquad t = -1$$
$$\qquad\qquad \text{reject}$$

The object reaches the ground after 3 seconds.

# CHAPTER 4
# Rational Expressions, Equations, and Functions

# CHAPTER 4
# Rational Expressions, Equations, and Functions

## Section 4.1  Integer Exponents and Scientific Notation

Solutions to Even-Numbered Exercises

**2.** $2^{-4} = \dfrac{1}{2^4} = \dfrac{1}{16}$

**4.** $-20^{-2} = -\dfrac{1}{20^2} = -\dfrac{1}{400}$

**6.** $25^0 = 1$

**8.** $\dfrac{1}{-8^{-2}} = \dfrac{8^2}{-1} = -64$

**10.** $-\dfrac{1}{6^{-2}} = -6^2 = -36$

**12.** $\left(\dfrac{4}{5}\right)^{-3} = \left(\dfrac{5}{4}\right)^3 = \dfrac{125}{64}$

**14.** $\left(-\dfrac{5}{8}\right)^{-2} = \left(-\dfrac{8}{5}\right)^2 = \dfrac{64}{25}$

**16.** $4^2 \cdot 4^{-3} = 4^{2+(-3)} = 4^{-1} = \dfrac{1}{4}$

**18.** $\dfrac{5^{-1}}{5^2} = \dfrac{1}{5^{2-(-1)}} = \dfrac{1}{5^3} = \dfrac{1}{125}$

**20.** $\dfrac{10^{-5}}{10^{-6}} = 10^{-5-(-6)} = 10^{-1} = 10$

**22.** $(5^3 \cdot 5^{-4})^{-3} = (5^{3+(-4)})^{-3}$
$= (5^{-1})^{-3}$
$= 5^3$
$= 125$

**24.** $(-4^{-1})^{-2} = (-4)^2 = 16$

**26.** $4 - 3^{-2} = 4 - \dfrac{1}{3^2}$
$= 4 - \dfrac{1}{9}$
$= \dfrac{36}{9} - \dfrac{1}{9}$
$= \dfrac{35}{9}$

**28.** $\left(\dfrac{1}{2} - \dfrac{2}{3}\right)^{-1} = \left(\dfrac{1(3)}{2(3)} - \dfrac{2(2)}{3(2)}\right)^{-1}$
$= \left(\dfrac{3}{6} - \dfrac{4}{6}\right)^{-1}$
$= \left(-\dfrac{1}{6}\right)^{-1}$
$= (-6)^1$
$= -6$

**30.** $(32 + 4^{-3})^0 = 1$

**32.** $x^{-2} \cdot x^{-5} = x^{-2+(-5)}$
$= x^{-7}$
$= \dfrac{1}{x^7}$

**34.** $t^{-1} \cdot t^{-6} = t^{-1+(-6)}$
$= t^{-7}$
$= \dfrac{1}{t^7}$

**36.** $3y^{-3} = \dfrac{3}{y^3}$

**38.** $(5u)^{-2} = \dfrac{1}{(5u)^2} = \dfrac{1}{25u^2}$

**40.** $\dfrac{4}{y^{-1}} = 4y$

**42.** $\dfrac{6u^{-2}}{15u^{-1}} = \dfrac{2 \cdot 3}{3 \cdot 5u^{-1-(-2)}}$
$= \dfrac{2}{5u^1}$
$= \dfrac{2}{5u}$

**44.** $\dfrac{(5u)^{-4}}{(5u)^0} = (5u)^{-4-0} = (5u)^{-4} = \dfrac{1}{(5u)^4} = \dfrac{1}{625u^4}$

**46.** $(4a^{-2})^{-3} = 4^{-3}a^6 = \dfrac{a^6}{4^3} = \dfrac{a^6}{64}$

**48.** $(5s^5t^{-5})(-6s^{-2}t^4) = -30s^{5+(-2)}t^{(-5)+4}$

$\qquad\qquad\qquad = -30s^3t^{-1}$

$\qquad\qquad\qquad = -\dfrac{30s^3}{t}$

**50.** $(-4y^{-3}z)^{-3} = \dfrac{1}{(-4y^{-3}z)^3}$

$\qquad\qquad = \dfrac{1}{-64y^{-9}z^3}$

$\qquad\qquad = \dfrac{y^9}{-64z^3}$

**52.** $\left(\dfrac{4}{z}\right)^{-2} = \dfrac{4^{-4}}{z^{-2}}$

$\qquad = \dfrac{z^2}{4^2}$

$\qquad = \dfrac{z^2}{16}$

**54.** $\dfrac{2y^{-1}z^{-3}}{4yz^{-3}} = \dfrac{2z^3}{4yyz^3}$

$\qquad = \dfrac{1z^{3-3}}{2y^2}$

$\qquad = \dfrac{1z^0}{2y^2}$

$\qquad = \dfrac{1}{2y^2}$

**56.** $\left(\dfrac{5^2x^3y^{-3}}{125xy}\right)^{-1} = \left(\dfrac{5^2x^{3-1}y^{-3-1}}{5^3}\right)^{-1}$

$\qquad\qquad = \left(\dfrac{x^2y^{-4}}{5^{3-2}}\right)^{-1}$

$\qquad\qquad = \left(\dfrac{x^2}{5y^4}\right)^{-1}$

$\qquad\qquad = \dfrac{5y^4}{x^2}$

**58.** $\left(\dfrac{a^{-3}}{b^{-3}}\right)\left(\dfrac{b}{a}\right)^3 = \left(\dfrac{b^3}{a^3}\right)\left(\dfrac{b^3}{a^3}\right)$

$\qquad = \dfrac{b^{3+3}}{a^{3+3}}$

$\qquad = \dfrac{b^6}{a^6}$

**60.** $(ab)^{-2}(a^2b^2)^{-1} = a^{-2}b^{-2}a^{-2}b^{-2}$

$\qquad\qquad = a^{-2+(-2)}b^{-2+(-2)}$

$\qquad\qquad = a^{-4}b^{-4}$

$\qquad\qquad = \dfrac{1}{a^4b^4}$

**62.** $x^5(3x^0y^4)(7y)^0 = x^5(3y^4)(1)$

$\qquad\qquad = 3x^5y^4$

**64.** $[(2x^{-3}y^{-2})^2]^{-2} = [(2x^{-3}y^{-2})^{-4}]$

$\qquad\qquad = 2^{-4}x^{12}y^8$

$\qquad\qquad = \dfrac{x^{12}y^8}{2^4}$

$\qquad\qquad = \dfrac{x^{12}y^8}{16}$

**66.** $\dfrac{(5x^2y^{-5})^{-1}}{2x^{-5}y^4} = \dfrac{5^{-1}x^{-2}y^5}{2x^{-5}y^4}$

$\qquad = \dfrac{x^{-2-(-5)}y^{5-4}}{5\cdot 2}$

$\qquad = \dfrac{x^3y}{10}$

**68.** $x^{-2}(x^2 + y^2) = x^{-2+2} + x^{-2}y^2$

$\qquad\qquad = x^0 + \dfrac{y^2}{x^2}$

$\qquad\qquad = 1 + \dfrac{y^2}{x^2}$

**70.** $\dfrac{u^{-1} - v^{-1}}{u^{-1} + v^{-1}} = \dfrac{\dfrac{1}{u} - \dfrac{1}{v}}{\dfrac{1}{u} + \dfrac{1}{v}}$

$\qquad = \dfrac{\dfrac{1}{u} - \dfrac{1}{v}}{\dfrac{1}{u} + \dfrac{1}{v}} \cdot \dfrac{uv}{uv}$

$\qquad = \dfrac{v - u}{v + u}$

**72.** $98,100,000 = 9.81 \times 10^7$

**74.** $956,300,000 = 9.563 \times 10^8$

**76.** $0.00625 = 6.25 \times 10^{-3}$

**78.** $0.0007384 = 7.384 \times 10^{-4}$

**80.** $139,400,000 = 1.394 \times 10^8$

**82.** $0.0000001 = 1.0 \times 10^{-7}$

**84.** $0.00003937 = 3.937 \times 10^{-5}$

**86.** $5.05 \times 10^{12} = 5,050,000,000,000$

**88.** $8.6 \times 10^{-9} = 0.0000000086$

**90.** $3.5 \times 10^8 = 350,000,000$

**92.** $9.0 \times 10^{-9} = 0.000000009$

**94.** $9.0 \times 10^{-4} = 0.0009$

**96.** $(6.5 \times 10^6)(2 \times 10^4) = (6.5)(2) \times 10^{6+4}$

$= 13.0 \times 10^{10}$

$= 1.3 \times 10^{11}$

**98.** $(4 \times 10^6)^3 = 4^3 \times 10^{6 \cdot 3}$

$= 64 \times 10^{18}$

$= 6.4 \times 10^{19}$

**100.** $\dfrac{2.5 \times 10^{-3}}{5 \times 10^2} = 0.5 \times 10^{-3-2}$

$= 0.5 \times 10^{-5}$

$= 5.0 \times 10^{-6}$

**102.** $(62,000,000)(0.0002) = (6.2 \times 10^7)(2 \times 10^{-4})$

$= 12.4 \times 10^{7+(-4)}$

$= 12.4 \times 10^3$

$= 1.24 \times 10^4$

**104.** $\dfrac{72,000,000,000}{0.00012} = \dfrac{7.2 \times 10^{10}}{1.2 \times 10^{-4}}$

$= 6.0 \times 10^{10-(-4)}$

$= 6.0 \times 10^{14}$

**106.** $\dfrac{(3,450,000,000)(0.000125)}{(52,000,000)(0.000003)} = \dfrac{(3.45 \times 10^9)(1.25 \times 10^{-4})}{(5.2 \times 10^7)(3 \times 10^{-6})}$

$= \dfrac{(3.45 \times 1.25) \times 10^{9+(-4)}}{(5.2 \times 3) \times 10^{7+(-6)}}$

$= \dfrac{4.3125 \times 10^5}{15.6 \times 10^1}$

$\approx 0.276 \times 10^{5-1}$

$\approx 0.276 \times 10^4$

$\approx 2.76 \times 10^3$

**108.** $\dfrac{(3.82 \times 10^5)^2}{(8.5 \times 10^4)(5.2 \times 10^{-3})} = \dfrac{14.5924 \times 10^{10}}{44.2 \times 10^1}$

$\approx 0.3301447964 \times 10^9$

$\approx 3.301447964 \times 10^8$

$\approx 3.30 \times 10^8$

**110.** $(8.67 \times 10^4)^7 = (8.67)^7 \times 10^{28}$

$\approx 3682423.07 \times 10^{28}$

$\approx 3.68 \times 10^{34}$

**112.** $\dfrac{(6,200,000)(0.005)^3}{(0.00035)^5} = \dfrac{(6.2 \times 10^6)(5.0 \times 10^{-3})^3}{(3.5 \times 10^{-4})^5}$

$= \dfrac{(6.2 \times 10^6)(5.0)^3 \times 10^{-9}}{(3.5)^5 \times 10^{-20}}$

$= \dfrac{(6.2)(125) \times 10^{-3}}{(3.5)^5 \times 10^{-20}}$

$= \dfrac{(6.2)(125) \times 10^{-3-(-20)}}{(3.5)^5}$

$= \dfrac{(6.2)(125) \times 10^{17}}{(3.5)^5}$

$= \dfrac{775 \times 10^{17}}{525.21875}$

$\approx 1.48 \times 10^{17}$

**114.** $8 \times 10^{22} = 80,000,000,000,000,000,000,000$

**116.** $(90)(9.45 \times 10^{15}) = 850.5 \times 10^{15}$

$= 8.505 \times 10^{17}$ meters or

$850,500,000,000,000,000$ meters

**118.** $75(200)(10 \times 10^{-6}) = 150{,}000 \times 10^{-6} = 0.15$ foot

**120.** Mercury:   $K = \dfrac{(0.241)^2}{(0.387)^3} \approx 1.002$       Saturn:   $K = \dfrac{(29.457)^2}{(9.541)^3} \approx 0.999$

Venus:   $K = \dfrac{(0.615)^2}{(0.723)^3} \approx 1.001$       Uranus:   $K = \dfrac{(84.008)^2}{(19.190)^3} \approx 0.999$

Earth:   $K = \dfrac{(1.000)^2}{(1.000)^3} = 1$       Neptune:   $K = \dfrac{(164.784)^2}{(30.086)^3} = 0.997$

Mars:   $K = \dfrac{(1.881)^2}{(1.523)^3} \approx 1.002$       Pluto:   $K = \dfrac{(248.350)^2}{(39.508)^3} \approx 1.000$

Jupiter:   $K = \dfrac{(11.861)^2}{(5.203)^3} \approx 0.999$       Yes

**122.** $(-2x)^{-4} = \dfrac{1}{(-2)^4 x^4} = \dfrac{1}{16x^4}$       Both the $-2$ and $x$ are raised to the $-4$ power.

$-2x^{-4} = -\dfrac{2}{x^4}$       Only the $x$ is raised to the $-4$ power.

**124.** $32.5 \times 10^5$ is not in scientific notation because $32.5$ is not in the interval $[1, 10)$.

# Section 4.2    Rational Expressions and Functions

**2.** $x - 13 \neq 0$

$x \neq 13$

$D = (-\infty, 13) \cup (13, \infty)$

**4.** $6 - y \neq 0$

$y \neq 6$

$D = (-\infty, 6) \cup (6, \infty)$

**6.** $7 \neq 0$

$D = (-\infty, \infty)$

**8.** $t^3 - 4t^2 + 1 \neq 0$

$D = (-\infty, \infty)$

**10.** $x^2 + 16 \neq 0$

$D = (-\infty, \infty)$

**12.**   $z(z - 4) \neq 0$

$z \neq 0$   $z - 4 \neq 0$

$z \neq 4$

$D = (-\infty, 0) \cup (0, 4) \cup (4, \infty)$

**14.**   $x^2 - 4 \neq 0$

$(x - 2)(x + 2) \neq 0$

$x - 2 \neq 0$   $x + 2 \neq 0$

$x \neq 2$       $x \neq -2$

$D = (-\infty, -2) \cup (-2, 2) \cup (2, \infty)$

**16.**   $t^2 + 5t \neq 0$

$t(t + 5) \neq 0$

$t \neq 0$   $t \neq -5$

$D = (-\infty, -5) \cup (-5, 0) \cup (0, \infty)$

**18.**   $t^2 - 2t - 3 \neq 0$

$(t - 3)(t + 1) \neq 0$

$t - 3 \neq 0$   $t + 1 \neq 0$

$t \neq 3$       $t \neq -1$

$D = (-\infty, -1) \cup (-1, 3) \cup (3, \infty)$

**20.**   $4y^2 - 5y - 6 \neq 0$

$(4y + 3)(y - 2) \neq 0$

$4y + 3 \neq 0$      $y - 2 \neq 0$

$4y \neq -3$        $y \neq 2$

$y \neq -\frac{3}{4}$

$D = \left(-\infty, -\frac{3}{4}\right) \cup \left(-\frac{3}{4}, 2\right) \cup (2, \infty)$

**22.** (a) $f(10) = \dfrac{10 - 10}{4(10)} = \dfrac{0}{40} = 0$

(b) $f(0) = \dfrac{0 - 10}{4(0)} = \dfrac{-10}{0}$; not possible; undefined

(c) $f(-2) = \dfrac{-2 - 10}{4(-2)} = \dfrac{-12}{-8} = \dfrac{3}{2}$

(d) $f(12) = \dfrac{12 - 10}{4(12)} = \dfrac{2}{48} = \dfrac{1}{24}$

**24.** (a) $g(2) = \dfrac{2 - 2}{2(2) - 5} = \dfrac{0}{4 - 5} = \dfrac{0}{-1} = 0$

(b) $g\left(\dfrac{5}{2}\right) = \dfrac{\frac{5}{2} - 2}{2\left(\frac{5}{2}\right) - 5} = \dfrac{\frac{5}{2} - \frac{4}{2}}{5 - 5} = \dfrac{\frac{1}{2}}{0}$; not possible; undefined

(c) $g(-2) = \dfrac{-2 - 2}{2(-2) - 5} = \dfrac{-4}{-4 - 5} = \dfrac{-4}{-9} = \dfrac{4}{9}$

(d) $g(0) = \dfrac{0 - 2}{2(0) - 5} = \dfrac{-2}{0 - 5} = \dfrac{-2}{-5} = \dfrac{2}{5}$

**26.** (a) $f(-1) = \dfrac{(-1)^3 + 1}{(-1)^2 - 6(-1) + 9}$

$= \dfrac{-1 + 1}{1 + 6 + 9}$

$= \dfrac{0}{16}$

$= 0$

(c) $f(-2) = \dfrac{(-2)^3 + 1}{(-2)^2 - 6(-2) + 9}$

$= \dfrac{-8 + 1}{4 + 12 + 9}$

$= \dfrac{-7}{25}$

(b) $f(3) = \dfrac{(3)^3 + 1}{(3)^2 - 6(3) + 9}$

$= \dfrac{27 + 1}{9 - 18 + 9}$

$= \dfrac{28}{0}$; not possible; undefined

(d) $f(2) = \dfrac{(2)^3 + 1}{(2)^2 - 6(2) + 9}$

$= \dfrac{8 + 1}{4 - 12 + 9}$

$= \dfrac{9}{1}$

$= 9$

**28.** Since $p$ is the percent of a certain illegal drug, $p \geq 0$. Since

$$\dfrac{258p}{100 - p}$$

must be defined, $100 - p \neq 0$. Thus, $p \neq 100$. Since $p$ is a percent of a certain illegal drug, $p \leq 100$. Therefore, the domain is $[0, 100)$.

**30.** $x =$ units of a product

$D = \{1, 2, 3, 4, \ldots\}$

**32.** $x =$ units of a product

$D = \{1, 2, 3, 4, \ldots\}$

**34.** $\dfrac{7}{15} = \dfrac{7[3(x - 10)^2]}{45(x - 10)^2}$,   $x \neq 10$

**36.** $\dfrac{5x}{12} = \dfrac{25x^2(x - 10)}{12[5x(x - 10)]}$,   $x \neq 10$

**38.** $\dfrac{3y - 7}{y + 2} = \dfrac{(3y - 7)(y - 2)}{y^2 - 4}$,   $y \neq 2$

**40.** $\dfrac{3 - z}{z^2} = \dfrac{(3 - z)(z + 2)}{z^3 + 2z^2}$,   $z \neq -2$

**42.** $\dfrac{32y}{24} = \dfrac{4 \cdot 8y}{3 \cdot 8} = \dfrac{4y}{3}$

**44.** $\dfrac{15z^3}{15z^3} = 1$,   $z \neq 0$

**46.** $\dfrac{16y^2z^2}{60y^5z} = \dfrac{4 \cdot 4 \cdot y^2 \cdot z \cdot z}{4 \cdot 15 \cdot y^2 \cdot y^3 \cdot z} = \dfrac{4z}{15y^3}$,   $y \neq 0, z \neq 0$

**48.** $\dfrac{8x^3 + 4x^2}{20x} = \dfrac{4x^2(2x + 1)}{20x} = \dfrac{x(2x + 1)}{5}$,   $x \neq 0$

**50.** $\dfrac{a^2b(b - 3)}{b^3(b - 3)^2} = \dfrac{a^2b(b - 3)}{b \cdot b^2(b - 3)(b - 3)}$

$= \dfrac{a^2}{b^2(b - 3)}$,   $b \neq 0, 3$

**52.** $\dfrac{y^2 - 81}{2y - 18} = \dfrac{(y + 9)(y - 9)}{2(y - 9)} = \dfrac{y + 9}{2}$,   $y \neq 9$

**54.** $\dfrac{x^2 - 36}{6 - x} = \dfrac{(x + 6)(x - 6)}{-(x - 6)}$

$\qquad = \dfrac{(x + 6)}{-1}$

$\qquad = -(x + 6) \text{ or } -x - 6, \quad x \neq 6$

**56.** $\dfrac{u^2 - 12u + 36}{u - 6} = \dfrac{(u - 6)(u - 6)}{u - 6}$

$\qquad = u - 6, \quad u \neq 6$

**58.** $\dfrac{z^2 + 22z + 121}{3z + 33} = \dfrac{(z + 11)(z + 11)}{3(z + 11)}$

$\qquad = \dfrac{z + 11}{3}, \quad z \neq -11$

**60.** $\dfrac{x^2 - 7x}{x^2 - 4x - 21} = \dfrac{x(x - 7)}{(x + 3)(x - 7)}$

$\qquad = \dfrac{x}{x + 3}, \quad x \neq 7$

**62.** $\dfrac{x^4 - 25x^2}{x^2 + 2x - 15} = \dfrac{x^2(x^2 - 25)}{(x + 5)(x - 3)}$

$\qquad = \dfrac{x^2(x - 5)(x + 5)}{(x + 5)(x - 3)}$

$\qquad = \dfrac{x^2(x - 5)}{x - 3}, \quad x \neq -5$

**64.** $\dfrac{2x^2 + 3x - 5}{7 - 6x - x^2} = \dfrac{(2x + 5)(x - 1)}{-1(x^2 + 6x - 7)}$

$\qquad = \dfrac{(2x + 5)(x - 1)}{-1(x + 7)(x - 1)}$

$\qquad = -\dfrac{2x + 5}{x + 7}, \quad x \neq 1$

**66.** $\dfrac{2y^2 + 13y + 20}{2y^2 + 17y + 30} = \dfrac{(2y + 5)(y + 4)}{(2y + 5)(y + 6)}$

$\qquad = \dfrac{y + 4}{y + 6}, \quad y \neq -\dfrac{5}{2}$

**68.** $\dfrac{56z^2 - 3z - 20}{49z^2 - 16} = \dfrac{(7z + 4)(8z - 5)}{(7z + 4)(7z - 4)}$

$\qquad = \dfrac{8z - 5}{7z - 4}, \quad z \neq -\dfrac{4}{7}$

**70.** $\dfrac{x + 3x^2y}{3xy + 1} = \dfrac{x(1 + 3xy)}{1 + 3xy} = x, \quad 3xy \neq -1$

**72.** $\dfrac{x^2 - 25z^2}{x + 5z} = \dfrac{(x + 5z)(x - 5z)}{x + 5z} = x - 5z, \quad x \neq -5z$

**74.** $\dfrac{4u^2v - 12uv^2}{18uv} = \dfrac{4uv(u - 3v)}{18uv}$

$\qquad = \dfrac{2 \cdot 2 \cdot u \cdot v \cdot (u - 3v)}{2 \cdot 9 \cdot u \cdot v}$

$\qquad = \dfrac{2(u - 3v)}{9}, \quad u \neq 0, v \neq 0$

**76.** $\dfrac{x^2 + 4xy}{x^2 - 16y^2} = \dfrac{x(x + 4y)}{(x + 4y)(x - 4y)}$

$\qquad = \dfrac{x}{x - 4y}, \quad x \neq -4y$

**78.** $\dfrac{x^2 + xy - 2y^2}{x^2 + 3xy + 2y^2} = \dfrac{(x + 2y)(x - y)}{(x + 2y)(x + y)}$

$\qquad = \dfrac{x - y}{x + y}, \quad x \neq -2y$

**80.** $\dfrac{x - 4}{x} \neq -4$

$\dfrac{10 - 4}{10} \neq -4$

Choose a value such as 10 for $x$ and evaluate both sides.

$\dfrac{6}{10} \neq -4$

$\dfrac{3}{5} \neq -4$

**82.** $\dfrac{1-x}{2-x} \neq \dfrac{1}{2}$

$\dfrac{1-1}{2-1} \neq \dfrac{1}{2}$

$\dfrac{0}{1} \neq \dfrac{1}{2}$

Choose a value such as 1 for $x$ and evaluate both sides.

**84.**

| $x$ | $-2$ | $-1$ | 0 | 1 | 2 | 3 | 4 |
|---|---|---|---|---|---|---|---|
| $\dfrac{x^2 + 5x}{x}$ | 3 | 4 | undefined | 6 | 7 | 8 | 9 |
| $x + 5$ | 3 | 4 | 5 | 6 | 7 | 8 | 9 |

$\dfrac{x^2 + 5x}{x} = \dfrac{x(x+5)}{x} = x + 5, \quad x \neq 0$

Domain of $\dfrac{x^2 + 5x}{x}$ is $(-\infty, 0) \cup (0, \infty)$.

Domain of $x + 5$ is $(-\infty, \infty)$.

The two expressions are equal for all replacements of the variable except $x = 0$.

**86.** $\dfrac{\text{Area of shaded portion}}{\text{Area of total figure}} = \dfrac{\frac{1}{2}x(0.6x)}{\frac{1}{2}(3x)(1.8x)} = \dfrac{0.6x^2}{5.4x^2} = \dfrac{0.6}{5.4} = \dfrac{6}{54} = \dfrac{1}{9}, \quad x > 0$

**88.** (a) *Verbal Model:* $\boxed{\text{Total cost}} = \boxed{\text{Number of units}} \cdot \boxed{\text{Cost per unit}} + \boxed{\text{Initial cost}}$

*Labels:* Total cost $= C$

Number of units $= x$

*Equation:* $C = 6.50x + 60,000$

(b) *Verbal Model:* $\boxed{\text{Average cost}} = \boxed{\text{Total cost}} \div \boxed{\text{Number of units}}$

*Label:* Average cost $= \overline{C}$

*Equation:* $\overline{C} = \dfrac{C}{x}$

$\overline{C} = \dfrac{6.50x + 60,000}{x}$

(c) $D = \{1, 2, 3, 4, \ldots\}$

(d) $\overline{C}(11,000) = \dfrac{6.50(11,000) + 60,000}{11,000} \approx \$11.95$

**90.** (a) *Verbal Model:* $\boxed{\text{Distance}} = \boxed{\text{Rate}} \cdot \boxed{\text{Time}}$

*Labels:* Car 1: $55(t + 2)$

Car 2: $65t$

(b) Distance between car 1 and car 2:

$d = |55(t + 2) - 65t|$

$= |55t + 110 - 65t|$

$= |110 - 10t|$

$= |10(11 - t)|$

(c) $\dfrac{\text{Distance of car 2}}{\text{Distance of car 1}} = \dfrac{65t}{55(t + 2)} = \dfrac{13t}{11(t + 2)}$

**92.** $\dfrac{\text{Rectangular pool volume}}{\text{Circular pool volume}} = \dfrac{d(6d - 2)(3d + 4)}{\pi(5d)^2 d}$

$= \dfrac{2(3d - 1)(3d + 4)}{25\pi d^2}$

**94.**

| Year | 1990 | 1991 | 1992 | 1993 |
|---|---|---|---|---|
| Average cost | $3132 | $3506 | $3866 | $4213 |

| Year | 1994 | 1995 | 1996 |
|---|---|---|---|
| Average cost | $4549 | $4872 | $5184 |

**96.** $\dfrac{1}{x^2 + 1}$

There are many correct answers.

**98.** No. You can cancel only common factors.

**100.** True.

$\dfrac{6x - 5}{5 - 6x} = \dfrac{-(5 - 6x)}{5 - 6x} = -1$

# Section 4.3   Multiplying and Dividing Rational Expressions

**2.** (a) $x = 0$

$$\frac{0^2 - 4(0)}{(0)^2 - 9} = \frac{0 - 0}{0 - 9} = \frac{0}{-9} = 0$$

(b) $x = 4$

$$\frac{4^2 - 4(4)}{4^2 - 9} = \frac{16 - 16}{16 - 9} = \frac{0}{7} = 0$$

(c) $x = 3$

$$\frac{3^2 - 4(3)}{3^2 - 9} = \frac{9 - 12}{9 - 9} = \frac{-3}{0} \text{ is undefined.}$$

(d) $x = -3$

$$\frac{(-3)^2 - 4(-3)}{(-3)^2 - 9} = \frac{9 + 12}{9 - 9} = \frac{21}{0} \text{ is undefined.}$$

**4.** $\dfrac{2x}{x - 3} = \dfrac{14x(x - 3)^2}{(x - 3)(7(x - 3)^2)}, \quad x \neq 0$

**6.** $\dfrac{x + 1}{x} = \dfrac{(x + 1)^3}{x(x + 1)^2}, \quad x \neq -1$

**8.** $\dfrac{3t + 5}{t} = \dfrac{(3t + 5)[5t(3t - 5)]}{5t^2(3t - 5)}, \quad t \neq \dfrac{5}{3}$

**10.** $\dfrac{x^2}{10 - x} = \dfrac{x^2(-x)}{x^2 - 10x}, \quad x \neq 0$

**12.** $24\left(-\dfrac{7}{18}\right) = \dfrac{6 \cdot 4 \cdot -7}{6 \cdot 3} = -\dfrac{28}{3}$

**14.** $\dfrac{6}{5a} \cdot (25a) = \dfrac{6 \cdot 5 \cdot 5 \cdot a}{5a} = 30, \quad a \neq 0$

**16.** $\dfrac{3x^4}{7x} \cdot \dfrac{8x^2}{9} = \dfrac{3 \cdot x \cdot x^3 \cdot 8 \cdot x^2}{7 \cdot x \cdot 3 \cdot 3} = \dfrac{8x^3x^2}{7 \cdot 3} = \dfrac{8x^5}{21}, \quad x \neq 0$

**18.** $25x^3 \cdot \dfrac{8}{35x} = \dfrac{5 \cdot 5 \cdot x \cdot x \cdot x \cdot 8}{7 \cdot 5 \cdot x} = \dfrac{40x^2}{7}, \quad x \neq 0$

**20.** $(6 - 4x) \cdot \dfrac{10}{3 - 2x} = 2(3 - 2x) \cdot \dfrac{10}{3 - 2x} = 20, \quad x \neq \dfrac{3}{2}$

**22.** $\dfrac{1 - 3xy}{4x^2y} \cdot \dfrac{46x^4y^2}{15 - 45xy} = \dfrac{1 - 3xy}{4x^2y} \cdot \dfrac{2(23)x^4y^2}{15(1 - 3xy)} = \dfrac{23x^2y}{30}$

$x \neq 0, y \neq 0, 1 - 3xy \neq 0$

**24.** $\dfrac{8 - z}{8 + z} \cdot \dfrac{z + 8}{z - 8} = \dfrac{-(z - 8)(z + 8)}{(z + 8)(z - 8)} = -1, \quad z \neq -8, 8$

**26.** $\dfrac{x + 14}{x^3(10 - x)} \cdot \dfrac{x(x - 10)}{5} = \dfrac{(x + 14)(x)(x - 10)}{-x \cdot x^2(x - 10)(5)}$

$$= \dfrac{x + 14}{-5x^2}$$

$$= -\dfrac{x + 14}{5x^2}, \quad x \neq 10$$

**28.** $\dfrac{5y - 20}{5y + 15} \cdot \dfrac{2y + 6}{y - 4} = \dfrac{5(y - 4) \cdot 2(y + 3)}{5(y + 3)(y - 4)} = 2, \quad y \neq -3, 4$

**30.** $\dfrac{y^2 - 16}{y^2 + 8y + 16} \cdot \dfrac{3y^2 - 5y - 2}{y^2 - 6y + 8} = \dfrac{(y - 4)(y + 4)}{(y + 4)(y + 4)} \cdot \dfrac{(3y + 1)(y - 2)}{(y - 4)(y - 2)} = \dfrac{3y + 1}{y + 4}, \quad y \neq 2, 4$

**32.** $(u - 2v)^2 \cdot \dfrac{u + 2v}{u - 2v} = \dfrac{(u - 2v)(u - 2v)(u + 2v)}{u - 2v} = (u - 2v)(u + 2v), \quad u \neq 2v$

**34.** $\dfrac{(x - 2y)^2}{x + 2y} \cdot \dfrac{x^2 + 7xy + 10y^2}{x^2 - 4y^2} = \dfrac{(x - 2y)^2}{x + 2y} \cdot \dfrac{(x + 5y)(x + 2y)}{(x - 2y)(x + 2y)} = \dfrac{(x - 2y)(x + 5y)}{x + 2y}, \quad x \neq 2y$

**36.** $\dfrac{t^2 + 4t + 3}{2t^2 - t - 10} \cdot \dfrac{t}{t^2 + 3t + 2} \cdot \dfrac{2t^2 + 4t^3}{t^2 + 3t} = \dfrac{(t + 3)(t + 1)}{(2t - 5)(t + 2)} \cdot \dfrac{t}{(t + 2)(t + 1)} \cdot \dfrac{2t^2(1 + 2t)}{t(t + 3)}$

$$= \dfrac{2t^2(1 + 2t)}{(2t - 5)(t + 2)(t + 2)}, \quad t \neq -1, -3, 0$$

**38.** $\dfrac{16x^2 - 1}{4x^2 + 9x + 5} \cdot \dfrac{5x^2 - 9x - 18}{x^2 - 12x + 36} \cdot \dfrac{12 + 4x - x^2}{4x^2 - 13x + 3} = \dfrac{\cancel{(4x - 1)}(4x + 1)}{4x^2 + 9x + 5} \cdot \dfrac{(5x + 6)\cancel{(x - 3)}}{(x - 6)(x - 6)} \cdot \dfrac{(6 - x)(2 + x)}{\cancel{(4x - 1)}\cancel{(x - 3)}}$

$$= \dfrac{(4x + 1)(5x + 6) \cdot -1(x - 6)(2 + x)}{(4x^2 + 9x + 5)(x - 6)(x - 6)}$$

$$= \dfrac{(4x + 1)(5x + 6)(2 + x)(-1)}{(4x^2 + 9x + 5)(x - 6)}$$

$$= \dfrac{(4x + 1)(5x + 6)(2 + x)}{(4x^2 + 9x + 5)(6 - x)}, \quad x \neq \dfrac{1}{4}, 3$$

**40.** $\dfrac{xu - yu + xv - yv}{xu + yu - xv - yv} \cdot \dfrac{xu + yu + xv + yv}{xu - yu - xv + yv} = \dfrac{u(x - y) + v(x - y)}{u(x + y) - v(x + y)} \cdot \dfrac{u(x + y) + v(x + y)}{u(x - y) - v(x - y)}$

$$= \dfrac{(x - y)(u + v)}{(x + y)(u - v)} \cdot \dfrac{(x + y)(u + v)}{(x - y)(u - v)}$$

$$= \dfrac{(x - y)(u + v)(x + y)(u + v)}{(x + y)(u - v)(x - y)(u - v)}$$

$$= \dfrac{(u + v)(u + v)}{(u - v)(u - v)} \quad \text{or} \quad \dfrac{(u + v)^2}{(u - v)^2}, \quad x \neq y, x \neq -y$$

**42.** $-\dfrac{7}{15} \div \left( \dfrac{-14}{25} \right) = -\dfrac{7}{15} \cdot \dfrac{25}{-14}$

$$= \dfrac{-7 \cdot 5 \cdot 5}{5 \cdot 3 \cdot -7 \cdot 2} = \dfrac{5}{6}$$

**44.** $\dfrac{u}{10} \div u^2 = \dfrac{u}{10} \cdot \dfrac{1}{u^2} = \dfrac{1}{10u}$

**46.** $\dfrac{25x^2y}{60x^3y^2} \div \dfrac{5x^4y^3}{16x^2y} = \dfrac{25x^2y}{60x^3y^2} \cdot \dfrac{16x^2y}{5x^4y^3}$

$$= \dfrac{5 \cdot 5 \cdot x^2 \cdot y \cdot 4 \cdot 4 \cdot x^2 \cdot y}{3 \cdot 4 \cdot 5 \cdot x^2 \cdot x \cdot y \cdot y \cdot 5 \cdot x^2 \cdot x^2 \cdot y \cdot y^2}$$

$$= \dfrac{4}{3 \cdot x \cdot yx^2y^2}$$

$$= \dfrac{4}{3x^3y^3}$$

**48.** $\dfrac{x^2 + 9}{5(x + 2)} \div \dfrac{x + 3}{5(x^2 - 4)} = \dfrac{x^2 + 9}{5(x + 2)} \cdot \dfrac{5(x - 2)(x + 2)}{x + 3}$

$$= \dfrac{(x - 2)(x^2 + 9)}{(x + 3)}, \quad x \neq \pm 2$$

**50.** $\dfrac{x^2 - y^2}{2x^2 - 8x} \div \dfrac{(x - y)^2}{2xy} = \dfrac{x^2 - y^2}{2x^2 - 8x} \cdot \dfrac{2xy}{(x - y)^2}$

$$= \dfrac{(x - y)(x + y) \cdot 2 \cdot x \cdot y}{2 \cdot x(x - 4)(x - y)(x - y)}$$

$$= \dfrac{y(x + y)}{(x - 4)(x - y)}, \quad x \neq 0, y \neq 0$$

**52.** $\dfrac{\left(\dfrac{3u^2}{6v^3}\right)}{\left(\dfrac{u}{3v}\right)} = \dfrac{3u^2}{6v^3} \div \dfrac{u}{3v} = \dfrac{3u^2}{6v^3} \cdot \dfrac{3v}{u} = \dfrac{3u}{2v^2}, \quad u \neq 0$

**54.** $\dfrac{\left(\dfrac{5x}{x+7}\right)}{\left(\dfrac{10}{x^2+8x+7}\right)} = \dfrac{5x}{x+7} \cdot \dfrac{x^2+8x+7}{10}$

$\qquad = \dfrac{5x}{x+7} \cdot \dfrac{(x+7)(x+1)}{5 \cdot 2}$

$\qquad = \dfrac{x(x+1)}{2}, \quad x \neq -7, -1$

**56.** $\dfrac{9x^2 - 24x + 16}{x^2 + 10x + 25} \div \dfrac{6x^2 - 5x - 4}{2x^2 + 3x - 35} = \dfrac{9x^2 - 24x + 16}{x^2 + 10x + 25} \cdot \dfrac{2x^2 + 3x - 35}{6x^2 - 5x - 4}$

$\qquad = \dfrac{(3x-4)(3x-4)}{(x+5)(x+5)} \cdot \dfrac{(2x-7)(x+5)}{(3x-4)(2x+1)}$

$\qquad = \dfrac{(3x-4)(2x-7)}{(x+5)(2x+1)}, \quad x \neq \dfrac{4}{3}, \dfrac{7}{2}$

**58.** $\dfrac{t^3 + t^2 - 9t - 9}{t^2 - 5t + 6} \div \dfrac{t^2 + 6t + 9}{t - 2} = \dfrac{t^3 + t^2 - 9t - 9}{t^2 - 5t + 6} \cdot \dfrac{t - 2}{t^2 + 6t + 9}$

$\qquad = \dfrac{t^2(t+1) - 9(t+1)}{(t-3)(t-2)} \cdot \dfrac{t-2}{(t+3)(t+3)}$

$\qquad = \dfrac{(t^2-9)(t+1)}{(t-3)(t+3)(t+3)}$

$\qquad = \dfrac{(t-3)(t+3)(t+1)}{(t-3)(t+3)(t+3)}$

$\qquad = \dfrac{t+1}{t+3}, \quad t \neq 3, 2$

**60.** $\dfrac{\left(\dfrac{x^2 + 5x + 6}{4x^2 - 20x + 25}\right)}{\left(\dfrac{x^2 - 5x - 24}{4x^2 - 25}\right)} = \dfrac{x^2 + 5x + 6}{4x^2 - 20x + 25} \div \dfrac{x^2 - 5x - 24}{4x^2 - 25}$

$\qquad = \dfrac{(x+3)(x+2)}{(2x-5)(2x-5)} \cdot \dfrac{(2x-5)(2x+5)}{(x-8)(x+3)}$

$\qquad = \dfrac{(x+2)(2x+5)}{(2x-5)(x-8)}, \quad x \neq -3, -\dfrac{5}{2}$

**62.** $\left(\dfrac{x^2 + 6x + 9}{x^2} \cdot \dfrac{2x+1}{x^2-9}\right) \div \dfrac{4x^2 + 4x + 1}{x^2 - 3x} = \left(\dfrac{(x+3)^2}{x^2} \cdot \dfrac{2x+1}{(x-3)(x+3)}\right) \div \dfrac{(2x+1)^2}{x(x-3)}$

$\qquad = \dfrac{(x+3)(2x+1)}{x^2(x-3)} \cdot \dfrac{x(x-3)}{(2x+1)^2}$

$\qquad = \dfrac{x+3}{x(2x+1)}, \quad x \neq 3, -3$

**64.** $\dfrac{3u^2 - u - 4}{u^2} \div \dfrac{3u^2 + 12u + 4}{u^4 - 3u^3} = \dfrac{(3u - 4)(u + 1)}{u^2} \div \dfrac{3u^2 + 12u + 4}{u^3(u - 3)}$

$$= \dfrac{(3u - 4)(u + 1)}{u^2} \cdot \dfrac{u^3(u - 3)}{3u^2 + 12u + 4}$$

$$= \dfrac{u(u - 3)(3u - 4)(u + 1)}{3u^2 + 12u + 4}, \quad u \neq 3, 0$$

**66.** $\dfrac{t^2 - 100}{4t^2} \cdot \dfrac{t^3 - 5t^2 - 50t}{t^4 + 10t^3} \div \dfrac{(t - 10)^2}{5t} = \dfrac{(t - 10)(t + 10)}{4t^2} \cdot \dfrac{t(t^2 - 5t - 50)}{t^3(t + 10)} \cdot \dfrac{5t}{(t - 10)^2}$

$$= \dfrac{(t - 10)(t + 10)t(t - 10)(t + 5)5t}{4t^2 \cdot t^3(t + 10)(t - 10)^2}$$

$$= \dfrac{5(t + 5)}{4t^3}, \quad t \neq -10, 10$$

**68.** $\dfrac{x^{n+1} - 8x}{x^{2n} + 2x^n + 1} \cdot \dfrac{x^{2n} - 4x^n - 5}{x} \div x^n = \dfrac{x(x^n - 8)}{(x^n + 1)(x^n + 1)} \cdot \dfrac{(x^n - 5)(x^n + 1)}{x} \cdot \dfrac{1}{x^n}$

$$= \dfrac{(x^n - 8)(x^n - 5)}{x^n(x^n + 1)}$$

**70.** *Keystrokes:*

$y_1$: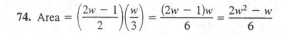

    ×〔〔 X,T,θ ＋ 5 〕÷ 2 〕 ENTER

$y_2$: 〔 X,T,θ － 5 〕÷ 2 GRAPH

$$\dfrac{x^2 - 10x + 25}{x^2 - 25} \cdot \dfrac{x + 5}{2} = \dfrac{(x - 5)^2}{(x - 5)(x + 5)} \cdot \dfrac{x + 5}{2} = \dfrac{x - 5}{2}$$

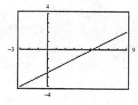

**72.** *Keystrokes:*

$y_1$: 〔 X,T,θ $x^2$ ＋ 6 X,T,θ ＋ 9 〕 ×

    〔3 ÷ 〔2 X,T,θ 〕× 〔 X,T,θ ＋ 3 〕〕〕 ENTER

$y_2$: 〔3〔 X,T,θ ＋ 3 〕〕÷〔2 X,T,θ 〕 GRAPH

$$(x^2 + 6x + 9) \cdot \dfrac{3}{2x(x + 3)} = (x + 3)^2 \cdot \dfrac{3}{2x(x + 3)} = \dfrac{3(x + 3)}{2x}, \quad x \neq -3$$

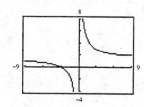

**74.** Area $= \left(\dfrac{2w - 1}{2}\right)\left(\dfrac{w}{3}\right) = \dfrac{(2w - 1)w}{6} = \dfrac{2w^2 - w}{6}$

**76.** $\dfrac{\text{Unshaded area}}{\text{Total area}} = \dfrac{\frac{1}{2}(x)\left(\frac{x}{2}\right)}{x(2x+1)}$

$$= \dfrac{\frac{x^2}{4}}{x(2x+1)}$$

$$= \dfrac{x^2}{4} \cdot \dfrac{1}{x(2x+1)}$$

$$= \dfrac{x \cdot x}{4 \cdot x \cdot (2x+1)}$$

$$= \dfrac{x}{4(2x+1)}$$

**78.** $\dfrac{\text{Unshaded area}}{\text{Total area}} = \dfrac{\pi\left(\frac{x}{2}\right)^2}{x(2x+1)}$

$$= \dfrac{\pi x^2}{4} \cdot \dfrac{1}{x(2x+1)}$$

$$= \dfrac{\pi x}{4(2x+1)}$$

**80.** (a) $\dfrac{15 \text{ gallons}}{1 \text{ minute}} = \dfrac{15 \text{ gallons}}{60 \text{ seconds}} = \dfrac{1 \text{ gallon}}{4 \text{ seconds}}, \quad t = 4 \text{ seconds or } \dfrac{1}{15} \text{ minute}$

(b) $\dfrac{4 \text{ seconds}}{1 \text{ gallon}} \cdot x \text{ gallons} = 4x \text{ seconds or } \dfrac{1}{15}x \text{ minutes}$

(c) $\dfrac{4 \text{ seconds}}{1 \text{ gallon}} \cdot 130 \text{ gallons} = 520 \text{ seconds or } \dfrac{520}{60} = \dfrac{26}{3} \text{ minutes}$

**82.** To divide rational expressions, invert the divisor and multiply.

**84.** A complex fraction is a fraction involving the division of two rational expressions. Example:

$$\dfrac{\frac{x^2}{6}}{\frac{2x}{9}} = \dfrac{x^2}{6} \cdot \dfrac{9}{2x} = \dfrac{3x}{4}$$

# Section 4.4    Adding and Subtracting Rational Expressions

**2.** $\dfrac{7}{12} - \dfrac{5}{12} = \dfrac{7-5}{12} = \dfrac{2}{12} = \dfrac{1}{6}$

**4.** $\dfrac{7y}{12} + \dfrac{9y}{12} = \dfrac{16y}{12} = \dfrac{4y}{3}$

**6.** $\dfrac{6x}{13} - \dfrac{7x}{13} = -\dfrac{x}{13}$

**8.** $\dfrac{4-y}{4} + \dfrac{3y}{4} = \dfrac{4-y+3y}{4}$

$$= \dfrac{4+2y}{4} = \dfrac{2(2+y)}{4} = \dfrac{2+y}{2}$$

**10.** $\dfrac{10x^2+1}{3} - \dfrac{10x^2}{3} = \dfrac{10x^2+1-10x^2}{3} = \dfrac{1}{3}$

**12.** $\dfrac{16+z}{5z} - \dfrac{11-z}{5z} = \dfrac{16+z-(11-z)}{5z}$

$$= \dfrac{16+z-11+z}{5z} = \dfrac{2z+5}{5z}$$

**14.** $\dfrac{-16u}{9} - \dfrac{27-16u}{9} + \dfrac{2}{9} = \dfrac{-16u-(27-16u)+2}{9}$

$$= \dfrac{-16u-27+16u+2}{9}$$

$$= -\dfrac{25}{9}$$

**16.** $\dfrac{5x-1}{x+4} + \dfrac{5-4x}{x+4} = \dfrac{5x-1+5-4x}{x+4}$

$\qquad\qquad = \dfrac{x+4}{x+4} = 1, \quad x \neq -4$

**18.** $\dfrac{7s-5}{2s+5} + \dfrac{3(s+10)}{2s+5} = \dfrac{7s-5+3(s+10)}{2s+5}$

$\qquad\qquad = \dfrac{7s-5+3s+30}{2s+5}$

$\qquad\qquad = \dfrac{10s+25}{2s+5}$

$\qquad\qquad = \dfrac{5(2s+5)}{2s+5}$

$\qquad\qquad = 5, \quad s \neq -\dfrac{5}{2}$

**20.** $14t^2 = 2 \cdot 7 \cdot t \cdot t$

$\quad 42t^5 = 2 \cdot 3 \cdot 7 \cdot t \cdot t \cdot t \cdot t \cdot t$

$\quad$ LCM $= 42t^5$

**22.** $44m^2 = 2 \cdot 2 \cdot 11 \cdot m \cdot m$

$\quad 10m = 2 \cdot 5 \cdot m$

$\quad$ LCM $= 220m^2$

**24.** $6x^2 = 2 \cdot 3 \cdot x \cdot x$

$\quad 15x(x-1) = 5 \cdot 3 \cdot x \cdot (x-1)$

$\quad$ LCM $= 2 \cdot 3 \cdot 5 \cdot x \cdot x \cdot (x-1) = 30x^2(x-1)$

**26.** $18y^3 = 2 \cdot 3 \cdot 3 \cdot y \cdot y \cdot y$

$\quad 27y(y-3)^2 = 3 \cdot 3 \cdot 3 \cdot y \cdot (y-3)(y-3)$

$\quad$ LCM $= 54y^3(y-3)^2$

**28.** $2y^2 + y - 1 = (2y-1)(y+1)$

$\quad 4y^2 - 2y = 2y(2y-1)$

$\quad$ LCM $= 2y(2y-1)(y+1)$

**30.** $t^3 + 3t^2 + 9t = t(t^2 + 3t + 9)$

$\quad 2t^2(t^2 - 9) = 2 \cdot t \cdot t(t-3)(t+3)$

$\quad$ LCM $= 2t^2(t-3)(t+3)(t^2 + 3t + 9)$

**32.** $\dfrac{3y(x-3)^2}{(x-3)\dfrac{(x-3)^2}{7}} = \dfrac{21y}{x-3}, \quad x \neq 3$

**34.** $\dfrac{(3t+5)(5t(3t-5))}{10t^2(3t-5)} = \dfrac{3t+5}{2t}, \quad t \neq \dfrac{5}{3}$

**36.** $\dfrac{4x^2(-x)}{x^2 - 10x} = \dfrac{4x^2}{10-x}, \quad x \neq 0$

**38.** $\dfrac{8s}{(s+2)^2} = \dfrac{8s \cdot s(s-1)}{s(s+2)^2(s-1)} = \dfrac{8s^2(s-1)}{s(s+2)^2(s-1)}$

$\quad \dfrac{3}{s^3 + s^2 - 2s} = \dfrac{3}{s(s^2 + s - 2)}$

$\qquad\qquad = \dfrac{3}{s(s+2)(s-1)} = \dfrac{3(s+2)}{s(s+2)^2(s-1)}$

$\quad$ LCD $= s(s+2)^2(s-1)$

**40.** $\dfrac{5t}{2t(t-3)^2} = \dfrac{5t}{2t(t-3)^2}$

$\quad \dfrac{4}{t(t-3)} = \dfrac{4(2)(t-3)}{t(t-3) \cdot 2(t-3)} = \dfrac{8(t-3)}{2t(t-3)^2}$

$\quad$ LCD $= 2t(t-3)^2$

**42.** $\dfrac{4x}{(x+5)^2} = \dfrac{4x(x-5)}{(x+5)^2(x-5)} = \dfrac{4x^2 - 20x}{(x+5)^2(x-5)}$

$\quad \dfrac{x-2}{x^2 - 25} = \dfrac{x-2}{(x-5)(x+5)}$

$\qquad\qquad = \dfrac{(x-2)(x+5)}{(x-5)(x+5)(x+5)} = \dfrac{x^2 + 3x - 10}{(x+5)^2(x-5)}$

$\quad$ LCD $= (x+5)^2(x-5)$

**44.** $\dfrac{3y}{y^2 - y - 12} = \dfrac{3y}{(y-4)(y+3)}$

$\qquad\qquad = \dfrac{3y \cdot y}{(y-4)(y+3) \cdot y} = \dfrac{3y^2}{y(y+3)(y-4)}$

$\quad \dfrac{y-4}{y^2 + 3y} = \dfrac{y-4}{y(y+3)} = \dfrac{(y-4)^2}{y(y+3)(y-4)}$

$\quad$ LCD $= y(y+3)(y-4)$

**46.** $\dfrac{10}{b} + \dfrac{1}{10b} = \dfrac{10(10)}{b(10)} + \dfrac{1}{10b}$

$\qquad\qquad = \dfrac{100}{10b} + \dfrac{1}{10b}$

$\qquad\qquad = \dfrac{100 + 1}{10b}$

$\qquad\qquad = \dfrac{101}{10b}$

**48.** $\dfrac{1}{6u^2} - \dfrac{2}{9u} = \dfrac{1(3)}{6u^2(3)} - \dfrac{2(2u)}{9u(2u)}$

$\qquad\qquad = \dfrac{3}{18u^2} - \dfrac{4u}{18u^2}$

$\qquad\qquad = \dfrac{3 - 4u}{18u^2}$

**50.** $\dfrac{15}{2 - t} - \dfrac{7}{t - 2} = \dfrac{15(-1)}{(2 - t)(-1)} - \dfrac{7}{t - 2}$

$\qquad\qquad = \dfrac{-15}{t - 2} - \dfrac{7}{t - 2}$

$\qquad\qquad = \dfrac{-15 - 7}{t - 2}$

$\qquad\qquad = \dfrac{-22}{t - 2}$

$\qquad\qquad = \dfrac{22}{2 - t}$

**52.** $\dfrac{1}{y - 6} + \dfrac{y}{6 - y} = \dfrac{1}{y - 6} + \dfrac{y(-1)}{(6 - y)(-1)}$

$\qquad\qquad = \dfrac{1}{y - 6} - \dfrac{y}{y - 6}$

$\qquad\qquad = \dfrac{1 - y}{y - 6}$

**54.** $\dfrac{100}{x - 10} - 8 = \dfrac{100}{x - 10} - \dfrac{8(x - 10)}{x - 10}$

$\qquad\qquad = \dfrac{100 - 8(x - 10)}{x - 10}$

$\qquad\qquad = \dfrac{100 - 8x + 80}{x - 10}$

$\qquad\qquad = \dfrac{-8x + 180}{x - 10}$

$\qquad\qquad = \dfrac{-4(2x - 45)}{-1(10 - x)} = \dfrac{4(2x - 45)}{10 - x} = \dfrac{4(45 - 2x)}{x - 10}$

**56.** $\dfrac{y}{5y - 3} - \dfrac{3}{3 - 5y} = \dfrac{y}{5y - 3} - \dfrac{3(-1)}{(3 - 5y)(-1)}$

$\qquad\qquad = \dfrac{y}{5y - 3} + \dfrac{3}{5y - 3}$

$\qquad\qquad = \dfrac{y + 3}{5y - 3}$

**58.** $\dfrac{3}{t(t + 1)} + \dfrac{4}{t} = \dfrac{3}{t(t + 1)} + \dfrac{4t(t + 1)}{t(t + 1)}$

$\qquad\qquad = \dfrac{3 + 4(t + 1)}{t(t + 1)}$

$\qquad\qquad = \dfrac{3 + 4t + 4}{t(t + 1)}$

$\qquad\qquad = \dfrac{4t + 7}{t(t + 1)}$

**60.** $\dfrac{1}{x + 4} - \dfrac{1}{x + 2} = \dfrac{1(x + 2)}{x + 4(x + 2)} - \dfrac{1(x + 4)}{x + 2(x + 4)}$

$\qquad\qquad = \dfrac{x + 2 - x - 4}{(x + 4)(x + 2)}$

$\qquad\qquad = \dfrac{-2}{(x + 4)(x + 2)}$

**62.** $\dfrac{5}{x-4} - \dfrac{3}{x} = \dfrac{5(x)}{(x-4)(x)} - \dfrac{3(x-4)}{x(x-4)}$

$\qquad = \dfrac{5x}{x(x-4)} - \dfrac{3(x-4)}{x(x-4)}$

$\qquad = \dfrac{5x - 3(x-4)}{x(x-4)}$

$\qquad = \dfrac{5x - 3x + 12}{x(x-4)}$

$\qquad = \dfrac{2x + 12}{x(x-4)}$

$\qquad = \dfrac{2(x+6)}{x(x-4)}$

**64.** $\dfrac{7}{2x-3} + \dfrac{3}{2x+3} = \dfrac{7(2x+3)}{(2x-3)(2x+3)} + \dfrac{3(2x-3)}{(2x+3)(2x-3)}$

$\qquad = \dfrac{7(2x+3) + 3(2x-3)}{(2x-3)(2x+3)}$

$\qquad = \dfrac{14x + 21 + 6x - 9}{(2x-3)(2x+3)}$

$\qquad = \dfrac{20x + 12}{(2x-3)(2x+3)}$

$\qquad = \dfrac{4(5x+3)}{(2x-3)(2x+3)}$

**66.** $\dfrac{2}{y^2+2} + \dfrac{1}{2y^2} = \dfrac{2(2y^2)}{(y^2+2)(2y^2)} + \dfrac{1(y^2+2)}{2y^2(y^2+2)}$

$\qquad = \dfrac{4y^2 + y^2 + 2}{(y^2+2)(2y^2)}$

$\qquad = \dfrac{5y^2 + 2}{(2y^2)(y^2+2)}$

**68.** $\dfrac{x}{x^2-x-30} - \dfrac{1}{x+5} = \dfrac{x}{(x+5)(x-6)} - \dfrac{(x-6)}{(x+5)(x-6)}$

$\qquad = \dfrac{x-(x-6)}{(x+5)(x-6)}$

$\qquad = \dfrac{x-x+6}{(x+5)(x-6)}$

$\qquad = \dfrac{6}{(x+5)(x-6)}$

**70.** $\dfrac{3}{x-2} - \dfrac{1}{(x-2)^2} = \dfrac{3(x-2)}{(x-2)(x-2)} - \dfrac{1}{(x-2)^2}$

$\qquad = \dfrac{3(x-2)}{(x-2)^2} - \dfrac{1}{(x-2)^2}$

$\qquad = \dfrac{3(x-2)-1}{(x-2)^2}$

$\qquad = \dfrac{3x-6-1}{(x-2)^2}$

$\qquad = \dfrac{3x-7}{(x-2)^2}$

**72.** $\dfrac{5}{x+y} + \dfrac{5}{x^2-y^2} = \dfrac{5(x-y)}{(x+y)(x-y)} + \dfrac{5}{(x+y)(x-y)}$

$\qquad = \dfrac{5(x-y)+5}{(x+y)(x-y)}$

$\qquad = \dfrac{5x-5y+5}{(x+y)(x-y)}$

$\qquad = \dfrac{5(x-y+1)}{(x+y)(x-y)}$

**74.** $\dfrac{5}{2} - \dfrac{1}{2x} - \dfrac{3}{(x+1)} = \dfrac{5(x+1)(x)}{2(x+1)(x)} - \dfrac{1(x+1)}{2x(x+1)} - \dfrac{3}{2(x+1)(x)}$

$\qquad = \dfrac{5x(x+1)}{2x(x+1)} - \dfrac{(x+1)}{2x(x+1)} - \dfrac{6}{2x(x+1)}$

$\qquad = \dfrac{5x(x+1) - (x+1) - 6x}{2x(x+1)}$

$\qquad = \dfrac{5x^2 + 5x - x - 1 - 6x}{2x(x+1)}$

$\qquad = \dfrac{5x^2 - 2x - 1}{2x(x+1)}$

**76.** $\dfrac{1}{x-y} - \dfrac{3}{x+y} + \dfrac{3x-y}{x^2-y^2} = \dfrac{x+y}{(x-y)(x+y)} - \dfrac{3(x-y)}{(x+y)(x-y)} + \dfrac{3x-y}{(x+y)(x-y)}$

$$= \frac{x+y-3(x-y)+3x-y}{(x-y)(x+y)}$$

$$= \frac{x+y-3x+3y+3x-y}{(x-y)(x+y)}$$

$$= \frac{x+3y}{(x-y)(x+y)}$$

**78.** $\dfrac{x}{x^2+15x+50} + \dfrac{7}{x+10} - \dfrac{x-1}{x+5} = \dfrac{x}{(x+10)(x+5)} + \dfrac{7(x+5)}{(x+10)(x+5)} - \dfrac{(x-1)(x+10)}{(x+10)(x+5)}$

$$= \frac{x+7(x+5)-(x+10)(x-1)}{(x+10)(x+5)}$$

$$= \frac{x+7x+35-(x^2+9x-10)}{(x+10)(x+5)}$$

$$= \frac{x+7x+35-x^2-9x+10}{(x+10)(x+5)}$$

$$= \frac{-x^2-x+45}{(x+10)(x+5)}$$

$$= -\frac{x^2+x-45}{(x+10)(x+5)}$$

**80.** *Keystrokes:*

$y_1$: 3 $\boxed{-}$ $\boxed{(}$ 1 $\boxed{\div}$ $\boxed{(}$ $\boxed{X,T,\theta}$ $\boxed{-}$ 1 $\boxed{)}$ $\boxed{)}$ $\boxed{\text{ENTER}}$

$y_2$: $\boxed{(}$ 3 $\boxed{X,T,\theta}$ $\boxed{-}$ 4 $\boxed{)}$ $\boxed{\div}$ $\boxed{(}$ $\boxed{X,T,\theta}$ $\boxed{-}$ 1 $\boxed{)}$ $\boxed{\text{GRAPH}}$

$3 - \dfrac{1}{x-1} = \dfrac{3(x-1)}{x-1} - \dfrac{1}{x-1}$

$$= \frac{3(x-1)-1}{x-1}$$

$$= \frac{3x-3-1}{x-1}$$

$$= \frac{3x-4}{x-1}$$

$y_1 = y_2$

**82.** $\dfrac{\frac{2}{3}}{\left(4-\frac{1}{x}\right)} = \dfrac{\frac{2}{3}}{\left(4-\frac{1}{x}\right)} \cdot \dfrac{3x}{3x}$

$$= \frac{2x}{12x-3} = \frac{2x}{3(4x-1)}, \quad x \neq 0$$

**84.** $\dfrac{\left(\frac{1}{t}-1\right)}{\left(\frac{1}{t}+1\right)} = \dfrac{\left(\frac{1}{t}-1\right)}{\left(\frac{1}{t}+1\right)} \cdot \dfrac{t}{t} = \dfrac{1-t}{1+t}, \quad t \neq 0$

**86.** $\dfrac{\left(\dfrac{36}{y}-y\right)}{6+y} = \dfrac{\left(\dfrac{36}{y}-y\right)}{6+y}\cdot\dfrac{y}{y}$

$\qquad = \dfrac{36-y^2}{y(6+y)}$

$\qquad = \dfrac{(6+y)(6-y)}{y(6+y)}$

$\qquad = \dfrac{6-y}{y},\quad y\neq -6$

**88.** $\dfrac{\left(x+\dfrac{2}{x-3}\right)}{\left(x+\dfrac{6}{x-3}\right)} = \dfrac{\left(x+\dfrac{2}{x-3}\right)}{\left(x+\dfrac{6}{x-3}\right)}\cdot\dfrac{x-3}{x-3}$

$\qquad = \dfrac{x(x-3)+2}{x(x-3)+6}$

$\qquad = \dfrac{x^2-3x+2}{x^2-3x+6}$

$\qquad = \dfrac{(x-2)(x-1)}{x^2-3x+6},\quad x\neq 3$

**90.** $\dfrac{\left(16-\dfrac{1}{x^2}\right)}{\left(\dfrac{1}{4x^2}-4\right)} = \dfrac{\left(16-\dfrac{1}{x^2}\right)}{\left(\dfrac{1}{4x^2}-4\right)}\cdot\dfrac{4x^2}{4x^2}$

$\qquad = \dfrac{64x^2-4}{1-16x^2}$

$\qquad = \dfrac{4(16x^2-1)}{1-16x^2}$

$\qquad = \dfrac{4(16x^2-1)}{-1(16x^2-1)}$

$\qquad = \dfrac{4}{-1}$

$\qquad = -4,\quad x\neq 0,\ -\dfrac{1}{4},\dfrac{1}{4}$

**92.** $\dfrac{\left(x-\dfrac{2y^2}{x-y}\right)}{x-2y} = \dfrac{\left(x-\dfrac{2y^2}{x-y}\right)}{x-2y}\cdot\dfrac{(x-y)}{(x-y)}$

$\qquad = \dfrac{x(x-y)-2y^2}{(x-2y)(x-y)}$

$\qquad = \dfrac{x^2-xy-2y^2}{(x-2y)(x-y)}$

$\qquad = \dfrac{(x-2y)(x+y)}{(x-2y)(x-y)}$

$\qquad = \dfrac{x+y}{x-y},\quad x\neq 2y$

**94.** $\dfrac{\left(\dfrac{x+1}{x+2}-\dfrac{1}{x}\right)}{\left(\dfrac{2}{x+2}\right)} = \dfrac{\left(\dfrac{x+1}{x+2}-\dfrac{1}{x}\right)}{\left(\dfrac{2}{x+2}\right)}\cdot\dfrac{x(x+2)}{x(x+2)}$

$\qquad = \dfrac{(x+1)x-1(x+2)}{2(x)}$

$\qquad = \dfrac{x^2+x-x-2}{2x}$

$\qquad = \dfrac{x^2-2}{2x},\quad x\neq -2$

**96.** $\dfrac{\left(\dfrac{1}{2x}-\dfrac{6}{x+5}\right)}{\left(\dfrac{x}{x-5}+\dfrac{1}{x}\right)} = \dfrac{\left(\dfrac{1}{2x}-\dfrac{6}{x+5}\right)}{\left(\dfrac{x}{x-5}+\dfrac{1}{x}\right)}\cdot\dfrac{2x(x+5)(x-5)}{2x(x+5)(x-5)}$

$\qquad = \dfrac{(x+5)(x-5)-6(2x)(x-5)}{x(2x)(x+5)+1(2)(x+5)(x-5)}$

$\qquad = \dfrac{x^2-25-12x(x-5)}{2x^2(x+5)+2(x+5)(x-5)}$

$\qquad = \dfrac{x^2-25-12x^2+60x}{(x+5)(2x^2+2(x-5))}$

$\qquad = \dfrac{-11x^2+60x-25}{(x+5)(2x^2+2x-10)}$

$\qquad = -\dfrac{(11x^2-60x+25)}{(x+5)(2x^2+2x-10)}$

$\qquad = -\dfrac{(11x-5)(x-5)}{(x+5)(2x^2+2x-10)}$

$\qquad = \dfrac{(5-x)(11x-5)}{2(x+5)(x^2+x-5)},\quad x\neq 0,5$

**98.** $\dfrac{f(2+h)-f(2)}{h} = \dfrac{\dfrac{2+h}{2+h-1}-\dfrac{2}{2-1}}{h}$

$= \dfrac{\dfrac{2+h}{1+h}-\dfrac{2}{1}}{h}$

$= \dfrac{\dfrac{2+h}{1+h}-2}{h} \cdot \dfrac{1+h}{1+h}$

$= \dfrac{2+h-2(1+h)}{h(1+h)}$

$= \dfrac{2+h-2-2h}{h(1+h)}$

$= \dfrac{-h}{h(1+h)}$

$= -\dfrac{1}{1+h}$

**100.**

| $x$ | $-3$ | $-2$ | $-1$ | $0$ | $1$ | $2$ | $3$ |
|---|---|---|---|---|---|---|---|
| $\dfrac{\left(1+\dfrac{4}{x}+\dfrac{4}{x^2}\right)}{\left(1-\dfrac{4}{x^2}\right)}$ | $\dfrac{1}{5}$ | Undefined | $-\dfrac{1}{3}$ | Undefined | $-3$ | Undefined | $5$ |
| $\dfrac{x+2}{x-2}$ | $\dfrac{1}{5}$ | $0$ | $-\dfrac{1}{3}$ | $-1$ | $-3$ | Undefined | $5$ |

*Keystrokes:*

$y_1$: `(` 1 `+` 4 `÷` `X,T,θ` `+` 4 `÷` `X,T,θ` `x²` `)` `÷` `(` 1 `-` 4 `÷` `X,T,θ` `x²` `)` `ENTER`

$y_2$: `X,T,θ` `+` 2 `)` `÷` `(` `X,T,θ` `-` 2 `)` `GRAPH`

$0, -2,$ and $2$ are not in the domain of $\dfrac{\left(1+\dfrac{4}{x}+\dfrac{4}{x^2}\right)}{\left(1-\dfrac{4}{x^2}\right)}$, but $-2$ and $0$ are in the domain of $\dfrac{x+2}{x-2}$.

$y_1 = y_2$ except for $x = -2, 0$.

**102.** $\dfrac{t}{3}+\dfrac{t}{5} = \dfrac{t(5)}{3(5)}+\dfrac{t(3)}{5(3)}$

$= \dfrac{5t}{15}+\dfrac{3t}{15} = \dfrac{5t+3t}{15} = \dfrac{8t}{15}$

**104.** $\dfrac{x+\dfrac{x}{2}+\dfrac{x}{3}}{3} = \dfrac{x+\dfrac{x}{2}+\dfrac{x}{3}}{3}\cdot\dfrac{6}{6}$

$= \dfrac{6x+3x+2x}{18} = \dfrac{11x}{18}$

**106.** (a) $r = \dfrac{\left[\dfrac{24(48(300)-10,000)}{48}\right]}{\left(10,000+\dfrac{300(48)}{12}\right)} = 19.6\%$

(b) $r = \dfrac{\dfrac{24(NM-P)}{N}}{P+\dfrac{MN}{12}}\cdot\dfrac{12N}{12N} = \dfrac{288(NM-P)}{12NP+MN^2} = \dfrac{288(NM-P)}{N(MN+12P)}$

$r = \dfrac{288[(48)(300)-10,000]}{12(48)(10,000)+(300)(48)^2}$

$r = 19.6\%$

**108.** (a) *Verbal Model:* $\boxed{\text{Total daily circulation}} = \boxed{\text{Morning newspapers}} + \boxed{\text{Evening newspapers}}$

(b)

*Labels:*  Total daily circulation $= T$

Morning newspapers $= M$

Evening newspapers $= E$

*Equation:* $T = M + E$

$$T = (41.1 + 0.61t) + \left(\frac{20.675 - 1.675t}{1 - 0.023t}\right)$$

$$= \frac{(41.1 + 0.61t)(1 - 0.023t) + (20.675 - 1.675t)}{1 - 0.023t}$$

$$= \frac{41.1 - 0.9453t + 0.61t - 0.01403t^2 + 20.675 - 1.675t}{1 - 0.023t}$$

$$= \frac{-0.01403t^2 - 2.0103t + 61.775}{1 - 0.023t}$$

$$= \frac{-1(0.01403t^2 + 2.0103t - 61.775)}{-1(0.023t - 1)} \cdot \frac{1000}{1000}$$

$$= \frac{14.03t^2 + 2010.3t - 61775}{23t - 1000}$$

(c) Total circulation is decreasing.

(d) $T(1) = \dfrac{14.03(1)^2 + 2010.3(1) - 61775}{23(1) - 1000} \approx 61.16$ million

**110.** Add or subtract the numerator and place the result over the common denominator.

**112.** Yes. $\dfrac{3}{2}(x+2) + \dfrac{x}{x+2}$

**114.** The simplification is not correct.

$$\frac{2}{x} - \frac{3}{x+1} + \frac{x+1}{x^2} = \frac{2x(x+1) - 3x^2 + (x+1)^2}{x^2(x+1)}$$

$$= \frac{2x^2 + 2x - 3x^2 + x^2 + 2x + 1}{x^2(x+1)}$$

$$= \frac{4x+1}{x^2(x+1)}$$

Two errors occurred, both in step 2. The $2x$ was not distributed properly, $2x(x+1) \neq 2x^2 + x$. The product $(x+1)^2$ was not multiplied correctly, $(x+1)^2 \neq x^2 + 1$.

# Section 4.5   Dividing Polynomials

**2.** $\dfrac{9x+12}{3} = \dfrac{9x}{3} + \dfrac{12}{3}$

$= 3x + 4$

**4.** $\dfrac{4u^2 + 8u - 24}{16} = \dfrac{4u^2}{16} + \dfrac{8u}{16} - \dfrac{24}{16}$

$= \dfrac{u^2}{4} + \dfrac{u}{2} - \dfrac{3}{2}$

**6.** $(6a^2 + 7a) \div a = \dfrac{6a^2 + 7a}{a}$

$\qquad\qquad\qquad = \dfrac{6a^2}{a} + \dfrac{7a}{a}$

$\qquad\qquad\qquad = 6a + 7, \quad a \neq 0$

**8.** $\dfrac{l^2 - 8l + 4}{-l} = \dfrac{l^2}{-l} - \dfrac{8l}{-l} + \dfrac{4}{-l}$

$\qquad\qquad\quad = -l + 8 - \dfrac{4}{l}$

**10.** $\dfrac{18c^4 - 24c^2}{-6c} = \dfrac{18c^4}{-6c} - \dfrac{24c^2}{-6c}$

$\qquad\qquad\qquad = -3c^3 + 4c, \quad c \neq 0$

**12.** $\dfrac{6x^4 + 8x^3 - 18x^2}{3x^2} = \dfrac{6x^4}{3x^2} + \dfrac{8x^3}{3x^2} - \dfrac{18x^3}{3x^2}$

$\qquad\qquad\qquad = 2x^2 + \dfrac{8x}{3} - 6, \quad x \neq 0$

**14.** $(-14s^4t^2 + 7s^2t^2 - 18t) \div 2s^2t = \dfrac{-14s^4t^2 + 7s^2t^2 - 18t}{2s^2t}$

$\qquad\qquad\qquad\qquad\qquad\quad = \dfrac{-14s^4t^2}{2s^2t} + \dfrac{7s^2t^2}{2s^2t} - \dfrac{18t}{2s^2t}$

$\qquad\qquad\qquad\qquad\qquad\quad = -7s^2t + \dfrac{7t}{2} - \dfrac{9}{s^2}, \quad t \neq 0$

**16.**
$$
\begin{array}{r}
t - 6\;)\overline{\;t^2 - 18t + 72} \\[2pt]
\underline{t^2 - \phantom{0}6t} \phantom{+ 72} \\[2pt]
-12t + 72 \\[2pt]
\underline{-12t + 72}
\end{array}
$$

**18.**
$$
\begin{array}{r}
y - \phantom{0}8, \quad y \neq -2 \\[2pt]
y + 2\;)\overline{\;y^2 - \phantom{0}6y - 16} \\[2pt]
\underline{y^2 + 2y} \phantom{- 16} \\[2pt]
-8y - 16 \\[2pt]
\underline{-8y - 16}
\end{array}
$$

**20.**
$$
\begin{array}{r}
x + 13 + \dfrac{30}{x - 3} \\[2pt]
x - 3\;)\overline{\;x^2 + 10x - \phantom{0}9} \\[2pt]
\underline{x^2 - \phantom{0}3x} \phantom{- 9} \\[2pt]
13x - \phantom{0}9 \\[2pt]
\underline{13x - 39} \\[2pt]
30
\end{array}
$$

**22.**
$$
\begin{array}{r}
-x + 5, \quad x \neq -1 \\[2pt]
x + 1\;)\overline{\;-x^2 + 4x + 5} \\[2pt]
\underline{-x^2 - \phantom{0}x} \phantom{+ 5} \\[2pt]
5x + 5 \\[2pt]
\underline{5x + 5}
\end{array}
$$

**24.**
$$
\begin{array}{r}
2x - \phantom{0}3 + \dfrac{14}{x + 4} \\[2pt]
x + 4\;)\overline{\;2x^2 + 5x + \phantom{0}2} \\[2pt]
\underline{2x^2 + 8x} \phantom{+ 2} \\[2pt]
-3x + \phantom{0}2 \\[2pt]
\underline{-3x - 12} \\[2pt]
14
\end{array}
$$

**26.**
$$
\begin{array}{r}
2x + 1 + \dfrac{4}{4x - 1} \\[2pt]
4x - 1\;)\overline{\;8x^2 + 2x + 3} \\[2pt]
\underline{8x^2 - 2x} \phantom{+ 3} \\[2pt]
4x + 3 \\[2pt]
\underline{4x - 1} \\[2pt]
4
\end{array}
$$

**28.**
$$
\begin{array}{r}
-4u + \phantom{0}3, \quad u \neq -\tfrac{5}{2} \\[2pt]
2u + 5\;)\overline{\;-8u^2 - 14u + 15} \\[2pt]
\underline{-8u^2 - 20u} \phantom{+ 15} \\[2pt]
6u + 15 \\[2pt]
\underline{6u + 15}
\end{array}
$$

**30.**
$$
\begin{array}{r}
5t + \phantom{0}4, \quad t \neq \tfrac{3}{2} \\[2pt]
2t - 3\;)\overline{\;10t^2 - \phantom{0}7t - 12} \\[2pt]
\underline{10t^2 - 15t} \phantom{- 12} \\[2pt]
8t - 12 \\[2pt]
\underline{8t - 12}
\end{array}
$$

**32.**

$$
\begin{array}{r}
x^2 \phantom{xxxx} + 7, \quad x \neq -4 \\
x + 4 \overline{\smash{)}\, x^3 + 4x^2 + 7x + 28} \\
\underline{x^3 + 4x^2 \phantom{xxxxxxxxx}} \\
7x + 28 \\
\underline{7x + 28}
\end{array}
$$

**34.**

$$
\begin{array}{r}
5x^2 - 2x + 14 + \dfrac{6}{x+1} \\
x + 1 \overline{\smash{)}\, 5x^3 + 3x^2 + 12x + 20} \\
\underline{5x^3 + 5x^2 \phantom{xxxxxxxxxx}} \\
-2x^2 + 12x \phantom{xxxx} \\
\underline{-2x^2 - 2x \phantom{xxxx}} \\
14x + 20 \\
\underline{14x + 14} \\
6
\end{array}
$$

**36.**

$$
\begin{array}{r}
6 + \dfrac{-23}{2x+3} \\
2x + 3 \overline{\smash{)}\, 12x - 5} \\
\underline{12x + 18} \\
-23
\end{array}
$$

**38.**

$$
\begin{array}{r}
y - 2 + \dfrac{12}{y+2} \\
y + 2 \overline{\smash{)}\, y^2 + 0y + 8} \\
\underline{y^2 + 2y \phantom{xxxx}} \\
-2y + 8 \\
\underline{-2y - 4} \\
12
\end{array}
$$

**40.**

$$
\begin{array}{r}
\dfrac{8}{3}y - \dfrac{46}{9} + \dfrac{230}{9(3y+5)} \\
3y + 5 \overline{\smash{)}\, 8y^2 - 2y + 0} \\
\underline{8y^2 + \dfrac{40}{3}y \phantom{xxxxx}} \\
-\dfrac{46}{3}y + 0 \\
\underline{-\dfrac{46}{3}y - \dfrac{230}{9}} \\
\dfrac{230}{9}
\end{array}
$$

**42.**

$$
\begin{array}{r}
9y + 5, \quad y \neq \tfrac{5}{9} \\
9y - 5 \overline{\smash{)}\, 81y^2 + 0y - 25} \\
\underline{81y^2 - 45y \phantom{xxxx}} \\
45y - 25 \\
\underline{45y - 25}
\end{array}
$$

**44.**

$$
\begin{array}{r}
x^2 + 3x + 9, \quad x \neq 3 \\
x - 3 \overline{\smash{)}\, x^3 + 0x^2 + 0x - 27} \\
\underline{x^3 - 3x^2 \phantom{xxxxxxxxx}} \\
3x^2 + 0x \phantom{xxxx} \\
\underline{3x^2 - 9x \phantom{xxxx}} \\
9x - 27 \\
\underline{9x - 27}
\end{array}
$$

**46.**

$$
\begin{array}{r}
x - 1 + \dfrac{-3x - 10}{2x^2 + 4x + 5} \\
2x^2 + 4x + 5 \overline{\smash{)}\, 2x^3 + 2x^2 - 2x - 15} \\
\underline{2x^3 + 4x^2 + 5x \phantom{xxxxx}} \\
-2x^2 - 7x - 15 \\
\underline{-2x^2 - 4x - 5} \\
-3x - 10
\end{array}
$$

**48.**

$$2x^3 - x^2 - 3 \overline{)\,8x^5 + 6x^4 - x^3 + 0x^2 + 0x + 1\,}$$

with quotient $4x^2 + 5x + 2 + \dfrac{14x^2 + 15x + 7}{2x^3 - x^2 - 3}$

$$
\begin{aligned}
&\underline{8x^5 - 4x^4 \qquad\quad - 12x^2} \\
&\quad 10x^4 - x^3 + 12x^2 + 0x \\
&\quad \underline{10x^4 - 5x^3 \qquad\quad - 15x} \\
&\qquad\quad 4x^3 + 12x^2 + 15x + 1 \\
&\qquad\quad \underline{4x^3 - 2x^2 \qquad\quad - 6} \\
&\qquad\qquad\quad 14x^2 + 15x + 7
\end{aligned}
$$

**50.**

$$x - 1 \overline{)\,x^3 + 0x^2 + 0x + 0\,}$$

with quotient $x^2 + x + 1 + \dfrac{1}{x - 1}$

$$
\begin{aligned}
&\underline{x^3 - x^2} \\
&\quad x^2 + 0x \\
&\quad \underline{x^2 - x} \\
&\qquad x + 0 \\
&\qquad \underline{x - 1} \\
&\qquad\quad 1
\end{aligned}
$$

**52.**

$$x - 2 \overline{)\,x^4 + 0x^3 + 0x^2 + 0x + 0\,}$$

with quotient $x^3 + 2x^2 + 4x + 8 + \dfrac{16}{x - 2}$

$$
\begin{aligned}
&\underline{x^4 - 2x^3} \\
&\quad 2x^3 + 0x^2 \\
&\quad \underline{2x^3 - 4x^2} \\
&\qquad 4x^2 + 0x \\
&\qquad \underline{4x^2 - 8x} \\
&\qquad\quad 8x + 0 \\
&\qquad\quad \underline{8x - 16} \\
&\qquad\qquad 16
\end{aligned}
$$

**54.**

$$\frac{15x^3 y}{10x^2} + \frac{3xy^2}{2y} = \frac{3xy}{2} + \frac{3}{2}xy$$

$$= \frac{6}{2}xy$$

$$= 3xy, \quad x \neq 0, y \neq 0$$

**56.**

$$\frac{x^2 + 2x - 3}{x - 1} - (3x - 4) = \frac{(x + 3)\cancel{(x - 1)}}{\cancel{x - 1}} - (3x - 4)$$

$$= x + 3 - 3x + 4$$

$$= -2x + 7, \quad x \neq 1$$

**58.** $\dfrac{x^3 + 6x^2 + 8x - 2}{x + 3}$

$$
\begin{array}{r|rrrr}
-3 & 1 & 6 & 8 & -2 \\
   &   & -3 & -9 & 3 \\
\hline
   & 1 & 3 & -1 & 1
\end{array}
$$

$$\frac{x^3 + 6x^2 + 8x - 2}{x + 3} = x^2 + 3x - 1 + \frac{1}{x + 3}$$

**60.** $\dfrac{x^4 - 4x^2 + 6}{x - 4}$

$$
\begin{array}{r|rrrrr}
4 & 1 & 0 & -4 & 0 & 6 \\
  &   & 4 & 16 & 48 & 192 \\
\hline
  & 1 & 4 & 12 & 48 & 198
\end{array}
$$

$$\frac{x^4 - 4x^2 + 6}{x - 4} = x^3 + 4x^2 + 12x + 48 + \frac{198}{x - 4}$$

**62.** $\dfrac{2x^5 - 3x^3 + x}{x - 3}$

$$
\begin{array}{r|rrrrrr}
3 & 2 & 0 & -3 & 0 & 1 & 0 \\
  &   & 6 & 18 & 45 & 135 & 408 \\
\hline
  & 2 & 6 & 15 & 45 & 136 & 408
\end{array}
$$

$$\frac{2x^5 - 3x^3 + x}{x - 3} = 2x^4 + 6x^3 + 15x^2 + 45x + 136 + \frac{408}{x - 3}$$

**64.** $\dfrac{5x^3 + 6x + 8}{x + 2}$

$$
\begin{array}{r|rrrr}
-2 & 5 & 0 & 6 & 8 \\
   &   & -10 & 20 & -52 \\
\hline
   & 5 & -10 & 26 & -44
\end{array}
$$

$$\frac{5x^3 + 6x + 8}{x + 2} = 5x^2 - 10x + 26 - \frac{44}{x + 2}$$

**66.** $\dfrac{x^5 - 13x^4 - 120x + 80}{x + 3}$

$$
\begin{array}{r|rrrrrr}
-3 & 1 & -13 & 0 & 0 & -120 & 80 \\
 &  & -3 & 48 & -144 & 432 & -936 \\
\hline
 & 1 & -16 & 48 & -144 & 312 & -856
\end{array}
$$

$$\dfrac{x^5 - 13x^4 - 120x + 80}{x + 3} = x^4 - 16x^3 + 48x^2 - 144x + 312 - \dfrac{856}{x + 3}$$

**68.** $\dfrac{x^3 - 0.8x + 2.4}{x + 0.1}$

$$
\begin{array}{r|rrrr}
-0.1 & 1 & 0 & -0.8 & 2.4 \\
 &  & -0.1 & 0.01 & 0.079 \\
\hline
 & 1 & -0.1 & -0.79 & 2.479
\end{array}
$$

$$\dfrac{x^3 - 0.8x + 2.4}{x + 0.1} = x^2 - 0.1x - 0.79 + \dfrac{2.479}{x + 0.1}$$

**70.**
$$
\begin{array}{r|rrrr}
-5 & 1 & 1 & -32 & -60 \\
 &  & -5 & 20 & 60 \\
\hline
 & 1 & -4 & -12 & 0
\end{array}
$$

$$x^2 - 4x - 12 = (x - 6)(x + 2)$$
$$x^3 + x^2 - 32x - 60 = (x + 5)(x - 6)(x + 2)$$

**72.**
$$
\begin{array}{r|rrrr}
3 & 9 & -3 & -56 & -48 \\
 &  & 27 & 72 & 48 \\
\hline
 & 9 & 24 & 16 & 0
\end{array}
$$

$$9x^2 + 24x + 16 = (3x + 4)^2$$
$$9x^3 - 3x^2 - 56x - 48 = (x - 3)(3x + 4)^2$$

**74.**
$$
\begin{array}{r|rrrr}
-2 & 4 & 8 & -25 & -50 \\
 &  & -8 & 0 & 50 \\
\hline
 & 4 & 0 & -25 & 0
\end{array}
$$

$$4x^2 - 25 = (2x - 5)(2x + 5)$$
$$4x^3 + 8x^2 - 25x - 50 = (x + 2)(2x - 5)(2x + 5)$$

**76.**
$$
\begin{array}{r|rrrrr}
4 & 1 & -6 & -8 & 96 & -128 \\
 &  & 4 & -8 & -64 & 128 \\
\hline
 & 1 & -2 & -16 & 32 & 0
\end{array}
$$

$$
\begin{aligned}
x^3 - 2x^2 - 16x + 32 &= x^2(x - 2) - 16(x - 2) \\
&= (x - 2)(x^2 - 16) \\
&= (x - 2)(x - 4)(x + 4)
\end{aligned}
$$

$$x^4 + 7x^3 + 3x^2 - 63x - 108 = (x - 4)^2(x - 2)(x + 4)$$

**78.** $\dfrac{18x^2 - 9x - 20}{x + \frac{5}{6}}$

$$
\begin{array}{r|rrr}
-\frac{5}{6} & 18 & -9 & -20 \\
 &  & -15 & 20 \\
\hline
 & 18 & -24 & 0
\end{array}
$$

$$18x - 24 = 6(3x - 4)$$

$$(18x^2 - 9x - 20) = 6\left(x + \dfrac{5}{6}\right)(3x - 4)$$

**80.** $\dfrac{x^4 - 3x^2 + c}{x + 6}$

$$
\begin{array}{r|rrrrr}
-6 & 1 & 0 & -3 & 0 & c \\
 &  & -6 & 36 & -198 & 1188 \\
\hline
 & 1 & -6 & 33 & -198 & 0
\end{array}
$$

$$c + 1188 = 0$$
$$c = -1188$$

**82.** *Keystrokes:*

$y_1$ [Y=] [(] [X,T,θ] [x²] [+] 2 [)] [÷] [(] [X,T,θ] [+] 1 [)] [ENTER]

$y_2$ [X,T,θ] [−] 1 [+] 3 [÷] [(] [X,T,θ] [+] 1 [)] [GRAPH]

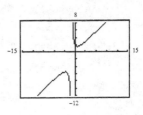

$$
\begin{array}{r|rrr}
-1 & 1 & 0 & 2 \\
   &   & -1 & 1 \\
\hline
   & 1 & -1 & 3
\end{array}
$$

Thus, $\dfrac{x^2 + 2}{x + 1} = x - 1 + \dfrac{3}{x + 1}$.

**84.** *Keystrokes:*

$y_1$ [Y=] [X,T,θ] [^] 3 [÷] [(] [X,T,θ] [x²] [+] 1 [)] [ENTER]

$y_2$ [X,T,θ] [−] [X,T,θ] [÷] [(] [X,T,θ] [x²] [+] 1 [)] [GRAPH]

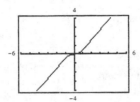

$$
\dfrac{x^3}{x^2 + 1} = x^2 + 1 \overline{\smash{)}\begin{array}{l} x \\ x^3 \phantom{xxxx} \end{array}} = x - \dfrac{x}{x^2 + 1}
$$

$$
\begin{array}{r}
\underline{x^3 + x} \\
-x
\end{array}
$$

**86.** $x^n - 1 \overline{\smash{)}\, x^{3n} - x^{2n} + 5x^n - 5}$    $x^{2n} \quad\quad + 5, \quad x^n \neq 1$

$$
\begin{array}{r}
\underline{x^{3n} - x^{2n}} \\
0x^{2n} + 5x^n - 5 \\
\underline{5x^n - 5}
\end{array}
$$

**88.** Dividend = Divisor · Quotient + Remainder

$= (x + 3)(x^3 + x^2 - 4) + 8$

$= x^4 + x^3 - 4x + 3x^3 + 3x^2 - 12 + 8$

$= x^4 + 4x^3 + 3x^2 - 4x - 4$

**90.**

| $x$-values | Polynomial values | Divisors | Remainders |
|---|---|---|---|
| $-2$ | $-15$ | $x + 2$ | $-15$ |
| $-1$ | $0$ | $x + 1$ | $0$ |
| $0$ | $1$ | $x$ | $1$ |
| $\frac{1}{2}$ | $0$ | $x - \frac{1}{2}$ | $0$ |
| $1$ | $0$ | $x - 1$ | $0$ |
| $2$ | $9$ | $x - 2$ | $9$ |

$f(-2) = 2(-2)^3 - (-2)^2 - 2(-2) + 1$

$\quad = -16 - 4 + 4 + 1$

$\quad = -15$

$$
\begin{array}{r|rrrr}
-2 & 2 & -1 & -2 & 1 \\
   &   & -4 & 10 & -16 \\
\hline
   & 2 & -5 & 8 & -15
\end{array}
$$

$f(-1) = 2(-1)^3 - (-1)^2 - 2(-1) + 1$

$\quad = -2 - 1 + 2 + 1$

$\quad = 0$

$$
\begin{array}{r|rrrr}
-1 & 2 & -1 & -2 & 1 \\
   &   & -2 & 3 & -1 \\
\hline
   & 2 & -3 & 1 & 0
\end{array}
$$

$f(0) = 2(0)^3 - (0)^2 - 2(0) + 1$

$\quad = 0 - 0 - 0 + 1$

$\quad = 1$

$$
\begin{array}{r|rrrr}
0 & 2 & -1 & -2 & 1 \\
  &   & 0 & 0 & 0 \\
\hline
  & 2 & -1 & -2 & 1
\end{array}
$$

**—CONTINUED—**

**90.** —CONTINUED—

$$f\left(\tfrac{1}{2}\right) = 2\left(\tfrac{1}{2}\right)^3 - \left(\tfrac{1}{2}\right)^2 - 2\left(\tfrac{1}{2}\right) + 1$$
$$= \tfrac{1}{4} - \tfrac{1}{4} - 1 + 1$$
$$= 0$$

$$\begin{array}{c|cccc} \tfrac{1}{2} & 2 & -1 & -2 & 1 \\ & & 1 & 0 & -1 \\ \hline & 2 & 0 & -2 & 0 \end{array}$$

$$f(1) = 2(1)^3 - (1)^2 - 2(1) + 1$$
$$= 2 - 1 - 2 + 1$$
$$= 0$$

$$\begin{array}{c|cccc} 1 & 2 & -1 & -2 & 1 \\ & & 2 & 1 & -1 \\ \hline & 2 & 1 & -1 & 0 \end{array}$$

$$f(2) = 2(2)^3 - (2)^2 - 2(2) + 1$$
$$= 16 - 4 - 4 + 1$$
$$= 9$$

$$\begin{array}{c|cccc} 2 & 2 & -1 & -2 & 1 \\ & & 4 & 6 & 8 \\ \hline & 2 & 3 & 4 & 9 \end{array}$$

The polynomial values equal the remainders.

**92.** Volume = Length · Width · Height

Volume = Area of first floor · Height

$$\text{Area of first floor} = \frac{\text{Volume}}{\text{Height}} = \frac{x^3 + 55x^2 + 650x + 2000}{x + 5}$$

$$\begin{array}{c|cccc} -5 & 1 & 55 & 650 & 2000 \\ & & -5 & -250 & -2000 \\ \hline & 1 & 50 & 400 & 0 \end{array}$$

Area of first floor = $x^2 + 50x + 400$ (square feet)

**94.** Volume = Length · Width · Height

$$\text{Length} = \frac{\text{Volume}}{\text{Width} \cdot \text{Height}}$$
$$= \frac{h^4 + 3h^3 + 2h^2}{h(h + 1)}$$
$$= \frac{h^2(h^2 + 3h + 2)}{h(h + 1)}$$
$$= \frac{h^2(h + 2)(h + 1)}{h(h + 1)}$$
$$= h(h + 2)$$
$$= h^2 + 2h$$

**96.** $\dfrac{x^2 + 4}{x + 1} = x - 1 + \dfrac{5}{x + 1}$

Divisor: $x + 1$

Dividend: $x^2 + 4$

Quotient: $x - 1$

Remainder: $5$

**98.** Check polynomial division by multiplication. Using Exercise 96 as an example:

$$(x + 1)\left(x - 1 + \frac{5}{x + 1}\right) = (x + 1)(x - 1) + (x - 1)\frac{5}{x + 1}$$
$$= x^2 - 1 + 5$$
$$= x^2 + 4$$

**100.** For synthetic division, the divisor must be of the form $x - k$.

# Section 4.6 Solving Rational Equations

**2.** (a) $x = 0$

$0 \overset{?}{=} 4 + \dfrac{21}{0}$

$\dfrac{21}{0}$ is undefined.

Not a solution

(b) $x = -3$

$-3 \overset{?}{=} 4 + \dfrac{21}{-3}$

$-3 \overset{?}{=} 4 + (-7)$

$-3 = -3$

Solution

(c) $x = 7$

$7 \overset{?}{=} 4 + \dfrac{21}{7}$

$7 \overset{?}{=} 4 + 3$

$7 = 7$

Solution

(d) $x = -1$

$-1 \overset{?}{=} 4 + \dfrac{21}{-1}$

$-1 \overset{?}{=} 4 - 21$

$-1 \neq -17$

Not a solution

**4.** (a) $x = \dfrac{10}{3}$

$5 - \dfrac{1}{\frac{10}{3} - 3} \overset{?}{=} 2$

$5 - \dfrac{1}{\frac{10}{3} - \frac{9}{3}} \overset{?}{=} 2$

$5 - \dfrac{1}{\frac{1}{3}} \overset{?}{=} 2$

$5 - 3 \overset{?}{=} 2$

$2 = 2$

Solution

(b) $x = -\dfrac{1}{3}$

$5 - \dfrac{1}{-\frac{1}{3} - 3} \overset{?}{=} 2$

$5 - \dfrac{1}{-\frac{1}{3} - \frac{9}{3}} \overset{?}{=} 2$

$5 - \dfrac{1}{-\frac{10}{3}} \overset{?}{=} 2$

$5 - \left(-\dfrac{3}{10}\right) \overset{?}{=} 2$

$\dfrac{50}{10} + \dfrac{3}{10} \overset{?}{=} 2$

$\dfrac{53}{10} \neq 2$

Not a solution

(c) $x = 0$

$5 - \dfrac{1}{0 - 3} \overset{?}{=} 2$

$5 - \left(-\dfrac{1}{3}\right) \overset{?}{=} 2$

$\dfrac{15}{3} + \dfrac{1}{3} \overset{?}{=} 2$

$\dfrac{16}{3} \neq 2$

Not a solution

(d) $x = 1$

$5 - \dfrac{1}{1 - 3} \overset{?}{=} 2$

$5 - \dfrac{1}{-2} \overset{?}{=} 2$

$\dfrac{10}{2} + \dfrac{1}{2} \overset{?}{=} 2$

$\dfrac{11}{2} \neq 2$

Not a solution

**6.** $\dfrac{y}{8} + 7 = -\dfrac{1}{2}$

$8\left(\dfrac{y}{8} + 7\right) = \left(-\dfrac{1}{2}\right)8$

$y + 56 = -4$

$y = -60$

**Check:**

$\dfrac{-60}{8} + 7 \overset{?}{=} -\dfrac{1}{2}$

$\dfrac{-15}{2} + \dfrac{14}{2} \overset{?}{=} -\dfrac{1}{2}$

$-\dfrac{1}{2} = -\dfrac{1}{2}$

**8.** $\dfrac{x - 5}{5} + 3 = -\dfrac{x}{4}$

$20\left(\dfrac{x - 5}{5} + 3\right) = \left(-\dfrac{x}{4}\right)20$

$4(x - 5) + 60 = -5x$

$4x - 20 + 60 = -5x$

$4x + 40 = -5x$

$40 = -9x$

$\dfrac{40}{-9} = x$

**Check:**

$\dfrac{-\frac{40}{9} - 5}{5} + 3 \overset{?}{=} -\dfrac{-\frac{40}{9}}{4}$

$\dfrac{-\frac{40}{9} - \frac{45}{9}}{5} + 3 \overset{?}{=} \dfrac{40}{9} \cdot \dfrac{1}{4}$

$\dfrac{1}{5}\left(-\dfrac{85}{9}\right) + 3 \overset{?}{=} \dfrac{10}{9}$

$-\dfrac{17}{9} + \dfrac{27}{9} \overset{?}{=} \dfrac{10}{9}$

$\dfrac{10}{9} = \dfrac{10}{9}$

**10.**  $\dfrac{4x-2}{7} - \dfrac{5}{14} = 2x$

$14\left(\dfrac{4x-2}{7} - \dfrac{5}{14}\right) = (2x)14$

$2(4x-2) - 5 = 28x$

$8x - 4 - 5 = 28x$

$-9 = 20x$

$\dfrac{-9}{20} = x$

**Check:**

$\dfrac{4\left(\frac{-9}{20}\right)-2}{7} - \dfrac{5}{14} \stackrel{?}{=} 2\left(\dfrac{-9}{20}\right)$

$\dfrac{\frac{-9}{5} - \frac{10}{5}}{7} - \dfrac{5}{14} \stackrel{?}{=} \dfrac{-9}{10}$

$\dfrac{-19}{5} \cdot \dfrac{1}{7} - \dfrac{5}{14} \stackrel{?}{=} \dfrac{-9}{10}$

$\dfrac{-19}{35} - \dfrac{5}{14} \stackrel{?}{=} \dfrac{-9}{10}$

$\dfrac{-38}{70} - \dfrac{25}{70} \stackrel{?}{=} \dfrac{-9}{10}$

$\dfrac{-63}{70} \stackrel{?}{=} \dfrac{-9}{10}$

$\dfrac{-9}{10} = \dfrac{-9}{10}$

**12.**  $\dfrac{x}{3} + \dfrac{x}{6} = 10$

$6\left(\dfrac{x}{3} + \dfrac{x}{6}\right) = (10)6$

$2x + x = 60$

$3x = 60$

$x = 20$

**Check:**  $\dfrac{20}{3} + \dfrac{20}{6} \stackrel{?}{=} 10$

$\dfrac{20}{3} + \dfrac{10}{3} \stackrel{?}{=} 10$

$\dfrac{30}{3} \stackrel{?}{=} 10$

$10 = 10$

**14.**  $\dfrac{z-4}{9} - \dfrac{3z+1}{18} = \dfrac{3}{2}$

$18\left(\dfrac{z-4}{9} - \dfrac{3z+1}{18}\right) = \left(\dfrac{3}{2}\right)18$

$2(z-4) - (3z+1) = 27$

$2z - 8 - 3z - 1 = 27$

$-z - 9 = 27$

$-z = 36$

$z = -36$

**Check:**  $\dfrac{-36-4}{9} - \dfrac{3(-36)+1}{18} \stackrel{?}{=} \dfrac{3}{2}$

$\dfrac{-40}{9} - \dfrac{-107}{18} \stackrel{?}{=} \dfrac{3}{2}$

$\dfrac{-80}{18} + \dfrac{107}{18} \stackrel{?}{=} \dfrac{3}{2}$

$\dfrac{27}{18} \stackrel{?}{=} \dfrac{3}{2}$

$\dfrac{3}{2} = \dfrac{3}{2}$

**16.**  $\dfrac{u-2}{6} + \dfrac{2u+5}{15} = 3$

$30\left(\dfrac{u-2}{6} + \dfrac{2u+5}{15}\right) = (3)30$

$5(u-2) + 2(2u+5) = 90$

$5u - 10 + 4u + 10 = 90$

$9u = 90$

$u = 10$

**Check:**  $\dfrac{10-2}{6} + \dfrac{2(10)+5}{15} \stackrel{?}{=} 3$

$\dfrac{8}{6} + \dfrac{25}{15} \stackrel{?}{=} 3$

$\dfrac{4}{3} + \dfrac{5}{3} \stackrel{?}{=} 3$

$\dfrac{9}{3} \stackrel{?}{=} 3$

$3 = 3$

**18.**  $\dfrac{2x-7}{10} - \dfrac{3x+1}{5} = \dfrac{6-x}{5}$

$10\left(\dfrac{2x-7}{10} - \dfrac{3x+1}{5}\right) = \left(\dfrac{6-x}{5}\right)10$

$2x - 7 - 2(3x+1) = 2(6-x)$

$2x - 7 - 6x - 2 = 12 - 2x$

$-4x - 9 = 12 - 2x$

$-2x = 21$

$x = \dfrac{-21}{2}$

**Check:**  $\dfrac{2\left(\frac{-21}{2}\right)-7}{10} - \dfrac{3\left(\frac{-21}{2}\right)+1}{5} \stackrel{?}{=} \dfrac{6-\left(\frac{-21}{2}\right)}{5}$

$\dfrac{-21-7}{10} - \dfrac{\frac{-63}{2}+\frac{2}{2}}{5} \stackrel{?}{=} \dfrac{\frac{12}{2}+\frac{21}{2}}{5}$

$\dfrac{-28}{10} - \dfrac{\frac{-61}{2}}{5} \stackrel{?}{=} \dfrac{\frac{33}{2}}{5}$

$\dfrac{-28}{10} + \dfrac{61}{10} \stackrel{?}{=} \dfrac{33}{10}$

$\dfrac{33}{10} = \dfrac{33}{10}$

**20.**
$$\frac{2}{u+4} = \frac{5}{8}$$

$$8(u+4)\left(\frac{2}{u+4}\right) = \left(\frac{5}{8}\right)8(u+4)$$

$$8(2) = 5(u+4)$$

$$16 = 5u + 20$$

$$-4 = 5u$$

$$-\frac{4}{5} = u$$

Check: $\dfrac{2}{-\frac{4}{5}+4} \overset{?}{=} \dfrac{5}{8}$

$$\frac{2}{-\frac{4}{5}+\frac{20}{5}} \overset{?}{=} \frac{5}{8}$$

$$\frac{2}{\frac{16}{5}} \overset{?}{=} \frac{5}{8}$$

$$\frac{10}{16} \overset{?}{=} \frac{5}{8}$$

$$\frac{5}{8} = \frac{5}{8}$$

**22.** $\dfrac{6}{b} + 22 = 24$

$$\frac{6}{b} = 2$$

$$b\left(\frac{6}{b}\right) = (2)b$$

$$6 = 2b$$

$$3 = b$$

Check: $\dfrac{6}{3} + 22 \overset{?}{=} 24$

$$2 + 22 \overset{?}{=} 24$$

$$24 = 24$$

**24.** $\dfrac{5}{3} = \dfrac{6}{7x} + \dfrac{2}{x}$

$$21x\left(\frac{5}{3}\right) = \left(\frac{6}{7x} + \frac{2}{x}\right)21x$$

$$35x = 18 + 42$$

$$35x = 60$$

$$x = \frac{60}{35}$$

$$x = \frac{12}{7}$$

Check: $\dfrac{5}{3} \overset{?}{=} \dfrac{6}{7\left(\frac{12}{7}\right)} + \dfrac{2}{\frac{12}{7}}$

$$\frac{5}{3} \overset{?}{=} \frac{6}{12} + \frac{14}{12}$$

$$\frac{5}{3} \overset{?}{=} \frac{20}{12}$$

$$\frac{5}{3} = \frac{5}{3}$$

**26.**
$$\frac{7}{8} - \frac{16}{t-2} = \frac{3}{4}$$

$$8(t-2)\left(\frac{7}{8} - \frac{16}{t-2}\right) = \left(\frac{3}{4}\right)8(t-2)$$

$$7(t-2) - 16(8) = 6(t-2)$$

$$7t - 14 - 128 = 6t - 12$$

$$7t - 142 = 6t - 12$$

$$7t = 6t + 130$$

$$t = 130$$

Check: $\dfrac{7}{8} - \dfrac{16}{130-2} \overset{?}{=} \dfrac{3}{4}$

$$\frac{7}{8} - \frac{16}{128} \overset{?}{=} \frac{3}{4}$$

$$\frac{7}{8} - \frac{1}{8} \overset{?}{=} \frac{3}{4}$$

$$\frac{6}{8} \overset{?}{=} \frac{3}{4}$$

$$\frac{3}{4} = \frac{3}{4}$$

**28.**
$$\frac{10}{x+4} = \frac{15}{4(x+1)}$$

$$4(x+1)(x+4)\left(\frac{10}{x+4}\right) = \left(\frac{15}{4(x+1)}\right)4(x+1)(x+4)$$

$$4 \cdot 10(x+1) = 15(x+4)$$

$$40(x+1) = 15(x+4)$$

$$40x + 40 = 15x + 60$$

$$25x + 40 = 60$$

$$25x = 20$$

$$x = \frac{20}{25}$$

$$x = \frac{4}{5}$$

Check: $\dfrac{10}{\frac{4}{5}+4} \overset{?}{=} \dfrac{15}{4\left(\frac{4}{5}+1\right)}$

$$\frac{10}{\frac{4}{5}+\frac{20}{5}} \overset{?}{=} \frac{15}{4\left(\frac{4}{5}+\frac{5}{5}\right)}$$

$$\frac{10}{\frac{24}{5}} \overset{?}{=} \frac{15}{4\left(\frac{9}{5}\right)}$$

$$\frac{50}{24} \overset{?}{=} \frac{15}{\frac{36}{5}}$$

$$\frac{50}{24} \overset{?}{=} \frac{75}{36}$$

$$\frac{25}{12} = \frac{25}{12}$$

**30.**

$$\frac{500}{3x + 5} = \frac{50}{x - 3}$$

$$(x - 3)(3x + 5)\left(\frac{500}{3x + 5}\right) = \left(\frac{50}{x - 3}\right)(x - 3)(3x + 5)$$

$$500(x - 3) = 50(3x + 5)$$

$$500x - 1500 = 150x + 250$$

$$350x - 1500 = 250$$

$$350x = 1750$$

$$x = \frac{1750}{350}$$

$$x = 5$$

Check: $\dfrac{500}{3(5) + 5} \stackrel{?}{=} \dfrac{50}{5 - 3}$

$$\frac{500}{15 + 5} \stackrel{?}{=} \frac{50}{2}$$

$$\frac{500}{20} \stackrel{?}{=} 25$$

$$25 = 25$$

**32.**

$$\frac{12}{x \div 5} + \frac{5}{x} = \frac{20}{x}$$

$$x(x + 5)\left(\frac{12}{x + 5} + \frac{5}{x}\right) = \left(\frac{20}{x}\right)x(x + 5)$$

$$12x + 5(x + 5) = 20(x + 5)$$

$$12x + 5x + 25 = 20x + 100$$

$$17x + 25 = 20x + 100$$

$$25 = 3x + 100$$

$$-75 = 3x$$

$$-25 = x$$

Check: $\dfrac{12}{-25 + 5} + \dfrac{5}{-25} \stackrel{?}{=} \dfrac{20}{-25}$

$$\frac{12}{-20} + \frac{5}{-25} \stackrel{?}{=} \frac{20}{-25}$$

$$\frac{3}{-5} + \frac{1}{-5} \stackrel{?}{=} -\frac{4}{5}$$

$$-\frac{4}{5} = -\frac{4}{5}$$

**34.**

$$\frac{1}{4} = \frac{16}{z^2}$$

$$4\left(\frac{1}{4}\right) = \left(\frac{16}{z^2}\right)4$$

$$z^2 = 64$$

$$z^2 - 64 = 0$$

$$(z - 8)(z + 8) = 0$$

$$z - 8 = 0 \qquad z + 8 = 0$$

$$z = 8 \qquad z = -8$$

Check: $\dfrac{1}{4} \stackrel{?}{=} \dfrac{16}{(8)^2}$

$$\frac{1}{4} \stackrel{?}{=} \frac{16}{64}$$

$$\frac{1}{4} = \frac{1}{4}$$

Check: $\dfrac{1}{4} \stackrel{?}{=} \dfrac{16}{(-8)^2}$

$$\frac{1}{4} \stackrel{?}{=} \frac{16}{64}$$

$$\frac{1}{4} = \frac{1}{4}$$

**36.**

$$\frac{20}{u} = \frac{u}{5}$$

$$5u\left(\frac{20}{u}\right) = \left(\frac{u}{5}\right)5u$$

$$100 = u^2$$

$$0 = u^2 - 100$$

$$0 = (u - 10)(u + 10)$$

$$u - 10 = 0 \qquad u + 10 = 0$$

$$u = 10 \qquad u = -10$$

Check: $\dfrac{20}{10} \stackrel{?}{=} \dfrac{10}{5}$

$$2 = 2$$

Check: $\dfrac{20}{-10} \stackrel{?}{=} \dfrac{-10}{5}$

$$-2 = -2$$

**38.**  $\dfrac{48}{x} = x - 2$

$x\left(\dfrac{48}{x}\right) = (x - 2)x$

$48 = x^2 - 2x$

$0 = x^2 - 2x - 48$

$0 = (x - 8)(x + 6)$

$x = 8 \qquad x = -6$

**Check:** $\dfrac{48}{8} \overset{?}{=} 8 - 2$

$6 = 6$

**Check:** $\dfrac{48}{-6} \overset{?}{=} -6 - 2$

$-8 = -8$

**40.**  $x - \dfrac{24}{x} = 5$

$x\left(x - \dfrac{24}{x}\right) = (5)x$

$x^2 - 24 = 5x$

$x^2 - 5x - 24 = 0$

$(x - 8)(x + 3) = 0$

$x = 8$

$x = -3$

**Check:** $8 - \dfrac{24}{8} \overset{?}{=} 5$

$8 - 3 = 5$

$5 = 5$

**Check:** $-3 - \dfrac{24}{-3} \overset{?}{=} 5$

$-3 + 8 = 5$

$5 = 5$

**42.**  $\dfrac{x + 42}{x} = x$

$x\left(\dfrac{x + 42}{x}\right) = (x)x$

$x + 42 = x^2$

$0 = x^2 - x - 42$

$0 = (x - 7)(x + 6)$

$x - 7 = 0 \qquad x + 6 = 0$

$x = 7 \qquad x = -6$

**Check:** $\dfrac{7 + 42}{7} \overset{?}{=} 7$

$\dfrac{49}{7} \overset{?}{=} 7$

$7 = 7$

**Check:** $\dfrac{-6 + 42}{-6} \overset{?}{=} -6$

$\dfrac{36}{-6} \overset{?}{=} -6$

$-6 = -6$

**44.**  $\dfrac{3x}{4} = \dfrac{x^2 + 3x}{8x}$

$8x\left(\dfrac{3x}{4}\right) = \left(\dfrac{x^2 + 3x}{8x}\right)8x$

$6x^2 = x^2 + 3x$

$5x^2 - 3x = 0$

$x(5x - 3) = 0$

$x = 0 \qquad 5x - 3 = 0$

$x = \dfrac{3}{5}$

**Check:** $\dfrac{3(0)}{4} \overset{?}{=} \dfrac{0^2 + 3(0)}{8(0)}$

Division by zero is undefined.

Solution $x = 0$ is extraneous.

**Check:** $\dfrac{3\left(\frac{3}{5}\right)}{4} \overset{?}{=} \dfrac{\left(\frac{3}{5}\right)^2 + 3\left(\frac{3}{5}\right)}{8\left(\frac{3}{5}\right)}$

$\dfrac{9}{5} \cdot \dfrac{1}{4} \overset{?}{=} \dfrac{\frac{9}{25} + \frac{9}{5}}{\frac{24}{5}}$

$\dfrac{9}{20} \overset{?}{=} \dfrac{54}{25} \cdot \dfrac{5}{24}$

$\dfrac{9}{20} \overset{?}{=} \dfrac{18}{5} \cdot \dfrac{1}{8}$

$\dfrac{9}{20} = \dfrac{9}{20}$

**46.**  $\dfrac{10}{x(x - 2)} + \dfrac{4}{x} = \dfrac{5}{x - 2}$

$x(x - 2)\left(\dfrac{10}{x(x - 2)} + \dfrac{4}{x}\right) = \left(\dfrac{5}{x - 2}\right)x(x - 2)$

$10 + 4(x - 2) = 5x$

$10 + 4x - 8 = 5x$

$2 = x$

**Check:** $\dfrac{10}{2(2 - 2)} + \dfrac{4}{2} \overset{?}{=} \dfrac{5}{2 - 2}$

$\dfrac{10}{0} + \dfrac{4}{2} \neq \dfrac{5}{0}$

Division by zero is undefined. Solution is extraneous, so equation has no solution.

**48.**
$$\frac{1}{x-1} + \frac{3}{x+1} = 2$$

$$(x-1)(x+1)\left(\frac{1}{x-1} + \frac{3}{x+1}\right) = (2)(x-1)(x+1)$$

$$x + 1 + 3(x-1) = 2(x^2-1)$$

$$x + 1 + 3x - 3 = 2x^2 - 2$$

$$4x - 2 = 2x^2 - 2$$

$$0 = 2x^2 - 4x$$

$$0 = 2x(x-2)$$

$$x = 0 \qquad x = 2$$

**Check:** $\dfrac{1}{0-1} + \dfrac{3}{0+1} \stackrel{?}{=} 2$

$$-1 + 3 = 2$$

$$2 = 2$$

**Check:** $\dfrac{1}{2-1} + \dfrac{3}{2+1} \stackrel{?}{=} 2$

$$1 + 1 = 2$$

$$2 = 2$$

**50.**
$$\frac{5}{x+2} + \frac{2}{x^2 - 6x - 16} = \frac{-4}{x-8}$$

$$(x+2)(x-8)\left(\frac{5}{x+2} + \frac{2}{(x+2)(x-8)}\right) = \left(\frac{-4}{x-8}\right)(x+2)(x-8)$$

$$5(x-8) + 2 = -4(x+2)$$

$$5x - 40 + 2 = -4x - 8$$

$$5x - 38 = -4x - 8$$

$$9x = 30$$

$$x = \frac{30}{9}$$

$$x = \frac{10}{3}$$

**Check:** $\dfrac{5}{\left(\frac{10}{3}\right)+2} + \dfrac{2}{\left(\frac{10}{3}\right)^2 - 6\left(\frac{10}{3}\right) - 16} \stackrel{?}{=} \dfrac{-4}{\left(\frac{10}{3}\right) - 8}$

$$\frac{5}{\frac{10}{3} + \frac{6}{3}} + \frac{2}{\frac{100}{9} - \frac{60}{3} - 16} \stackrel{?}{=} \frac{-4}{\frac{10}{3} - \frac{24}{3}}$$

$$\frac{5}{\frac{16}{3}} + \frac{2}{\frac{100}{9} - \frac{180}{9} - \frac{144}{9}} \stackrel{?}{=} \frac{-4}{-\frac{14}{3}}$$

$$\frac{15}{16} + \frac{2}{-\frac{224}{9}} \stackrel{?}{=} \frac{12}{14}$$

$$\frac{15}{16} - \frac{18}{224} \stackrel{?}{=} \frac{12}{14}$$

$$\frac{210}{224} - \frac{18}{224} \stackrel{?}{=} \frac{192}{224}$$

$$\frac{192}{224} = \frac{192}{224}$$

**52.**
$$1 - \frac{6}{4-x} = \frac{x+2}{x^2 - 16}$$

$$1 + \frac{6}{x-4} = \frac{x+2}{(x-4)(x+4)}$$

$$(x-4)(x+4)\left(1 + \frac{6}{x-4}\right) = \left(\frac{x+2}{(x-4)(x+4)}\right)(x-4)(x+4)$$

$$(x^2 - 16) + 6(x+4) = x + 2$$

$$x^2 - 16 + 6x + 24 = x + 2$$

$$x^2 + 5x + 6 = 0$$

$$(x+3)(x+2) = 0$$

$$x + 3 = 0 \qquad x + 2 = 0$$

$$x = -3 \qquad x = -2$$

**Check:** $1 - \dfrac{6}{4-(-3)} \stackrel{?}{=} \dfrac{-3+2}{(-3)^2 - 16}$

$$1 - \frac{6}{7} \stackrel{?}{=} \frac{-1}{9 - 16}$$

$$\frac{1}{7} = \frac{1}{7}$$

**Check:** $1 - \dfrac{6}{4-(-2)} \stackrel{?}{=} \dfrac{-2+2}{(-2)^2 - 16}$

$$1 - \frac{6}{6} \stackrel{?}{=} \frac{0}{4 - 16}$$

$$0 = 0$$

**54.**
$$\frac{2(x+1)}{x^2-4x+3} + \frac{6x}{x-3} = \frac{3x}{x-1}$$

$$(x-3)(x-1)\left(\frac{2(x+1)}{(x-3)(x-1)}\right) + (x-3)(x-1)\left(\frac{6x}{x-3}\right) = \left(\frac{3x}{x-1}\right)(x-3)(x-1)$$

$$2(x+1) + 6x(x-1) = 3x(x-3)$$

$$2x + 2 + 6x^2 - 6x = 3x^2 - 9x$$

$$3x^2 + 5x + 2 = 0$$

$$(3x+2)(x+1) = 0$$

$$3x + 2 = 0 \qquad x + 1 = 0$$

$$x = -\frac{2}{3} \qquad x = -1$$

**Check:** $\dfrac{2\left(-\frac{2}{3}+1\right)}{\left(-\frac{2}{3}\right)^2 - 4\left(-\frac{2}{3}\right) + 3} + \dfrac{6\left(-\frac{2}{3}\right)}{-\frac{2}{3}-3} \overset{?}{=} \dfrac{3\left(-\frac{2}{3}\right)}{-\frac{2}{3}-1}$

$$\frac{\frac{2}{3}}{\frac{4}{9}+\frac{8}{3}+3} + \frac{-4}{-\frac{11}{3}} \overset{?}{=} \frac{-2}{-\frac{5}{3}}$$

$$\frac{2}{3}\cdot\frac{9}{55} + 4\cdot\frac{3}{11} \overset{?}{=} 2\cdot\frac{3}{5}$$

$$\frac{6}{55} + \frac{12}{11} \overset{?}{=} \frac{6}{5}$$

$$\frac{66}{55} \overset{?}{=} \frac{6}{5}$$

$$\frac{6}{5} = \frac{6}{5}$$

**Check:** $\dfrac{2(-1+1)}{(-1)^2 - 4(-1) + 3} + \dfrac{6(-1)}{-1-3} \overset{?}{=} \dfrac{3(-1)}{-1-1}$

$$0 + \frac{6}{4} \overset{?}{=} \frac{-3}{-2}$$

$$\frac{3}{2} = \frac{3}{2}$$

**56.**
$$\frac{2x^2-5}{x^2-4} + \frac{6}{x+2} = \frac{4x-7}{x-2}$$

$$(x^2-4)\left(\frac{2x^2-5}{x^2-4} + \frac{6}{x+2}\right) = \left(\frac{4x-7}{x-2}\right)(x^2-4)$$

$$2x^2 - 5 + 6(x-2) = (4x-7)(x+2)$$

$$2x^2 - 5 + 6x - 12 = 4x^2 + x - 14$$

$$0 = 2x^2 - 5x + 3$$

$$0 = (2x-3)(x-1)$$

$$2x - 3 = 0 \qquad x - 1 = 0$$

$$x = \frac{3}{2} \qquad x = 1$$

**Check:** $\dfrac{2\left(\frac{3}{2}\right)^2 - 5}{\left(\frac{3}{2}\right)^2 - 4} + \dfrac{6}{\frac{3}{2}+2} \overset{?}{=} \dfrac{4\left(\frac{3}{2}\right) - 7}{\frac{3}{2}-2}$

$$\frac{-\frac{1}{2}}{-\frac{7}{4}} + \frac{6}{\frac{7}{2}} \overset{?}{=} \frac{-1}{-\frac{1}{2}}$$

$$\frac{2}{7} + \frac{12}{7} \overset{?}{=} 2$$

$$\frac{14}{7} \overset{?}{=} 2$$

$$2 = 2$$

**Check:** $\dfrac{2(1)^2 - 5}{1^2 - 4} + \dfrac{6}{1+2} \overset{?}{=} \dfrac{4(1) - 7}{1-2}$

$$\frac{-3}{-3} + 2 \overset{?}{=} \frac{-3}{-1}$$

$$3 = 3$$

**58.**
$$\frac{2x}{3} = \frac{1 + \dfrac{2}{x}}{1 + \dfrac{1}{x}}$$

$$\frac{2x}{3} = \frac{\left(1 + \dfrac{2}{x}\right)x}{\left(1 + \dfrac{1}{x}\right)x}$$

$$\frac{2x}{3} = \frac{(x + 2)}{(x + 1)}$$

$$3(x + 1)\frac{2x}{3} = 3(x + 1)\frac{(x + 2)}{(x + 1)}$$

$$2x^2 + 2x = 3x + 6$$

$$2x^2 - x - 6 = 0$$

$$(2x + 3)(x - 2) = 0$$

$$2x + 3 = 0 \qquad x - 2 = 0$$

$$x = -\frac{3}{2} \qquad x = 2$$

**Check:** $\dfrac{2\left(-\dfrac{3}{2}\right)}{3} \stackrel{?}{=} \dfrac{1 + \dfrac{2}{(-3/2)}}{1 + \dfrac{1}{(-3/2)}}$

$$\frac{-3}{3} \stackrel{?}{=} \frac{1 - \dfrac{4}{3}}{1 - \dfrac{2}{3}}$$

$$-1 \stackrel{?}{=} \frac{-\dfrac{1}{3}}{\dfrac{1}{3}}$$

$$-1 = -1$$

**Check:** $\dfrac{2(2)}{3} \stackrel{?}{=} \dfrac{1 + \dfrac{2}{2}}{1 + \dfrac{1}{2}}$

$$\frac{4}{3} \stackrel{?}{=} \frac{2}{\dfrac{3}{2}}$$

$$\frac{4}{3} = \frac{4}{3}$$

**60.** (a) *x*-intercept: $(0, 0)$

(b)
$$0 = \frac{2x}{x + 4}$$

$$(x + 4)(0) = \left(\frac{2x}{x + 4}\right)(x + 4)$$

$$0 = 2x$$

$$0 = x$$

(a) and (b) $(0, 0)$

**62.** (a) *x*-intercepts: $(-1, 0)$ and $(2, 0)$

(b)
$$0 = x - \frac{2}{x} - 1$$

$$x(0) = \left(x - \frac{2}{x} - 1\right)x$$

$$0 = x^2 - 2 - x$$

$$0 = x^2 - x - 2$$

$$0 = (x - 2)(x + 1)$$

$$x - 2 = 0 \qquad x + 1 = 0$$

$$x = 2, \qquad x = -1$$

(a) and (b) $(-1, 0), (2, 0)$

**64.** (a) *Keystrokes:*

$\boxed{Y=}\ 1\ \boxed{\div}\ \boxed{X,T,\theta}\ \boxed{-}\ 3\ \boxed{\div}\ \boxed{(}\ \boxed{X,T,\theta}\ \boxed{+}\ 4\ \boxed{)}\ \boxed{GRAPH}$

*x*-intercept: $(2, 0)$

(b)
$$0 = \frac{1}{x} - \frac{3}{x + 4}$$

$$x(x + 4)(0) = \left(\frac{1}{x} - \frac{3}{x + 4}\right)x(x + 4)$$

$$0 = x + 4 - 3x$$

$$0 = -2x + 4$$

$$2x = 4$$

$$x = 2$$

$(2, 0)$

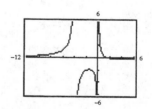

**66.** (a) *Keystrokes:*

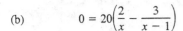 20 [(] 2 [÷] [X,T,θ] [–] 3 [÷] [(] [X,T,θ] [–] 1 [)] [)] [GRAPH]

*x*-intercept: $(-2, 0)$

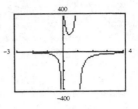

(b)
$$0 = 20\left(\frac{2}{x} - \frac{3}{x-1}\right)$$

$$x(x-1)(0) = \left[20\left(\frac{2}{x} - \frac{3}{x-1}\right)\right]x(x-1)$$

$$0 = 40(x-1) - 60x$$

$$0 = 40x - 40 - 60x$$

$$0 = -20x - 40$$

$$20x = -40$$

$$x = -2$$

$$(-2, 0)$$

**68.** (a) *Keystrokes:*

[=] [(] [X,T,θ] [x²] [–] 4 [)] [÷] [X,T,θ] [GRAPH]

*x*-intercepts: $(-2, 0)$ and $(2, 0)$

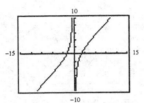

(b)
$$0 = \frac{x^2 - 4}{x}$$

$$x(0) = \left(\frac{x^2 - 4}{x}\right)x$$

$$0 = x^2 - 4$$

$$0 = (x-2)(x+2)$$

$$x - 2 = 0 \qquad x + 2 = 0$$

$$x = 2 \qquad\quad x = -2$$

$$(2, 0) \qquad\qquad (-2, 0)$$

**70.** (a) *Keystrokes:*

[Y=] [X,T,θ] [÷] 2 [–] 4 [÷] [X,T,θ] [–] 1 [GRAPH]

*x*-intercepts: $(4, 0)$ and $(-2, 0)$

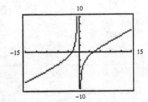

(b)
$$0 = \frac{x}{2} - \frac{4}{x} - 1$$

$$2x(0) = \left(\frac{x}{2} - \frac{4}{x} - 1\right)(2x)$$

$$0 = x^2 - 8 - 2x$$

$$0 = x^2 - 2x - 8$$

$$0 = (x-4)(x+2)$$

$$x - 4 = 0 \qquad x + 2 = 0$$

$$x = 4 \qquad\quad x = -2$$

$$(4, 0) \qquad\qquad (-2, 0)$$

**72.** *Verbal Model:*    $\boxed{\text{Twice a number}} + \boxed{\text{3 times the reciprocal}} = \dfrac{97}{4}$

     *Labels:*      $x$ = a number

$$\frac{1}{x} = \text{reciprocal of number}$$

     *Equation:*

$$2x + \frac{3}{x} = \frac{97}{4}$$

$$4x\left(2x + \frac{3}{x}\right) = \left(\frac{97}{4}\right)4x$$

$$8x^2 + 12 = 97x$$

$$8x^2 - 97x + 12 = 0$$

$$(8x - 1)(x - 12) = 0$$

$$8x - 1 = 0 \qquad x - 12 = 0$$

$$8x = 1 \qquad\qquad x = 12$$

$$x = \frac{1}{8}$$

**74.** *Verbal Model:*    $\boxed{\dfrac{\text{Distance first part}}{\text{Speed first part}}} + \boxed{\dfrac{\text{Distance second part}}{\text{Speed second part}}} = \boxed{\begin{array}{c}\text{Total}\\\text{time}\end{array}}$

     *Labels:*      $r$ = speed first part

$$r + 10 = \text{speed second part}$$

     *Equation:*

$$\frac{240}{r} + \frac{72}{r + 10} = 6$$

$$r(r + 10)\left(\frac{240}{r}\right) + r(r + 10)\left(\frac{72}{r + 10}\right) = r(r + 10)6$$

$$(r + 10)240 + r(72) = 6r^2 + 60r$$

$$312 + 2400 = 6r^2 + 60r$$

$$6r^2 - 252r - 2400 = 0$$

$$3r^2 - 126r - 1200 = 0$$

$$(r - 50)(3r + 24) = 0$$

$$r - 50 = 0 \qquad 3r + 24 = 0$$

$$r = 50 \qquad\qquad r = -8$$

Choose the positive value of $r$. The two average speeds are 50 miles per hour and 60 miles per hour.

**76.** *Verbal Model:*    $\boxed{\dfrac{\text{Distance traveled by commuter plane}}{\text{Rate of commuter plane}}} = \boxed{\dfrac{\text{Distance traveled by jet}}{\text{Rate of jet}}}$

     *Labels:*      $x$ = rate of jet

$$x - 150 = \text{rate of commuter plane}$$

     *Equation:*

$$\frac{450}{x - 150} = \frac{1150}{x}$$

$$x(x - 150)\left(\frac{450}{x - 150}\right) = \left(\frac{1150}{x}\right)x(x - 150)$$

$$450x = 1150(x - 150)$$

$$450x = 1150x - 172{,}500$$

$$-700x = -172{,}500$$

$$x \approx 246.43$$

Thus, the speed of the jet is approximately 246 miles per hour and the speed of the commuter plane is approximately $246 - 150 = 96$ miles per hour.

**78.** *Verbal Model:* $\boxed{\dfrac{\text{Distance}}{\text{slower speed}}} = \boxed{\dfrac{\text{Distance}}{\text{faster speed}}} + \dfrac{1}{6}$

*Labels:*    $r = $ slower speed

$r + 6 = $ faster speed

*Equation:*

$$\frac{72}{r} = \frac{72}{r+6} + \frac{1}{6}$$

$$6r(r+6)\left(\frac{72}{r}\right) = \left(\frac{72}{r+6} + \frac{1}{6}\right)6r(r+6)$$

$$432(r+6) = 432r + r(r+6)$$

$$432r + 2592 = 432r + r^2 + 6r$$

$$0 = r^2 + 432r + 6r - 432r - 2592$$

$$0 = r^2 + 6r - 2592$$

$$0 = (r-48)(r+54)$$

$$r - 48 = 0 \qquad r + 54 = 0$$

$$r = 48 \qquad r = -54$$

Not a solution

Thus, your speed was 48 miles per hour.

**80.** *Verbal Model:* $\boxed{\dfrac{\text{Total cost}}{\text{Current group}}} - \boxed{\dfrac{\text{Total cost}}{\text{New group}}} = 6250$

*Labels:*    $x = $ number of persons in current group

$x + 4 = $ number of persons in new group

*Equation:*

$$\frac{150{,}000}{x} - \frac{150{,}000}{x+4} = 6250$$

$$x(x+4)\left(\frac{150{,}000}{x} - \frac{150{,}000}{x+4}\right) = (6250)x(x+4)$$

$$150{,}000(x+4) - 150{,}000x = 6250x^2 + 25{,}000x$$

$$150{,}000x + 600{,}000 - 150{,}000x = 6250x^2 + 25{,}000x$$

$$0 = 6250x^2 + 25{,}000x - 600{,}000$$

$$0 = 6250(x^2 + 4x - 96)$$

$$0 = 6250(x+12)(x-8)$$

$$x + 12 = 0 \qquad x - 8 = 0$$

$$x = -12 \qquad x = 8$$

There are presently 8 people in the group.

**82.**

$$1000 = \frac{500(1+3t)}{5+t}$$

$$(5+t)(1000) = \left(\frac{500(1+3t)}{5+t}\right)(5+t)$$

$$5000 + 1000t = 500(1+3t)$$

$$5000 + 1000t = 500 + 1500t$$

$$4500 + 1000t = 1500t$$

$$4500 = 500t$$

$$9 = t$$

Thus, it takes 9 hours for the population to reach 1000.

**84.**

$$2.90 = 1.50 + \frac{4200}{x}$$

$$1.40x = \left(\frac{4200}{x}\right)x$$

$$1.40x = 4200$$

$$x = \frac{4200}{1.4}$$

$$x = 3000$$

3000 units must be produced to have an average cost of $2.90.

**86.** $\dfrac{1}{4} + \dfrac{1}{4} = \dfrac{1}{t}$

$\dfrac{1}{2} = \dfrac{1}{t}$

$t = 2$ days

| Person #1 | Person #2 | Together |
|---|---|---|
| 4 days | 4 days | 2 days |
| $5\frac{1}{2}$ hours | 3 hours | $\dfrac{33}{17}$ hours |
| $a$ days | $b$ days | $\dfrac{ab}{a+b}$ days |

$\dfrac{1}{5\frac{1}{2}} + \dfrac{1}{3} = \dfrac{1}{t}$

$\dfrac{1}{\frac{11}{2}} + \dfrac{1}{3} = \dfrac{1}{t}$

$\dfrac{2}{11} + \dfrac{1}{3} = \dfrac{1}{t}$

$\dfrac{6}{33} + \dfrac{11}{33} = \dfrac{1}{t}$

$\dfrac{17}{33} = \dfrac{1}{t}$

$17t = 33$

$t = \dfrac{33}{17}$ hours

$\dfrac{1}{a} + \dfrac{1}{b} = \dfrac{1}{t}$

$abt\left(\dfrac{1}{a} + \dfrac{1}{b}\right) = \left(\dfrac{1}{t}\right)abt$

$bt + at = ab$

$t(b + a) = ab$

$t = \dfrac{ab}{a+b}$ days

**88.** *Verbal Model:* $\boxed{\text{Work of 1 person}} + \boxed{\text{Work of 2 people}} = \boxed{\text{Job completed}}$

*Equation:* $\dfrac{1}{x}(14) + \dfrac{1}{\frac{3}{2}x}(4) = 1$

$\dfrac{14}{x} + \dfrac{8}{3x} = 1$

$3x\left(\dfrac{14}{x} + \dfrac{8}{3x}\right) = 1(3x)$

$42 + 8 = 3x$

$50 = 3x$

$\dfrac{50}{3} = x$

Thus, the rate of the slower worker was $\frac{50}{3} = 16\frac{2}{3}$ hours and the rate of the faster worker was $\frac{3}{2}\left(\frac{50}{3}\right) = 25$ hours.

**90.** *Verbal Model:* $\boxed{\text{Work pipe 1}} + \boxed{\text{Work pipe 2}} = \boxed{\text{Job together}}$

*Equation:* $\dfrac{1}{x}(1) + \dfrac{1}{\frac{5}{4}x}(11) = 1$

$5x\left(\dfrac{1}{x} + \dfrac{44}{5x}\right) = (1)5x$

$5 + 44 = 5x$

$49 = 5x$

$9\frac{4}{5}$ hours $= \dfrac{49}{5} = x$

$12\frac{1}{4}$ hours $= \dfrac{49}{4} = \dfrac{5}{4}x$

**92.** *Keystrokes:* $\boxed{Y=}$ $\boxed{(}$ 87709 $\boxed{-}$ 1236 $\boxed{X,T,\theta}$ $\boxed{)}$ $\boxed{\div}$ $\boxed{(}$ 1000 $\boxed{-}$ 93 $\boxed{X,T,\theta}$ $\boxed{)}$ $\boxed{GRAPH}$

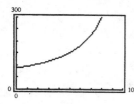

(a) Revenue will exceed $250 billion in the year 1998.

(b) The model fails after the year 2000 because if $t \geq 11$, $y$ (revenue) becomes a negative quantity.

**94.** (a)
$$\frac{16}{x^2 - 16} + \frac{x}{2x - 8} = \frac{1}{2} \rightarrow \text{equation}$$

$$2(x - 4)(x + 4)\left(\frac{16}{x^2 - 16} + \frac{x}{2(x - 4)}\right) = \left(\frac{1}{2}\right)2(x - 4)(x + 4)$$

$$32 + x(x + 4) = x^2 - 16$$

$$32 + x^2 + 4x = x^2 - 16$$

$$4x = -48$$

$$x = -12$$

(b) $\dfrac{16}{x^2 - 16} + \dfrac{x}{2x - 8} + \dfrac{1}{2} \rightarrow \text{expression}$

$$\frac{16}{(x - 4)(x + 4)} + \frac{x}{2(x - 4)} + \frac{1}{2} = \frac{16(2)}{2(x - 4)(x + 4)} + \frac{x(x + 4)}{2(x - 4)(x + 4)} + \frac{1(x - 4)(x + 4)}{2(x - 4)(x + 4)}$$

$$= \frac{32 + x^2 + 4x + x^2 - 16}{2(x - 4)(x + 4)}$$

$$= \frac{2x^2 + 4x + 16}{2(x - 4)(x + 4)}$$

$$= \frac{2(x^2 + 2x + 8)}{2(x - 4)(x + 4)}$$

$$= \frac{x^2 + 2x + 8}{(x - 4)(x + 4)}$$

(c) $\dfrac{5}{x + 3} + \dfrac{5}{3} + 3 \rightarrow \text{expression}$

$$\frac{3(5)}{3(x + 3)} + \frac{5(x + 3)}{3(x + 3)} + \frac{3(3)(x + 3)}{3(x + 3)} = \frac{15 + 5x + 15 + 9x + 27}{3(x + 3)} = \frac{14x + 57}{3(x + 3)} .$$

(d)
$$\frac{5}{x + 3} + \frac{5}{3} = 3 \rightarrow \text{equation}$$

$$3(x + 3)\left(\frac{5}{x + 3} + \frac{5}{3}\right) = (3)3(x + 3)$$

$$15 + 5(x + 3) = 9(x + 3)$$

$$15 + 5x + 15 = 9x + 27$$

$$3 = 4x$$

$$\frac{3}{4} = x$$

**96.** (a) An equation is a statement of equality of two expressions.

(b) When solving a rational equation, multiply the equation by the lowest common denominator. When adding or subtracting rational expressions, rewrite each fraction in terms of the lowest common denominator and combine the terms.

**98.** Extraneous solution is an extra solution found by multiplying both sides of the original equation by an expression containing the variable. It is identified by checking all solutions in the original equation.

**100.** Graph the rational equation and approximate any $x$-intercepts of the graph.

# Review Exercises for Chapter 4

**2.** $(2^{-2} \cdot 5^2)^{-2} = \left(\dfrac{1}{2^2} \cdot 5^2\right)^{-2}$

$\qquad = \left(\dfrac{25}{4}\right)^{-2}$

$\qquad = \left(\dfrac{4}{25}\right)^{2}$

$\qquad = \dfrac{16}{625}$

**4.** $\left(\dfrac{1}{3^{-2}}\right)^2 = (3^2)^2 = 3^4 = 81$

**6.** $(3 \times 10^{-3})(8 \times 10^7) = 24 \times 10^{-3+7}$

$\qquad\qquad\qquad\qquad = 24 \times 10^4$

$\qquad\qquad\qquad\qquad = 2.4 \times 10^5$

**8.** $\dfrac{1}{(6 \times 10^{-3})^2} = \dfrac{1}{36 \times 10^{-6}}$

$\qquad\qquad = \dfrac{10^6}{36}$

$\qquad\qquad = \dfrac{1,000,000}{36}$

$\qquad\qquad = \dfrac{2,500,000}{9}$

**10.** $30,296,000,000 = 3.0296 \times 10^{10}$

**12.** $2.74 \times 10^{-4} = 0.000274$

**14.** $4(-3x)^{-3} = \dfrac{4}{(-3x)^3}$

$\qquad\qquad = -\dfrac{4}{27x^3}$

**16.** $\dfrac{15t^5}{24t^{-3}} = \dfrac{5}{8}t^{5-(-3)}$

$\qquad\quad = \dfrac{5}{8}t^8$

**18.** $5yx^0 = 5y(1) = 5y$

**20.** $\dfrac{a^5 \cdot a^{-3}}{a^{-2}} = \dfrac{a^{5+(-3)}}{a^{-2}}$

$\qquad = \dfrac{a^2}{a^{-2}}$

$\qquad = a^{2-(-2)}$

$\qquad = a^4$

**22.** $(2x^2y^4)^4(2x^2y^4)^{-4} = (2x^2y^4)^{4+(-4)}$

$\qquad\qquad\qquad\qquad = (2x^2y^4)^0$

$\qquad\qquad\qquad\qquad = 1$

**24.** $t + 12 \neq 0$

$\qquad t \neq -12$

$\qquad D = (-\infty, -12) \cup (-12, \infty)$

**26.** $\qquad x(x^2 - 16) \neq 0$

$x(x - 4)(x + 4) \neq 0$

$x \neq 0 \quad x - 4 \neq 0 \quad x + 4 \neq 0$

$\qquad\qquad x \neq 4 \qquad\quad x \neq -4$

$D = (-\infty, -4) \cup (-4, 0) \cup (0, 4) \cup (4, \infty)$

**28.** $\dfrac{2(y^3z)^2}{28(yz^2)^2} = \dfrac{2y^6z^2}{28y^2z^4}$

$\qquad\qquad = \dfrac{2 \cdot y^2 \cdot y^4 \cdot z^2}{2 \cdot 14 \cdot y^2 \cdot z^2 \cdot z^2}$

$\qquad\qquad = \dfrac{y^4}{14z^2}, \quad y \neq 0$

**30.** $\dfrac{4a}{10a^2 + 26a} = \dfrac{4a}{2a(5a + 13)}$

$\qquad\qquad = \dfrac{2 \cdot 2 \cdot a}{2 \cdot a(5a + 13)}$

$\qquad\qquad = \dfrac{2}{5a + 13}, \quad a \neq 0$

**32.** $\dfrac{x + 3}{x^2 - x - 12} = \dfrac{x + 3}{(x - 4)(x + 3)}$

$\qquad\qquad = \dfrac{1}{x - 4}, \quad x \neq -3$

**34.** $\dfrac{x^2 + 3x + 9}{x^3 - 27} = \dfrac{x^2 + 3x + 9}{(x - 3)(x^2 + 3x + 9)}$

$\qquad\qquad = \dfrac{1}{x - 3}$

**36.** $2b(-3b)^3 = (2b)(-27b^3) = -54b^4$

**38.** $\dfrac{8u^2v}{6v} = \dfrac{4u^2}{3}$

**40.** $\dfrac{15(x^2y)^3}{3y^3} \cdot \dfrac{12y}{x} = \dfrac{3 \cdot 5 \cdot x \cdot x \cdot x \cdot x \cdot x \cdot x \cdot y \cdot y \cdot y \cdot 2 \cdot 2 \cdot 3 \cdot y}{3 \cdot y \cdot y \cdot y \cdot x}$

$= 5x^5y \cdot 2 \cdot 2 \cdot 3$

$= 60x^5y, \quad x \neq 0, y \neq 0$

**42.** $\dfrac{1}{6}(x^2 - 16) \cdot \dfrac{3}{x^2 - 8x + 16} = \dfrac{(x-4)(x+4) \cdot 3}{2 \cdot 3 \cdot (x-4)(x-4)}$

$= \dfrac{x+4}{2(x-4)}$

**44.** $x^2 \cdot \dfrac{x+1}{x^2 - x} \cdot \dfrac{(5x-5)^2}{x^2 + 6x + 5} = \dfrac{x^2 \cdot (x+1)(5)^2(x-1)^2}{1 \cdot x(x-1)(x+5)(x+1)}$

$= \dfrac{25x(x-1)}{x+5}, \quad x \neq -1, 1, 0$

**46.** $\dfrac{0}{\frac{5x^2}{2y}} = 0, \quad x \neq 0, y \neq 0$

**48.** $\dfrac{6}{z^2} \div 4z^2 = \dfrac{6}{z^2} \cdot \dfrac{1}{4z^2}$

$= \dfrac{2 \cdot 3}{z^2 \cdot 2 \cdot 2 \cdot z^2}$

$= \dfrac{3}{2z^4}$

**50.** $\left(\dfrac{6x}{y^2}\right)^2 \div \left(\dfrac{3x}{y}\right)^3 = \dfrac{36x^2}{y^4} \div \dfrac{27x^3}{y^3}$

$= \dfrac{36x^2}{y^4} \cdot \dfrac{y^3}{27x^3}$

$= \dfrac{2 \cdot 2 \cdot 3 \cdot 3 \cdot x^2 \cdot y^3}{3 \cdot 3 \cdot 3x^2xy^3y}$

$= \dfrac{4}{3xy}$

**52.** $\dfrac{\left[\frac{24 - 18x}{(2-x)^2}\right]}{\left(\frac{60 - 45x}{x^2 - 4x - 4}\right)} = \dfrac{\left[\frac{6(4-3x)}{(2-x)^2}\right]}{\left[\frac{15(4-3x)}{(x-2)^2}\right]}$

$= \dfrac{6(4-3x)}{(2-x)^2} \cdot \dfrac{(x-2)^2}{15(4-3x)}$

$= \dfrac{2 \cdot 3(4-3x)(x-2)^2}{[-1(x-2)]^2 \cdot 3 \cdot 5 \cdot (4-3x)}$

$= \dfrac{2 \cdot 3(4-3x)(x-2)^2}{(-1)^2(x-2)^2 \cdot 3 \cdot 5(4-3x)}$

$= \dfrac{2 \cdot 3 \cdot (4-3x)(x-2)^2}{1 \cdot 3 \cdot 5 \cdot (x-2)^2(4-3x)}$

$= \dfrac{2}{5}, \quad x \neq 2, \dfrac{4}{3}$

**54.** $-\dfrac{3}{8} + \dfrac{7}{6} - \dfrac{1}{12} = \dfrac{-3(3)}{8(3)} + \dfrac{7(4)}{6(4)} - \dfrac{1(2)}{12(2)}$

$= -\dfrac{9}{24} + \dfrac{28}{24} - \dfrac{2}{24}$

$= \dfrac{-9 + 28 - 2}{24}$

$= \dfrac{17}{24}$

**56.** $\dfrac{2(3y+4)}{2y+1} + \dfrac{3-y}{2y+1} = \dfrac{2(3y+4) + 3 - y}{2y+1}$

$= \dfrac{6y + 8 + 3 - y}{2y+1}$

$= \dfrac{5y + 11}{2y+1}$

**58.** $\dfrac{2}{x-10} + \dfrac{3}{4-x} = \dfrac{2(4-x)}{(x-10)(4-x)} + \dfrac{3(x-10)}{(x-10)(4-x)}$

$$= \dfrac{2(4-x) + 3(x-10)}{(x-10)(4-x)}$$

$$= \dfrac{8 - 2x + 3x - 30}{(x-10)(4-x)}$$

$$= \dfrac{x - 22}{(x-10)(4-x)}$$

**60.** $4 - \dfrac{4x}{x+6} + \dfrac{7}{x-5} = \dfrac{4(x+6)(x-5)}{(x+6)(x-5)} - \dfrac{4x(x-5)}{(x+6)(x-5)} + \dfrac{7(x+6)}{(x+6)(x-5)}$

$$= \dfrac{4(x+6)(x-5) - 4x(x-5) + 7(x+6)}{(x+6)(x-5)}$$

$$= \dfrac{4(x^2 + x - 30) - 4x^2 + 20x + 7x + 42}{(x+6)(x-5)}$$

$$= \dfrac{4x^2 + 4x - 120 - 4x^2 + 20x + 7x + 42}{(x+6)(x-5)}$$

$$= \dfrac{31x - 78}{(x+6)(x-5)}$$

**62.** $\dfrac{5}{x+2} + \dfrac{25-x}{x^2 - 3x - 10} = \dfrac{5}{x+2} + \dfrac{25-x}{(x+2)(x-5)}$

$$= \dfrac{5(x-5)}{(x+2)(x-5)} + \dfrac{25-x}{(x+2)(x-5)}$$

$$= \dfrac{5(x-5) + 25 - x}{(x+2)(x-5)}$$

$$= \dfrac{5x - 25 + 25 - x}{(x+2)(x-5)}$$

$$= \dfrac{4x}{(x+2)(x-5)}$$

**64.** $\dfrac{8}{y} - \dfrac{3}{y+5} + \dfrac{4}{y-2} = \dfrac{8(y+5)(y-2)}{y(y+5)(y-2)} - \dfrac{3y(y-2)}{y(y+5)(y-2)} + \dfrac{4y(y+5)}{y(y+5)(y-2)}$

$$= \dfrac{8(y+5)(y-2) - 3y(y-2) + 4y(y+5)}{y(y+5)(y-2)}$$

$$= \dfrac{8(y^2 + 3y - 10) - 3y^2 + 6y + 4y^2 + 20y}{y(y+5)(y-2)}$$

$$= \dfrac{8y^2 + 24y - 80 - 3y^2 + 6y + 4y^2 + 20y}{y(y+5)(y-2)}$$

$$= \dfrac{9y^2 + 50y - 80}{y(y+5)(y-2)}$$

**66.** $\dfrac{\left(x - 3 + \dfrac{2}{x}\right)}{\left(1 - \dfrac{2}{x}\right)} = \dfrac{\left(x - 3 + \dfrac{2}{x}\right)}{\left(1 - \dfrac{2}{x}\right)} \cdot \dfrac{x}{x}$

$= \dfrac{x^2 - 3x + 2}{x - 2}, \quad x \neq 0$

$= \dfrac{(x - 2)(x - 1)}{x - 2}, \quad x \neq 0$

$= x - 1, \quad x \neq 0, x \neq 2$

**68.** $\dfrac{\left(\dfrac{1}{x^2} - \dfrac{1}{y^2}\right)}{\left(\dfrac{1}{x} + \dfrac{1}{y}\right)} = \dfrac{\left(\dfrac{1}{x^2} - \dfrac{1}{y^2}\right)}{\left(\dfrac{1}{x} + \dfrac{1}{y}\right)} \cdot \dfrac{x^2 y^2}{x^2 y^2}$

$= \dfrac{y^2 - x^2}{xy^2 + x^2 y}$

$= \dfrac{(y - x)(y + x)}{xy(y + x)}$

$= \dfrac{y - x}{xy}, \quad x \neq -y$

**70.** Keystrokes:

$y_1$ Y= 1 ÷ X,T,θ − 3 ÷ ( X,T,θ + 3 ) ENTER

$y_2$ ( 3 − 2 X,T,θ ) ÷ ( X,T,θ ( X,T,θ + 3 ) ) GRAPH

$\dfrac{1}{x} - \dfrac{3}{x + 3} = \dfrac{x + 3}{x(x + 3)} - \dfrac{3x}{x(x + 3)}$

$= \dfrac{x + 3 - 3x}{x(x + 3)}$

$= \dfrac{3 - 2x}{x(x + 3)}$

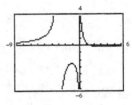

**72.** Keystrokes:

$y_1$ Y= ( X,T,θ ^ 3 − 2 X,T,θ $x^2$ − 7 ) ÷ ( X,T,θ − 2 ) ENTER

$y_2$ X,T,θ $x^2$ − 7 ÷ ( X,T,θ − 2 ) GRAPH

$$
\begin{array}{r|rrrr}
2 & 1 & -2 & 0 & -7 \\
  &   & 2 & 0 & 0 \\
\hline
  & 1 & 0 & 0 & -7
\end{array}
$$

$\dfrac{x^3 - 2x^2 - 7}{x - 2} = x^2 - \dfrac{7}{x - 2}$

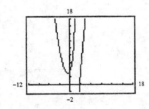

**74.** 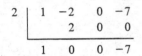 $\dfrac{10x + 15}{5x} = \dfrac{10x}{5x} + \dfrac{15}{5x}$

$= 2 + \dfrac{3}{x}$

**76.**

$$
\begin{array}{r}
4x^3 + 7x^2 + 7x + 32 + \dfrac{64}{x - 2} \\
x - 2 \overline{\smash{)}\, 4x^4 - x^3 - 7x^2 + 18x + 0} \\
\underline{4x^4 - 8x^3} \phantom{aaaaaaaaaaaaaaaaaa} \\
7x^3 - 7x^2 \phantom{aaaaaaaaaaaaa} \\
\underline{7x^3 - 14x^2} \phantom{aaaaaaaaaa} \\
7x^2 + 18x \phantom{aaaaaa} \\
\underline{7x^2 - 14x} \phantom{aaaa} \\
32x + 0 \\
\underline{32x - 64} \\
64
\end{array}
$$

**78.**

$$x^2 - 4x + 1 - \frac{1}{x+1}, \quad x \neq 1$$

$$
\begin{array}{r}
x^2 - 4x + 1 \\
x^2 - 1 \overline{\smash{)}\; x^4 - 4x^3 + 0x^2 + 3x + 0} \\
\underline{x^4 \qquad\; -\; x^2} \\
-4x^3 + x^2 + 3x \\
\underline{-4x^3 \qquad + 4x} \\
x^2 - x + 0 \\
\underline{x^2 \qquad -1} \\
-x + 1
\end{array}
$$

$$\frac{-x + 1}{x^2 - 1} = \frac{-1(\cancel{x - 1})}{(\cancel{x - 1})(x + 1)} = \frac{-1}{x + 1}$$

**80.**

$$
\begin{array}{r}
x^3 + 3x^2 + x + 8 - \dfrac{16x^2 - 34x + 24}{x^3 + x^2 - 4x + 3} \\
x^3 + x^2 - 4x + 3 \overline{\smash{)}\; x^6 + 4x^5 + 0x^4 + 0x^3 - 3x^2 + 5x + 0} \\
\underline{x^6 + x^5 - 4x^4 + 3x^3} \\
3x^5 + 4x^4 - 3x^3 - 3x^2 \\
\underline{3x^5 + 3x^4 - 12x^3 + 9x^2} \\
x^4 + 9x^3 - 12x^2 + 5x \\
\underline{x^4 + x^3 - 4x^2 + 3x} \\
8x^3 - 8x^2 + 2x + 0 \\
\underline{8x^3 + 8x^2 - 32x + 24} \\
-16x^2 + 34x - 24
\end{array}
$$

**82.**

$$
\begin{array}{r|rrrrr}
5 & 1 & -2 & -15 & -2 & 10 \\
& & 5 & 15 & 0 & -10 \\
\hline
& 1 & 3 & 0 & -2 & 0
\end{array}
$$

$$\frac{x^4 - 2x^3 - 15x^2 - 2x + 10}{x - 5} = x^3 + 3x^2 - 2, \; x \neq 5$$

**84.**

$$
\begin{array}{r|rrrr}
-\frac{1}{2} & 2 & 0 & 5 & -2 \\
& & -1 & \frac{1}{2} & -\frac{11}{4} \\
\hline
& 2 & -1 & \frac{11}{2} & -\frac{19}{4}
\end{array}
$$

$$(2x^3 + 5x - 2) \div \left(x + \frac{1}{2}\right) = 2x^2 - x + \frac{11}{2} - \frac{\frac{19}{4}}{x + \frac{1}{2}}$$

$$= 2x^2 - x + \frac{11}{2} - \frac{19}{4x - 2}$$

**86.**

$$\frac{t + 1}{6} = \frac{1}{2} - 2t$$

$$6\left(\frac{t + 1}{6}\right) = \left(\frac{1}{2} - 2t\right)6$$

$$t + 1 = 3 - 12t$$

$$13t = 2$$

$$t = \frac{2}{13}$$

**88.**

$$5 + \frac{2}{x} = \frac{1}{4}$$

$$4x\left(5 + \frac{2}{x}\right) = \left(\frac{1}{4}\right)4x$$

$$20x + 8 = x$$

$$19x = -8$$

$$x = -\frac{8}{19}$$

**90.**

$$\frac{7}{4x} - \frac{6}{8x} = 1$$

$$8x\left(\frac{7}{4x} - \frac{6}{8x}\right) = (1)8x$$

$$14 - 6 = 8x$$

$$8 = 8x$$

$$1 = x$$

**92.**   $\dfrac{2}{x} - \dfrac{x}{6} = \dfrac{2}{3}$

$6x\left(\dfrac{2}{x} - \dfrac{x}{6}\right) = \left(\dfrac{2}{3}\right)6x$

$12 - x^2 = 4x$

$0 = x^2 + 4x - 12$

$0 = (x - 2)(x + 6)$

$x = 2, \quad x = -6$

**94.**   $\dfrac{3}{y + 1} - \dfrac{8}{y} = 1$

$y(y + 1)\left(\dfrac{3}{y + 1} - \dfrac{8}{y}\right) = (1)y(y + 1)$

$3y - 8(y + 1) = y(y + 1)$

$3y - 8y - 8 = y^2 + y$

$-5y - 8 = y^2 + y$

$0 = y^2 + 6y + 8$

$0 = (y + 4)(y + 2)$

$y + 4 = 0 \qquad y + 2 = 0$

$y = -4 \qquad\quad y = -2$

**96.**   $\dfrac{2x}{x - 3} - \dfrac{3}{x} = 0$

$x(x - 3)\left(\dfrac{2x}{x - 3} - \dfrac{3}{x}\right) = (0)x(x - 3)$

$2x(x) - 3(x - 3) = 0$

$2x^2 - 3x + 9 = 0$

No real solution

**98.**   $\dfrac{3}{x - 1} + \dfrac{6}{x^2 - 3x + 2} = 2$

$\dfrac{3}{x - 1} + \dfrac{6}{(x - 1)(x - 2)} = 2$

$(x - 1)(x - 2)\left(\dfrac{3}{x - 1} + \dfrac{6}{(x - 1)(x - 2)}\right) = 2(x - 1)(x - 2)$

$3(x - 2) + 6 = 2(x - 1)(x - 2)$

$3x - 6 + 6 = 2(x^2 - 3x + 2)$

$3x = 2x^2 - 6x + 4$

$0 = 2x^2 - 9x + 4$

$0 = (2x - 1)(x - 4)$

$2x - 1 = 0 \qquad x - 4 = 0$

$2x = 1 \qquad\qquad x = 4$

$x = \dfrac{1}{2}$

**100.**

$$\frac{3}{x^2 - 9} + \frac{4}{x + 3} = 1$$

$$\frac{3}{(x - 3)(x + 3)} + \frac{4}{x + 3} = 1$$

$$(x + 3)(x - 3)\left(\frac{3}{(x + 3)(x - 3)} + \frac{4}{x + 3}\right) = (1)(x + 3)(x - 3)$$

$$3 + 4(x - 3) = (x + 3)(x - 3)$$

$$3 + 4x - 12 = x^2 - 9$$

$$4x - 9 = x^2 - 9$$

$$0 = x^2 - 4x$$

$$0 = x(x - 4)$$

$$x = 0 \qquad x - 4 = 0$$

$$x = 4$$

**102.** (a) Keystrokes:

$$\boxed{Y=}\ \boxed{X,T,\theta}\ \boxed{\div}\ 4\ \boxed{-}\ 2\ \boxed{\div}\ \boxed{X,T,\theta}\ \boxed{-}\ 1\ \boxed{\div}\ 2\ \boxed{GRAPH}$$

*x*-intercepts: $(-2, 0)$ and $(4, 0)$

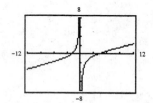

(b)

$$0 = \frac{x}{4} - \frac{2}{x} - \frac{1}{2}$$

$$4x(0) = \left(\frac{x}{4} - \frac{2}{x} - \frac{1}{2}\right)4x$$

$$0 = x^2 - 8 - 2x$$

$$0 = x^2 - 2x - 8$$

$$0 = (x - 4)(x + 2)$$

$$x - 4 = 0 \qquad x + 2 = 0$$

$$x = 4 \qquad\qquad x = -2$$

**104.** Domain of $\overline{C} = \dfrac{15{,}000 + 0.75x}{x}$ is $\{1, 2, 3, 4, \ldots\}$.

**106.** *Verbal Model:*

$$\boxed{\text{Distance}} = \boxed{\text{Rate}} \cdot \boxed{\text{Time}}$$

$$\boxed{\begin{array}{c}\text{Original}\\ \text{trip time}\end{array}} = \boxed{\begin{array}{c}\text{Return}\\ \text{trip time}\end{array}} + \frac{1}{3}$$

*Labels:*   Speed of return trip $= x$

Speed of original trip $= x - 5$

*Equation:*

$$\frac{220}{x - 5} = \frac{220}{x} + \frac{1}{3}$$

$$3x(x - 5)\left(\frac{220}{x - 5}\right) = \left(\frac{220}{x} + \frac{1}{3}\right)3x(x - 5)$$

$$660x = 660(x - 5) + x(x - 5)$$

$$660x = 660x - 3300 + x^2 - 5x$$

$$0 = x^2 - 5x - 3300$$

$$0 = (x - 60)(x + 55)$$

$$x = 60, \quad x = -55$$

$$x = 60 \text{ miles per hour}$$

**108.** *Verbal Model:*  $\boxed{\dfrac{\text{Hits}}{\text{At bats}}} = 0.350$

*Label:*  Consecutive times $= x$

*Equation:*
$$\frac{23 + x}{75 + x} = 0.350$$

$$(75 + x)\left(\frac{23 + x}{75 + x}\right) = (0.350)(75 + x)$$

$$23 + x = 26.25 + 0.350x$$

$$0.65x = 3.25$$

$$x = 5$$

**110.** *Verbal Model:*  $\boxed{\begin{array}{c}\text{Share per}\\\text{person now}\end{array}} = \boxed{\begin{array}{c}\text{Share per}\\\text{person later}\end{array}} + 1200$

*Labels:*  People presently in group $= x$
People in new group $= x + 3$

*Equation:*
$$\frac{28{,}000}{x} = \frac{28{,}000}{x + 3} + 1200$$

$$x(x + 3)\left(\frac{28{,}000}{x}\right) = \left(\frac{28{,}000}{x + 3} + 1200\right)x(x + 3)$$

$$28{,}000(x + 3) = 28{,}000x + 1200x(x + 3)$$

$$28{,}000x + 84{,}000 = 28{,}000x + 1200x^2 + 3600x$$

$$0 = 1200x^2 + 3600x - 84{,}000$$

$$0 = x^2 + 3x - 70$$

$$0 = (x + 10)(x - 7)$$

$$x = -10, \quad x = 7 \text{ people}$$

**112.** *Verbal Model:*  $\boxed{\begin{array}{c}\text{Rate of}\\\text{person 1}\end{array}} + \boxed{\begin{array}{c}\text{Rate of}\\\text{person 2}\end{array}} = \boxed{\begin{array}{c}\text{Rate}\\\text{together}\end{array}}$

*Labels:*  Supervisor's time $= 21$ minutes
Your time $= 24$ minutes
Time together $= x$

*Equation:*
$$\frac{1}{21} + \frac{1}{24} = \frac{1}{x}$$

$$168x\left(\frac{1}{21} + \frac{1}{24}\right) = \left(\frac{1}{x}\right)168x$$

$$8x + 7x = 168$$

$$15x = 168$$

$$x = 11 \text{ minutes, } 12 \text{ seconds}$$

# CHAPTER 5
# Radicals and Complex Numbers

# CHAPTER 5
# Radicals and Complex Numbers

## Section 5.1    Radicals and Rational Exponents

**Solutions to Even-Numbered Exercises**

**2.** $-\sqrt{100} = -10$
because $10 \cdot 10 = 100$

**4.** $\sqrt{-25} =$ not a real number
because no real number multiplied
by itself yields $-25$.

**6.** $\sqrt[3]{-64} = -4$
because $-4 \cdot -4 \cdot -4 = -64$

**8.** $-\sqrt[3]{1} = -1$
because $1 \cdot 1 \cdot 1 = 1$

**10.** Because $24.5^2 = 600.25$,
24.5 is a square root.

**12.** Because $6^4 = 1296$,
6 is a fourth root of 1296.

**14.** Because $12^3 = 1728$, 12 is the
cube root of 1728.

**16.** $-\sqrt{10^2} = -10$

(index is even)

**18.** $\sqrt{(-12)^2} = |-12| = 12$

(index is even)

**20.** $\sqrt{-12^2} =$ not a real number

(even root of a negative number)

**22.** $\sqrt{\left(\frac{3}{4}\right)^2} = \frac{3}{4}$

(index is even)

**24.** $\sqrt{\left(-\frac{3}{5}\right)^2} = \left|-\frac{3}{5}\right| = \frac{3}{5}$

(index is even)

**26.** $(\sqrt{10})^2 = -10$

(inverse property
of powers and roots)

**28.** $\left(-\sqrt{18}\right)^2 = 18$

(inverse property
of powers and roots

**30.** $\sqrt[3]{(-2)^3} = -2$

(index is odd)

**32.** $\sqrt[3]{4^3} = 4$

(index is odd)

**34.** $-\sqrt[3]{9^3} = -9$

(inverse property
of powers and roots)

**36.** $-\sqrt[3]{\left(\frac{1}{5}\right)^3} = \frac{-1}{5}$

(index is odd)

**38.** $\left(\sqrt[3]{-6}\right)^3 = -6$

(inverse property
of powers and roots)

**40.** $\left(\sqrt[3]{21}\right)^3 = 21$

(inverse property
of powers and roots)

**42.** $\sqrt[5]{(-2)^5} = -2$

(inverse property
of powers and roots)

**44.** $-\sqrt[4]{24} = -2$

(inverse property
of powers and roots)

**46.** $\sqrt{\frac{9}{16}}$ is rational, because
$\frac{3}{4} \cdot \frac{3}{4} = \frac{9}{16}$ is a perfect square.

**48.** $\sqrt{72}$ is not rational because
72 is not a perfect square.

**50.** *Radical Form*      *Rational Exponent Form*
$\sqrt[4]{81} = 3$           $81^{1/4} = 3$

**52.** *Radical Form*     *Rational Exponent Form*
$\sqrt[3]{125} = 5$          $125^{1/3} = 5$

**54.** *Radical Form*     *Rational Exponent Form*
$\sqrt[3]{27^2} = 9$          $27^{2/3} = 9$

**56.** $49^{1/2} = \sqrt{49} = 7$    Root is 2. Power is 1.

**58.** $-121^{1/2} = -\sqrt{121} = -11$

Root is 2. Power is 1.

**60.** $-(125)^{2/3} = -\left(\sqrt[3]{125}\right)^2 = -5^2 = -25$

Root is 3. Power is 2.

**62.** $81^{-3/4} = \dfrac{1}{\left(\sqrt[4]{81}\right)^3} = \dfrac{1}{3^3} = \dfrac{1}{27}$

Root is 4. Power is 3.

**64.** $(-243)^{-3/5} = \dfrac{1}{(-243)^{3/5}} = \dfrac{1}{\left(\sqrt[5]{-243}\right)^3} = \dfrac{1}{-3^3} = \dfrac{1}{-27}$

Root is 5. Power is 3.

**66.** $\left(\dfrac{256}{625}\right)^{1/4} = \dfrac{\sqrt[4]{256}}{\sqrt[4]{625}} = \dfrac{4}{5}$

Root is 4. Power is 1.

**68.** $\left(\dfrac{27}{1000}\right)^{-4/3} = \left(\dfrac{1000}{27}\right)^{4/3} = \dfrac{\left(\sqrt[3]{1000}\right)^4}{\left(\sqrt[3]{27}\right)^4} = \dfrac{10^4}{3^4} = \dfrac{10{,}000}{81}$

Root is 3. Power is 4.

**70.** $(8^2)^{3/2} = 8^{2 \cdot 3/2} = 8^3 = 512$

Root is 2. Power is 3.

**72.** $(-2^3)^{5/3} = (-2)^{3 \cdot 5/3} = (-2)^5 = -32$

Root is 3. Power is 5.

**74.** $\left(\dfrac{4}{6^2}\right)^{-3/2} = \dfrac{1}{\left(\dfrac{4}{6^2}\right)^{3/2}} = \left(\dfrac{6^2}{4}\right)^{3/2} = \dfrac{6^{2 \cdot 3/2}}{4^{3/2}} = \dfrac{6^3}{\left(\sqrt{4}\right)^3} = \dfrac{216}{8} = 27$

**76.** $\sqrt[3]{x} = x^{1/3}$

Root is 3. Power is 1.

**78.** $t\sqrt[5]{t^2} = t \cdot t^{2/5} = t^{1+2/5} = t^{7/5}$

Root is 5. Power is 2.

**80.** $y\sqrt[4]{y^2} = y \cdot y^{2/4} = y^{1+2/4} = y^{3/2}$

Root is 4. Power is 2.

**82.** $n^3\sqrt[4]{n^6} = n^3 \cdot n^{6/4} = n^{3+6/4}$

$= n^{6/2+3/2} = n^{9/2}$

Root is 4. Power is 6.

**84.** $\dfrac{\sqrt[3]{x^2}}{\sqrt[3]{x^4}} = \dfrac{x^{2/3}}{x^{4/3}} = x^{2/3-4/3} = x^{-2/3} = \dfrac{1}{x^{2/3}}$

**86.** $\dfrac{\sqrt[3]{x^4}}{\sqrt{x^3}} = \dfrac{x^{4/3}}{x^{3/2}} = x^{4/3-3/2} = x^{8/6-9/6} = x^{-1/6} = \dfrac{1}{x^{1/6}}$

**88.** $\sqrt[5]{z^3} \cdot \sqrt[5]{z^2} \cdot = z^{3/5} \cdot z^{2/5} = z^{3/5+2/5} = z^{5/5} = z$

**90.** $\sqrt[6]{x^5} \cdot \sqrt[3]{x^4} = x^{5/6} \cdot x^{4/3} = x^{5/6+4/3} = x^{5/6+8/6} = x^{13/6}$

**92.** $\sqrt[3]{u^4v^2} = (u^4v^2)^{1/3} = u^{4/3}v^{2/3}$

**94.** $x^2\sqrt[3]{xy^4} = x^2 \cdot (xy^4)^{1/3} = x^2 \cdot x^{1/3} \cdot y^{4/3} = x^{2+1/3}y^{4/3} = x^{7/3}y^{4/3}$

**96.** $2^{2/5} \cdot 2^{3/5} = 2^{2/5+3/5} = 2^{5/5} = 2$

**98.** $(4^{1/3})^{9/4} = 4^{3/4}$

**100.** $\dfrac{5^{-3/4}}{5} = \dfrac{1}{5^{1-(-3/4)}} = \dfrac{1}{5^{1+3/4}} = \dfrac{1}{5^{7/4}}$

**102.** $(k^{-1/3})^{3/2} = k^{-1/2} = \dfrac{1}{k^{1/2}} = \dfrac{1}{\sqrt{k}}$

**104.** $\dfrac{a^{3/4} \cdot a^{1/2}}{a^{5/2}} = \dfrac{a^{3/4+1/2}}{a^{5/2}} = \dfrac{a^{3/4+2/4}}{a^{5/2}} = \dfrac{a^{5/4}}{a^{5/2}} = \dfrac{1}{a^{5/2-5/4}} = \dfrac{1}{a^{10/4-5/4}} = \dfrac{1}{a^{5/4}}$

**106.** $(-2u^{3/5}v^{-1/5})^3 = (-2)^3u^{9/5}v^{-3/5} = -\dfrac{8u^{9/5}}{v^{3/5}}$

**108.** $\left(\dfrac{3m^{1/6}n^{1/3}}{4n^{-2/3}}\right)^2 = \dfrac{3^2m^{1/3}n^{2/3}}{4^2n^{-4/3}} = \dfrac{9m^{1/3}n^{2/3-(-4/3)}}{16} = \dfrac{9m^{1/3}n^{6/3}}{16} = \dfrac{9m^{1/3}n^2}{16}$

**110.** $\sqrt[3]{\sqrt{2x}} = ((2x)^{1/2})^{1/3} = (2x)^{1/6} = \sqrt[6]{2x}$

**112.** $\sqrt[5]{\sqrt[3]{y^4}} = \sqrt[5]{y^{4/3}} = (y^{4/3})^{1/5} = y^{4/3\cdot 1/5} = y^{4/15}$

**114.** $\dfrac{(a-b)^{1/3}}{\sqrt[3]{a-b}} = \dfrac{(a-b)^{1/3}}{(a-b)^{1/3}} = (a-b)^{1/3-1/3} = (a-b)^0 = 1$

**116.** $\dfrac{\sqrt[4]{2x+y}}{(2x+y)^{3/2}} = \dfrac{(2x+y)^{1/4}}{(2x+y)^{3/2}} = (2x+y)^{1/4-3/2} = (2x+y)^{1/4-6/4} = (2x+y)^{-5/4} = \dfrac{1}{(2x+y)^{5/4}}$

**118.** $\sqrt{-532}$ is not a real number, because the even root of a negative number is not real.

**120.** $962^{2/3} \approx 97.4503$

Scientific: 962 $\boxed{y^x}$ $\boxed{(}$ 2 $\boxed{\div}$ 3 $\boxed{)}$ $\boxed{=}$

Graphing: 962 $\boxed{\wedge}$ $\boxed{(}$ 2 $\boxed{\div}$ 3 $\boxed{)}$ $\boxed{\text{ENTER}}$

**122.** $382.5^{-3/2} \approx 0.0001$

Scientific: 382.5 $\boxed{y^x}$ $\boxed{(}$ 3 $\boxed{\div}$ 2 $\boxed{+/-}$ $\boxed{)}$ $\boxed{=}$

Graphing: 382.5 $\boxed{\wedge}$ $\boxed{(}$ $\boxed{(-)}$ 3 $\boxed{\div}$ 2 $\boxed{)}$ $\boxed{\text{ENTER}}$

**124.** $\sqrt[3]{159} \approx 5.4175$

$(159)^{1/3}$

Scientific: 159 $\boxed{y^x}$ $\boxed{(}$ 1 $\boxed{\div}$ 3 $\boxed{)}$ $\boxed{=}$

Graphing: 159 $\boxed{\wedge}$ $\boxed{(}$ 1 $\boxed{\div}$ 3 $\boxed{)}$ $\boxed{\text{ENTER}}$

**126.** $\sqrt[5]{-35^3} \approx -8.4419$

$\sqrt[5]{-1} \cdot \sqrt[5]{35^3} = -1 \cdot 35^{3/5}$

Scientific: 35 3 $\boxed{y^x}$ $\boxed{(}$ 5 $\boxed{\div}$ $\boxed{)}$ $\boxed{+/-}$ $\boxed{=}$

Graphing: 5 $\boxed{\sqrt[x]{\phantom{x}}}$ $\boxed{(-)}$ 35 $\boxed{\wedge}$ 3 $\boxed{\text{ENTER}}$

**128.** $\dfrac{-5+\sqrt{3215}}{10} \approx 5.1701$

Scientific: $\boxed{(}$ 5 $\boxed{+/-}$ $\boxed{+}$ 3215 $\boxed{\sqrt{\phantom{x}}}$ $\boxed{)}$ $\boxed{\div}$ 10 $\boxed{=}$

Graphing: $\boxed{(}$ $\boxed{(-)}$ 5 $\boxed{+}$ $\boxed{\sqrt{\phantom{x}}}$ 3215 $\boxed{)}$ $\boxed{\div}$ 10 $\boxed{\text{ENTER}}$

**130.** $\dfrac{7-\sqrt{241}}{12} \approx -0.7103$

Scientific: $\boxed{(}$ 7 $\boxed{-}$ 241 $\boxed{\sqrt{\phantom{x}}}$ $\boxed{)}$ $\boxed{\div}$ 12

Graphing: $\boxed{(}$ 7 $\boxed{-}$ $\boxed{\sqrt{\phantom{x}}}$ 241 $\boxed{)}$ $\boxed{\div}$ 12

**132.** The domain of $h(x) = \sqrt[4]{x}$ is the set of all nonnegative real numbers or $[0, \infty)$.

**134.**    $\sqrt[3]{x} \neq 0$

$(\sqrt[3]{x})^3 \neq 0^3$

$x \neq 0$

Domain: $(-\infty, 0) \cup (0, \infty)$

**136.** The domain of $f(x) = \sqrt[3]{x^4}$ is all reals $(-\infty, \infty)$.

**138.** *Keystrokes:*

$\boxed{Y=}$ 4 $\boxed{\text{MATH}}$ 4 $\boxed{X,T,\theta}$ $\boxed{\text{GRAPH}}$

Domain is $(-\infty, \infty)$ so the graphing utility did complete the graph.

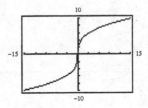

**140.** *Keystrokes:*

$\boxed{Y=}$ 5 $\boxed{X,T,\theta}$ $\boxed{\wedge}$ $\boxed{(}$ 2 $\boxed{\div}$ 3 $\boxed{)}$ $\boxed{GRAPH}$

Domain is $(-\infty, \infty)$ so the graphing utility did complete the graph.

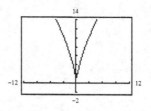

**142.** $x^{4/3}(3x^2 - 4x + 5) = 3x^{2+4/3} - 4x^{1+4/3} + 5x^{4/3}$

$\qquad\qquad = 3x^{6/3+4/3} - 4x^{2/3+4/3} + 5x^{4/3}$

$\qquad\qquad = 3x^{10/3} - 4x^{7/3} + 5x^{4/3}$

**144.** $(x^{1/2} - 3)(x^{1/2} + 3) = x^{1/2+1/2} + 3x^{1/2} - 3x^{1/2} - 9$

$\qquad\qquad = x^1 - 9$

$\qquad\qquad = x - 9$

**146.** $r = 1 - \left(\dfrac{s}{c}\right)^{1/n}$

$\quad r = 1 - \left(\dfrac{25{,}000}{125{,}000}\right)^{1/10}$

$\quad\ \ = 1 - \left(\dfrac{1}{5}\right)^{1/10}$

$\quad\ \ \approx 0.149$

$\quad\ \ \approx 14.9\%$

**148.** *Verbal model:* $\boxed{\text{Area}} = \boxed{\text{Side}} \cdot \boxed{\text{Side}}$

*Labels:* $\qquad$ Area = 1024

$\qquad\qquad\qquad$ Side = $x$

*Equation:* $\qquad 1024 = x \cdot x$

$\qquad\qquad\qquad 1024 = x^2$

$\qquad\qquad\qquad \sqrt{1024} = x$

$\qquad\qquad\qquad\quad 32 = x$

32 inches $\times$ 32 inches

**150.** (a) particle size = $0.03\sqrt{v}$ inches

$\qquad$ particle size = $0.03\sqrt{\dfrac{3}{4}}$

$\qquad$ particle size = $0.0259808$ inch $\approx 0.026$ inch

(b) particle size = $0.03\sqrt{v}$ inch

$\qquad$ particle size = $0.03\sqrt{\dfrac{3}{16}}$

$\qquad\qquad\qquad = 0.0129903811$ inch

$\qquad\qquad\qquad \approx 0.013$ inch

**152.** If $a$ and $b$ are real numbers, $n$ is an integer greater than or equal to 2, and $a = b^n$, then $b$ is the $n$th root of $a$.

**154.** If the $n$th root is a real number and $n$ is even, the radicand must be nonnegative.

**156.** (a) $\sqrt[n]{x^n} = x$

$\qquad$ $n$ is odd.

(b) $\sqrt[n]{x^n} = |x|$

$\qquad$ $n$ is even because $x$ cannot be negative.

## Section 5.2    Simplifying Radical Expressions

**2.** $\sqrt{27} = \sqrt{9 \cdot 3} = \sqrt{3^2 \cdot 3} = 3\sqrt{3}$

**4.** $\sqrt{125} = \sqrt{5^2 \cdot 5} = 5\sqrt{5}$

**6.** $\sqrt{84} = \sqrt{4 \cdot 21} = \sqrt{2^2 \cdot 21}$

$\qquad\ = 2\sqrt{21}$

**8.** $\sqrt{147} = \sqrt{49 \cdot 3} = \sqrt{7^2 \cdot 3} = 7\sqrt{3}$

**10.** $\sqrt{1176} = \sqrt{49 \cdot 4 \cdot 6} = \sqrt{7^2 \cdot 2^2 \cdot 6}$

$\qquad\quad = 14\sqrt{6}$

**12.** $\sqrt{0.25} = \sqrt{25 \cdot .01} = \sqrt{5^2 \cdot .01} = 5 \cdot .1 = 0.5$

**14.** $\sqrt{0.0027} = \sqrt{9 \cdot 3 \cdot .0001} = \sqrt{3^2 \cdot 3 \cdot .0001} = 3 \cdot .01\sqrt{3} = 0.03\sqrt{3}$

**16.** $\sqrt{9.8} = \sqrt{49 \cdot \dfrac{2}{10} \cdot \dfrac{10}{10}} = \dfrac{7\sqrt{4 \cdot 5}}{\sqrt{100}} = \dfrac{14}{10}\sqrt{5} = 1.4\sqrt{5}$

**18.** $\sqrt{\dfrac{5}{36}} = \dfrac{\sqrt{5}}{\sqrt{36}} = \dfrac{\sqrt{5}}{\sqrt{6^2}} = \dfrac{\sqrt{5}}{6}$

**20.** $\sqrt{\dfrac{15}{36}} = \dfrac{\sqrt{15}}{\sqrt{36}} = \dfrac{\sqrt{15}}{6}$

**22.** $\sqrt{64x^3} = \sqrt{8^2 \cdot x^2 \cdot x} = 8|x|\sqrt{x} = 8x\sqrt{x}$

**24.** $\sqrt{32x} = \sqrt{16 \cdot 2x}$
$= 4\sqrt{2x}$

**26.** $\sqrt{160x^8} = \sqrt{16 \cdot 10 \cdot x^8} = 4\sqrt{10}x^4$

**28.** $\sqrt{125u^4v^6} = \sqrt{25 \cdot 5u^4v^6} = 5\sqrt{5}u^2|v^3|$

**30.** $\sqrt{363x^{10}y^9} = \sqrt{121 \cdot 3 \cdot x^{10} \cdot y^8 \cdot y} = 11|x^5|y^4\sqrt{3y}$

**32.** $\sqrt[3]{81} = \sqrt[3]{27 \cdot 3} = 3\sqrt[3]{3}$

**34.** $\sqrt[4]{112} = \sqrt[4]{16 \cdot 7} = 2\sqrt[4]{7}$

**36.** $\sqrt[3]{54z^7} = \sqrt[3]{27 \cdot 2 \cdot z^6 \cdot z} = 3z^2\sqrt[3]{2z}$

**38.** $\sqrt[5]{160x^8} = \sqrt[5]{32 \cdot 5 \cdot x^5 \cdot x^3} = 2|x|\sqrt[5]{5x^3}$

**40.** $\sqrt[3]{a^5b^6} = \sqrt[3]{a^3 \cdot a^2 \cdot b^3 \cdot b^3} = ab^2\sqrt[3]{a^2}$

**42.** $\sqrt[4]{128u^4v^7} = \sqrt[4]{2^4 \cdot 8u^4 \cdot v^4 \cdot v^3} = 2|uv|\sqrt[4]{8v^3}$

**44.** $\sqrt[3]{16x^4y^5} = \sqrt[3]{2^3 \cdot 2 \cdot x^3 \cdot x \cdot y^3 \cdot y^2} = 2xy\sqrt[3]{2xy^2}$

**46.** $\sqrt[4]{\dfrac{5}{16}} = \dfrac{\sqrt[4]{5}}{\sqrt[4]{16}} = \dfrac{\sqrt[4]{5}}{2}$

**48.** $\sqrt[3]{\dfrac{1}{1000}} = \dfrac{\sqrt[3]{1}}{\sqrt[3]{1000}} = \dfrac{1}{10}$

**50.** $\sqrt[3]{\dfrac{16z^3}{y^6}} = \sqrt[3]{\dfrac{2^3 \cdot 2 \cdot z^3}{y^3 \cdot y^3}} = \dfrac{\sqrt[3]{2^3 \cdot 2 \cdot z^3}}{\sqrt[3]{y^3 \cdot y^3}} = \dfrac{2z\sqrt[3]{2}}{y^2}$

**52.** $\sqrt[4]{\dfrac{3u^2}{16v^8}} = \dfrac{\sqrt[4]{3u^2}}{\sqrt[4]{16u^8}} = \dfrac{\sqrt[4]{3u^2}}{\sqrt[4]{2^4 \cdot v^4 \cdot v^4}} = \dfrac{\sqrt[4]{3u^2}}{2v^2}$

**54.** $\sqrt{\dfrac{18x^2}{z^6}} = \dfrac{\sqrt{3^2 \cdot 2x^2}}{\sqrt{z^2 \cdot z^2 \cdot z^2}} = \dfrac{3|x|\sqrt{2}}{z^3}$

**56.** $\sqrt[5]{96x^5} = \sqrt[5]{32 \cdot 3 \cdot x^5} = 2x\sqrt[5]{3}$

**58.** $\sqrt{\dfrac{1}{5}} = \dfrac{\sqrt{1}}{\sqrt{5}} = \dfrac{1}{\sqrt{5}} = \dfrac{1}{\sqrt{5}} \cdot \dfrac{\sqrt{5}}{\sqrt{5}} = \dfrac{\sqrt{5}}{5}$

**60.** $\dfrac{1}{\sqrt{15}} = \dfrac{1}{\sqrt{15}} \cdot \dfrac{\sqrt{15}}{\sqrt{15}} = \dfrac{\sqrt{15}}{15}$

**62.** $\dfrac{5}{\sqrt{10}} = \dfrac{5}{\sqrt{10}} \cdot \dfrac{\sqrt{10}}{\sqrt{10}} = \dfrac{5\sqrt{10}}{10} = \dfrac{\sqrt{10}}{2}$

**64.** $\sqrt[3]{\dfrac{9}{25}} = \dfrac{\sqrt[3]{9}}{\sqrt[3]{25}} = \dfrac{\sqrt[3]{9}}{\sqrt[3]{5^2}} = \dfrac{\sqrt[3]{9}}{\sqrt[3]{5^2}} \cdot \dfrac{\sqrt[3]{5}}{\sqrt[3]{5}} = \dfrac{\sqrt[3]{45}}{5}$

**66.** $\dfrac{10}{\sqrt[5]{16}} = \dfrac{10}{\sqrt[5]{2^4}} = \dfrac{10}{\sqrt[5]{2^4}} \cdot \dfrac{\sqrt[5]{2}}{\sqrt[5]{2}} = \dfrac{10\sqrt[5]{2}}{2} = 5\sqrt[5]{2}$

**68.** $\sqrt{\dfrac{5}{c}} = \dfrac{\sqrt{5}}{\sqrt{c}} = \dfrac{\sqrt{5}}{\sqrt{c}} \cdot \dfrac{\sqrt{c}}{\sqrt{c}} = \dfrac{\sqrt{5c}}{c}$

**70.** $\sqrt{\dfrac{4}{x^3}} = \dfrac{\sqrt{4}}{\sqrt{x^3}} \cdot \dfrac{\sqrt{x}}{\sqrt{x}} = \dfrac{\sqrt{4x}}{\sqrt{x^4}} = \dfrac{\sqrt{4x}}{x^2} = \dfrac{2\sqrt{x}}{x^2}$

**72.** $\dfrac{5}{\sqrt{8x^5}} = \dfrac{5}{\sqrt{8x^5}} \cdot \dfrac{\sqrt{2x}}{\sqrt{2x}} = \dfrac{5\sqrt{2x}}{\sqrt{16x^6}} = \dfrac{5\sqrt{2x}}{4x^3}$

**74.** $\dfrac{1}{\sqrt{xy}} = \dfrac{1}{\sqrt{xy}} \cdot \dfrac{\sqrt{xy}}{\sqrt{xy}} = \dfrac{\sqrt{xy}}{xy}$

**76.** $\sqrt[3]{\dfrac{20x^2}{9y^2}} = \dfrac{\sqrt[3]{20x^2}}{\sqrt[3]{9y^2}} = \dfrac{\sqrt[3]{20x^2}}{\sqrt[3]{3^2y^2}} = \dfrac{\sqrt[3]{20x^2}}{\sqrt[3]{3^2y^2}} \cdot \dfrac{\sqrt[3]{3y}}{\sqrt[3]{3y}} = \dfrac{\sqrt[3]{60x^2y}}{3y}$

**78.** $\dfrac{3u^2}{\sqrt[4]{8u^3}} = \dfrac{3u^2}{\sqrt[4]{2^3u^3}} = \dfrac{3u^2}{\sqrt[4]{2^3u^3}} \cdot \dfrac{\sqrt[4]{2u}}{\sqrt[4]{2u}} = \dfrac{3u^2\sqrt[4]{2u}}{2u} = \dfrac{3u\sqrt[4]{2u}}{2}$

**80.** $6\sqrt{5} - 2\sqrt{5} = (6-2)\sqrt{5}$
$= 4\sqrt{5}$

**82.** $4\sqrt{32} + 7\sqrt{32} = 11\sqrt{32} = 11\sqrt{16\cdot 2} = 11(4)\sqrt{2} = 44\sqrt{2}$

**84.** $9\sqrt[3]{17} + 7\sqrt[3]{2} - 4\sqrt[3]{17} + \sqrt[3]{2} = 5\sqrt[3]{17} + 8\sqrt[3]{2}$

**86.** $4\sqrt[4]{48} - \sqrt[4]{243} = 4\sqrt[4]{2^4\cdot 3} - \sqrt[4]{3^4\cdot 3}$
$= 4(2)\sqrt[4]{3} - 3\sqrt[4]{3}$
$= 8\sqrt[4]{3} - 3\sqrt[4]{3}$
$= 5\sqrt[4]{3}$

**88.** $3\sqrt{x+1} + 10\sqrt{x+1} = 13\sqrt{x+1}$

**90.** $\sqrt[3]{16t^4} - \sqrt[3]{54t^4} = \sqrt[3]{2^3\cdot 2t^3\cdot t} - \sqrt[3]{3^3\cdot 2t^3\cdot t}$
$= 2t\sqrt[3]{2t} - 3t\sqrt[3]{2t}$
$= -t\sqrt[3]{2t}$

**92.** $5\sqrt[3]{24u^2} + 2\sqrt[3]{81u^5} = 5\sqrt[3]{2^3\cdot 3u^2} + 2\sqrt[3]{3^3\cdot 3u^3\cdot u^2}$
$= 5(2)\sqrt[3]{3u^2} + 2(3u)\sqrt[3]{3u^2}$
$= 10\sqrt[3]{3u^2} + 6u\sqrt[3]{3u^2}$
$= (10 + 6u)\sqrt[3]{3u^2}$

**94.** $\sqrt{10} + \dfrac{5}{\sqrt{10}} = \sqrt{10} + \dfrac{5}{\sqrt{10}}\cdot\dfrac{\sqrt{10}}{\sqrt{10}}$
$= \sqrt{10} + \dfrac{5\sqrt{10}}{10}$
$= \dfrac{10\sqrt{10}}{10} + \dfrac{5\sqrt{10}}{10}$
$= \dfrac{15\sqrt{10}}{10}$
$= \dfrac{3\sqrt{10}}{2}$

**96.** $\dfrac{x}{\sqrt{3x}} + \sqrt{27x} = \dfrac{x}{\sqrt{3x}} + \sqrt{3^2\cdot 3x}$
$= \dfrac{x}{\sqrt{3x}} + 3\sqrt{3x}$
$= \dfrac{x\sqrt{3x}}{3x} + 3\sqrt{3x}$
$= \dfrac{\sqrt{3x}}{3} + 3\sqrt{3x}$
$= \dfrac{\sqrt{3x}}{3} + \dfrac{9\sqrt{3x}}{3}$
$= \dfrac{10\sqrt{3x}}{3}$

**98.** $\sqrt{\dfrac{4}{3x^3}} + \sqrt{3x^3} = \dfrac{\sqrt{4}}{\sqrt{3x^3}} + \sqrt{3x^3} = \dfrac{\sqrt{4}}{\sqrt{3x^3}}\cdot\dfrac{\sqrt{3x}}{\sqrt{3x}} + \sqrt{3x^3} = \dfrac{2\sqrt{3x}}{3x^2} + \sqrt{3x^3}$
$= \dfrac{2\sqrt{3x}}{3x^2} + x\sqrt{3x}\cdot\dfrac{3x^2}{3x^2} = \dfrac{2\sqrt{3x}}{3x^2} + \dfrac{3x^3\sqrt{3x}}{3x^2} = \dfrac{2\sqrt{3x} + 3x^3\sqrt{3x}}{3x^2}$
$= \dfrac{(2 + 3x^3)\sqrt{3x}}{3x^2}$

**100.** $\sqrt{10} - \sqrt{6} < \sqrt{10-6}$

**102.** $5 = \sqrt{3^2 + 4^2}$

**104.** $c = \sqrt{a^2 + b^2}$

$c = \sqrt{(6)^2 + (4)^2}$

$c = \sqrt{36 + 16}$

$c = \sqrt{52}$

$c = \sqrt{4 \cdot 13}$

$c = 2\sqrt{13}$

**106.** $c = \sqrt{a^2 + b^2}$

$c = \sqrt{(10)^2 + (5)^2}$

$c = \sqrt{100 + 25}$

$c = \sqrt{125}$

$c = \sqrt{25 \cdot 5}$

$c = 5\sqrt{5}$

**108.** Hypotenuse of right triangles cut off $= \sqrt{(2)^2 + (2)^2}$

$= \sqrt{4 + 4}$

$= \sqrt{8}$

$= \sqrt{4 \cdot 2}$

$= 2\sqrt{2}$

Perimeter $= 4 + 2\sqrt{2} + 2\sqrt{2} + 4 + 2\sqrt{2} + 2\sqrt{2}$

$= 8 + 8\sqrt{2}$

$= 8\left(1 + \sqrt{2}\right)$ (feet) $\approx 19.3$ feet

**110.** $T = 2\pi\sqrt{\frac{4}{32}}$

$= 2\pi\sqrt{\frac{1}{8}}$

$= 2.2214415$

$\approx 2.22$ seconds

**112.** The display appears to be approaching 1.

**114.** (a) $1^2 + 1^2 = r_1^2$

$1 + 1 = r_1^2$

$2 = r_1^2$

$\sqrt{2} = r_1$

(b) $r_2^2 = 1^2 + \sqrt{2}^2$

$r_2^2 = 1 + 2$

$r_2^2 = 3$

$r_2 = \sqrt{3}$

$r_3^2 = 1^2 + \sqrt{3}^2$

$r_3^2 = 1 + 3$

$r_3^2 = 4$

$r_3 = \sqrt{4}$

$r_4^2 = 1^2 + \sqrt{4}^2$

$r_4^2 = 1 + 4$

$r_4^2 = 5$

$r_4 = \sqrt{5}$

$r_5^2 = 1^2 + \sqrt{5}^2$

$r_5^2 = 1 + 5$

$r_5^2 = 6$

$r_5 = \sqrt{6}$

$r_6^2 = 1^2 + \sqrt{6}^2$

$r_6^2 = 1 + 6$

$r_6^2 = 7$

$r_6 = \sqrt{7}$

(c) $r_n^2 = 1^2 + \sqrt{n}^2$

$r_n^2 = 1 + n$

$r_n^2 = (1 + n)$

$r_n = \sqrt{1 + n}$

**116.** A simplified radical expression:

(a) All possible factors have been removed from the radical.

(b) No radical contains a fraction.

(c) No denominator of a fraction contains a radical.

**118.** To simplify

$\frac{1}{\sqrt{3}}$

multiply numerator and denominator by $\sqrt{3}$.

**120.** Two radical expressions are alike if they have the same index and the same radicand.

# Section 5.3    Multiplying and Dividing Radical Expressions

**2.** $\sqrt{6} \cdot \sqrt{18} = \sqrt{6 \cdot 18} = \sqrt{108} = \sqrt{36 \cdot 3} = 6\sqrt{3}$

**4.** $\sqrt{5} \cdot \sqrt{10} = \sqrt{5 \cdot 10} = \sqrt{50} = \sqrt{25 \cdot 2} = 5\sqrt{2}$

**6.** $\sqrt[3]{9} \cdot \sqrt[3]{9} = \sqrt[3]{9 \cdot 9} = \sqrt[3]{81} = \sqrt[3]{27 \cdot 3} = 3\sqrt[3]{3}$

**8.** $\sqrt[4]{54} \cdot \sqrt[4]{3} = \sqrt[4]{54 \cdot 3} = \sqrt[4]{81 \cdot 2} = 3\sqrt[4]{2}$

**10.** $\sqrt{11}\left(\sqrt{5} - 3\right) = \sqrt{11} \cdot \sqrt{5} - 3\sqrt{11} = \sqrt{11 \cdot 5} - 3\sqrt{11} = \sqrt{55} - 3\sqrt{11}$

**12.** $\sqrt{7}\left(\sqrt{14} + 3\right) = \sqrt{7} \cdot \sqrt{14} + \sqrt{7}(3) = \sqrt{98} + 3\sqrt{7} = \sqrt{49 \cdot 2} + 3\sqrt{7} = 7\sqrt{2} + 3\sqrt{7}$

**14.** $\sqrt{10}\left(\sqrt{5} + \sqrt{6}\right) = \sqrt{10} \cdot \sqrt{5} + \sqrt{10} \cdot \sqrt{6} = \sqrt{50} + \sqrt{60} = \sqrt{25 \cdot 2} + \sqrt{4 \cdot 15}$
$$= 5\sqrt{2} + 2\sqrt{15}$$

**16.** $\sqrt{5}\left(\sqrt{15} + \sqrt{5}\right) = \sqrt{5} \cdot \sqrt{15} + \sqrt{5} \cdot \sqrt{5} = \sqrt{75} + 5 = \sqrt{25 \cdot 3} + 5 = 5\sqrt{3} + 5$

**18.** $\sqrt{x}\left(5 - \sqrt{x}\right) = 5\sqrt{x} - \left(\sqrt{x}\right)^2 = 5\sqrt{x} - x$

**20.** $\sqrt{z}\left(\sqrt{z} + 5\right) = \sqrt{z} \cdot \sqrt{z} + \sqrt{z} \cdot 5$
$$= z + 5\sqrt{z}$$

**22.** $\sqrt[3]{9}\left(\sqrt[3]{3} + 2\right) = \sqrt[3]{9} \cdot \sqrt[3]{3} + \sqrt[3]{9} \cdot 2 = \sqrt[3]{27} + 2\sqrt[3]{9} = 3 + 2\sqrt[3]{9}$

**24.** $\left(3 - \sqrt{5}\right)\left(3 + \sqrt{5}\right) = 3^2 - \left(\sqrt{5}\right)^2$
$$= 9 - 5$$
$$= 4$$

**26.** $\left(\sqrt{7} + 6\right)\left(\sqrt{2} + 6\right) = \sqrt{7} \cdot \sqrt{2} + 6\sqrt{7} + 6\sqrt{2} + 36$
$$= \sqrt{14} + 6\sqrt{7} + 6\sqrt{2} + 36$$

**28.** $\left(4 - \sqrt{20}\right)^2 = 4^2 - 2(4)\sqrt{20} + \left(\sqrt{20}\right)^2$
$$= 16 - 8\sqrt{20} + 20$$
$$= 36 - 8\sqrt{4 \cdot 5}$$
$$= 36 - 16\sqrt{5}$$

**30.** $\left(\sqrt[3]{9} + 5\right)\left(\sqrt[3]{5} - 5\right) = \sqrt[3]{9} \cdot \sqrt[3]{5} - 5\sqrt[3]{9} + 5\sqrt[3]{5} - 25$
$$= \sqrt[3]{45} - 5\sqrt[3]{9} + 5\sqrt[3]{5} - 25$$

**32.** $\left(5 - \sqrt{3v}\right)^2 = 5^2 - 2(5)\sqrt{3v} + \left(\sqrt{3v}\right)^2$
$$= 25 - 10\sqrt{3v} + 3v$$

**34.** $\left(16\sqrt{u} - 3\right)\left(\sqrt{u} - 1\right) = 16\sqrt{u} \cdot \sqrt{u} - 16\sqrt{u} - 3\sqrt{u} + 3$
$$= 16u - 19\sqrt{u} + 3$$

**36.** $\left(7 - 3\sqrt{3t}\right)\left(7 + 3\sqrt{3t}\right) = (7)^2 - \left(3\sqrt{3t}\right)^2 = 49 - 27t$

**38.** $\left(\sqrt[3]{3x} - 4\right)^2 = \left(\sqrt[3]{3x}\right)^2 - 4\sqrt[3]{3x} - 4\sqrt[3]{3x} + (-4)^2$
$$= \sqrt[3]{9x^2} - 8\sqrt[3]{3x} + 16$$

**40.** $\left(\sqrt[3]{2y} + 10\right)\left(\sqrt[3]{4y^2} - 10\right) = \sqrt[3]{2y} \cdot \sqrt[3]{4y^2} - 10\sqrt[3]{2y} + 10\sqrt[3]{4y^2} - 100$
$$= \sqrt[3]{8y^3} - 10\sqrt[3]{2y} + 10\sqrt[3]{4y^2} - 100$$
$$= 2y - 10\sqrt[3]{2y} + 10\sqrt[3]{4y^2} - 100$$

**42.** $\left(\sqrt{x} - 2\right)\left(\sqrt{x^3} - 2\sqrt{x^2} + 1\right) = \sqrt{x} \cdot \sqrt{x^3} - \sqrt{x} \cdot 2\sqrt{x^2} + \sqrt{x} - 2\sqrt{x^3} + 4\sqrt{x^2} - 2$
$$= \sqrt{x^4} - 2\sqrt{x^3} + \sqrt{x} - 2\sqrt{x^3} + 4\sqrt{x^2} - 2$$
$$= x^2 - 4\sqrt{x^3} + 4\sqrt{x^2} + \sqrt{x} - 2$$
$$= x^2 - 4|x|\sqrt{x} + 4|x| + \sqrt{x} - 2$$

**44.** $x\sqrt{7} - x^2\sqrt{7} = x\sqrt{7}(1 - x)$

**46.** $5\sqrt{50} + 10y\sqrt{8} = 5\sqrt{25 \cdot 2} + 10y\sqrt{4 \cdot 2}$
$$= 25\sqrt{2} + 20y\sqrt{2}$$
$$= 5\sqrt{2}(5 + 4y)$$

**48.** $12s^3 - \sqrt{32s^4} = 12s^3 - \sqrt{16 \cdot 2 \cdot s^2 \cdot s^2}$
$$= 12s^3 - 4s^2\sqrt{2}$$
$$= 4s^2(3s - \sqrt{2})$$

**50.** $\sqrt{2} - 9$, conjugate $= \sqrt{2} + 9$
$$\text{product} = (\sqrt{2} - 9)(\sqrt{2} + 9)$$
$$= (\sqrt{2})^2 - (9)^2$$
$$= 2 - 81$$
$$= -79$$

**52.** $\sqrt{10} + \sqrt{7}$, conjugate $= \sqrt{10} - \sqrt{7}$
$$\text{product} = (\sqrt{10} + \sqrt{7})(\sqrt{10} - \sqrt{7})$$
$$= (\sqrt{10})^2 - (\sqrt{7})^2$$
$$= 10 - 7$$
$$= 3$$

**54.** $\sqrt{11} + 3$, conjugate $= \sqrt{11} - 3$
$$\text{product} = (\sqrt{11} + 3)(\sqrt{11} - 3)$$
$$= (\sqrt{11})^2 - (3)^2$$
$$= 11 - 9 = 2$$

**56.** $\sqrt{t} + 7$, conjugate $= \sqrt{t} - 7$
$$\text{product} = (\sqrt{t} + 7)(\sqrt{t} - 7)$$
$$= (\sqrt{t})^2 - (7)^2$$
$$= t - 49$$

**58.** $\sqrt{5a} + \sqrt{2}$, conjugate $= \sqrt{5a} - \sqrt{2}$
$$\text{product} = (\sqrt{5a} + \sqrt{2})(\sqrt{5a} - \sqrt{2})$$
$$= (\sqrt{5a})^2 - (\sqrt{2})^2$$
$$= 5a - 2$$

**60.** $4\sqrt{3} + \sqrt{2}$, conjugate $= 4\sqrt{3} - \sqrt{2}$
$$\text{product} = (4\sqrt{3} + \sqrt{2})(4\sqrt{3} - \sqrt{2})$$
$$= (4\sqrt{3})^2 - (\sqrt{2})^2$$
$$= 48 - 2 = 46$$

**62.** $3\sqrt{u} + \sqrt{3v}$, conjugate $= 3\sqrt{u} - \sqrt{3v}$
$$\text{product} = (3\sqrt{u} + \sqrt{3v})(3\sqrt{u} - \sqrt{3v})$$
$$= (3\sqrt{u})^2 - (\sqrt{3v})^2$$
$$= 9u - 3v$$

**64.** $\dfrac{-3 + 27\sqrt{2y}}{18} = \dfrac{-3(1 - 9\sqrt{2y})}{18} = \dfrac{-(1 - 9\sqrt{2y})}{6}$

**66.** $\dfrac{-t^2 - \sqrt{2t^3}}{3t} = \dfrac{-t^2 - t\sqrt{2t}}{3t} = \dfrac{-t(t + \sqrt{2t})}{3t} = \dfrac{-(t + \sqrt{2t})}{3}$

**68.** $g(x) = x^2 + 8x + 11$

(a) $g(-4 + \sqrt{5}) = (-4 + \sqrt{5})^2 + 8(-4 + \sqrt{5}) + 11$
$$= (-4)^2 + 2(-4)(\sqrt{5}) + (\sqrt{5})^2 - 32 + 8\sqrt{5} + 11$$
$$= 16 - 8\sqrt{5} + 5 - 32 + 8\sqrt{5} + 11$$
$$= 0$$

(b) $g(-4\sqrt{2}) = (-4\sqrt{2})^2 + 8(-4\sqrt{2}) + 11$
$$= 32 - 32\sqrt{2} + 11$$
$$= 43 - 32\sqrt{2}$$

**70.** $g(x) = x^2 - 4x + 1$

(a) $g(1 + \sqrt{5}) = (1 + \sqrt{5})^2 - 4(1 + \sqrt{5}) + 1$

$= 1^2 + 2(1)(\sqrt{5}) + (\sqrt{5})^2 - 4 - 4\sqrt{5} + 1$

$= 1 + 2\sqrt{5} + 5 - 4 - 4\sqrt{5} + 1$

$= 3 - 2\sqrt{5}$

(b) $g(2 - \sqrt{3}) = (2 - \sqrt{3})^2 - 4(2 - \sqrt{3}) + 1$

$= 2^2 - 2(2)(\sqrt{3}) + (\sqrt{3})^2 - 8 + 4\sqrt{3} + 1$

$= 4 - 4\sqrt{3} + 3 - 8 + 4\sqrt{3} + 1$

$= 0$

**72.** $\dfrac{8}{\sqrt{7} + 3} = \dfrac{8}{\sqrt{7} + 3} \cdot \dfrac{\sqrt{7} - 3}{\sqrt{7} - 3} = \dfrac{8(\sqrt{7} - 3)}{(\sqrt{7})^2 - 3^2} = \dfrac{8(\sqrt{7} - 3)}{7 - 9} = \dfrac{8(\sqrt{7} - 3)}{-2} = -4(\sqrt{7} - 3) = -4\sqrt{7} + 12$

**74.** $\dfrac{5}{9 - \sqrt{6}} = \dfrac{5}{9 - \sqrt{6}} \cdot \dfrac{9 + \sqrt{6}}{9 + \sqrt{6}} = \dfrac{5(9 + \sqrt{6})}{9^2 - (\sqrt{6})^2} = \dfrac{5(9 + \sqrt{6})}{81 - 6} = \dfrac{5(9 + \sqrt{6})}{75} = \dfrac{9 + \sqrt{6}}{15}$

**76.** $\dfrac{4}{3\sqrt{5} - 1} = \dfrac{4}{3\sqrt{5} - 1} \cdot \dfrac{3\sqrt{5} + 1}{3\sqrt{5} + 1}$

$= \dfrac{4(3\sqrt{5} + 1)}{45 - 1}$

$= \dfrac{4(3\sqrt{5} + 1)}{44}$

$= \dfrac{3\sqrt{5} + 1}{11}$

**78.** $\dfrac{10}{\sqrt{9} + \sqrt{5}} = \dfrac{10}{3 + \sqrt{5}}$

$= \dfrac{10}{3 + \sqrt{5}} \cdot \dfrac{3 - \sqrt{5}}{3 - \sqrt{5}}$

$= \dfrac{10(3 - \sqrt{5})}{3^2 - (\sqrt{5})^2}$

$= \dfrac{10(3 - \sqrt{5})}{9 - 5}$

$= \dfrac{10(3 - \sqrt{5})}{4}$

$= \dfrac{5(3 - \sqrt{5})}{2}$

**80.** $\dfrac{12}{\sqrt{5} + \sqrt{8}} = \dfrac{12}{\sqrt{5} + \sqrt{8}} \cdot \dfrac{\sqrt{5} - \sqrt{8}}{\sqrt{5} - \sqrt{8}} = \dfrac{12(\sqrt{5} - \sqrt{8})}{(\sqrt{5})^2 - (\sqrt{8})^2} = \dfrac{12(\sqrt{5} - \sqrt{8})}{5 - 8} = \dfrac{12(\sqrt{5} - \sqrt{8})}{-3}$

$= -4(\sqrt{5} - \sqrt{8})$

$= 4(2\sqrt{2} - \sqrt{5})$

**82.** $\left(5 - \sqrt{3}\right) \div \left(3 + \sqrt{3}\right) = \dfrac{5 - \sqrt{3}}{3 + \sqrt{3}}$

$= \dfrac{5 - \sqrt{3}}{3 + \sqrt{3}} \cdot \dfrac{3 - \sqrt{3}}{3 - \sqrt{3}}$

$= \dfrac{\left(5 - \sqrt{3}\right)\left(3 - \sqrt{3}\right)}{9 - 3}$

$= \dfrac{15 - 5\sqrt{3} - 3\sqrt{3} + 3}{6}$

$= \dfrac{18 - 8\sqrt{3}}{6}$

$= \dfrac{9 - 4\sqrt{3}}{3}$

**84.** $\left(2\sqrt{t} + 1\right) \div \left(2\sqrt{t} - 1\right) = \dfrac{2\sqrt{t} + 1}{2\sqrt{t} - 1}$

$= \dfrac{2\sqrt{t} + 1}{2\sqrt{t} - 1} \cdot \dfrac{2\sqrt{t} + 1}{2\sqrt{t} + 1}$

$= \dfrac{4t + 2\sqrt{t} + 2\sqrt{t} + 1}{4t - 1}$

$= \dfrac{4t + 4\sqrt{t} + 1}{4t - 1}$

**86.** $\dfrac{5y}{\sqrt{12} + \sqrt{10}} = \dfrac{5y}{\sqrt{12} + \sqrt{10}} \cdot \dfrac{\sqrt{12} - \sqrt{10}}{\sqrt{12} - \sqrt{10}}$

$= \dfrac{5y\left(\sqrt{12} - \sqrt{10}\right)}{\left(\sqrt{12}\right)^2 - \left(\sqrt{10}\right)^2}$

$= \dfrac{5y\left(2\sqrt{3} - \sqrt{10}\right)}{12 - 10}$

$= \dfrac{5y\left(2\sqrt{3} - \sqrt{10}\right)}{2}$

**88.** $\dfrac{5x}{\sqrt{x} - \sqrt{2}} = \dfrac{5x}{\sqrt{x} - \sqrt{2}} \cdot \dfrac{\sqrt{x} + \sqrt{2}}{\sqrt{x} + \sqrt{2}}$

$= \dfrac{5x\left(\sqrt{x} + \sqrt{2}\right)}{\left(\sqrt{x}\right)^2 - \left(\sqrt{2}\right)^2}$

$= \dfrac{5x\left(\sqrt{x} + \sqrt{2}\right)}{x - 2}$

**90.** $\dfrac{7z}{\sqrt{5z} - \sqrt{z}} = \dfrac{7z}{\sqrt{5z} - \sqrt{z}} \cdot \dfrac{\sqrt{5z} + \sqrt{z}}{\sqrt{5z} + \sqrt{z}}$

$= \dfrac{7z\left(\sqrt{5z} + \sqrt{z}\right)}{\left(\sqrt{5z}\right)^2 - \left(\sqrt{z}\right)^2}$

$= \dfrac{7z\left(\sqrt{5z} + \sqrt{z}\right)}{5z - z} = \dfrac{7z\left(\sqrt{5z} + \sqrt{z}\right)}{4z} = \dfrac{7\left(\sqrt{5z} + \sqrt{z}\right)}{4}$

**92.** $\dfrac{6(y + 1)}{y^2 + \sqrt{y}} = \dfrac{6(y + 1)}{y^2 + \sqrt{y}} \cdot \dfrac{y^2 - \sqrt{y}}{y^2 - \sqrt{y}}$

$= \dfrac{6(y + 1)\left(y^2 - \sqrt{y}\right)}{y^4 - y}$

$= \dfrac{6(y + 1)\left(y^2 - \sqrt{y}\right)}{y(y^3 - 1)}$

$= \dfrac{6(y + 1)\left(y^2 - \sqrt{y}\right)}{y(y - 1)(y^2 + y + 1)}$

**94.** $\dfrac{z}{\sqrt{u + z} - \sqrt{u}} = \dfrac{z}{\sqrt{u + z} - \sqrt{u}} \cdot \dfrac{\sqrt{u + z} + \sqrt{u}}{\sqrt{u + z} - \sqrt{u}}$

$= \dfrac{z\left(\sqrt{u + z} + \sqrt{u}\right)}{\left(\sqrt{u + z}\right)^2 - \left(\sqrt{u}\right)^2}$

$= \dfrac{z\left(\sqrt{u + z} + \sqrt{u}\right)}{u + z - u}$

$= \dfrac{z\left(\sqrt{u + z} + \sqrt{u}\right)}{z}$

$= \sqrt{u + z} + \sqrt{u}$

**96.** *Keystrokes:*

$y_1$ Y= 4 X,T,θ ÷ ( ( √ X,T,θ + 4 ) ENTER

$y_2$ ( 4 X,T,θ ( ( √ X,T,θ − 4 ) ) ) ÷ ( X,T,θ − 16 ) GRAPH

$y_1 = y_2$, except at $x = 16$

$$y_1 = \frac{4x}{\sqrt{x}+4} = \frac{4x}{\sqrt{x}+4} \cdot \frac{\sqrt{x}-4}{\sqrt{x}-4} = \frac{4x(\sqrt{x}-4)}{x-16}, x \neq 16$$

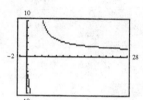

**98.** *Keystrokes:*

$y_1$ Y= ( ( √ ( 2 X,T,θ ) + 6 ) ÷ ( ( √ ( 2 X,T,θ ) − 2 ENTER

$y_2$ ( X,T,θ + 6 + 4 √ ( 2 X,T,θ ) ) ÷ ( X,T,θ − 2 ) GRAPH

$y_1 = y_2$

$$\frac{\sqrt{2x}+6}{\sqrt{2x}-2} = \frac{\sqrt{2x}+6}{\sqrt{2x}-2} \cdot \frac{\sqrt{2x}+2}{\sqrt{2x}+2}$$

$$= \frac{2x + 2\sqrt{2x} + 6\sqrt{2x} + 12}{2x - 4}$$

$$= \frac{2x + 8\sqrt{2x} + 12}{2x - 4}$$

$$= \frac{2(x + 4\sqrt{2x} + 6)}{2(x - 2)} = \frac{x + 4\sqrt{2x} + 6}{x - 2}$$

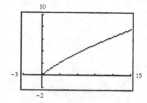

**100.** $\dfrac{\sqrt{10}}{5} = \dfrac{\sqrt{10}}{5} \cdot \dfrac{\sqrt{10}}{\sqrt{10}} = \dfrac{10}{5\sqrt{10}} = \dfrac{2}{\sqrt{10}}$

**102.** $\dfrac{\sqrt{x}+6}{\sqrt{2}} = \dfrac{\sqrt{x}+6}{\sqrt{2}} \cdot \dfrac{\sqrt{x}-6}{\sqrt{x}-6} = \dfrac{(\sqrt{x})^2 - 6^2}{\sqrt{2}(\sqrt{x}-6)} = \dfrac{x - 36}{\sqrt{2}(\sqrt{x}-6)}$

**104.** $\boxed{\begin{array}{c}\text{radius of}\\\text{small circle}\end{array}} : \boxed{\begin{array}{c}\text{radius of}\\\text{large circle}\end{array}} = \sqrt{\dfrac{15}{\pi}} : \sqrt{\dfrac{20}{\pi}}$

$$= \frac{\sqrt{\dfrac{15}{\pi}}}{\sqrt{\dfrac{20}{\pi}}} = \sqrt{\frac{15/\pi}{20/\pi}} = \sqrt{\frac{15}{20}} = \sqrt{\frac{3}{4}} = \frac{\sqrt{3}}{\sqrt{4}} = \frac{\sqrt{3}}{2}$$

**106.**

| $v/T$ | 0° | 5° | 10° | 15° | 20° | 25° |
|---|---|---|---|---|---|---|
| 10 mi/hr | −20.9 | −14.8 | −8.6 | −2.5 | 3.7 | 9.8 |
| 20 mi/hr | −38.5 | −31.4 | −24.3 | −17.2 | −10.1 | −3.0 |
| 30 mi/hr | −47.7 | −40.1 | −32.5 | −24.9 | −17.3 | −9.7 |
| 40 mi/hr | −52.5 | −44.7 | −36.8 | −28.9 | −21.0 | −13.2 |

**108.** The "First, Outer, Inner, Last" is the same for both. For polynomial expressions, use the properties of positive integer exponents to find the individual products. For radical expressions, find the products by using the properties of radicals.

**110.** No. Multiply the numerator and denominator by the conjugate of the denominator, $1 - \sqrt{5}$.

# Section 5.4    Solving Radical Equations

**2.** (a) $x = \frac{2}{3}$      $\sqrt{3\left(\frac{2}{3}\right)} - 6 \neq 0$      not a solution

   (b) $x = 2$      $\sqrt{3(2)} - 6 \neq 0$      not a solution

   (c) $x = 12$      $\sqrt{3(12)} - 6 = 0$      a solution

   (d) $x = -\frac{1}{3}\sqrt{6}$      $\sqrt{3\left(-\frac{1}{3}\right)\sqrt{6}} - 6 \neq 0$      not a solution

**4.** (a) $x = 128$      $\sqrt[4]{2(128)} + 2 = 6$      a solution

   (b) $x = 2$      $\sqrt[4]{2(2)} + 2 \neq 6$      not a solution

   (c) $x = -2$      $\sqrt[4]{2(-2)} + 2 \neq 6$      not a solution

   (d) $x = 0$      $\sqrt[4]{2(0)} + 2 \neq 6$      not a solution

**6.**   $\sqrt{x} = 5$      **Check:**

$\left(\sqrt{x}\right)^2 = 5^2$      $\sqrt{25} \stackrel{?}{=} 5$

$x = 25$      $5 = 5$

**8.**   $\sqrt{t} = 4$      **Check:**

$\left(\sqrt{t}\right)^2 = 4^2$      $\sqrt{16} \stackrel{?}{=} 4$

$t = 16$      $4 = 4$

**10.**   $\sqrt[4]{x} = 2$      **Check:**

$\left(\sqrt[4]{x}\right)^4 = 2^4$      $\sqrt[4]{16} \stackrel{?}{=} 2$

$x = 16$      $2 = 2$

**12.** $\sqrt{t} - 13 = 0$      **Check:**

$\sqrt{t} = 13$      $\sqrt{169} - 13 \stackrel{?}{=} 0$

$\left(\sqrt{t}\right)^2 = 13^2$      $13 - 13 \stackrel{?}{=} 0$

$t = 169$      $0 = 0$

**14.** $\sqrt{y} + 15 = 0$      **Check:**

$\sqrt{y} = -15$      $\sqrt{225} + 15 \stackrel{?}{=} 0$

$\left(\sqrt{y}\right)^2 = (-15)^2$      $15 + 15 \stackrel{?}{=} 0$

$y = 225$      $30 \neq 0$

         No solution

**16.** $\sqrt{x} - 10 = 0$      **Check:**

$\sqrt{x} = 10$      $\sqrt{100} - 10 \stackrel{?}{=} 0$

$\left(\sqrt{x}\right)^2 = 10^2$      $10 - 10 \stackrel{?}{=} 0$

$x = 100$      $0 = 0$

**18.**   $\sqrt{8x} = 6$      **Check:**

$\left(\sqrt{8x}\right)^2 = 6^2$      $\sqrt{8\left(\frac{9}{2}\right)} \stackrel{?}{=} 6$

$8x = 36$      $\sqrt{36} \stackrel{?}{=} 6$

$x = \frac{36}{8}$      $6 = 6$

$x = \frac{9}{2}$

**20.** $\sqrt{-4y} = 4$

$\left(\sqrt{-4y}\right)^2 = 4^2$

$-4y = 16$

$y = -4$

**Check:**

$\sqrt{-4(-4)} \overset{?}{=} 4$

$\sqrt{16} \overset{?}{=} 4$

$4 = 4$

**22.** $6 - \sqrt{8x} = 0$

$6 = \sqrt{8x}$

$6^2 = \left(\sqrt{8x}\right)^2$

$36 = 8x$

$\frac{36}{8} = x$

$\frac{9}{2} = x$

**Check:**

$6 - \sqrt{8\left(\frac{9}{2}\right)} \overset{?}{=} 0$

$6 - \sqrt{36} \overset{?}{=} 0$

$6 - 6 \overset{?}{=} 0$

$0 = 0$

**24.** $\sqrt{3 - 2x} = 2$

$\left(\sqrt{3 - 2x}\right)^2 = 2^2$

$3 - 2x = 4$

$-2x = 1$

$x = -\frac{1}{2}$

**Check:**

$\sqrt{3 - 2\left(-\frac{1}{2}\right)} \overset{?}{=} 2$

$\sqrt{3 + 1} \overset{?}{=} 2$

$\sqrt{4} \overset{?}{=} 2$

$2 = 2$

**26.** $\sqrt{2t - 7} = -5$

$\left(\sqrt{2t - 7}\right)^2 = (-5)^2$

$2t - 7 = 25$

$2t = 32$

$t = 16$

**Check:**

$\sqrt{2(16) - 7} \overset{?}{=} -5$

$\sqrt{32 - 7} \overset{?}{=} -5$

$\sqrt{25} \overset{?}{=} -5$

$5 \neq -5$

No solution

**28.** $\sqrt{5z - 2} + 7 = 10$

$\sqrt{5z - 2} = 3$

$\left(\sqrt{5z - 2}\right)^2 = 3^2$

$5z - 2 = 9$

$5z = 11$

$z = \frac{11}{5}$

**Check:**

$\sqrt{5\left(\frac{11}{5}\right) - 2} + 7 \overset{?}{=} 10$

$\sqrt{11 - 2} + 7 \overset{?}{=} 10$

$\sqrt{9} + 7 \overset{?}{=} 10$

$3 + 7 \overset{?}{=} 10$

$10 = 10$

**30.** $2\sqrt{x + 4} = 7$

$\left(2\sqrt{x + 4}\right)^2 = 7^2$

$4(x + 4) = 49$

$4x + 16 = 49$

$4x = 33$

$x = \frac{33}{4}$

**Check:**

$2\sqrt{\frac{33}{4} + 4} \overset{?}{=} 7$

$2\sqrt{\frac{33 + 16}{4}} \overset{?}{=} 7$

$2\sqrt{\frac{49}{4}} \overset{?}{=} 7$

$2\left(\frac{7}{2}\right) \overset{?}{=} 7$

$7 = 7$

**32.** $\sqrt{1 - x} + 10 = 4$

$\sqrt{1 - x} = -6$

$\left(\sqrt{1 - x}\right)^2 = (-6)^2$

$1 - x = 36$

$1 = x + 36$

$-35 = x$

**Check:**

$\sqrt{1 - (-35)} + 10 \overset{?}{=} 4$

$\sqrt{36} + 10 \overset{?}{=} 4$

$6 + 10 \overset{?}{=} 4$

$16 \neq 4$

No solution

**34.** $\sqrt{3t + 1} = \sqrt{t + 15}$

$\left(\sqrt{3t + 1}\right)^2 = \left(\sqrt{t + 15}\right)^2$

$3t + 1 = t + 15$

$2t + 1 = 15$

$2t = 14$

$t = 7$

**Check:**

$\sqrt{3(7) + 1} \stackrel{?}{=} \sqrt{7 + 15}$

$\sqrt{21 + 1} \stackrel{?}{=} \sqrt{22}$

$\sqrt{22} = \sqrt{22}$

**36.** $\sqrt{2u + 10} - 2\sqrt{u} = 0$

$\sqrt{2u + 10} = 2\sqrt{u}$

$\left(\sqrt{2u + 10}\right)^2 = \left(2\sqrt{u}\right)^2$

$2u + 10 = 4u$

$10 = 2u$

$5 = u$

**Check:**

$\sqrt{2(5) + 10} - 2\sqrt{5} \stackrel{?}{=} 0$

$\sqrt{10 + 10} - 2\sqrt{5} \stackrel{?}{=} 0$

$\sqrt{20} - 2\sqrt{5} \stackrel{?}{=} 0$

$2\sqrt{5} - 2\sqrt{5} \stackrel{?}{=} 0$

$0 = 0$

**38.** $2\sqrt[3]{10 - 3x} = \sqrt[3]{2 - x}$

$\left(2\sqrt[3]{10 - 3x}\right)^3 = \left(\sqrt[3]{2 - x}\right)^3$

$8(10 - 3x) = 2 - x$

$80 - 24x = 2 - x$

$80 = 2 + 23x$

$78 = 23x$

$\dfrac{78}{23} = x$

**Check:**

$2\sqrt[3]{10 - 3\left(\dfrac{78}{23}\right)} \stackrel{?}{=} \sqrt[3]{2 - \dfrac{78}{23}}$

$2\sqrt[3]{10 - \dfrac{234}{23}} \stackrel{?}{=} \sqrt[3]{\dfrac{46}{23} - \dfrac{78}{23}}$

$2\sqrt[3]{\dfrac{230 - 234}{23}} \stackrel{?}{=} \sqrt[3]{-\dfrac{32}{23}}$

$2\sqrt[3]{-\dfrac{4}{23}} = 2\sqrt[3]{-\dfrac{4}{23}}$

**40.** $\sqrt[4]{2x} + \sqrt[4]{x + 3} = 0$

$\sqrt[4]{2x} = -\sqrt[4]{x + 3}$

$\left(\sqrt[4]{2x}\right)^4 = \left(-\sqrt[4]{x + 3}\right)^4$

$2x = x + 3$

$x = 3$

**Check:**

$\sqrt[4]{2(3)} + \sqrt[4]{3 + 3} \stackrel{?}{=} 0$

$\sqrt[4]{6} + \sqrt[4]{6} \stackrel{?}{=} 0$

$2\sqrt[4]{6} \neq 0$

No solution

**42.** $\sqrt{x^2 - 4} = x - 2$

$\left(\sqrt{x^2 - 4}\right)^2 = (x - 2)^2$

$x^2 - 4 = x^2 - 2(x)(2) + (2)^2$

$x^2 - 4 = x^2 - 4x + 4$

$-4 = -4x + 4$

$-8 = -4x$

$2 = x$

**Check:**

$\sqrt{(2)^2 - 4} \stackrel{?}{=} 2 - 2$

$\sqrt{4 - 4} \stackrel{?}{=} 0$

$\sqrt{0} \stackrel{?}{=} 0$

$0 = 0$

**44.**   $\sqrt{x} = x - 6$

$\left(\sqrt{x}\right)^2 = (x - 6)^2$

$x = x^2 - 2x(6) + 6^2$

$x = x^2 - 12x + 36$

$0 = x^2 - 13x + 36$

$0 = (x - 4)(x - 9)$

$x - 4 = 0 \quad x - 9 = 0$

$x = 4 \qquad x = 9$

Not a solution

**Check:**

$\sqrt{4} \overset{?}{=} 4 - 6$

$2 \neq -2$

$\sqrt{9} \overset{?}{=} 9 - 6$

$3 = 3$

**46.**   $\sqrt{3x + 7} = x + 3$

$\left(\sqrt{3x + 7}\right)^2 = (x + 3)^2$

$3x + 7 = x^2 + 2x(3) + 3^2$

$3x + 7 = x^2 + 6x + 9$

$0 = x^2 + 3x + 2$

$0 = (x + 2)(x + 1)$

$x + 2 = 0 \quad x + 1 = 0$

$x = -2 \qquad x = -1$

**Check:**

$\sqrt{3(-2) + 7} \overset{?}{=} -2 + 3$

$\sqrt{-6 + 7} \overset{?}{=} 1$

$\sqrt{1} \overset{?}{=} 1$

$1 = 1$

$\sqrt{3 - (-1) + 7} \overset{?}{=} -1 + 3$

$\sqrt{-3 + 7} \overset{?}{=} 2$

$\sqrt{4} \overset{?}{=} 2$

$2 = 2$

**48.**   $\sqrt{2x + 5} = 7 - \sqrt{2x}$

$\left(\sqrt{2x + 5}\right)^2 = \left(7 - \sqrt{2x}\right)^2$

$2x + 5 = 49 - 14\sqrt{2x} + 2x$

$-44 = -14\sqrt{2x}$

$22 = 7\sqrt{2x}$

$22^2 = \left(7\sqrt{2x}\right)^2$

$484 = 49 \cdot 2x$

$\frac{242}{49} = x$

**Check:**

$\sqrt{2\left(\frac{242}{49}\right) + 5} \overset{?}{=} 7 - \sqrt{2\left(\frac{242}{49}\right)}$

$\sqrt{\frac{484}{49} + \frac{245}{49}} \overset{?}{=} 7 - \sqrt{\frac{484}{49}}$

$\sqrt{\frac{729}{49}} \overset{?}{=} 7 = \frac{22}{7}$

$\frac{27}{7} \overset{?}{=} \frac{49}{7} - \frac{22}{7}$

$\frac{27}{7} = \frac{27}{7}$

**50.**   $\sqrt{x} + \sqrt{x + 2} = 2$

$\sqrt{x} = 2 - \sqrt{x + 2}$

$\left(\sqrt{x}\right)^2 = \left(2 - \sqrt{x + 2}\right)^2$

$x = 4 - 4\sqrt{x + 2} + x + 2$

$-6 = -4\sqrt{x + 2}$

$3 = 2\sqrt{x + 2}$

$3^2 = \left(2\sqrt{x + 2}\right)^2$

$9 = 4(x + 2)$

$9 = 4x + 8$

$1 = 4x$

$\frac{1}{4} = x$

**Check:**

$\sqrt{\frac{1}{4}} + \sqrt{\frac{1}{4} + 2} \overset{?}{=} 2$

$\frac{1}{2} + \sqrt{\frac{9}{4}} \overset{?}{=} 2$

$\frac{1}{2} + \frac{3}{2} \overset{?}{=} 2$

$2 = 2$

**52.** $\sqrt{x+3} - \sqrt{x-1} = 1$

$$\sqrt{x+3} = 1 + \sqrt{x-1}$$

$$\left(\sqrt{x}+3\right)^2 = \left(1 + \sqrt{x-1}\right)^2$$

$$x + 3 = 1 + 2\sqrt{x-1} + x - 1$$

$$3 = 2\sqrt{x-1}$$

$$3^2 = \left(2\sqrt{x-1}\right)^2$$

$$9 = 4(x-1)$$

$$9 = 4x - 4$$

$$13 = 4x$$

$$\tfrac{13}{4} = x$$

**Check:**

$$\sqrt{\tfrac{13}{4}+3} - \sqrt{\tfrac{13}{4}-1} \overset{?}{=} 1$$

$$\sqrt{\tfrac{13}{4}+\tfrac{12}{4}} - \sqrt{\tfrac{13}{4}-\tfrac{4}{4}} \overset{?}{=} 1$$

$$\sqrt{\tfrac{25}{4}} - \sqrt{\tfrac{9}{4}} \overset{?}{=} 1$$

$$\tfrac{5}{2} - \tfrac{3}{2} \overset{?}{=} 1$$

$$\tfrac{2}{2} \overset{?}{=} 1$$

$$1 = 1$$

**54.** $v^{2/3} = 25$

$$\sqrt[3]{v^2} = 25$$

$$\left(\sqrt[3]{v^2}\right)^3 = 25^3$$

$$v^2 = 15,625$$

$$v = 125, -125$$

**Check:**

$$125^{2/3} \overset{?}{=} 25$$

$$25 = 25$$

$$(-125)^{2/3} \overset{?}{=} 25$$

$$25 = 25$$

**56.** $2x^{3/4} = 54$

$$x^{3/4} = 27$$

$$\sqrt[4]{x^3} = 27$$

$$\left(\sqrt[4]{x^3}\right)^4 = (27)^4$$

$$x^3 = 531,441$$

$$x = \sqrt[3]{531,441}$$

$$x = 81$$

**Check:**

$$2(81)^{3/4} \overset{?}{=} 54$$

$$2(3)^3 \overset{?}{=} 54$$

$$2(27) \overset{?}{=} 54$$

$$54 = 54$$

**58.** $(u-2)^{4/3} = 81$

$$\sqrt[3]{(u-2)^4} = 81$$

$$\left(\sqrt[3]{(u-2)^4}\right)^3 = 81^3$$

$$(u-2)^4 = 531,441$$

$$u - 2 = \sqrt[4]{531,441}$$

$$u - 2 = 27$$

$$u = 29$$

**Check:**

$$(29-2)^{4/3} \overset{?}{=} 81$$

$$27^{4/3} \overset{?}{=} 81$$

$$\left(\sqrt[3]{27}\right)^4 \overset{?}{=} 81$$

$$3^4 \overset{?}{=} 81$$

$$81 = 81$$

**60.** $(x-6)^{3/2} - 27 = 0$

$$(x-6)^{3/2} = 27$$

$$\sqrt{(x-6)^3} = 27$$

$$\left(\sqrt{(x-6)^3}\right)^2 = 27^2$$

$$(x-6)^3 = 729$$

$$x - 6 = \sqrt[3]{729}$$

$$x - 6 = 9$$

$$x = 15$$

**Check:**

$$(15-6)^{3/2} - 27 \overset{?}{=} 0$$

$$9^{3/2} - 27 \overset{?}{=} 0$$

$$\sqrt{9^3} - 27 \overset{?}{=} 0$$

$$3^3 - 27 \overset{?}{=} 0$$

$$27 - 27 \overset{?}{=} 0$$

$$0 = 0$$

**62.** $\sqrt{2x + 3} = 4x - 3$

*Keystrokes:*

$y_1$ [Y=] [√] [(] 2 [X,T,θ] [+] 3 [)] [ENTER]

$y_2$ 4 [X,T,θ] [−] 3 [GRAPH]

Approximate Solution: $x \approx 1.347$

*Check algebraically:* $\sqrt{2(1.347) + 3} \overset{?}{=} 4(1.347) - 3$

$$2.386 \approx 2.388$$

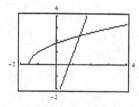

**64.** $\sqrt{8 - 3x} = x$

*Keystrokes:*

$y_1$ [Y=] [√] [(] 8 [−] 3 [X,T,θ] [)] [ENTER]

$y_2$ [X,T,θ] [GRAPH]

Approximate solution: $x \approx 1.702$

*Check algebraically:* $\sqrt[3]{8 - 3(1.702)} \overset{?}{=} 1.702$

$$1.701 \approx 1.702$$

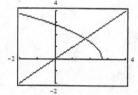

**66.** $\sqrt[3]{5x - 8} = 4 - \sqrt[3]{x}$

*Keystrokes:*

$y_1$ [Y=] [MATH] 4 [(] 5 [X,T,θ] [−] 8 [)] [ENTER]

$y_2$ 4 [−] [MATH] 4 [X,T,θ] [GRAPH]

Approximate Solution: $x \approx 4.283$

*Check algebraically:* $\sqrt[3]{5(4.283) - 8} \overset{?}{=} 4 - \sqrt[3]{4.283}$

$$2.376 = 2.376$$

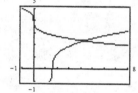

**68.** $\sqrt[3]{x + 4} = \sqrt{6 - x}$

*Keystrokes:*

$y_1$ [Y=] [MATH] 4 [(] [X,T,θ] [+] 4 [)] [ENTER]

$y_2$ [√] [(] 6 [−] [X,T,θ] [)] [GRAPH]

Approximate solution: $x \approx 2.513$

*Check algebraically:* $\sqrt[3]{2.513 + 4} \overset{?}{=} \sqrt{6 - 2.513}$

$$1.867 = 1.867$$

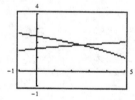

**70.** $\dfrac{4}{\sqrt{x}} = 3\sqrt{x} - 4$

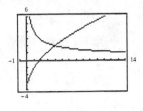

*Keystrokes:*

$y_1$ [Y=] 4 [÷] [√] [X,T,θ] [ENTER]

$y_2$ 3 [√] [X,T,θ] [−] 4 [GRAPH]

Solution: $x = 4$

*Check algebraically:* $\dfrac{4}{\sqrt{4}} \overset{?}{=} 3\sqrt{4} - 4$

$$\dfrac{4}{2} \overset{?}{=} 3(2) - 4$$

$$2 = 2$$

**72.** (e)          **74.** (a)          **76.** (b)

**78.**  $26^2 = x^2 + 10^2$

$676 = x^2 + 100$

$576 = x^2$

$\sqrt{576} = x$

$24 = x$

**80.** $x^2 = 4^2 + 6^2$

$x^2 = 16 + 36$

$x^2 = 52$

$x = \sqrt{52}$

$x = 2\sqrt{13} \approx 7.211$

**82.** $c = \sqrt{94^2 + 50^2}$

$= \sqrt{8836 + 2500}$

$= \sqrt{11{,}336}$

$\approx 106.47$ feet

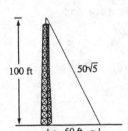

50 ft

94 ft

**84.** $x^2 = (5)^2 + (100)^2$

$x^2 = 2500 + 10{,}000$

$x^2 = 12{,}500$

$x = \sqrt{12{,}500}$

$x \approx 111.8$ feet

100 ft          $50\sqrt{5}$

$\mid\!\leftarrow$ 50 ft $\rightarrow\!\mid$

**86.** $x^2 + 6^2 = 10^2$

$x^2 = 100 - 36$

$x = \sqrt{100 - 36}$

$x = \sqrt{64} = 8$

$m = \dfrac{6}{8} = \dfrac{3}{4} = 0.75$

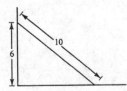

10

6

**88.** Perimeter = 2 · Length + 2 · Width

$$68 = 2l + 2w$$
$$68 = 2(l + w)$$
$$34 = l + w$$
$$34 - w = l$$

$$26^2 = w^2 + (34 - w)^2$$
$$676 = w^2 + 34^2 - 2(34)w + w^2$$
$$676 = w^2 + 1156 - 68w + w^2$$
$$0 = 2w^2 - 68w + 480$$
$$0 = 2(w^2 - 34w + 240)$$
$$0 = 2(2 - 24)(w - 10)$$
$$w - 24 = 0 \qquad w - 10 = 0$$
$$w = 24 \qquad w = 10$$

If $w = 24$ in., $l = 34 - 24 = 10$ in.

If $w = 10$ in., $l = 34 - 10 = 24$ in.

Thus, the rectangle is 10 in. × 24 in. or 24 in. × 10 in.

**90.** Area of circle = $\pi r^2$

$$A = \pi r^2$$
$$\frac{A}{\pi} = r^2$$
$$\sqrt{\frac{A}{\pi}} = r$$

*Keystrokes:*

$\boxed{Y=}\ \boxed{\sqrt{\phantom{x}}}\ \boxed{(}\ \boxed{X,T,\theta}\ \boxed{\div}\ \boxed{\pi}\ \boxed{)}\ \boxed{GRAPH}$

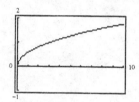

**92.**
$$t = \sqrt{\frac{d}{16}}$$
$$3 = \sqrt{\frac{d}{16}}$$
$$3^2 = \left(\sqrt{\frac{d}{16}}\right)^2$$
$$9 = \frac{d}{16}$$
$$d = 144 \text{ feet}$$

**94.**
$$v = \sqrt{2(32)(200)}$$
$$v = \sqrt{12{,}800}$$
$$v = 80\sqrt{2}$$
$$v \approx 113.14 \text{ feet per second}$$

**96.**
$$120 = \sqrt{2(32)h}$$
$$120 = \sqrt{64h}$$
$$120^2 = \left(\sqrt{64h}\right)^2$$
$$14{,}400 = 64h$$
$$\frac{14{,}400}{64} = h$$
$$225 \text{ feet} = h$$

**98.**
$$t = 2\pi\sqrt{\frac{L}{32}}$$
$$0.75 = 2\pi\sqrt{\frac{L}{32}}$$
$$\frac{0.75}{2\pi} = \sqrt{\frac{L}{32}}$$
$$\left(\frac{0.75}{2\pi}\right)^2 = \left(\sqrt{\frac{L}{32}}\right)^2$$
$$0.014248 = \frac{L}{32}$$
$$0.4559 = L$$
$$L \approx 0.46 \text{ foot}$$

**100.** Looking at the graph, we see that there are approximately 30,000 passengers.

**Check:**

$$2.5 = \sqrt{0.2x + 1}$$
$$2.5^2 = \left(\sqrt{0.2x + 1}\right)^2$$
$$6.25 = 0.2x + 1$$
$$5.25 = 0.2x$$
$$\frac{5.25}{0.2} = x$$

26.25 thousand = number of passengers

26,250 = number of passengers

**102.** $x + \sqrt{x - a} = b$

If $x = 20$    $20 + \sqrt{20 - a} = b$

then

$$\sqrt{20 - a} = b - 20$$

$$\left(\sqrt{20 - a}\right)^2 = (b - 20)^2$$

$$20 - a = b^2 - 40b + 400$$

$$a = -b^2 + 40b - 380$$

Now substitute values for $b$ and solve for $a$ such as

$$a = -24^2 + 40(24) - 380$$

$$a = -576 + 960 - 380$$

$$a = -4, \ b = 24$$

**104.** Isolate a radical on one side of the equation and then raise both sides of the equation to the power necessary to eliminate the radical. If there are more radicals, continue the process. When the radicals have been eliminated, solve the resulting equation and check your results.

**106.** Check for extraneous solutions.

# Section 5.5    Complex Numbers

**2.** $\sqrt{-9} = \sqrt{-1 \cdot 9} = \sqrt{-1} \cdot \sqrt{9} = 3i$

**4.** $\sqrt{-49} = \sqrt{49 \cdot -1} = \sqrt{49} \cdot \sqrt{-1} = 7i$

**6.** $-\sqrt{\dfrac{36}{121}} = -\sqrt{\dfrac{36}{121} \cdot -1} = -\sqrt{\dfrac{36}{121}} \cdot \sqrt{-1} = -\dfrac{6}{11}i$

**8.** $\sqrt{-0.0004} = \sqrt{0.0004 \cdot -1} = \sqrt{0.0004} \cdot \sqrt{-1} = 0.02i$

**10.** $\sqrt{-75} = \sqrt{75 \cdot -1} = \sqrt{75}\sqrt{-1} = \sqrt{25 \cdot 3}\sqrt{-1} = 5\sqrt{3}i$

**12.** $\sqrt{-80} = \sqrt{16 \cdot 5 \cdot -1}$
$= 4i\sqrt{5}$

**14.** $\sqrt{-15} = \sqrt{-1 \cdot 15} = \sqrt{-1} \cdot \sqrt{15} = i\sqrt{15}$

**16.** $\dfrac{\sqrt{-45}}{\sqrt{-5}} = \sqrt{\dfrac{-45}{-5}} = \sqrt{9} = 3$

**18.** $\dfrac{\sqrt{72}}{\sqrt{-2}} = \sqrt{\dfrac{72}{-2}} = \sqrt{-36} = \sqrt{36 \cdot -1} = -6i$

**20.** $\sqrt{\dfrac{-8}{25}} = \sqrt{-1 \cdot \dfrac{8}{25}} = \sqrt{-1} \cdot \sqrt{\dfrac{8}{25}} = \dfrac{2\sqrt{2}}{5}i$

**22.** $\sqrt{-25} - \sqrt{-9} = \sqrt{25 \cdot -1} - \sqrt{9 \cdot -1}$
$= 5i - 3i$
$= 2i$

**24.** $\sqrt{-500} + \sqrt{-45} = i\sqrt{500} + i\sqrt{45}$
$= 10i\sqrt{5} + 3i\sqrt{5}$
$= \left(10\sqrt{5} + 3\sqrt{5}\right)i$
$= 13i\sqrt{5}$

**26.** $\sqrt{-32} - \sqrt{-18} + \sqrt{-50} = \sqrt{16 \cdot 2 \cdot -1} - \sqrt{9 \cdot 2 \cdot -1} + \sqrt{25 \cdot 2 \cdot -1}$
$= 4i\sqrt{2} - 3i\sqrt{2} + 5i\sqrt{2}$
$= (4i - 3i + 5i)\sqrt{2}$
$= 6i\sqrt{2}$

**28.** $\sqrt{-25}\sqrt{-6} = \sqrt{-1\cdot 25}\cdot\sqrt{-1\cdot 6}$

$= \sqrt{-1}\cdot 5\cdot\sqrt{-1}\cdot\sqrt{6}$

$= 5i\cdot i\sqrt{6}$

$= i^2 5\sqrt{6}$

$= (-1)5\sqrt{6}$

$= -5\sqrt{6}$

**30.** $\sqrt{-7}\sqrt{-7} = i\sqrt{7}\cdot i\sqrt{7}$

$= i\cdot i\cdot\left(\sqrt{7}\right)^2$

$= i^2\cdot 7$

$= -7$

**32.** $\sqrt{-0.49}\sqrt{-1.44} = (0.7i)(1.2i)$

$= 0.84i^2$

$= -0.84$

**34.** $\sqrt{-12}\left(\sqrt{-3} - \sqrt{-12}\right) = 2i\sqrt{3}\left(\sqrt{3}i - 2i\sqrt{3}\right)$

$= 2i\sqrt{3}\left(-i\sqrt{3}\right)$

$= -6i^2$

$= 6$

**36.** $\sqrt{-24}\left(\sqrt{-9} + \sqrt{-4}\right) = 2i\sqrt{6}(3i + 2i)$

$= 2i\sqrt{6}(5i)$

$= 10i^2\sqrt{6}$

$= -10\sqrt{6}$

**38.** $\sqrt{-9}\left(1 + \sqrt{-16}\right) = 3i(1 + 4i)$

$= 3i + 12i^2$

$= -12 + 3i$

**40.** $\left(\sqrt{-2}\right)^2 = \left(i\sqrt{2}\right)^2 = i^2\cdot 2 = -2$

**42.** $\left(\sqrt{-5}\right)^3 = \left(i\sqrt{5}\right)^3 = i^3\cdot 5\sqrt{5} = -5i\sqrt{5}$

**44.** $-8 + 6i = a + bi$

$a = -8 \quad b = 6$

**46.** $-10 + 12i = 2a + (5b - 3)i$

$2a = -10 \quad 5b - 3 = 12$

$a = -5 \qquad 5b = 15$

$b = 3$

**48.** $\sqrt{-36} - 3 = a + bi$

$6i - 3 = a + bi$

$-3 + 6i = a + bi$

$a = -3 \quad b = 6$

**50.** $(2a + 1) + (2b + 3)i = 5 + 12i$

$2a + 1 = 5 \quad 2b + 3 = 12$

$2a = 4 \qquad 2b = 9$

$a = 2 \qquad b = \frac{9}{2}$

**52.** $(-10 + 2i) + (4 - 7i) = (-10 + 4) + (2 - 7)i = -6 - 5i$

**54.** $(15 + 10i) - (2 + 10i) = (15 - 2) + (10 - 10)i$

$= 13 + 0i = 13$

**56.** $(-21 - 50i) + (21 - 20i) = (-21 + 21) + (-50 - 20)i = 0 - 70i = -70i$

**58.** $(4 + 6i) + (15 + 24i) - (1 - i) = (4 + 15 - 1) + (6 + 24 - (-1))i$

$= 18 + 31i$

**60.** $22 + (-5 + 8i) + 10i = (22 - 5) + (8 + 10)i = 17 + 18i$

**62.** $(0.05 + 2.50i) - (6.2 + 11.8i) = (0.05 - 6.2) + (2.50 - 11.8)i$
$$= -6.15 - 9.3i$$

**64.** $(-1 + i) - \sqrt{2} - \sqrt{-2} = -1 + i - \sqrt{2} - i\sqrt{2}$
$$= \left(-1 - \sqrt{2}\right) + \left(1 - \sqrt{2}\right)i$$

**66.** $\left(7 - \sqrt{-96}\right) - (-8 + 10i) - 3i = \left(7 - \sqrt{16 \cdot 6 \cdot -1}\right) - (-8 + 10i) - 3i$
$$= 7 - 4i\sqrt{6} + 8 - 10i - 3i$$
$$= 15 - \left(13 + 4\sqrt{6}\right)i$$

**68.** $(-5i)(4i) = -20i^2 = 20$      **70.** $(2i)(-10i) = 20i^2 = -20$      **72.** $(10i)(12i)(-3i) = -360i^3$
$$= -360i^2 \cdot i$$
$$= 360i$$

**74.** $(8i)^2 = 64i^2 = -64$      **76.** $(2i)^4 = 16i^4 = 16(1) = 16$      **78.** $10(8 - 6i) = 80 - 60i$

**80.** $-3i(10 - 15i) = -30i + 45i^2 = -30i - 45 = -45 - 30i$

**82.** $(11 + 3i)\left(\sqrt{-25}\right) = (11 + 3i)(5i)$      **84.** $\sqrt{-24}\left(-3\sqrt{6} - 4i\right) = 2i\sqrt{6}\left(-3\sqrt{6} - 4i\right)$
$$= 55i + 15i^2 \qquad\qquad\qquad\qquad\qquad = -6i\left(\sqrt{6}\right)^2 - 8\sqrt{6}i^2$$
$$= 55i - 15 \qquad\qquad\qquad\qquad\qquad\quad = -36i + 8\sqrt{6}$$
$$= -15 + 55i \qquad\qquad\qquad\qquad\qquad\; = 8\sqrt{6} - 36i$$

**86.** $(3 + 5i)(2 + 15i) = 6 + 45i + 10i + 75i^2$      **88.** $(3 + 5i)(2 - 15i) = 6 - 45i + 10i - 75i^2$
$$= 6 + 45i + 10i - 75 \qquad\qquad\qquad\qquad = 6 - 35i + 75$$
$$= -69 + 55i \qquad\qquad\qquad\qquad\qquad\qquad\; = 81 - 35i$$

**90.** $\left(-3 - \sqrt{-12}\right)\left(4 - \sqrt{-12}\right) = \left(-3 - 2i\sqrt{3}\right)\left(4 - 2i\sqrt{3}\right)$
$$= -12 + 6i\sqrt{3} - 8i\sqrt{3} + \left(2i\sqrt{3}\right)^2$$
$$= -12 - 2i\sqrt{3} + 12i^2$$
$$= -24 - 2i\sqrt{3}$$

**92.** $(7 + i)^2 = 7^2 + 2(7)i + i^2 = 49 + 14i - 1 = 48 + 14i$      **94.** $(8 - 3i)^2 = 8^2 + 2(8)(-3i) + (-3i)$
$$= 64 - 48i + 9i^2$$
$$= 64 - 9 - 48i$$
$$= 55 - 48i$$

**96.** $(3 - 2i)^3 = (3 - 2i)(3 - 2i)(3 - 2i)$

$\qquad = (9 - 2(3)(2i) + 4i^2)(3 - 2i)$

$\qquad = (9 - 12i - 4)(3 - 2i)$

$\qquad = (5 - 12i)(3 - 2i)$

$\qquad = 15 - 10i - 36i + 24i^2$

$\qquad = 15 - 46i - 24$

$\qquad = -9 - 46i$

**98.** $3 + 2i$, conjugate $= 3 - 2i$

$\qquad$ product $= (3 + 2i)(3 - 2i)$

$\qquad = 3^2 - (2i)^2$

$\qquad = 9 - 4i^2$

$\qquad = 9 + 4$

$\qquad = 13$

**100.** $10 - 3i$, conjugate $= 10 + 3i$

$\qquad$ product $= (10 - 3i)(10 + 3i)$

$\qquad = 10^2 - (3i)^2$

$\qquad = 100 - 9i^2$

$\qquad = 100 + 9$

$\qquad = 109$

**102.** $-4 + \sqrt{2}i$, conjugate $= -4 - \sqrt{2}i$

$\qquad$ product $= \left(-4 + \sqrt{2}i\right)\left(-4 - \sqrt{2}i\right)$

$\qquad = (-4)^2 - \left(\sqrt{2}i\right)^2$

$\qquad = 16 - 2i^2$

$\qquad = 16 + 2$

$\qquad = 18$

**104.** $20$, conjugate $= 20$

$\qquad$ product $= 20 \cdot 20$

$\qquad = 400$

**106.** $-3 - \sqrt{-5} = -3 - i\sqrt{5}$, conjugate $= -3 + i\sqrt{5}$

$\qquad$ product $= \left(-3 - i\sqrt{5}\right)\left(-3 + i\sqrt{5}\right)$

$\qquad = (-3)^2 - \left(i\sqrt{5}\right)^2$

$\qquad = 9 - 5i^2$

$\qquad = 9 + 5$

$\qquad = 14$

**108.** $3.2 - \sqrt{-0.04} = 3.2 - 0.2i$, conjugate $= 3.2 + 0.2i$

$\qquad$ product $= (3.2 - 0.2i)(3.2 + 0.2i)$

$\qquad = 3.2^2 - (0.2i)^2$

$\qquad = 10.24 - 0.04i^2$

$\qquad = 10.24 + 0.04$

$\qquad = 10.28$

**110.** $\dfrac{1+i}{3i} = \dfrac{1+i}{3i} \cdot \dfrac{-3i}{-3i}$

$= \dfrac{-3i(1+i)}{9}$

$= \dfrac{-i(1+i)}{3}$

$= \dfrac{-i+1}{3}$

$= -\dfrac{1}{3}i + \dfrac{1}{3}$

$= \dfrac{1}{3} - \dfrac{1}{3}i$

**112.** $\dfrac{20}{3+i} = \dfrac{20}{3+i} \cdot \dfrac{3-i}{3-i}$

$= \dfrac{20(3-i)}{9+1}$

$= \dfrac{20(3-i)}{10}$

$= 2(3-i)$

$= 6 - 2i$

**114.** $\dfrac{15}{2(1-i)} = \dfrac{15}{2(1-i)} \cdot \dfrac{1+i}{1+i}$

$= \dfrac{15(1+i)}{2(1+1)}$

$= \dfrac{15(1+i)}{2(2)}$

$= \dfrac{15(1+i)}{4}$

$= \dfrac{15+15i}{4}$

$= \dfrac{15}{4} + \dfrac{15}{4}i$

**116.** $\dfrac{17i}{5+3i} = \dfrac{17i}{5+3i} \cdot \dfrac{5-3i}{5-3i}$

$= \dfrac{17i(5-3i)}{25+9}$

$= \dfrac{85i - 51i^2}{34}$

$= \dfrac{85i}{34} + \dfrac{51}{34}$

$= \dfrac{5}{2}i + \dfrac{3}{2}$

$= \dfrac{3}{2} + \dfrac{5}{2}i$

**118.** $\dfrac{4-5i}{4+5i} = \dfrac{4-5i}{4+5i} \cdot \dfrac{4-5i}{4-5i}$

$= \dfrac{(4-5i)(4-5i)}{16+25}$

$= \dfrac{16 - 20i - 20i + 25i^2}{41}$

$= \dfrac{16 - 20i - 20i - 25}{41}$

$= \dfrac{-9 - 40i}{41}$

$= \dfrac{-9}{41} - \dfrac{40}{41}i$

**120.** $\dfrac{3i}{1+i} + \dfrac{2}{2+3i} = \dfrac{3i}{1+i} \cdot \dfrac{1-i}{1-i} + \dfrac{2}{2+3i} \cdot \dfrac{2-3i}{2-3i}$

$= \dfrac{3i(1-i)}{1+1} + \dfrac{2(2-3i)}{4+9}$

$= \dfrac{3i(1-i)}{2} + \dfrac{2(2-3i)}{13}$

$= \dfrac{3 \cdot 13(1-i) + 2 \cdot 2(2-3i)}{26}$

$= \dfrac{39i(1-i) + 4(2-3i)}{26}$

$= \dfrac{39i + 39 + 8 - 12i}{26}$

$= \dfrac{47 + 27i}{26}$

$= \dfrac{47}{26} + \dfrac{27}{26}i$

**122.** $\dfrac{1+i}{i} - \dfrac{3}{5-2i} = \dfrac{1+i}{i} \cdot \dfrac{-i}{-i} - \dfrac{3}{5-2i} \cdot \dfrac{5+2i}{5+2i}$

$= \dfrac{-i(1+i)}{1} - \dfrac{3(5+2i)}{25+4}$

$= \dfrac{-i(1+i)}{1} - \dfrac{3(5+2i)}{29}$

$= \dfrac{-29i(1+i) - 3(5+2i)}{29}$

$= \dfrac{-29i + 29 - 15 - 6i}{29}$

$= \dfrac{14 - 35i}{29}$

$= \dfrac{14}{29} - \dfrac{35}{29}i$

**124.** $x^2 - 4x + 13 = 0$

(a) $x = 2 - 3i$

$$(2 - 3i)^2 - 4(2 - 3i) + 13 \stackrel{?}{=} 0$$

$$4 - 12i + 9i^2 - 8 + 12i + 13 \stackrel{?}{=} 0$$

$$4 - 12i - 9 - 8 + 12i + 13 \stackrel{?}{=} 0$$

$$0 = 0$$

Solution

(b) $x = 2 + 3i$

$$(2 + 3i)^2 - 4(2 + 3i) + 13 \stackrel{?}{=} 0$$

$$4 + 12i + 9i^2 - 8 - 12i + 13 \stackrel{?}{=} 0$$

$$4 + 12i - 9 - 8 - 12i + 13 \stackrel{?}{=} 0$$

$$0 = 0$$

Solution

**126.** $x^3 - 8x^2 + 25x - 26 = 0$

(a) $x = 2$

$$2^3 - 8 \cdot 2^2 + 25 \cdot 2 - 26 \stackrel{?}{=} 0$$

$$8 - 32 + 50 - 26 \stackrel{?}{=} 0$$

$$0 = 0$$

Solution

(b) $x = 3 - 2i$

$$(3 - 2i)^3 - 8(3 - 2i)^2 + 25(3 - 2i) - 26 \stackrel{?}{=} 0$$

$$(9 - 12i + 4i^2)(3 - 2i) - 8(9 - 12i + 4i^2) + 25(3 - 2i) - 26 \stackrel{?}{=} 0$$

$$(5 - 12i)(3 - 2i) - 8(5 - 12i) + 25(3 - 2i) - 26 \stackrel{?}{=} 0$$

$$15 - 46i - 24 - 40 + 96i + 75 - 50i - 26 \stackrel{?}{=} 0$$

$$0 = 0$$

Solution

**128.** (a) $\left(\dfrac{-3 + 3\sqrt{3}i}{2}\right)^3 = \left(\dfrac{-3}{2} + \dfrac{3\sqrt{3}}{2}i\right)^2\left(\dfrac{-3}{2} + \dfrac{3\sqrt{3}}{2}i\right)$

$$= \left(\dfrac{9}{4} - \dfrac{9\sqrt{3}}{2}i + \dfrac{9}{4}(3)i^2\right)\left(\dfrac{-3}{2} + \dfrac{3\sqrt{3}}{2}i\right)$$

$$= \left(\dfrac{9}{4} - \dfrac{9\sqrt{3}}{2}i - \dfrac{27}{4}\right)\left(\dfrac{-3}{2} + \dfrac{3\sqrt{3}}{2}i\right)$$

$$= \left(\dfrac{-18}{4} - \dfrac{9\sqrt{3}}{2}i\right)\left(\dfrac{-3}{2} + \dfrac{3\sqrt{3}}{2}i\right)$$

$$= \left(\dfrac{-9}{2} - \dfrac{9\sqrt{3}}{2}i\right)\left(\dfrac{-3}{2} + \dfrac{3\sqrt{3}}{2}i\right)$$

$$= \dfrac{27}{4} - \dfrac{27}{4}\sqrt{3}i + \dfrac{27\sqrt{3}}{4}i - \dfrac{27(3)}{4}i^2$$

$$= \dfrac{27}{4} + \dfrac{81}{4} = \dfrac{108}{4} = 27$$

(b) Use same method as (a) part

$$\left(\dfrac{-3 - 3\sqrt{3}i}{2}\right)^3 = 27$$

**130.** Complex number $1 + 5i$

(a) Additional inverse is $-(1 + 5i)$ or $-1 - 5i$

(b) Multiplicative inverse is $\dfrac{1}{1 + 5i} = \dfrac{1}{1 + 5i} \cdot \dfrac{1 - 5i}{1 - 5i} = \dfrac{1 - 5i}{1 + 25}$

$$= \dfrac{1 - 5i}{26}$$

$$= \dfrac{1}{26} - \dfrac{5}{26}i$$

**132.** $(a + bi)(a - bi) = a^2 - abi + abi - b^2i^2$

$$= a^2 - b^2(-1)$$

$$= a^2 + b^2$$

**134.** $(a + bi)^2 + (a - bi)^2 = (a^2 + 2abi + b^2i^2) + (a^2 - 2abi + b^2i^2)$

$$= a^2 + 2abi - b^2 + a^2 - 2abi - b^2$$

$$= 2a^2 - 2b^2$$

**136.** The square of any real number is nonnegative.

**138.** False. A number cannot be both. The number 2 is real but not imaginary. The number $2i$ is imaginary but not real.

**140.** To add or subtract complex numbers, add or subtract the real and imaginary parts. To multiply complex numbers, use FOIL and simplify powers of $i$. To divide complex numbers, multiply the numerator and denominator by the complex conjugate of the denominator and simplify.

# Review Exercises for Chapter 5

**2.** $\sqrt{64} = 8$ because $8 \cdot 8 = 64$

**4.** $\sqrt{-16} = 4i$

**6.** $\sqrt[3]{-27} = -3$ because $-3 \cdot -3 \cdot -3 = -27$

**8.** $-\sqrt[3]{125} = -5$ because $5 \cdot 5 \cdot 5 = 125$

**10.** $\sqrt{(0.4)^2} = 0.4$   (inverse property of powers and roots)

**12.** $\sqrt{\left(\frac{8}{15}\right)^2} = \frac{8}{15}$   (inverse property of powers and roots)

**14.** $-\sqrt[3]{\left(\dfrac{-27}{64}\right)^3} = -\left(\dfrac{-27}{64}\right) = \dfrac{27}{64}$

(inverse property of powers and roots)

**16.** $\sqrt{-4^2} = 4i$

**18.** $(0.125)^{1/3} = 0.5$

**20.** $\sqrt[4]{16} = 2$

**22.** $16^{3/4} = \left(\sqrt[4]{16}\right)^3 = 2^3 = 8$

**24.** $(-9)^{5/2} = \left(\sqrt{(-9)}\right)^5 = (3i)^5 = 243i^5$

$$= 243i$$

**26.** $243^{-2/5} = \dfrac{1}{243^{2/5}} = \dfrac{1}{\left(\sqrt[5]{243}\right)^2} = \dfrac{1}{3^2} = \dfrac{1}{9}$

**28.** $\left(\dfrac{-8}{125}\right)^{1/3} = \sqrt[3]{\dfrac{-8}{125}} = \dfrac{-2}{5}$

**30.** $a^{2/3} \cdot a^{3/5} = a^{2/3+3/5}$

$= a^{10/15+9/15}$

$= a^{19/15}$

**32.** $x^2 \sqrt[4]{x^3} = x^2 \cdot x^{3/4}$

$= x^{2+3/4}$

$= x^{8/4+3/4}$

$= x^{11/4}$

**34.** $\dfrac{\sqrt{x^3}}{\sqrt[3]{x^2}} = \dfrac{x^{3/2}}{x^{2/3}} = x^{3/2-2/3} = x^{9/6-4/6}$

$= x^{5/6}$

**36.** $\sqrt[5]{x^6 y^2} = \sqrt[5]{x^5 \cdot x \cdot y^2}$

$= x \sqrt[5]{xy^2}$

**38.** $\sqrt{\sqrt[3]{x^4}} = \sqrt{x^{4/3}} = (x^{4/3})^{1/2}$

$= x^{2/3} = \sqrt[3]{x^2}$

**40.** $\dfrac{\sqrt[5]{3x+6}}{(3x+6)^{4/5}} = \dfrac{(3x+6)^{1/5}}{(3x+6)^{4/5}} = (3x+6)^{1/5-4/5} = (3x+6)^{-3/5} = \dfrac{1}{(3x+6)^{3/5}}$

$= \dfrac{1}{\sqrt[5]{(3x+6)^3}}$

**42.** $510^{5/3} \approx 32{,}554.94457 \approx 32{,}554.94$

**44.** $\dfrac{-3.7 + \sqrt{15.8}}{2(2.3)} \approx 0.059765518 \approx 0.06$

**46.** $y = \dfrac{10}{\sqrt[4]{x^2+1}}$

*Keystrokes:*

$\boxed{Y=}$ 10 $\boxed{\div}$ 4 $\boxed{\text{MATH}}$ 5 $\boxed{(}$ $\boxed{X,T,\theta}$ $\boxed{x^2}$ $\boxed{+}$ 1 $\boxed{)}$ $\boxed{\text{GRAPH}}$

Domain: $(-\infty, \infty)$

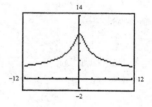

**48.** $h(x) = \tfrac{1}{2} x^{4/3}$

*Keystrokes:*

$\boxed{Y=}$ $\boxed{(}$ 1 $\boxed{\div}$ 2 $\boxed{)}$ $\boxed{\text{MATH}}$ 4 $\boxed{X,T,\theta}$ $\boxed{\wedge}$ 4 $\boxed{\text{GRAPH}}$

Domain: $(-\infty, \infty)$

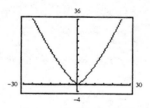

**50.** $\sqrt{\dfrac{50}{9}} = \dfrac{\sqrt{50}}{\sqrt{9}} = \dfrac{\sqrt{25 \cdot 2}}{\sqrt{9}} = \dfrac{5\sqrt{2}}{3} = \dfrac{5}{3}\sqrt{2}$

**52.** $\sqrt{24x^3 y^4} = \sqrt{4 \cdot 6 \cdot x \cdot x^2 \cdot y^4}$

$= 2xy^2 \sqrt{6x}$

**54.** $\sqrt{0.16s^6t^3} = \sqrt{16 \cdot 10^{-2}s^6t^2t}$

$$= \sqrt{\frac{16s^6t^2t}{10^2}}$$

$$= \frac{4s^3t\sqrt{t}}{10}$$

$$= \frac{2s^3t\sqrt{t}}{5}$$

$$= 0.4s^3t\sqrt{t}$$

**56.** $\sqrt{36x^3y^2} = \sqrt{36x^2 \cdot x \cdot y^2}$

$$= 6x|y|\sqrt{x}$$

**58.** $\sqrt[4]{32u^4v^5} = \sqrt[4]{2 \cdot 16 \cdot u^4v^4v} = 2uv\sqrt[4]{2v}$

**60.** $\sqrt{\dfrac{3}{20}} = \dfrac{\sqrt{3}}{\sqrt{20}} = \dfrac{\sqrt{3}}{2\sqrt{5}} = \dfrac{\sqrt{3}}{2\sqrt{5}} \cdot \dfrac{\sqrt{5}}{\sqrt{5}} = \dfrac{\sqrt{15}}{10}$

**62.** $\dfrac{4y}{\sqrt{10z}} = \dfrac{4y}{\sqrt{10z}} \cdot \dfrac{\sqrt{10z}}{\sqrt{10z}} = \dfrac{4y\sqrt{10z}}{10z} = \dfrac{2y\sqrt{10z}}{5z}$

**64.** $\sqrt[3]{\dfrac{16t}{s^2}} = \dfrac{\sqrt[3]{16t}}{\sqrt[3]{s^2}} = \dfrac{2\sqrt[3]{2t}}{\sqrt[3]{s^2}} \cdot \dfrac{\sqrt[3]{s}}{\sqrt[3]{s}} = \dfrac{2\sqrt[3]{2st}}{s}$

**66.** $3\sqrt{5} - 7\sqrt{5} + 2\sqrt{5} = (3 - 7 + 2)\sqrt{5}$

$$= -2\sqrt{5}$$

**68.** $9\sqrt{50} - 5\sqrt{8} + \sqrt{48} = 9\sqrt{25 \cdot 2} - 5\sqrt{4 \cdot 2} + \sqrt{16 \cdot 3}$

$$= 45\sqrt{2} - 10\sqrt{2} + 4\sqrt{3}$$

$$= 35\sqrt{2} + 4\sqrt{3}$$

**70.** $\sqrt{3x} - \sqrt[4]{6x^2} + 2\sqrt[4]{6x^2} - 4\sqrt{3x} = \sqrt{3x} - 4\sqrt{3x} - \sqrt[4]{6x^2} + 2\sqrt[4]{6x^2}$

$$= (1 - 4)\sqrt{3x} + (-1 + 2)\sqrt[4]{6x^2}$$

$$= -3\sqrt{3x} + \sqrt[4]{6x^2}$$

**72.** $5\sqrt[3]{x-3} + 4\sqrt[3]{x-3} = (5 + 4)\sqrt[3]{x-3}$

$$= 9\sqrt[3]{x-3}$$

**74.** $\sqrt[3]{81x^4} + \sqrt[3]{24x^4} - \sqrt{3x} = \sqrt[3]{27 \cdot 3 \cdot x^3 \cdot x} + \sqrt[3]{8 \cdot 3 \cdot x^3 \cdot x} - \sqrt{3x}$

$$= 3x\sqrt[3]{3x} + 2x\sqrt[3]{3x} - \sqrt{3x}$$

$$= (3x + 2x)\sqrt[3]{3x} - \sqrt{3x}$$

$$= 5x\sqrt[3]{3x} - \sqrt{3x}$$

**76.** $\dfrac{4}{\sqrt{2}} + 3\sqrt{2} = \dfrac{4}{\sqrt{2}} \cdot \dfrac{\sqrt{2}}{\sqrt{2}} + 3\sqrt{2} = \dfrac{4\sqrt{2}}{2} + 3\sqrt{2}$

$$= 2\sqrt{2} + 3\sqrt{2} = 5\sqrt{2}$$

**78.** $\sqrt{42} \cdot \sqrt{21} = \sqrt{42 \cdot 21} = \sqrt{882} = \sqrt{441 \cdot 2} = 21\sqrt{2}$

**80.** $\sqrt{6}(\sqrt{24} - 8) = \sqrt{6}\sqrt{24} - 8\sqrt{6} = \sqrt{144} - 8\sqrt{6} = 12 - 8\sqrt{6}$

**82.** $\sqrt{12}(\sqrt{6} - \sqrt{8}) = \sqrt{12}\sqrt{6} - \sqrt{12}\sqrt{8} = \sqrt{72} - \sqrt{96}$

$$= \sqrt{36 \cdot 2} - \sqrt{16 \cdot 6} = 6\sqrt{2} - 4\sqrt{6}$$

**84.** $\left(2 - 4\sqrt{3}\right)\left(7 + \sqrt{3}\right) = 14 + 2\sqrt{3} - 28\sqrt{3} = 4(3) = 2 - 26\sqrt{3}$

**86.** $\left(4 - 3\sqrt{2}\right)^2 = \left(16 - 2(4)3\sqrt{2} + 9(2)\right) = 16 - 24\sqrt{2} + 18 = 34 - 24\sqrt{2}$

**88.** $\left(2 + 3\sqrt{5}\right)\left(2 - 3\sqrt{5}\right) = 2^2 - \left(3\sqrt{5}\right)^2 = 4 - 45 = -41$

**90.** $\dfrac{\sqrt{5}}{\sqrt{10} + 3} = \dfrac{\sqrt{5}}{\sqrt{10} + 3} \cdot \dfrac{\sqrt{10} - 3}{\sqrt{10} - 3} = \dfrac{\sqrt{5}\left(\sqrt{10} - 3\right)}{\left(\sqrt{10}\right)^2 - 3^2} = \dfrac{\sqrt{50} - 3\sqrt{5}}{10 - 9} = \dfrac{5\sqrt{2} - 3\sqrt{5}}{1}$

$$= 5\sqrt{2} - 3\sqrt{5}$$

**92.** $\dfrac{7\sqrt{6}}{\sqrt{3} - 4\sqrt{2}} = \dfrac{7\sqrt{6}}{\sqrt{3} - 4\sqrt{2}} \cdot \dfrac{\sqrt{3} + 4\sqrt{2}}{\sqrt{3} + 4\sqrt{2}} = \dfrac{7\sqrt{18} + 28\sqrt{12}}{\left(\sqrt{3}\right)^2 - \left(4\sqrt{2}\right)^2}$

$$= \dfrac{7\sqrt{9 \cdot 2} + 28\sqrt{4 \cdot 3}}{3 - 32} = \dfrac{21\sqrt{2} + 56\sqrt{3}}{-29} = \dfrac{-7\left(3\sqrt{2} + 8\sqrt{3}\right)}{29}$$

**94.** $\dfrac{3 + \sqrt{3}}{5 - \sqrt{3}} = \dfrac{3 + \sqrt{3}}{5 - \sqrt{3}} \cdot \dfrac{5 + \sqrt{3}}{5 + \sqrt{3}} = \dfrac{15 + 3\sqrt{3} + 5\sqrt{3} + 3}{5^2 - \left(\sqrt{3}\right)^2} = \dfrac{18 + 8\sqrt{3}}{25 - 3} = \dfrac{18 + 8\sqrt{3}}{22}$

$$= \dfrac{9 + 4\sqrt{3}}{11}$$

**96.** $\left(3\sqrt{s} + 4\right) \div \left(\sqrt{s} + 2\right) = \dfrac{3\sqrt{s} + 4}{\sqrt{s} + 2} \cdot \dfrac{\sqrt{s} - 2}{\sqrt{s} - 2} = \dfrac{3s - 6\sqrt{s} + 4\sqrt{s} - 8}{\left(\sqrt{s}\right)^2 - 2^2}$

$$= \dfrac{3s - 2\sqrt{s} - 8}{s - 4}$$

**98.** *Keystrokes:*

$y_1$ $\boxed{\text{Y=}}$ $\boxed{\text{X,T,}\theta}$ $\boxed{\div}$ $\boxed{(}$ $\boxed{1}$ $\boxed{+}$ $\boxed{\sqrt{\ }}$ $\boxed{\text{X,T,}\theta}$ $\boxed{)}$ $\boxed{\text{ENTER}}$

$y_2$ $\boxed{(}$ $\boxed{\text{X,T,}\theta}$ $\boxed{(}$ $\boxed{1}$ $\boxed{-}$ $\boxed{\sqrt{\ }}$ $\boxed{\text{X,T,}\theta}$ $\boxed{)}$ $\boxed{)}$ $\boxed{\div}$ $\boxed{(}$ $\boxed{1}$ $\boxed{-}$ $\boxed{\text{X,T,}\theta}$ $\boxed{)}$ $\boxed{\text{GRAPH}}$

$\dfrac{x}{1 + \sqrt{x}} = \dfrac{x}{1 + \sqrt{x}} \cdot \dfrac{1 - \sqrt{x}}{1 - \sqrt{x}}$

$\qquad = \dfrac{x\left(1 - \sqrt{x}\right)}{1 - x}$

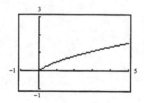

**100.** *Keystrokes:*

$y_1$ $\boxed{\text{Y=}}$ $\boxed{(-)}$ $\boxed{2}$ $\boxed{\sqrt{\ }}$ $\boxed{(}$ $\boxed{9}$ $\boxed{\text{X,T,}\theta}$ $\boxed{)}$ $\boxed{+}$ $\boxed{10}$ $\boxed{\sqrt{\ }}$ $\boxed{\text{X,T,}\theta}$ $\boxed{\text{ENTER}}$

$y_2$ $\boxed{4}$ $\boxed{\sqrt{\ }}$ $\boxed{\text{X,T,}\theta}$ $\boxed{\text{GRAPH}}$

$-2\sqrt{9x} + 10\sqrt{x} = -6\sqrt{x} + 10\sqrt{x}$

$\qquad\qquad = 4\sqrt{x}$

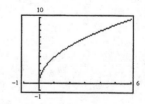

**102.** $\sqrt{x} - 3 = 0$

$\sqrt{x} = 3$

$(\sqrt{x})^2 = 3^2$

$x = 9$

**Check:**

$\sqrt{9} - 3 \overset{?}{=} 0$

$3 - 3 \overset{?}{=} 0$

$0 = 0$

**104.** $\sqrt{4x} + 6 = 9$

$\sqrt{4x} = 3$

$(\sqrt{4x})^2 = (3)^2$

$4x = 9$

$x = \frac{9}{4}$

**Check:**

$\sqrt{4 \cdot \frac{9}{4}} + 6 \overset{?}{=} 9$

$\sqrt{9} + 6 \overset{?}{=} 9$

$3 + 6 \overset{?}{=} 9$

$9 = 9$

**106.** $\sqrt{5(4 - 3x)} = 10$

$\left(\sqrt{5(4 - 3x)}\right)^2 = 10^2$

$5(4 - 3x) = 100$

$4 - 3x = 20$

$-3x = 16$

$x = \frac{-16}{3}$

**Check:**

$\sqrt{5\left(4 - 3\left(\frac{-16}{3}\right)\right)} \overset{?}{=} 10$

$\sqrt{5(4 + 16)} \overset{?}{=} 10$

$\sqrt{100} \overset{?}{=} 10$

$10 = 10$

**108.** $\sqrt[4]{2x + 3} + 4 = 5$

$\sqrt[4]{2x + 3} = 1$

$\left(\sqrt[4]{2x + 3}\right)^4 = 1^4$

$2x + 3 = 1$

$2x = -2$

$x = -1$

**Check:**

$\sqrt[4]{2(-1) + 3} + 4 \overset{?}{=} 5$

$\sqrt[4]{1} + 4 \overset{?}{=} 5$

$1 + 4 \overset{?}{=} 5$

$5 = 5$

**110.** $\sqrt[4]{9x - 2} - \sqrt[4]{8x} = 0$

$\sqrt[4]{9x - 2} = \sqrt[4]{8x}$

$\left(\sqrt[4]{9x - 2}\right)^4 = \left(\sqrt[4]{8x}\right)^4$

$9x - 2 = 8x$

$x - 2 = 0$

$x = 2$

**Check:**

$\sqrt[4]{9 \cdot 2 - 2} - \sqrt[4]{8 \cdot 2} \overset{?}{=} 0$

$\sqrt[4]{18 - 2} - \sqrt[4]{16} \overset{?}{=} 0$

$\sqrt[4]{16} - \sqrt[4]{16} \overset{?}{=} 0$

$2 - 2 \overset{?}{=} 0$

$0 = 0$

**112.** $y - 2 = \sqrt{y + 4}$

$(y - 2)^2 = \left(\sqrt{y + 4}\right)^2$

$y^2 - 4y + 4 = y + 4$

$y^2 - 5y = 0$

$y(y - 5) = 0$

$y = 0 \qquad y = 5$

Not a solution

**Check:**

$0 - 2 \overset{?}{=} \sqrt{0 + 4}$

$-2 \overset{?}{=} \sqrt{4}$

$-2 \neq 2$

**Check:**

$5 - 2 \overset{?}{=} \sqrt{5 + 4}$

$3 \overset{?}{=} \sqrt{9}$

$3 = 3$

**114.**

$$\sqrt{5t} = 1 + \sqrt{5(t-1)}$$
$$\sqrt{5t} - 1 = \sqrt{5(t-1)}$$
$$\left(\sqrt{5t} - 1\right)^2 = \left(\sqrt{5(t-1)}\right)^2$$
$$\left(\sqrt{5t}\right)^2 - 2\sqrt{5t} + 1 = 5(t-1)$$
$$5t - 2\sqrt{5t} + 1 = 5t - 5$$
$$-2\sqrt{5t} + 1 = -5$$
$$1 = 2\sqrt{5t} - 5$$
$$6 = 2\sqrt{5t}$$
$$3 = \sqrt{5t}$$
$$3^2 = \left(\sqrt{5t}\right)^2$$
$$9 = 5t$$
$$\tfrac{9}{5} = t$$

**Check:**

$$\sqrt{5\left(\tfrac{9}{5}\right)} \overset{?}{=} 1 + \sqrt{5\left(\tfrac{9}{5} - 1\right)}$$
$$\sqrt{9} \overset{?}{=} 1 + \sqrt{5\left(\tfrac{4}{5}\right)}$$
$$3 \overset{?}{=} 1 + \sqrt{4}$$
$$3 \overset{?}{=} 1 + 2$$
$$3 = 3$$

**116.** $\sqrt{2 + 9b} + 1 = 3\sqrt{b}$

$$\sqrt{2 + 9b} = -1 + 3\sqrt{b}$$
$$\left(\sqrt{2 + 9b}\right)^2 = \left(-1 + 3\sqrt{b}\right)^2$$
$$2 + 9b = 1 - 2(1)\left(3\sqrt{b}\right) + \left(3\sqrt{b}\right)^2$$
$$2 + 9b = 1 - 6\sqrt{b} + 9b$$
$$2 = 1 - 6\sqrt{b}$$
$$1 = -6\sqrt{b}$$
$$(1)^2 = \left(-6\sqrt{b}\right)^2$$
$$1 = 36b$$
$$\tfrac{1}{36} = b$$

**Check:**

$$\sqrt{2 + 9\left(\tfrac{1}{36}\right)} + 1 \overset{?}{=} 3\sqrt{\tfrac{1}{36}}$$
$$\sqrt{2 + \tfrac{1}{4}} + 1 \overset{?}{=} 3\left(\tfrac{1}{6}\right)$$
$$\sqrt{\tfrac{9}{4}} + 1 \overset{?}{=} \tfrac{1}{2}$$
$$\tfrac{3}{2} + 1 \overset{?}{=} \tfrac{1}{2}$$
$$\tfrac{5}{2} \neq \tfrac{1}{2}$$

No solution

**118.** $\sqrt{-0.16} = \sqrt{0.16 \cdot -1} = \sqrt{0.16}\,i = 0.4i$

**120.** $3 + 2\sqrt{-500} = 3 + 2\sqrt{500 \cdot -1} = 3 + 2\sqrt{500}\,i = 3 + 2\sqrt{100 \cdot 5}\,i$
$$= 3 + 20\sqrt{5}\,i$$

**122.** $-0.5 + 3\sqrt{-1.21} = -0.5 + 3\sqrt{1.21 \cdot -1}$
$$= -0.5 + 3\sqrt{1.21}\,i$$
$$= -0.5 + 3(1.1)i$$
$$= -0.5 + 3.3i$$

**124.** $\sqrt{-49} + \sqrt{-1} = 7i + i = 8i$

**126.** $\sqrt{-169} - \sqrt{-4} = 13i - 2i = 11i$

**128.** $\sqrt{-24}\sqrt{-6} = 2i\sqrt{6} \cdot i\sqrt{6} = 2i^2\left(\sqrt{6}\right)^2 = 2(-1)(6) = -12$

**130.** $\sqrt{-5}\left(\sqrt{-10} + \sqrt{-15}\right) = i\sqrt{5}\left(i\sqrt{10} + i\sqrt{15}\right)$

$$= i^2\sqrt{50} + i^2\sqrt{75}$$

$$= -\sqrt{25 \cdot 2} + -\sqrt{25 \cdot 3}$$

$$= -5\sqrt{2} - 5\sqrt{3}$$

**132.** $5x + \sqrt{-81} = 25 + 3yi$ $\quad\quad 5x = 25$ $\quad\quad\quad\quad 9 = 3y$

$\quad\quad 5x + 9i = 25 + 3yi$ $\quad\quad\quad x = 5$ $\quad\quad\quad\quad 3 = y$

**134.** $10 - \sqrt{-4y} = 2x - 16i$ $\quad\quad 10 = 2x$ $\quad\quad\quad\quad -2\sqrt{y} = -16$

$\quad\quad 10 - 2i\sqrt{y} = 2x - 16i$ $\quad\quad\quad 5 = x$ $\quad\quad\quad\quad\quad \sqrt{y} = 8$

$$\left(\sqrt{y}\right)^2 = 8^2$$

$$y = 64$$

**136.** $(-8 + 3i) - (6 + 7i) = -8 + 3i - 6 - 7i$

$$= -8 - 6 + 3i - 7i$$

$$= -14 - 4i$$

**138.** $(-6 + 3i) + (-1 + i) = -6 + 3i - 1 + i$

$$= -6 - 1 + 3i + i$$

$$= -7 + 4i$$

**140.** $(12 - 5i)(2 + 7i) = 24 + 84i - 10i - 35i^2$

$$= 24 + 74i + 35$$

$$= 59 + 74i$$

**142.** $(2 - 9i)^2 = 2^2 - 2(2)(9i) + (-9i)^2$

$$= 4 - 36i + 81i^2$$

$$= 4 - 36i - 81$$

$$= -77 - 36i$$

**144.** $\dfrac{4}{5i} = \dfrac{4}{5i} \cdot \dfrac{-5i}{-5i} = \dfrac{-20i}{-25i^2} = \dfrac{-20i}{25} = \dfrac{-4}{5}i$

**146.** $\dfrac{5i}{2 + 9i} = \dfrac{5i}{2 + 9i} \cdot \dfrac{2 - 9i}{2 - 9i} = \dfrac{10i - 45i^2}{4 + 81}$

$$= \dfrac{45 + 10i}{85}$$

$$= \dfrac{45}{85} + \dfrac{10}{85}i$$

$$= \dfrac{9}{17} + \dfrac{2}{17}i$$

**148.** $\dfrac{2 + i}{1 - 9i} = \dfrac{2 + i}{1 - 9i} \cdot \dfrac{1 + 9i}{1 + 9i}$

$$= \dfrac{2 + 18i + i + 9i^2}{1 + 81}$$

$$= \dfrac{2 + 19i - 9}{82}$$

$$= \dfrac{-7 + 19i}{82}$$

$$= -\dfrac{7}{82} + \dfrac{19}{82}i$$

**150.**

$$P = 2l + 2w \qquad 30^2 = l^2 + w^2$$

$$84 = 2l + 2w \qquad 900 = l^2 + w^2$$

$$42 = l + w \qquad 900 = l^2 + (42 - l)^2$$

$$42 - l = w \qquad 900 = l^2 + 1764 - 2(42)l + l^2$$

$$900 = l^2 + 1764 - 84l + l^2$$

$$0 = 2l^2 - 84l + 864$$

$$0 = 2(l^2 - 42l + 432)$$

$$0 = 2(l - 24)(l - 18)$$

$$l - 24 = 0 \qquad l - 18 = 0$$

$$l = 24 \qquad l = 18$$

When $l = 24$ inches, $w = 42 - 24 = 18$ inches.

When $l = 18$ inches, $w = 42 - 18 = 24$ inches.

**152.**

$$t = \sqrt{\frac{d}{16}}$$

$$4 = \sqrt{\frac{d}{16}}$$

$$4^2 = \left( \sqrt{\frac{d}{16}} \right)^2$$

$$16 = \frac{d}{16}$$

$$16(16) = d$$

$$256 = d$$

Thus, the height of the bridge is approximately 256 feet.

**154.**

$$I = \sqrt{\frac{P}{R}}$$

$$10 = \sqrt{\frac{P}{20}}$$

$$10^2 = \left( \sqrt{\frac{P}{20}} \right)^2$$

$$100 = \frac{P}{20}$$

$$2000 \text{ watts} = P$$

**156.**

$$I = \sqrt{\frac{P}{R}}$$

$$15 = \sqrt{\frac{P}{20}}$$

$$15^2 = \left( \sqrt{\frac{P}{20}} \right)^2$$

$$225 = \frac{P}{20}$$

$$4500 \text{ watts} = P$$

# CHAPTER 6
## Quadratic Equations and Inequalities

# CHAPTER 6
# Quadratic Equations and Inequalities

## Section 6.1  Factoring and Extracting Square Roots
Solutions to Even-Numbered Exercises

**2.** $x^2 + 15x + 44 = 0$

$(x + 11)(x + 4) = 0$

$x = -11 \quad x = -4$

**4.** $x^2 - 2x - 48 = 0$

$(x - 8)(x + 6) = 0$

$x = 8 \quad x = -6$

**6.** $\quad\quad x^2 - 7x = 18$

$x^2 - 7x - 18 = 0$

$(x - 9)(x + 2) = 0$

$x = 9 \quad x = -2$

**8.** $x^2 + 60x + 900 = 0$

$(x + 30)(x + 30) = 0$

$x = -30 \quad x = -30$

**10.** $\quad 8x^2 - 10x + 3 = 0$

$(4x - 3)(2x - 1) = 0$

$4x - 3 = 0 \quad\quad 2x - 1 = 0$

$4x = 3 \quad\quad\quad 2x = 1$

$x = \frac{3}{4} \quad\quad\quad x = \frac{1}{2}$

**12.** $25y^2 - 75y = 0$

$25y(y - 3) = 0$

$25y = 0 \quad\quad y - 3 = 0$

$y = 0 \quad\quad\quad y = 3$

**14.** $16x(x - 8) - 12(x - 8) = 0$

$(x - 8)(16x - 12) = 0$

$4(x - 8)(4x - 3) = 0$

$x - 8 = 0 \quad\quad 4x - 3 = 0$

$x = 8 \quad\quad\quad 4x = 3$

$x = \frac{3}{4}$

**16.** $3(4 - x) - 2x(4 - x) = 0$

$(4 - x)(3 - 2x) = 0$

$4 - x = 0 \quad\quad 3 - 2x = 0$

$4 = x \quad\quad\quad 3 = 2x$

$\frac{3}{2} = x$

**18.** $\quad (6 + u)(1 - u) = 10$

$6 - 6u + u - u^2 = 10$

$6 - 5u - u^2 = 10$

$0 = u^2 + 5u + 4$

$0 = (u + 4)(u + 1)$

$u + 4 = 0 \quad\quad u + 1 = 0$

$u = -4 \quad\quad\quad u = -1$

**20.** $(2z + 1)(2z - 1) = -4z^2 - 5z + 2$

$4z^2 - 1 = -4z^2 - 5z + 2$

$8z^2 + 5z - 3 = 0$

$(8z - 3)(z + 1) = 0$

$8z - 3 = 0 \quad\quad z + 1 = 0$

$8z = 3 \quad\quad\quad z = -1$

$z = \frac{3}{8}$

**22.** $z^2 = 169$

$z = \pm\sqrt{169}$

$z = \pm 13$

**24.** $\quad\quad 5t^2 = 125$

$5t^2 - 125 = 0$

$5(t^2 - 25) = 0$

$5(t - 5)(t + 5) = 0$

$t - 5 = 0 \quad\quad t + 5 = 0$

$t = 5 \quad\quad\quad t = -5$

**26.** $9z^2 = 121$

$z^2 = \frac{121}{9}$

$z = \pm\sqrt{\frac{121}{9}}$

$z = \pm\frac{11}{3}$

**28.** $\frac{x^2}{6} = 24$

$x^2 = 24(6)$

$x^2 = 144$

$x = \pm\sqrt{144}$

$x = \pm 12$

**30.** $16y^2 - 121 = 0$

$16y^2 = 121$

$y^2 = \dfrac{121}{16}$

$y = \pm\sqrt{\dfrac{121}{16}}$

$y = \pm\dfrac{11}{4}$

**32.** $16x^2 - 1 = 0$

$16x^2 = 1$

$x^2 = \dfrac{1}{16}$

$x = \pm\sqrt{\dfrac{1}{16}}$

$x = \pm\dfrac{1}{4}$

**34.** $(y - 20)^2 = 625$

$y - 20 = \pm\sqrt{625}$

$y - 20 = \pm25$

$y = 20 \pm 25$

$y = 45 \quad y = -5$

**36.** $(x + 2)^2 = 0.81$

$x + 2 = \pm\sqrt{0.81}$

$x + 2 = \pm0.9$

$x = -2 \pm 0.9$

$x = -2.9 \quad x = -1.1$

**38.** $(x + 8)^2 = 28$

$x + 8 = \pm\sqrt{28}$

$x + 8 = \pm2\sqrt{7}$

$x = -8 \pm 2\sqrt{7}$

**40.** $(3x - 5)^2 = 48$

$3x - 5 = \pm\sqrt{48}$

$3x - 5 = \pm4\sqrt{3}$

$3x = 5 \pm 4\sqrt{3}$

$x = \dfrac{5 \pm 4\sqrt{3}}{3}$

**42.** $(5x + 11)^2 - 300 = 0$

$(5x + 11)^2 = 300$

$5x + 11 = \pm\sqrt{300}$

$5x + 11 = \pm10\sqrt{3}$

$5x = -11 \pm 10\sqrt{3}$

$x = \dfrac{-11 \pm 10\sqrt{3}}{5}$

**44.** $x^2 = -9$

$x = \pm\sqrt{-9}$

$x = \pm3i$

**46.** $y^2 + 16 = 0$

$y^2 = -16$

$y = \pm\sqrt{-16}$

$y = \pm4i$

**48.** $4v^2 + 9 = 0$

$4v^2 = -9$

$v^2 = -\dfrac{9}{4}$

$v = \pm\sqrt{-\dfrac{9}{4}}$

$v = \pm\dfrac{3}{2}i$

**50.** $(x + 5)^2 = -81$

$(x + 5)^2 + 81 = 0$

$(x + 5)^2 = -81$

$x + 5 = \pm\sqrt{-81}$

$x + 5 = \pm9i$

$x = -5 \pm 9i$

**52.** $(2y - 3)^2 + 25 = 0$

$(2y - 3)^2 = -25$

$2y - 3 = \pm\sqrt{-25}$

$2y = 3 \pm 5i$

$y = \dfrac{3}{2} \pm \dfrac{5}{2}i$

**54.** $(6y - 5)^2 = -8$

$6y - 5 = \pm\sqrt{-8}$

$6y = 5 \pm 2i\sqrt{2}$

$y = \dfrac{5 \pm 2i\sqrt{2}}{6}$

$y = \dfrac{5}{6} \pm \dfrac{\sqrt{2}}{3}i$

**56.** $4(x - 4)^2 = -169$

$(x - 4)^2 = \dfrac{-169}{4}$

$x - 4 = \pm\sqrt{\dfrac{-169}{4}}$

$x - 4 = \pm\dfrac{13}{2}i$

$x = 4 \pm \dfrac{13}{2}i$

**58.** $(2x + 3)^2 = -54$

$2x + 3 = \pm\sqrt{-54}$

$2x + 3 = \pm3\sqrt{6}i$

$2x = -3 \pm 3\sqrt{6}i$

$x = \dfrac{-3 \pm 3\sqrt{6}i}{2}$

$= \dfrac{-3}{2} \pm \dfrac{3\sqrt{6}}{2}i$

**60.** $(x - 3)^2 + 2.25 = 0$

$$(x - 3)^2 = -2.25$$
$$x - 3 = \pm\sqrt{-2.25}$$
$$x = 3 \pm 1.5i$$

**62.** $\left(u + \dfrac{5}{8}\right)^2 + \dfrac{49}{16} = 0$

$$\left(u + \dfrac{5}{8}\right)^2 = -\dfrac{49}{16}$$
$$u + \dfrac{5}{8} = \pm\sqrt{-\dfrac{49}{16}}$$
$$u + \dfrac{5}{8} = \pm\dfrac{7}{4}i$$
$$u = -\dfrac{5}{8} \pm \dfrac{7}{4}i$$

**64.** $\left(y - \dfrac{5}{6}\right)^2 = -\dfrac{4}{5}$

$$y - \dfrac{5}{6} = \pm\sqrt{-\dfrac{4}{5}}$$
$$y - \dfrac{5}{6} = \pm\dfrac{2}{\sqrt{5}}i$$
$$y = \dfrac{5}{6} \pm \dfrac{2}{\sqrt{5}}i$$
$$y = \dfrac{5}{6} \pm \dfrac{2}{\sqrt{5}} \cdot \dfrac{\sqrt{5}}{\sqrt{5}}i$$
$$y = \dfrac{5}{6} \pm \dfrac{2\sqrt{5}}{5}i$$

**66.** $3t^2 + 6t = 0$

$$3t(t + 2) = 0$$
$$3t = 0 \qquad t + 2 = 0$$
$$t = 0 \qquad\quad t = -2$$

**68.** $3x^2 + 8x - 16 = 0$

$$(3x - 4)(x + 4) = 0$$
$$3x - 4 = 0 \qquad x + 4 = 0$$
$$3x = 4 \qquad\quad x = -4$$
$$x = \dfrac{4}{3}$$

**70.** $y^2 - 225 = 0$

$$y^2 = 225$$
$$y = \pm\sqrt{225}$$
$$y = \pm 15$$

**72.** $y^2 + 225 = 0$

$$y^2 = -225$$
$$y = \pm\sqrt{-225}$$
$$y = \pm 15i$$

**74.** $\dfrac{1}{3}x^2 = 4$

$$x^2 = 12$$
$$x = \pm\sqrt{12}$$
$$x = \pm 2\sqrt{3}$$

**76.** $(y + 12)^2 - 400 = 0$

$$(y + 12)^2 = 400$$
$$y + 12 = \pm 20$$
$$y = -12 \pm 20$$
$$y = -32 \qquad y = 8$$

**78.** $(y + 12)^2 + 400 = 0$

$$(y + 12)^2 = -400$$
$$y + 12 = \pm 20i$$
$$y = -12 \pm 20i$$

**80.** $(x + 2)^2 - 18 = 0$

$$(x + 2)^2 = 18$$
$$x + 2 = \pm\sqrt{18}$$
$$x = -2 \pm 3\sqrt{2}$$

**82.** $y = 5x - x^2$

*Keystrokes:*

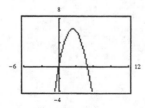

$0 = 5x - x^2$

$0 = x(5 - x)$

$0 = x \qquad 5 - x = 0$

$\qquad\qquad\quad 5 = x$

$x$-intercepts are 0 and 5.

$(0, 0), (5, 0)$

**84.** $y = 9 - 4(x - 3)^2$

*Keystrokes:*

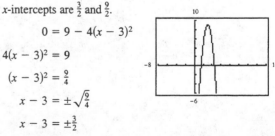

$x$-intercepts are $\dfrac{3}{2}$ and $\dfrac{9}{2}$.

$$0 = 9 - 4(x - 3)^2$$
$$4(x - 3)^2 = 9$$
$$(x - 3)^2 = \dfrac{9}{4}$$
$$x - 3 = \pm\sqrt{\dfrac{9}{4}}$$
$$x - 3 = \pm\dfrac{3}{2}$$
$$x = 3 \pm \dfrac{3}{2}$$
$$x = \dfrac{9}{2} \qquad x = \dfrac{3}{2}$$
$$\left(\dfrac{3}{2}, 0\right), \left(\dfrac{9}{2}, 0\right)$$

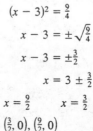

**86.** $y = 4(x + 1)^2 - 9$

*Keystrokes:*

[Y=] 4 [(] [X,T,θ] [+] 1 [)] [x²] [−] 9 [GRAPH]

$x$-intercepts are $-\frac{5}{2}$ and $\frac{1}{2}$.

$0 = 4(x + 1)^2 - 9$

$9 = 4(x + 1)^2$

$\frac{9}{4} = (x + 1)^2$

$\pm\sqrt{\frac{9}{4}} = x + 1$

$\pm\frac{3}{2} = x + 1$

$-1 \pm \frac{3}{2} = x$

$x = -\frac{5}{2} \qquad x = \frac{1}{2}$

$\left(-\frac{5}{2}, 0\right), \left(\frac{1}{2}, 0\right)$

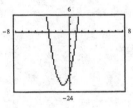

**88.** $y = 4x^2 - x - 14$

*Keystrokes:*

[Y=] 4 [X,T,θ] [x²] [−] [X,T,θ] [−] 14 [GRAPH]

$x$-intercepts are $-\frac{7}{4}$ and 2.

$0 = 4x^2 - x - 14$

$0 = (4x + 7)(x - 2)$

$x = -\frac{7}{4} \qquad x = 2$

$\left(-\frac{7}{4}, 0\right), (2, 0)$

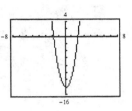

**90.** $y = 5x^2 + 9x - 18$

*Keystrokes:*

[Y=] 5 [X,T,θ] [x²] [+] 9 [X,T,θ] [−] 18 [GRAPH]

$x$-intercepts are $\frac{6}{5}$ and $-3$.

$0 = 5x^2 + 9x - 18$

$0 = (5x - 6)(x + 3)$

$x = \frac{6}{5} \qquad x = -3$

$(-3, 0), \left(\frac{6}{5}, 0\right)$

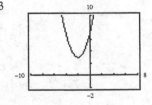

**92.** $y = x^2 + 5$

*Keystrokes:*

[Y=] [X,T,θ] [x²] [+] 5 [GRAPH]

$0 = x^2 + 5$

$-5 = x^2$

$\pm\sqrt{-5} = x$

$\pm i\sqrt{5} = x$

Not real, therefore, there are no $x$-intercepts.

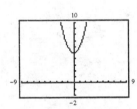

**94.** $y = (x + 2)^2 + 3$

*Keystrokes:*

[Y=] [(] [X,T,θ] [+] 2 [)] [x²] [+] 3 [GRAPH]

$0 = (x + 2)^2 + 3$

$-3 = (x + 2)^2$

$\pm\sqrt{-3} = x + 2$

$\pm\sqrt{3}i = x + 2$

$-2 \pm \sqrt{3}i = x$

Not real, therefore, there are no $x$-intercepts.

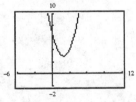

**96.** *Keystrokes:*

[Y=] [(] [X,T,θ] [−] 2 [)] [x²] [+] 3 [GRAPH]

$0 = (x - 2)^2 + 3$

$-3 = (x - 2)^2$

$\pm\sqrt{-3} = x - 2$

$2 \pm \sqrt{3}i = x$

Not real, therefore, there are no $x$-intercepts.

**98.** $x^2 - y^2 = 4$

$-y^2 = 4 - x^2$

$y^2 = x^2 - 4$

$y = \pm\sqrt{x^2 - 4}$

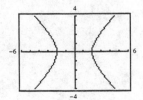

**100.** $x - y^2 = 0$

$-y^2 = -x$

$y^2 = x$

$y = \pm\sqrt{x}$

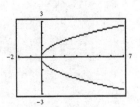

**102.**    $x^4 - 10x^2 + 25 = 0$

$$\text{let } u = x^2$$

$$(x^2)^2 - 10x^2 + 25 = 0$$

$$u^2 - 10u + 25 = 0$$

$$(u - 5)(u - 5) = 0$$

$$u - 5 = 0 \qquad u - 5 = 0$$

$$u = 5 \qquad u = 5$$

$$x^2 = 5 \qquad x^2 = 5$$

$$x = \pm\sqrt{5} \qquad x = \pm\sqrt{5}$$

**104.**    $x^4 - 11x^2 + 30 = 0$

$$\text{let } u = x^2$$

$$(x^2)^2 - 11x^2 + 30 = 0$$

$$u^2 - 11u + 30 = 0$$

$$(u - 5)(u - 6) = 0$$

$$u - 5 = 0 \qquad u - 6 = 0$$

$$u = 5 \qquad u = 6$$

$$x^2 = 5 \qquad x^2 = 6$$

$$x = \pm\sqrt{5} \qquad x = \pm\sqrt{6}$$

**106.**  $x^4 - x^2 - 6 = 0$

$$\text{let } u = x^2$$

$$(x^2)^2 - x^2 - 6 = 0$$

$$u^2 - u - 6 = 0$$

$$(u - 3)(u + 2) = 0$$

$$u - 3 = 0 \qquad u + 2 = 0$$

$$u = 3 \qquad u = -2$$

$$x^2 = 3 \qquad x^2 = -2$$

$$x = \pm\sqrt{3} \qquad x = \pm\sqrt{-2}$$

$$\qquad\qquad x = \pm\sqrt{2}\,i$$

**108.**  $(x^2 - 1)^2 + (x^2 - 1) - 6 = 0$

$$\text{let } u = x^2 - 1$$

$$(x^2 - 1)^2 + (x^2 - 1) - 6 = 0$$

$$u^2 + u - 6 = 0$$

$$(u + 3)(u - 2) = 0$$

$$u + 3 = 0 \qquad u - 2 = 0$$

$$u = -3 \qquad u = 2$$

$$x^2 - 1 = -3 \qquad x^2 - 1 = 2$$

$$x^2 = -2 \qquad x^2 = 3$$

$$x = \pm\sqrt{-2} \qquad x = \pm\sqrt{3}$$

$$x = \pm\sqrt{2}\,i$$

**110.**    $x - 11\sqrt{x} + 24 = 0$

$$\text{let } u = \sqrt{x}$$

$$\left(\sqrt{x}\right)^2 - 11\sqrt{x} + 24 = 0$$

$$u^2 - 11u + 24 = 0$$

$$(u - 8)(u - 3) = 0$$

$$u = 8 \qquad u = 3$$

$$\sqrt{x} = 8 \qquad \sqrt{x} = 3$$

$$\left(\sqrt{x}\right)^2 = 8^2 \qquad \left(\sqrt{x}\right)^2 = 3^2$$

$$x = 64 \qquad x = 9$$

**Check:**            **Check:**

$$64 - 11\sqrt{64} + 24 \overset{?}{=} 0 \qquad 9 - 11\sqrt{9} + 24 \overset{?}{=} 0$$

$$64 - 88 + 24 \overset{?}{=} 0 \qquad 9 - 33 + 24 \overset{?}{=} 0$$

$$0 = 0 \qquad\qquad 0 = 0$$

**112.**  $x^{2/3} + 3x^{1/3} - 10 = 0$

$$(x^{1/3} + 5)(x^{1/3} - 2) = 0$$

$$x^{1/3} = -5 \qquad x^{1/3} = 2$$

$$x = (-5)^3 \qquad x = 2^3$$

$$x = -125 \qquad x = 8$$

**114.**   $3x^{2/3} + 8x^{1/3} + 5 = 0$

$(3x^{1/3} + 5)(x^{1/3} + 1) = 0$

$3x^{1/3} + 5 = 0 \qquad x^{1/3} = -1$

$x^{1/3} = \dfrac{-5}{3} \qquad\qquad x = (-1)^3$

$\qquad\qquad\qquad\qquad x = -1$

$x = \left(\dfrac{-5}{3}\right)^3$

$x = \dfrac{-125}{27}$

**116.**   $x^{2/5} + 5x^{1/5} + 6 = 0$

$(x^{1/5} + 3)(x^{1/5} + 2) = 0$

$x^{1/5} = -3 \qquad x^{1/5} = -2$

$x = (-3)^5 \qquad x = (-2)^5$

$x = -243 \qquad x = -32$

**118.**   $2x^{2/5} + 3x^{1/5} + 1 = 0$

$(2x^{1/5} + 1)(x^{1/5} + 1) = 0$

$x^{1/5} = \dfrac{-1}{2} \qquad x^{1/5} = -1$

$x = \left(\dfrac{-1}{2}\right)^5 \qquad x = (-1)^5$

$x = \dfrac{-1}{32} \qquad\qquad x = -1$

**120.**   $3\left(\dfrac{x}{x+1}\right)^2 + 7\left(\dfrac{x}{x+1}\right) - 6 = 0$

$\left[3\left(\dfrac{x}{x+1}\right) - 2\right]\left[\left(\dfrac{x}{x+1}\right) + 3\right] = 0$

$\dfrac{3x}{x+1} - 2 = 0 \qquad\qquad \dfrac{x}{x+1} + 3 = 0$

$\dfrac{3x}{x+1} = 2 \qquad\qquad \dfrac{x}{x+1} = -3$

$3x = 2x + 2 \qquad\qquad x = -3x - 3$

$x = 2 \qquad\qquad\qquad 4x = -3$

$x = -\dfrac{3}{4}$

**122.**  $-2$ and $3$

$(x - (-2))(x - 3) = 0$

$(x + 2)(x - 3) = 0$

$x^2 - x - 6 = 0$

**124.**  $-3 + \sqrt{5}$ and $-3 - \sqrt{5}$

$\left(x - (-3 + \sqrt{5})\right)\left(x - (-3 - \sqrt{5})\right) = 0$

$(x + 3 - \sqrt{5})(x + 3 + \sqrt{5}) = 0$

$((x + 3) - \sqrt{5})((x + 3) + \sqrt{5}) = 0$

$(x + 3)^2 - (\sqrt{5})^2 = 0$

$x^2 + 6x + 9 - 5 = 0$

$x^2 + 6x + 4 = 0$

**126.**  $2i$ and $-2i$

$(x - 2i)(x - (-2i)) = 0$

$(x - 2i)(x + 2i) = 0$

$x^2 - 4i^2 = 0$

$x^2 + 4 = 0$

**128.**  $s_0 = 48$

$0 = -16t^2 + 48$

$16t^2 = 48$

$t^2 = 3$

$t = \pm\sqrt{3}$

$t = \sqrt{3} \approx 1.732$ seconds

**130.**   $s_0 = 500$

$0 = -16t^2 + 500$

$16t^2 = 500$

$t^2 = 31.25$

$t = \sqrt{31.25} \approx 5.59$ sec

**132.**
$$7000 = x\left(120 - \tfrac{1}{2}x\right)$$
$$7000 = 120x - \tfrac{1}{2}x^2$$
$$2(7000) = \left(120x - \tfrac{1}{2}x^2\right)2$$
$$14{,}000 = 240x - x^2$$
$$x^2 - 240x + 14{,}000 = 0$$
$$(x - 100)(x - 140) = 0$$
$$x - 100 = 0 \qquad x - 140 = 0$$
$$x = 100 \qquad\qquad x = 140$$

You must sell 100 units to produce a revenue of $7000. If you sell 140 units, you will also have a revenue of $7000.

**134.** $P = \$5000,\ A = \$5724.50$
$$A = P(1 + r)^2$$
$$5724.50 = 5000(1 + r)^2$$
$$\frac{5724.50}{5000} = (1 + r)^2$$
$$1.1449 = (1 + r)^2$$
$$\sqrt{1.1449} = 1 + r$$
$$1.07 - 1 = r$$
$$0.7 = r$$
$$7\% = r$$

**136.**
$$y = (26.6 + t)^2,\ 0 \le t \le 6$$
$$1000 = (26.6 + t)^2$$
$$\sqrt{1000} = 26.6 + t$$
$$\sqrt{1000} - 26.6 = t$$
$$5.02 \approx t$$
$$5 \approx t$$

year 1995

**138.** If $a = 0$, the equation would not be quadratic.

**140.** Yes. If $(x - 1)^2 = 0$, the only solution is $x = 1$.

**142.** Write the equation in the form $u^2 = d$, where $u$ is an algebraic expression and $d$ is a positive constant. Take the square roots of both sides to obtain the solutions $u = \pm\sqrt{d}$.

# Section 6.2    Completing the Square

**2.** $x^2 + 12x + 36$
$$\left[36 = \left(\frac{12}{2}\right)^2\right]$$

**4.** $y^2 - 2y + 1$
$$\left[1 = \left(\frac{-2}{2}\right)^2\right]$$

**6.** $x^2 + 18x + 81$
$$\left[81 = \left(\frac{18}{2}\right)^2\right]$$

**8.** $u^2 + 7u + \dfrac{49}{4}$
$$\left[\frac{49}{4} = \left(\frac{7}{2}\right)^2\right]$$

**10.** $y^2 - 11y + \dfrac{121}{4}$
$$\left[\frac{121}{4} = \left(\frac{-11}{2}\right)^2\right]$$

**12.** $y^2 + \dfrac{4}{3}y + \dfrac{4}{9}$
$$\left[\frac{4}{9} = \left[\left(\frac{4}{3}\right)\frac{1}{2}\right]^2\right]$$

**14.** $x^2 - \dfrac{6}{5}x + \dfrac{9}{25}$
$$\left[\frac{9}{25} = \left[\left(\frac{-6}{5}\right)\frac{1}{2}\right]^2\right]$$

**16.** $s^2 + 4.6s + 5.29$
$$\left[5.29 = \left(\frac{4.6}{2}\right)^2\right]$$

**18.** $x^2 + 32x = 0$

(a)
$$x^2 + 32x = 0$$
$$x^2 + 32x + 256 = 0 + 256$$
$$(x + 16)^2 = 256$$
$$x + 16 = \pm 16$$
$$x = -16 \pm 16$$
$$x = 0,\ -32$$

(b) $x^2 + 32x = 0$
$$x(x + 32) = 0$$
$$x = 0 \qquad x + 32 = 0$$
$$x = -32$$

**20.** $t^2 - 10t = 0$

   (a) $t^2 - 10t + 25 = 25$     (b) $t^2 - 10t = 0$

        $(t - 5)^2 = 25$          $t(t - 10) = 0$

         $t - 5 = \pm 5$            $t = 0, 10$

           $t = 5 \pm 5$

            $t = 10, 0$

**22.** $t^2 - 9t = 0$

   (a) $t^2 - 9t + \frac{81}{4} = 0 + \frac{81}{4}$    (b) $t^2 - 9t = 0$

        $\left(t - \frac{9}{2}\right)^2 = \frac{81}{4}$         $t(t - 9) = 0$

        $t - \frac{9}{2} = \pm\sqrt{\frac{81}{4}}$    $t = 0$    $t - 9 = 0$

        $t - \frac{9}{2} = \pm\frac{9}{2}$             $t = 9$

          $t = \frac{9}{2} \pm \frac{9}{2}$

           $t = 9, 0$

**24.** $y^2 - 8y + 12 = 0$

   (a) $y^2 - 8y + 12 = 0$

         $y^2 - 8y = -12$

     $y^2 - 8y + 16 = -12 + 16$

          $(y - 4)^2 = 4$

           $y - 4 = \pm\sqrt{4}$

           $y - 4 = \pm 2$

             $y = 4 \pm 2$

             $y = 6, 2$

   (b) $y^2 - 8y + 12 = 0$

       $(y - 6)(y - 2) = 0$

    $y - 6 = 0$    $y - 2 = 0$

      $y = 6$       $y = 2$

**26.** $x^2 + 12x + 27 = 0$

   (a) $x^2 + 12x + 27 = 0$

         $x^2 + 12x = -27$

     $x^2 + 12x + 36 = -27 + 36$

          $(x + 6)^2 = 9$

           $x + 6 = \pm\sqrt{9}$

           $x + 6 = \pm 3$

             $x = -6 \pm 3$

             $x = -9, -3$

   (b) $x^2 + 12x + 27 = 0$

       $(x + 9)(x + 3) = 0$

    $x + 9 = 0$    $x + 3 = 0$

      $x = -9$      $x = -3$

**28.** $z^2 + 3z - 10 = 0$

   (a) $z^2 + 3z - 10 = 0$

         $z^2 + 3z = 10$

    $z^2 + 3z + \frac{9}{4} = 10 + \frac{9}{4}$

        $\left(z + \frac{3}{2}\right)^2 = \frac{49}{4}$

        $z + \frac{3}{2} = \pm\sqrt{\frac{49}{4}}$

        $z + \frac{3}{2} = \pm\frac{7}{2}$

           $z = -\frac{3}{2} \pm \frac{7}{2}$

           $z = 2, -5$

   (b) $z^2 + 3z - 10 = 0$

     $(z + 5)(z - 2) = 0$

    $z + 5 = 0$    $z - 2 = 0$

      $z = -5$      $z = 2$

**30.** $t^2 - 5t - 36 = 0$

   (a) $t^2 - 5t - 36 = 0$

         $t^2 - 5t = 36$

    $t^2 - 5t + \frac{25}{4} = 36 + \frac{25}{4}$

        $\left(t - \frac{5}{2}\right)^2 = \frac{169}{4}$

        $t - \frac{5}{2} = \pm\sqrt{\frac{169}{4}}$

        $t - \frac{5}{2} = \pm\frac{13}{2}$

           $t = \frac{5}{2} \pm \frac{13}{2}$

           $t = 9, -4$

   (b) $t^2 - 5t - 36 = 0$

     $(t - 9)(t + 4) = 0$

    $t - 9 = 0$    $t + 4 = 0$

      $t = 9$      $t = -4$

**32.** $3x^2 - 3x - 6 = 0$

(a) $3x^2 - 3x - 6 = 0$

$x^2 - 1x - 2 = 0$

$x^2 - 1x + \frac{1}{4} = 2 + \frac{1}{4}$

$\left(x - \frac{1}{2}\right)^2 = \frac{9}{4}$

$x - \frac{1}{2} = \pm\frac{3}{2}$

$x = \frac{1}{2} \pm \frac{3}{2}$

$x = 2, -1$

(b) $3x^2 - 3x - 6 = 0$

$3(x^2 - x - 2) = 0$

$(x - 2)(x + 1) = 0$

$x = 2 \qquad x = -1$

**34.** $3x^2 - 13x + 12 = 0$

(a) $\qquad 3x^2 - 13x = -12$

$x^2 - \frac{13}{3}x + \frac{169}{36} = -4 + \frac{169}{36}$

$\left(x - \frac{13}{6}\right)^2 = \frac{25}{36}$

$x - \frac{13}{6} = \pm\frac{5}{6}$

$x = \frac{13}{6} \pm \frac{5}{6} = \frac{18}{6}, \frac{8}{6}$

$= 3, \frac{4}{3}$

(b) $3x^2 - 13x + 12 = 0$

$(3x - 4)(x - 3) = 0$

$x = \frac{4}{3} \qquad x = 3$

**36.** $x^2 - 6x + 7 = 0$

$x^2 - 6x = -7$

$x^2 - 6x + 9 = -7 + 9$

$(x - 3)^2 = 2$

$x - 3 = \pm\sqrt{2}$

$x = 3 \pm \sqrt{2}$

$x \approx 4.41, 1.59$

**38.** $x^2 + 6x + 7 = 0$

$x^2 + 6x + 9 = -7 + 9$

$(x + 3)^2 = 2$

$x + 3 = \pm\sqrt{2}$

$x = -3 \pm \sqrt{2}$

$x \approx -1.59, -4.41$

**40.** $a^2 - 10a - 15 = 0$

$a^2 - 10a + 25 = 15 + 25$

$(a - 5)^2 = 40$

$a - 5 = \pm\sqrt{40}$

$a = 5 \pm 2\sqrt{10}$

$a \approx 11.32, -1.32$

**42.** $x^2 - 6x + 12 = 0$

$x^2 - 6x = -12$

$x^2 - 6x + 9 = -12 + 9$

$(x - 3)^2 = -3$

$x - 3 = \pm\sqrt{-3}$

$x - 3 = \pm\sqrt{3}i$

$x = 3 \pm \sqrt{3}i$

$x \approx 3 + 1.73i, 3 - 1.73i$

**44.** $x^2 + 8x - 4 = 0$

$x^2 + 8x + 16 = 4 + 16$

$(x + 4)^2 = 20$

$x + 4 = \pm\sqrt{20}$

$x = -4 \pm 2\sqrt{5}$

$x \approx 0.47, -8.47$

**46.** $y^2 + 6y - 24 = 0$

$y^2 + 6y + 9 = 24 + 9$

$(y + 3)^2 = 33$

$y + 3 = \pm\sqrt{33}$

$y = -3 \pm \sqrt{33}$

$y \approx 2.74, -8.74$

**48.** $u^2 - 9u - 1 = 0$

$u^2 - 9u = 1$

$u^2 - 9u + \frac{81}{4} = 1 + \frac{81}{4}$

$\left(u - \frac{9}{2}\right)^2 = \frac{85}{4}$

$u - \frac{9}{2} = \pm\sqrt{\frac{85}{4}}$

$u = \frac{9}{2} \pm \frac{\sqrt{85}}{2}$

$u \approx 9.11, -0.11$

**50.** $z^2 - 7z + 9 = 0$

$z^2 - 7z + \frac{49}{4} = -9 + \frac{49}{4}$

$\left(z - \frac{7}{2}\right)^2 = \frac{13}{4}$

$z - \frac{7}{2} = \pm\sqrt{\frac{13}{2}}$

$z = \frac{7}{2} \pm \frac{\sqrt{13}}{2}$

$z \approx 5.30, 1.70$

**52.** $1 - x - x^2 = 0$

$0 = x^2 + x - 1$

$1 = x^2 + x$

$1 + \frac{1}{4} = x^2 + x + \frac{1}{4}$

$\frac{5}{4} = \left(x + \frac{1}{2}\right)^2$

$\pm\sqrt{\frac{5}{4}} = x + \frac{1}{2}$

$\pm\frac{\sqrt{5}}{2} - \frac{1}{2} = x$

$x \approx 0.62, -1.62$

**54.**  $y^2 + 5y + 9 = 0$

$$y^2 + 5y + \frac{25}{4} = -9 + \frac{25}{4}$$

$$\left(y + \frac{5}{2}\right)^2 = \frac{-36}{4} + \frac{25}{4}$$

$$\left(y + \frac{5}{2}\right)^2 = \frac{-11}{4}$$

$$y + \frac{5}{2} = \pm\sqrt{\frac{-11}{4}}$$

$$y = \frac{5}{2} \pm \frac{\sqrt{11}}{2}i$$

$$y \approx -2.50 + 1.66i$$

$$y \approx -2.50 - 1.66i$$

**56.**  $x^2 + \frac{4}{5}x - 1 = 0$

$$x^2 + \frac{4}{5}x = 1$$

$$x^2 + \frac{4}{5}x + \frac{4}{25} = 1 + \frac{4}{25}$$

$$\left(x + \frac{2}{5}\right)^2 = \frac{29}{25}$$

$$x + \frac{2}{5} = \pm\sqrt{\frac{29}{25}}$$

$$x + \frac{2}{5} = \pm\frac{\sqrt{29}}{5}$$

$$x = -\frac{2}{5} \pm \frac{\sqrt{29}}{5}$$

$$x \approx 0.68, -1.48$$

**58.**  $u^2 - \frac{2}{3}u + 5 = 0$

$$u^2 - \frac{2}{3}u + \frac{1}{9} = -5 + \frac{1}{9}$$

$$\left(u - \frac{1}{3}\right)^2 = \frac{-45}{9} + \frac{1}{9}$$

$$\left(u - \frac{1}{3}\right)^2 = \frac{-44}{9}$$

$$u - \frac{1}{3} = \pm\sqrt{\frac{-44}{9}}$$

$$u = \frac{1}{3} \pm \frac{2\sqrt{11}}{3}i$$

$$u \approx 0.33 + 2.11i$$

$$u \approx 0.33 - 2.11i$$

**60.**  $3x^2 - 24x - 5 = 0$

$$x^2 - 8x - \frac{5}{3} = 0$$

$$x^2 - 8x = \frac{5}{3}$$

$$x^2 - 8x + 16 = \frac{5}{3} + 16$$

$$(x - 4)^2 = \frac{53}{3}$$

$$x - 4 = \pm\sqrt{\frac{53}{3}}$$

$$x = 4 \pm \frac{\sqrt{53}}{\sqrt{3}}$$

$$x = 4 \pm \frac{\sqrt{159}}{3}$$

$$x \approx 8.20, -0.20$$

**62.**  $5x^2 - 15x + 7 = 0$

$$x^2 - 3x = -\frac{7}{5}$$

$$x^2 - 3x + \frac{9}{4} = -\frac{7}{5} + \frac{9}{4}$$

$$\left(x - \frac{3}{2}\right)^2 = \frac{17}{20}$$

$$x - \frac{3}{2} = \pm\sqrt{\frac{17}{20}}$$

$$x = \frac{3}{2} \pm \frac{\sqrt{17}}{2\sqrt{5}}$$

$$x = \frac{3}{2} \pm \frac{\sqrt{85}}{10}$$

$$x = \frac{15 \pm \sqrt{85}}{10}$$

$$x \approx 2.42, 0.58$$

**64.**  $4z^2 - 3z + 2 = 0$

$$z^2 - \frac{3}{4}z = \frac{-2}{4}$$

$$z^2 - \frac{3}{4}z + \frac{9}{64} = \frac{-1}{2} + \frac{9}{64}$$

$$\left(z - \frac{3}{8}\right)^2 = \frac{-23}{64}$$

$$z - \frac{3}{8} = \pm\sqrt{\frac{-23}{64}}$$

$$z = \frac{3}{8} \pm \frac{\sqrt{-23}}{8}$$

$$z = \frac{3}{8} \pm \frac{i\sqrt{23}}{8}$$

$$z \approx 0.38 + 0.60i$$

$$z \approx 0.38 - 0.60i$$

**66.**  $7x^2 + 4x + 3 = 0$

$$x^2 + \frac{4}{7}x + \frac{4}{49} = \frac{-3}{7} + \frac{4}{49}$$

$$\left(x + \frac{2}{7}\right)^2 = \frac{-21}{49} + \frac{4}{49}$$

$$\left(x + \frac{2}{7}\right)^2 = \frac{-17}{49}$$

$$x + \frac{2}{7} = \pm\sqrt{\frac{-17}{49}}$$

$$x = \frac{-2}{7} \pm \frac{\sqrt{17}i}{7}$$

$$x \approx -0.29 + 0.59i$$

$$x \approx -0.29 - 0.59i$$

**68.**    $2x\left(x + \dfrac{4}{3}\right) = 5$

$2x^2 + \dfrac{8}{3}x = 5$

$x^2 + \dfrac{4}{3}x = \dfrac{5}{2}$

$x^2 + \dfrac{4}{3}x + \dfrac{4}{9} = \dfrac{5}{2} + \dfrac{4}{9}$

$\left(x + \dfrac{2}{3}\right)^2 = \dfrac{53}{18}$

$x + \dfrac{2}{3} = \pm\sqrt{\dfrac{53}{18}}$

$x = -\dfrac{2}{3} \pm \dfrac{\sqrt{53}}{3\sqrt{2}}$

$x = -\dfrac{2}{3} \pm \dfrac{\sqrt{106}}{6}$

$x = -\dfrac{4}{6} \pm \dfrac{\sqrt{106}}{6}$

$x = \dfrac{-4 \pm \sqrt{106}}{6}$

$x \approx 1.05, \; -2.38$

**70.**    $0.1x^2 + 0.5x = -0.2$

$0.1x^2 + 0.5x + 0.2 = 0$

$x^2 + 5x + 2 = 0$

$x^2 + 5x + \dfrac{25}{4} = -2 + \dfrac{25}{4}$

$\left(x + \dfrac{5}{2}\right)^2 = \dfrac{-8 + 25}{4}$

$\left(x + \dfrac{5}{2}\right)^2 = \dfrac{17}{4}$

$x + \dfrac{5}{2} = \pm\sqrt{\dfrac{17}{4}}$

$x = -\dfrac{5}{2} \pm \dfrac{\sqrt{17}}{2}$

$x = \dfrac{-5 \pm \sqrt{17}}{2}$

$x \approx -0.44, \; -4.56$

**72.**    $0.02x^2 + 0.10x - 0.05 = 0$

$2x^2 + 10x - 5 = 0$

$x^2 + 5x - \dfrac{5}{2} = 0$

$x^2 + 5x = \dfrac{5}{2}$

$x^2 + 5x + \dfrac{25}{4} = \dfrac{5}{2} + \dfrac{25}{4}$

$\left(x + \dfrac{5}{2}\right)^2 = \dfrac{35}{4}$

$x + \dfrac{5}{2} = \pm\sqrt{\dfrac{35}{4}}$

$x = -\dfrac{5}{2} \pm \dfrac{\sqrt{35}}{2}$

$x \approx 0.46, \; -5.46$

**74.**    $\dfrac{x}{2} + \dfrac{5}{x} = 4$

$2x\left[\dfrac{x}{2} + \dfrac{5}{x}\right] = (4)2x$

$x^2 + 10 = 8x$

$x^2 - 8x = -10$

$x^2 - 8x + 16 = -10 + 16$

$(x - 4)^2 = 6$

$x - 4 = \pm\sqrt{6}$

$x = 4 \pm \sqrt{6}$

**76.**    $\dfrac{x^2 + 2}{24} = \dfrac{x - 1}{3}$

$$3x^2 + 6 = 24x - 24$$

$$3x^2 - 24x + 30 = 0$$

$$x^2 - 8x + 10 = 0$$

$$x^2 - 8x = -10$$

$$x^2 - 8x + 16 = -10 + 16$$

$$(x - 4)^2 = 6$$

$$x - 4 = \pm\sqrt{6}$$

$$x = 4 \pm \sqrt{6}$$

**78.**    $\sqrt{3x - 2} = x - 2$

$$\left(\sqrt{3x - 2}\right)^2 = (x - 2)^2$$

$$3x - 2 = x^2 - 2x(2) + 4$$

$$3x - 2 = x^2 - 4x + 4$$

$$0 = x^2 - 7x + 6$$

$$0 = (x - 6)(x - 1)$$

$$x - 6 = 0 \qquad x - 1 = 0$$

$$x = 6 \qquad\quad x = 1$$

Not a solution

**Check:**  $\sqrt{3 \cdot 6 - 2} \overset{?}{=} 6 - 2$

$$\sqrt{18 - 2} \overset{?}{=} 4$$

$$\sqrt{16} \overset{?}{=} 4$$

$$4 = 4$$

$$\sqrt{3 \cdot 1 - 2} \overset{?}{=} 1 - 2$$

$$\sqrt{3 - 2} \overset{?}{=} -1$$

$$\sqrt{1} \overset{?}{=} -1$$

$$1 \neq -1$$

**80.**  $y = x^2 + 6x - 4$

*Keystrokes:*

$\boxed{\text{Y=}}$ $\boxed{\text{X,T,}\theta}$ $\boxed{x^2}$ $\boxed{+}$ 6 $\boxed{\text{X,T,}\theta}$ $\boxed{-}$ 4 $\boxed{\text{GRAPH}}$

$$0 = x^2 + 6x - 4$$

$$4 = x^2 + 6x$$

$$4 + 9 = x^2 + 6x + 9$$

$$13 = (x + 3)^2$$

$$\pm\sqrt{13} = x + 3$$

$$-3 \pm \sqrt{13} = x$$

$$x \approx 0.61, -6.61$$

$$\left(-3 \pm \sqrt{13}, 0\right)$$

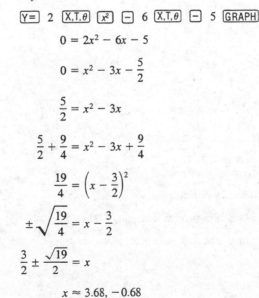

**82.**  $y = 2x^2 - 6x - 5$

*Keystrokes:*

$\boxed{\text{Y=}}$ 2 $\boxed{\text{X,T,}\theta}$ $\boxed{x^2}$ $\boxed{-}$ 6 $\boxed{\text{X,T,}\theta}$ $\boxed{-}$ 5 $\boxed{\text{GRAPH}}$

$$0 = 2x^2 - 6x - 5$$

$$0 = x^2 - 3x - \frac{5}{2}$$

$$\frac{5}{2} = x^2 - 3x$$

$$\frac{5}{2} + \frac{9}{4} = x^2 - 3x + \frac{9}{4}$$

$$\frac{19}{4} = \left(x - \frac{3}{2}\right)^2$$

$$\pm\sqrt{\frac{19}{4}} = x - \frac{3}{2}$$

$$\frac{3}{2} \pm \frac{\sqrt{19}}{2} = x$$

$$x \approx 3.68, -0.68$$

$$\left(\frac{3 \pm \sqrt{19}}{2}, 0\right)$$

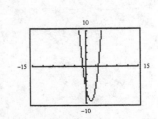

**84.** $y = \frac{1}{2}x^2 - 3x + 1$

*Keystrokes:*

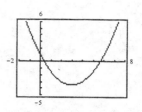

$\boxed{Y=}$ $\boxed{(}$ 1 $\boxed{\div}$ 2 $\boxed{)}$ $\boxed{X,T,\theta}$ $\boxed{x^2}$ $\boxed{-}$ 3 $\boxed{X,T,\theta}$ $\boxed{+}$ 1 $\boxed{GRAPH}$

$$0 = \frac{1}{2}x^2 - 3x + 1$$

$$0 = x^2 - 6x + 2$$

$$-2 = x^2 - 6x$$

$$-2 + 9 = x^2 - 6x + 9$$

$$7 = (x - 3)^2$$

$$\pm\sqrt{7} = x - 3$$

$$3 \pm \sqrt{7} = x$$

$$x \approx 5.65, 0.35$$

$$\left(3 \pm \sqrt{7}, 0\right)$$

**86.** $y = \sqrt{x} - x + 2$

*Keystrokes:*

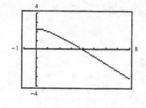

$\boxed{Y=}$ $\boxed{\sqrt{\ }}$ $\boxed{X,T,\theta}$ $\boxed{-}$ $\boxed{X,T,\theta}$ $\boxed{+}$ 2 $\boxed{GRAPH}$

$$0 = \sqrt{x} - x + 2$$

$$x - 2 = \sqrt{x}$$

$$(x - 2)^2 = \left(\sqrt{x}\right)^2$$

$$x^2 - 4x + 4 = x$$

$$x^2 - 5x = -4$$

$$x^2 - 5x + \frac{25}{4} = -4 + \frac{25}{4}$$

$$\left(x - \frac{5}{2}\right)^2 = \frac{9}{4}$$

$$x - \frac{5}{2} = \pm\sqrt{\frac{9}{4}}$$

$$x - \frac{5}{2} = \pm\frac{3}{2}$$

$$x = \frac{5}{2} \pm \frac{3}{2}$$

$$x = 4 \qquad\qquad x = 1$$

Not a solution

**Check:** $0 \overset{?}{=} \sqrt{4} - 4 + 2$

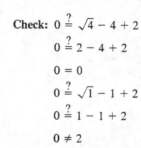

$$0 \overset{?}{=} 2 - 4 + 2$$

$$0 = 0$$

$$0 \overset{?}{=} \sqrt{1} - 1 + 2$$

$$0 \overset{?}{=} 1 - 1 + 2$$

$$0 \neq 2$$

**88.** (a) Area of square $= 3 \cdot 3 = 9$

Area of vertical rectangle $= 3x$

Area of horizontal rectangle $= 3x$

Total area $= 9 + 3x + 3x = 9 + 6x$

(b) Area of small square $= x \cdot x = x^2$

Total area $= x^2 + 6x + 9$

(c) $(x + 3)(x + 3) = x^2 + 3x + 3x + 9$

$$(x + 3)^2 = x^2 + 6x + 9$$

**90.** *Verbal model:* $\boxed{\text{Area}} = \boxed{\text{Length}} \cdot \boxed{\text{Width}}$

*Labels:* Length $= x$

Width $= \frac{1}{4}x + 3$

*Equations:* $160 = x\left(\frac{1}{4}x + 3\right)$

$$160 = \frac{1}{4}x^2 + 3x$$

$$0 = x^2 + 12x - 640$$

$$0 = (x + 32)(x - 20)$$

$$x + 32 = 0 \qquad x - 20 = 0$$

$$x = 20 \text{ length}$$

$$\frac{1}{4}x + 3 = 8 \text{ width}$$

**92.** *Verbal model:*    | Volume | = | Length | · | Width | · | Height |

*Labels:*    Length $= x + 4$          *Equation:*     $840 = (x + 4)(x)(6)$

Width $= x$                             $140 = x(x + 4)$

Height $= 6$                            $140 = x^2 + 4x$

$$0 = x^2 + 4x - 140$$

$$0 = (x + 14)(x - 10)$$

$x + 14 = 0 \qquad x - 10 = 0$

$\qquad x = -14 \qquad\quad x = 10$

Not a solution

Thus, the dimensions are:

length $= 10 + 4 = 14$ inches

width $= x = 10$ inches

height $= 6$ inches

**94.**
$$1218 = x\left(50 - \tfrac{1}{2}x\right)$$

$$1218 = 50x - \tfrac{1}{2}x^2$$

$$\tfrac{1}{2}x^2 - 50x = -1218$$

$$x^2 - 100x + 2500 = -2436 + 2500$$

$$(x - 50)^2 = 64$$

$$x - 50 = \pm 8$$

$$x = 50 \pm 8$$

$$x = 58 \text{ units}$$

$$x = 42 \text{ units}$$

**96.** A perfect square trinomial is one that can be written in the form $(x + k)^2$.

**98.** Use the method of completing the square to write the quadratic equation in the form $u^2 = d$, $d > 0$. This is the form for extracting square roots.

**100.** Divide both sides of the equation by the leading coefficient. Dividing both sides of an equation by a nonzero constant yields an equivalent equation.

**102.** (a) $d = 0$

(b) $d > 0$, and $d$ is a perfect square.

(c) $d > 0$, and $d$ is not a perfect square.

(d) $d < 0$

# Section 6.3    The Quadratic Formula

**2.**    $7x^2 + 15x = 5$

$7x^2 + 15x - 5 = 0$

**4.**    $x(3x + 8) = 15$

$3x^2 + 8x = 15$

$3x^2 + 8x - 15 = 0$

**6.** (a) $x^2 - 12x + 27 = 0$

$$x = \frac{12 \pm \sqrt{12^2 - 4(1)(27)}}{2(1)}$$

$$x = \frac{12 \pm \sqrt{144 - 108}}{2}$$

$$x = \frac{12 \pm \sqrt{36}}{2}$$

$$x = \frac{12 \pm 6}{2}$$

$$x = 9, 3$$

(b) $(x - 9)(x - 3) = 0$

$x - 9 = 0 \qquad x - 3 = 0$

$\quad x = 9 \qquad\quad x = 3$

**8.** (a) $x^2 + 9x + 14 = 0$

$$x = \frac{-9 \pm \sqrt{9^2 - 4(1)(14)}}{2(1)}$$

$$x = \frac{-9 \pm \sqrt{81 - 56}}{2}$$

$$x = \frac{-9 \pm \sqrt{25}}{2}$$

$$x = \frac{-9 \pm 5}{2}$$

$$x = -2, -7$$

(b) $(x + 7)(x + 2) = 0$

$x + 7 = 0 \qquad x + 2 = 0$

$\quad x = -7 \qquad\quad x = -2$

**10.** (a) $9x^2 + 12x + 4 = 0$

$$x = \frac{-12 \pm \sqrt{(12)^2 - 4(9)(4)}}{2(9)}$$

$$x = \frac{-12 \pm \sqrt{144 - 144}}{18}$$

$$x = \frac{-12 \pm 0}{18}$$

$$x = \frac{12}{18} = \frac{-2}{3}$$

(b) $(3x + 2)(3x + 2) = 0$

$3x + 2 = 0 \qquad 3x + 2 = 0$

$\quad x = \frac{-2}{3} \qquad\quad x = \frac{-2}{3}$

**12.** (a) $9x^2 - 30x + 25 = 0$

$$x = \frac{30 \pm \sqrt{(-30)^2 - 4(9)(25)}}{2(9)}$$

$$x = \frac{30 \pm \sqrt{900 - 900}}{18}$$

$$x = \frac{30 \pm \sqrt{0}}{18}$$

$$x = \frac{30}{18}$$

$$x = \frac{5}{3}$$

(b) $(3x - 5)(3x - 5) = 0$

$3x - 5 = 0 \qquad 3x - 5 = 0$

$3x = 5 \qquad\quad 3x = 5$

$\quad x = \frac{5}{3} \qquad\quad x = \frac{5}{3}$

**14.** (a) $10x^2 - 11x + 3 = 0$

$$x = \frac{11 \pm \sqrt{(-11)^2 - 4(10)(3)}}{2(10)}$$

$$x = \frac{11 \pm \sqrt{121 - 120}}{20}$$

$$x = \frac{11 \pm \sqrt{1}}{20}$$

$$x = \frac{11 \pm 1}{20}$$

$$x = \frac{3}{5}, \frac{1}{2}$$

(b) $(5x - 3)(2x - 1) = 0$

$5x - 3 = 0 \qquad 2x - 1 = 0$

$5x = 3 \qquad\quad 2x = 1$

$\quad x = \frac{3}{5} \qquad\quad x = \frac{1}{2}$

**16.** (a) $x^2 + 20x - 300 = 0$

$$x = \frac{-20 \pm \sqrt{20^2 - 4(1)(-300)}}{2(1)}$$

$$x = \frac{-20 \pm \sqrt{400 + 1200}}{2}$$

$$x = \frac{-20 \pm \sqrt{1600}}{2}$$

$$x = \frac{-20 \pm 40}{2} = 10, -30$$

(b) $(x + 30)(x - 10) = 0$

$x + 30 = 0 \qquad x - 10 = 0$

$\quad x = -30 \qquad\quad x = 10$

**18.** $x^2 - 2x - 6 = 0$

$$x = \frac{2 \pm \sqrt{(-2)^2 - 4(1)(-6)}}{2(1)}$$

$$x = \frac{2 \pm \sqrt{4 + 24}}{2}$$

$$x = \frac{2 \pm \sqrt{28}}{2}$$

$$x = \frac{2 \pm 2\sqrt{7}}{2}$$

$$x = \frac{2(1 \pm \sqrt{7})}{2}$$

$$x = 1 \pm \sqrt{7}$$

**20.** $y^2 + 6y + 4 = 0$

$$y = \frac{-6 \pm \sqrt{6^2 - 4(1)(4)}}{2(1)}$$

$$y = \frac{-6 \pm \sqrt{36 - 16}}{2}$$

$$y = \frac{-6 \pm \sqrt{20}}{2}$$

$$y = \frac{-6 \pm 2\sqrt{5}}{2}$$

$$y = \frac{2(-3 \pm \sqrt{5})}{2}$$

$$y = -3 \pm \sqrt{5}$$

**22.** $x^2 + 8x - 4 = 0$

$$x = \frac{-8 \pm \sqrt{8^2 - 4(1)(-4)}}{2(1)}$$

$$x = \frac{-8 \pm \sqrt{64 + 16}}{2}$$

$$x = \frac{-8 \pm \sqrt{80}}{2}$$

$$x = \frac{-8 \pm 4\sqrt{5}}{2}$$

$$x = \frac{2(-4 \pm 2\sqrt{5})}{2}$$

$$x = -4 \pm 2\sqrt{5}$$

**24.** $u^2 - 12u + 29 = 0$

$$u = \frac{12 \pm \sqrt{(-12)^2 - 4(1)(29)}}{2(1)}$$

$$u = \frac{12 \pm \sqrt{144 - 116}}{2}$$

$$u = \frac{12 \pm \sqrt{28}}{2}$$

$$u = \frac{12 \pm 2\sqrt{7}}{2}$$

$$u = \frac{2(6 \pm \sqrt{7})}{2}$$

$$u = 6 \pm \sqrt{7}$$

**26.** $2x^2 - x + 1 = 0$

$$x = \frac{1 \pm \sqrt{(-1)^2 - 4(2)(1)}}{2(2)}$$

$$x = \frac{1 \pm \sqrt{1 - 8}}{4}$$

$$x = \frac{1 \pm \sqrt{-7}}{4}$$

$$x = \frac{1 \pm \sqrt{7}i}{4}$$

$$x = \frac{1}{4} \pm \frac{\sqrt{7}i}{4}$$

**28.** $4x^2 + 6x + 1 = 0$

$$x = \frac{-6 \pm \sqrt{6^2 - 4(4)(1)}}{2(4)}$$

$$x = \frac{-6 \pm \sqrt{36 - 16}}{8}$$

$$x = \frac{-6 \pm \sqrt{20}}{8}$$

$$x = \frac{-6 \pm 2\sqrt{5}}{8}$$

$$x = \frac{2(-3 \pm \sqrt{5})}{8}$$

$$x = \frac{-3 \pm \sqrt{5}}{4}$$

**30.** $2x^2 + 3x + 3 = 0$

$$x = \frac{-3 \pm \sqrt{3^2 - 4(2)(3)}}{2(2)}$$

$$x = \frac{-3 \pm \sqrt{9 - 24}}{4}$$

$$x = \frac{-3 \pm \sqrt{-15}}{4}$$

$$x = \frac{-3 \pm \sqrt{15}i}{4}$$

$$x = \frac{-3}{4} \pm \frac{\sqrt{15}i}{4}$$

**32.** $8y^2 - 8y - 1 = 0$

$$y = \frac{8 \pm \sqrt{(-8)^2 - 4(8)(-1)}}{2(8)}$$

$$y = \frac{8 \pm \sqrt{64 + 32}}{16}$$

$$y = \frac{8 \pm \sqrt{96}}{16}$$

$$y = \frac{8 \pm 4\sqrt{6}}{16}$$

$$y = \frac{4(2 \pm \sqrt{6})}{16}$$

$$y = \frac{2 \pm \sqrt{6}}{4}$$

**34.** $-5x^2 - 15x + 10 = 0$

$$x = \frac{15 \pm \sqrt{(-15)^2 - 4(-5)(10)}}{2(-5)}$$

$$x = \frac{15 \pm \sqrt{225 + 200}}{-10}$$

$$x = \frac{-15 \pm \sqrt{425}}{10}$$

$$x = \frac{-15 \pm 5\sqrt{17}}{10}$$

$$x = \frac{-5(3 \pm \sqrt{17})}{10}$$

$$x = \frac{-3 \pm \sqrt{17}}{2}$$

**36.** $6x^2 + 3x - 9 = 0$

$$x = \frac{-3 \pm \sqrt{3^2 - 4(6)(-9)}}{2(6)}$$

$$x = \frac{-3 \pm \sqrt{9 + 216}}{12}$$

$$x = \frac{-3 \pm \sqrt{225}}{12}$$

$$x = \frac{-3 \pm 15}{12}$$

$$x = 1, \frac{-18}{12}$$

$$x = 1, \frac{-3}{2}$$

**38.** $-15x^2 - 10x + 25 = 0$    (Divide by $-5$)

$$3x^2 + 2x - 5 = 0$$

$$x = \frac{-2 \pm \sqrt{2^2 - 4(3)(-5)}}{2(3)}$$

$$x = \frac{-2 \pm \sqrt{4 + 60}}{6}$$

$$x = \frac{-2 \pm \sqrt{64}}{6}$$

$$x = \frac{-2 \pm 8}{6}$$

$$x = 1, \frac{-10}{6}$$

$$x = 1, \frac{-5}{3}$$

**40.** $7x^2 = 3 - 5x$

$7x^2 + 5x - 3 = 0$

$x = \dfrac{-5 \pm \sqrt{5^2 - 4(7)(-3)}}{2(7)}$

$x = \dfrac{-5 \pm \sqrt{25 + 84}}{14}$

$x = \dfrac{-5 \pm \sqrt{109}}{14}$

**42.** $x - x^2 = 1 - 6x^2$

$5x^2 + x - 1 = 0$

$x = \dfrac{-1 \pm \sqrt{1^2 - 4(5)(-1)}}{2(5)}$

$x = \dfrac{-1 \pm \sqrt{1 + 20}}{10}$

$x = \dfrac{-1 \pm \sqrt{21}}{10}$

**44.** $x^2 + 0.6x - 0.41 = 0$

$x = \dfrac{-0.6 \pm \sqrt{0.6^2 - 4(1)(-0.41)}}{2(1)}$

$x = \dfrac{-0.6 \pm \sqrt{0.36 + 1.64}}{2}$

$x = \dfrac{-0.6 \pm \sqrt{2.00}}{2}$

$x = \dfrac{-0.6 \pm \sqrt{2}}{2}$

**46.** $0.09x^2 - 0.12x - 0.26 = 0$

$9x^2 - 12x - 26 = 0$

$x = \dfrac{12 \pm \sqrt{(-12)^2 - 4(9)(-26)}}{2(9)}$

$x = \dfrac{12 \pm \sqrt{144 + 936}}{18}$

$x = \dfrac{12 \pm \sqrt{1080}}{18}$

$x = \dfrac{12 \pm 6\sqrt{30}}{18}$

$= \dfrac{6\left(2 \pm \sqrt{30}\right)}{18} = \dfrac{2 \pm \sqrt{30}}{3} = \dfrac{-0.02 \pm \sqrt{0.003}}{0.03}$

**48.** $x^2 + x - 1 = 0$

$b^2 - 4ac = 1^2 - 4(1)(-1)$

$= 1 + 4$

$= 5$

Two distinct irrational solutions.

**50.** $b^2 - 4ac = 5^2 - 4(10)(1)$

$= 25 - 40$

$= -15$

Two distinct imaginary solutions.

**52.** $b^2 - 4ac = (-2)^2 - 4(3)(-5)$

$= 4 + 60$

$= 64$

Two distinct rational solutions.

**54.** $b^2 - 4ac = 10^2 - 4(2)(6)$

$= 100 - 48$

$= 52$

Two distinct irrational solutions.

**56.** $b^2 - 4ac = (-24)^2 - 4(9)(16)$

$= 576 - 576$

$= 0$

One (repeated) rational solution.

**58.** $t^2 = 144$

$t = \pm\sqrt{144}$

$t = \pm 12$

**60.** $7u^2 + 49u = 0$

$7u(u + 7) = 0$

$7u = 0 \qquad u + 7 = 0$

$u = 0 \qquad\quad u = -7$

**62.** $9(x + 4)^2 + 16 = 0$

$9(x + 4)^2 = -16$

$(x + 4)^2 = \dfrac{-16}{9}$

$x + 4 = \pm\sqrt{\dfrac{-16}{9}}$

$x = -4 \pm \tfrac{4}{3}i$

**64.** $4y(y + 7) - 5(y + 7) = 0$

$(y + 7)(4y - 5) = 0$

$y + 7 = 0 \qquad 4y - 5 = 0$

$y = -7 \qquad\quad y = \tfrac{5}{4}$

**66.** $x^2 - 3x - 4 = 0$

$(x - 4)(x + 1) = 0$

$x - 4 = 0 \qquad x + 1 = 0$

$x = 4 \qquad\quad x = -1$

**68.** $y^2 + 21y + 108 = 0$

$(y + 12)(y + 9) = 0$

$y + 12 = 0 \qquad y + 9 = 0$

$y = -12 \qquad\quad y = -9$

**70.** $2x^2 - 15x + 225 = 0$

$$x = \frac{-(-15) \pm \sqrt{(-15)^2 - 4(2)(225)}}{2(2)}$$

$$x = \frac{15 \pm \sqrt{225 - 1800}}{4}$$

$$x = \frac{15 \pm \sqrt{-1575}}{4}$$

$$x = \frac{15}{4} \pm \frac{15\sqrt{7}i}{4}$$

**72.** $14x^2 + 11x - 40 = 0$

$$x = \frac{-11 \pm \sqrt{11^2 - 4(14)(-40)}}{2(14)}$$

$$x = \frac{-11 \pm \sqrt{121 + 2240}}{28}$$

$$x = \frac{-11 \pm \sqrt{2361}}{28}$$

**74.** $2x^2 + 8x + 4.5 = 0$

$$x = \frac{-8 \pm \sqrt{8^2 - 4(2)(4.5)}}{2(2)}$$

$$x = \frac{-8 \pm \sqrt{64 - 36}}{4}$$

$$x = \frac{-8 \pm \sqrt{28}}{4}$$

$$x = \frac{-8 \pm 2\sqrt{7}}{4}$$

$$x = \frac{2(-4 \pm \sqrt{7})}{4}$$

$$x = \frac{-4 \pm \sqrt{7}}{2}$$

$$x = -2 \pm \frac{1}{2}\sqrt{7}$$

**76.** $y = x^2 + x + 1$

*Keystrokes:*

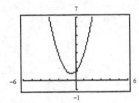

$0 = x^2 + x + 1$

$$x = \frac{(-1) \pm \sqrt{1^2 - 4(1)(1)}}{2(1)}$$

$$x = \frac{-1 \pm \sqrt{1 - 4}}{2}$$

$$x = \frac{-1 \pm \sqrt{-3}}{2}$$

$$x = \frac{-1 \pm \sqrt{3}i}{2}$$

No $x$-intercepts

**78.** $y = x^2 - 4x + 3$

*Keystrokes:*

$\boxed{Y=}\ \boxed{X,T,\theta}\ \boxed{x^2}\ \boxed{-}\ 4\ \boxed{X,T,\theta}\ \boxed{+}\ 3\ \boxed{GRAPH}$

$0 = x^2 - 4x + 3$

$0 = (x - 3)(x - 1)$

$x - 3 = 0 \qquad x - 1 = 0$

$\quad x = 3 \qquad\quad x = 1$

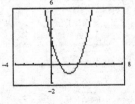

**80.** $y = 15x^2 + 3x - 105$

*Keystrokes:*

$\boxed{Y=}\ 15\ \boxed{X,T,\theta}\ \boxed{x^2}\ \boxed{+}\ 3\ \boxed{X,T,\theta}\ \boxed{-}\ 105\ \boxed{GRAPH}$

$$x = \frac{-3 \pm \sqrt{3^2 - 4(15)(-105)}}{2(15)}$$

$$x = \frac{-3 \pm \sqrt{9 + 6300}}{30}$$

$$x = \frac{-3 \pm \sqrt{6309}}{30}$$

$$x = \frac{-3 \pm 3\sqrt{701}}{30}$$

$$x = \frac{3(-1 \pm \sqrt{701})}{30}$$

$$x = \frac{-1 \pm \sqrt{701}}{10}$$

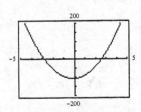

$x \approx 2.55, -2.75$

**82.** $y = 3.7x^2 - 10.2x + 3.2$

*Keystrokes:*

$\boxed{Y=}$ 3.7 $\boxed{X,T,\theta}$ $\boxed{x^2}$ $\boxed{-}$ 10.2 $\boxed{X,T,\theta}$ $\boxed{+}$ 3.2 $\boxed{GRAPH}$

$x = \dfrac{10.2 \pm \sqrt{(-10.2)^2 - 4(3.7)(3.2)}}{2(3.7)}$

$x = \dfrac{10.2 \pm \sqrt{104.04 - 47.36}}{7.4}$

$x = \dfrac{10.2 \pm \sqrt{56.68}}{7.4}$

$x \approx 2.40, 0.36$

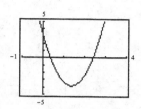

**84.** $2x^2 - x - 1 = 0$

*Keystrokes:*

$\boxed{Y=}$ 2 $\boxed{X,T,\theta}$ $\boxed{x^2}$ $\boxed{-}$ $\boxed{X,T,\theta}$ $\boxed{-}$ 1 $\boxed{GRAPH}$

$b^2 - 4ac = (-1)^2 - 4(2)(-1)$

$= 1 + 8$

$= 9$

Two real solutions.

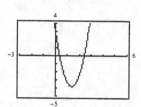

**86.** $\dfrac{1}{3}x^2 - 5x + 25 = 0$

Multiply by 3.

$x^2 - 15x + 75 = 0$

*Keystrokes:*

$\boxed{Y=}$ $\boxed{X,T,\theta}$ $\boxed{x^2}$ $\boxed{-}$ 15 $\boxed{X,T,\theta}$ $\boxed{+}$ 75 $\boxed{GRAPH}$

$b^2 - 4ac = (-15)^2 - 4(1)(75)$

$= 225 - 300$

$= -75$

No real solutions

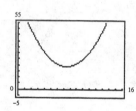

**88.** $\dfrac{x^2 - 9x}{6} = \dfrac{x - 1}{2}$

$2x^2 - 18x = 6x - 6$

$2x^2 - 24x + 6 = 0$

$x^2 - 12x + 3 = 0$

$x = \dfrac{-(-12) \pm \sqrt{(-12)^2 - 4(1)(3)}}{2(1)}$

$x = \dfrac{12 \pm \sqrt{144 - 12}}{2} = \dfrac{12 \pm \sqrt{132}}{2}$

$x = \dfrac{12 \pm 2\sqrt{33}}{2}$

$x = 6 \pm \sqrt{33}$

**90.**    $\sqrt{2x - 3} = x - 2$

$\left(\sqrt{2x - 3}\right)^2 = (x - 2)^2$

$2x - 3 = x^2 - 2(x)(2) + 2^2$

$2x - 3 = x^2 - 4x + 4$

$0 = x^2 - 6x + 7$

$x = \dfrac{6 \pm \sqrt{6^2 - 4(1)(7)}}{2(1)}$

$x = \dfrac{6 \pm \sqrt{36 - 28}}{2}$

$x = \dfrac{6 \pm \sqrt{8}}{2}$

$x = \dfrac{6 \pm 2\sqrt{2}}{2}$

$x = \dfrac{2(3 \pm \sqrt{2})}{2}$

$x = 3 \pm \sqrt{2}$

$x = 3 - \sqrt{2}$ does not check.

**Check:** $\sqrt{2(3 + \sqrt{2}) - 3} \overset{?}{=} 3 + \sqrt{2} - 2$

$\sqrt{6 + 2\sqrt{2} - 3} \overset{?}{=} 1 + \sqrt{2}$

$\sqrt{3 + 2\sqrt{2}} \overset{?}{=} 1 + \sqrt{2}$

$2.4142 = 2.4142$

$\sqrt{2(3 - \sqrt{2}) - 3} \overset{?}{=} 3 - \sqrt{2} - 2$

$\sqrt{6 - 2\sqrt{2} - 3} \overset{?}{=} 1 - \sqrt{2}$

$\sqrt{3 - 2\sqrt{2}} \overset{?}{=} 1 - \sqrt{2}$

$0.4142 \neq -0.4142$

**92.** $x^2 - 12x + c = 0$

(a) $b^2 - 4ac > 0$

$(-12)^2 - 4(1)c > 0$

$144 - 4c > 0$

$144 > 4c$

$36 > c$

(b) $b^2 - 4ac = 0$

$(-12)^2 - 4(1)c = 0$

$144 - 4c = 0$

$144 = 4c$

$36 = c$

(c) $b^2 - 4ac < 0$

$(-12)^2 - 4(1)c < 0$

$144 - 4c < 0$

$144 < 4c$

$36 < c$

**94.** $x^2 + 2x + c = 0$

(a) $b^2 - 4ac > 0$

$2^2 - 4(1)c > 0$

$4 - 4c > 0$

$4 > 4c$

$1 > c$

(b) $b^2 - 4ac = 0$

$2^2 - 4(1)c = 0$

$4 - 4c = 0$

$4 = 4c$

$1 = c$

(c) $b^2 - 4ac < 0$

$2^2 - 4(1)c < 0$

$4 - 4c < 0$

$4 < 4c$

$1 < c$

**96.** *Verbal model:* | Area | = | Length | · | Width |

*Labels:*  Length $= x + 1.5$

Width $= x$

*Equation:*  $18.36 = (x + 1.5) \cdot x$

$18.36 = x^2 + 1.5x$

$0 = x^2 + 1.5x - 18.36$

$$x = \frac{-1.5 \pm \sqrt{1.5^2 - 4(1)(-18.36)}}{2(1)}$$

$$x = \frac{-1.5 \pm \sqrt{2.25 + 73.44}}{2}$$

$$x = \frac{-1.5 \pm \sqrt{75.69}}{2} \approx 3.6$$

$x \approx 3.6$ inches

$x + 1.5 \approx 5.1$ inches

**98.** $h = -16t^2 + 20t + 40$

(a) $40 = -16t^2 + 20t + 40$

$0 = -16t^2 + 20t$

$0 = -4t(4t - 5)$

$-4t = 0$      $4t - 5 = 0$

$t = 0$      $t = \frac{5}{4} = 1.25$ seconds

(b) $0 = -16t^2 + 20t + 40$

$0 = -4(4t^2 - 5t - 10)$

$$t = \frac{-(-5) \pm \sqrt{(-5)^2 - 4(4)(-10)}}{2(4)}$$

$$t = \frac{5 \pm \sqrt{25 + 160}}{8}$$

$$t = \frac{5 \pm \sqrt{185}}{8}$$

$$t = \frac{5 + \sqrt{185}}{8} \approx 2.325$$ seconds

reject $\frac{5 - \sqrt{185}}{8}$

**100.** $s = 0.84t^2 + 1.51t + 4.70,\ -1 \leq t \leq 6$

(a) *Keystrokes:*

$\boxed{Y=}$ .84 $\boxed{X,T,\theta}$ $\boxed{x^2}$ $\boxed{+}$ 1.51 $\boxed{X,T,\theta}$ $\boxed{+}$ 4.70 $\boxed{GRAPH}$

(b) $10 = 0.84t^2 + 1.51t + 4.70$

$0 = 0.84t^2 + 1.51t - 5.3$

$t = \dfrac{-1.51 \pm \sqrt{1.51^2 - 4(.84)(-5.3)}}{2(.84)}$

$t = \dfrac{-1.51 \pm \sqrt{20.0881}}{1.68} \approx 1.769$

Year 1991

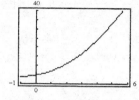

**102.** $ax^2 + bx + c = 0$

Solutions are $x_1$ and $x_2$

$$x_1 + x_2 = \frac{-b}{a} \qquad x_1 \cdot x_2 = \frac{c}{a}$$

**104.** Compute $-b$ plus or minus the square root of the quantity $b$ squared minus $4ac$. This quantity divided by the quantity $2a$ is the Quadratic Formula.

**106.** The Quadratic Formula is derived by solving the general quadratic equation $ax^2 + bx + c = 0$ by the method of completing the square.

## Section 6.4    Applications of Quadratic Equations

**2.** *Verbal model:* $\boxed{\begin{array}{c}\text{Selling price} \\ \text{per computer}\end{array}} = \boxed{\begin{array}{c}\text{Cost per} \\ \text{computer}\end{array}} + \boxed{\begin{array}{c}\text{Profit per} \\ \text{computer}\end{array}}$

*Equation:*

$$\frac{27{,}000}{x} = \frac{27{,}000}{x + 3} + 750$$

$$27{,}000(x + 3) = 27{,}000x + 750x(x + 3)$$

$$27{,}000x + 81{,}000 = 27{,}000x + 750x^2 + 2250x$$

$$0 = 750x^2 + 2250x - 81{,}000$$

$$0 = x^2 + 3x - 108$$

$$0 = (x + 12)(x - 9)$$

$$x = -12 \quad x = 9 \text{ computers}$$

$$\text{Selling price} = \frac{27{,}000}{9} = \$3000$$

**4.** *Verbal model:*  $\boxed{\begin{array}{c}\text{Selling price}\\\text{per sweatshirt}\end{array}} = \boxed{\begin{array}{c}\text{Cost per}\\\text{sweatshirt}\end{array}} + \boxed{\begin{array}{c}\text{Profit per}\\\text{sweatshirt}\end{array}}$

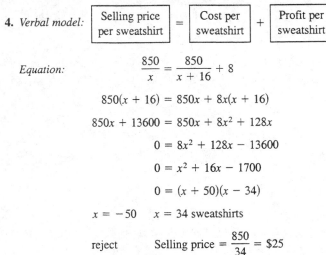

$\dfrac{1700}{2} = \dfrac{850}{2} - \dfrac{425}{5} - \dfrac{85}{5} - 17$

*Equation:*

$$\frac{850}{x} = \frac{850}{x + 16} + 8$$

$$850(x + 16) = 850x + 8x(x + 16)$$

$$850x + 13600 = 850x + 8x^2 + 128x$$

$$0 = 8x^2 + 128x - 13600$$

$$0 = x^2 + 16x - 1700$$

$$0 = (x + 50)(x - 34)$$

$x = -50$    $x = 34$ sweatshirts

reject        Selling price $= \dfrac{850}{34} = \$25$

**6.** *Verbal model:*  $2\,\boxed{\text{Length}} + 2\,\boxed{\text{Width}} = \boxed{\text{Perimeter}}$

*Equation:*  

$$2(1.5w) + 2w = 40$$
$$3w + 2w = 40$$
$$5w = 40$$
$$w = 8$$
$$l = 1.5w = 12$$

*Verbal model:*  $\boxed{\text{Length}} \cdot \boxed{\text{Width}} = \boxed{\text{Area}}$

*Equation:*  

$$12 \cdot 8 = A$$
$$96 \text{ square meters} = A$$

**8.** *Verbal model:*  $\boxed{\text{Length}} \cdot \boxed{\text{Width}} = \boxed{\text{Area}}$

*Equation:*  

$$1.5w \cdot w = 216$$
$$1.5w^2 = 216$$
$$w^2 = 144$$
$$w = 12$$
$$l = 1.5w = 12$$

*Verbal model:*  $2\,\boxed{\text{Length}} + 2\,\boxed{\text{Width}} = \boxed{\text{Perimeter}}$

*Equation:*  

$$2(18) + 2(12) = P$$
$$60 \text{ centimeters} = P$$

**10.** *Verbal model:*  $\boxed{\text{Length}} \cdot \boxed{\text{Width}} = \boxed{\text{Area}}$

*Equation:*  

$$l \cdot \tfrac{3}{4}l = 2700$$
$$\tfrac{3}{4}l^2 = 2700$$
$$l^2 = 3600$$
$$l = 60$$
$$w = \tfrac{3}{4}l = 45$$

*Verbal model:*  $2\,\boxed{\text{Length}} + 2\,\boxed{\text{Width}} = \boxed{\text{Perimeter}}$

*Equation:*  

$$2(60) + 2(45) = P$$
$$210 \text{ inches} = P$$

**12.** *Verbal model:*  $2\,\boxed{\text{Length}} + 2\,\boxed{\text{Width}} = \boxed{\text{Perimeter}}$

*Equation:*  

$$2l + 2(l - 6) = 108$$
$$2l + 2l - 12 = 108$$
$$4l = 120$$
$$l = 30$$
$$w = l - 6 = 24$$

*Verbal model:*  $\boxed{\text{Length}} \cdot \boxed{\text{Width}} = \boxed{\text{Area}}$

*Equation:*  

$$30 \cdot 24 = A$$
$$720 \text{ square feet} = A$$

**14.** *Verbal model:* | Area | = | Length | · | Width |

*Equation:* $500 = (w + 5) \cdot w$

$500 = w^2 + 5w$

$0 = w^2 + 5w$

$0 = (w + 25)(w - 20)$

$w = -25 \qquad w = 20$

reject $\qquad l = w + 5 = 25$

*Verbal model:* 2 | Length | + 2 | Width | = | Perimeter |

*Equation:* $2(25) + 2(20) = P$

$90 \text{ feet} = P$

**16.** *Verbal model:* | Area | = | Length | · | Width |

*Equation:* $10 = (x + 3) \cdot x$

$10 = x^2 + 3x$

$0 = x^2 + 3x - 10$

$0 = (x + 5)(x - 2)$

$x = -5 \qquad\qquad x = 2 \text{ feet}$

$\qquad\qquad x + 3 = 5 \text{ feet}$

**18.** *Verbal model:* | Area | $= \frac{1}{2} \cdot$ | Base | · | Height |

*Equation:* $625 = \frac{1}{2} \cdot x \cdot (x + 25)$

$1250 = x^2 + 25x$

$0 = x^2 + 25x - 1250$

$0 = (x + 50)(x - 25)$

$x = -50 \qquad\qquad x = 25 \text{ inches}$

$\qquad\qquad x + 25 = 50 \text{ inches}$

**20.** (a) *Verbal model:* | Area | = | Length | · | Width |

*Equation:*
$$630 = x \cdot (50 - x)$$
$$630 = 50x - x^2$$
$$x^2 - 50x + 630 = 0$$
$$x = \frac{-(-50) \pm \sqrt{(-50)^2 - 4(1)(630)}}{2(1)}$$
$$x = \frac{50 \pm \sqrt{2500 - 2520}}{2} = \frac{50 \pm \sqrt{-20}}{2}$$

Not real therefore cannot enclose a rectangular region.

$2x + 2w = 100$

$x + w = 50$

$w = 50 - x$

**—CONTINUED—**

**20.** —CONTINUED—

(b) *Verbal model:*   $\boxed{\text{Area}} = \pi \cdot \boxed{\text{Radius}}^2$

    *Equation:*    $A = \pi\left(\dfrac{50}{\pi}\right)^2$

              $A = \pi \cdot \dfrac{2500}{\pi^2}$

              $A = \dfrac{2500}{\pi}$

              $A \approx 796$ square feet

    Yes, can enclose a circular area.

$C = 2\pi r$

$100 = 2\pi r$

$\dfrac{100}{2\pi} = r$

$\dfrac{50}{\pi} = r$

**22.** *Verbal model:*   $\boxed{\text{Length}} \cdot \boxed{\text{Width}} = \boxed{\text{Area}}$

    *Equation:*         $(x + 20) \cdot x = 25{,}500$

                   $x^2 + 20x = 25{,}500$

              $x^2 + 20x - 25{,}500 = 0$

              $(x - 150)(x + 170) = 0$

        $x - 150 = 0 \qquad x + 170 = 0$

             $x = 150 \qquad\quad x = -170$

The lot is $150 + 20 = 170$ feet $\times$ 150 feet.

**24.**        $A = P(1 + r)^2$

    $11{,}990.25 = 10{,}000(1 + r)^2$

      $1.199025 = (1 + r)^2$

         $1.095 = 1 + r$

        $0.095 = r$ or $9.5\%$

**26.**      $A = P(1 + r)^2$

  $572.45 = 500(1 + r)^2$

  $\dfrac{572.45}{500} = (1 + r)^2$

  $1.1449 = (1 + r)^2$

    $1.07 = 1 + r$

    $0.07 = r$

     $7\% = r$

**28.**        $A = P(1 + r)^2$

   $7372.46 = 6500(1 + r)^2$

  $1.134224615 \approx (1 + r)^2$

  $1.064999819 \approx 1 + r$

  $0.064999819 \approx r$

      $0.065 \approx r$ or $6.5\%$

**30.** *Verbal model:*   $\boxed{\text{Cost per ticket}} \cdot \boxed{\begin{array}{c}\text{Number of}\\ \text{people going}\end{array}} = 210$

    *Equation:*       $\left(\dfrac{210}{x} - 3.50\right) \cdot (x + 3) = 210$

           $210 + \dfrac{630}{x} - 3.5x - 10.50 = 210$

        $210x + 630 - 3.5x^2 - 10.5x = 210x$

          $-3.5x^2 - 10.5x + 630 = 0$

                 $0 = 3.5x^2 + 10.5x - 630$

                 $0 = 3.5(x^2 + 3x - 180)$

                 $0 = 3.5(x - 12)(x + 15)$

    $x - 12 = 0 \qquad x + 15 = 0$

        $x = 12 \qquad\quad x = -15$

There are $12 + 3 = 15$ people going to the game.

**32.** *Verbal model:* $\boxed{\begin{array}{c}\text{Cost per}\\\text{member}\end{array}} \cdot \boxed{\begin{array}{c}\text{Number of}\\\text{members}\end{array}} = \boxed{\$480}$

*Equation:* $\left(\dfrac{480}{x} - 1\right) \cdot (x + 2) = 480$

$$\left(\dfrac{480 - x}{x}\right) \cdot (x + 2) = 480$$

$$(480 - x)(x + 2) = 480x$$

$$480x + 960 - x^2 - 2x = 480x$$

$$-x^2 - 2x + 960 = 0$$

$$x^2 + 2x - 960 = 0$$

$$x = -32 \qquad x = 30$$

$$x + 2 = 32$$

**34.** *Common formula:* $a^2 + b^2 = c^2$

*Equation:* $x^2 + (51 - x)^2 = 39^2$

$$x^2 + 2601 - 102x + x^2 = 1521$$

$$2x^2 - 102x + 1080 = 0$$

$$x^2 - 51x + 540 = 0$$

$$x = \dfrac{51 \pm \sqrt{51^2 - 4(1)(540)}}{2(1)}$$

$$x = \dfrac{51 \pm \sqrt{441}}{2}$$

$$x = 36, 15$$

15 inches $\times$ 36 inches

**36.** *Common formula:* $a^2 + b^2 + c^2$

*Equation:* $x^2 + (100 - x)^2 = 80^2$

$$x^2 + 10,000 - 200x + x^2 = 6400$$

$$2x^2 - 200x + 3600 = 0$$

$$x^2 - 100x + 1800 = 0$$

$$x = \dfrac{-(-100) \pm \sqrt{(-100)^2 - 4(1)(1800)}}{2(1)}$$

$$x = \dfrac{100 \pm \sqrt{10,000 - 7200}}{2} = \dfrac{100 \pm \sqrt{2800}}{2} \approx 76.5 \text{ yards}, 23.5 \text{ yards}$$

**38.** (a) $d = \sqrt{100^2 + h^2}$

$d = \sqrt{10,000 + h^2}$

(b) *Keystrokes:*

When $d = 200$ feet $h$ is approximately 173 feet.

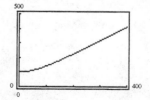

**—CONTINUED—**

**38. —CONTINUED—**

(c)

| $h$ | 0 | 100 | 200 | 300 |
|---|---|---|---|---|
| $d$ | 100 | 141.4 | 223.6 | 316.2 |

$d = \sqrt{10,000 + 0^2}$          $d = \sqrt{10,000 + 100^2}$          $d = \sqrt{10,000 + 200^2}$          $d = \sqrt{10,000 + 300^2}$

$\quad = \sqrt{10,000}$                  $\quad = \sqrt{20,000}$                  $\quad = \sqrt{10,000 + 40,000}$          $\quad = \sqrt{10,000 + 90,000}$

$\quad = 100$                             $\quad \approx 141.4$                   $\quad = \sqrt{50,000}$                  $\quad = \sqrt{100,000}$

$\quad\quad\quad\quad\quad\quad\quad\quad\quad\quad\quad\quad\quad\quad\quad\quad\quad\quad \approx 223.6$                  $\quad \approx 316.2$

**40.** *Verbal model:*    $\boxed{\begin{array}{c}\text{Work done by}\\\text{Machine A}\end{array}} + \boxed{\begin{array}{c}\text{Work done by}\\\text{Machine B}\end{array}} = \boxed{\begin{array}{c}\text{1 complete}\\\text{job}\end{array}}$

*Equation:*

$$\frac{1}{x}(6) + \frac{1}{x+3}(6) = 1$$

$$x(x+3)\left[\frac{1}{x}(6) + \frac{1}{x+3}(6)\right] = (1)x(x+3)$$

$$6(x+3) + 6x = x(x+3)$$

$$6x + 18 + 6x = x^2 + 3x$$

$$-x^2 + 9x + 18 = 0$$

$$x^2 - 9x - 18 = 0$$

$$x = \frac{9 \pm \sqrt{(-9)^2 - 4(1)(-18)}}{2(1)}$$

$$x = \frac{9 \pm \sqrt{81 + 72}}{2}$$

$$x = \frac{9 \pm \sqrt{153}}{2}$$

$$x = 10.684658, -1.6846584 \qquad\qquad x \approx 10.7 \text{ minutes}$$

$$x + 3 = 13.684658 \qquad\qquad x + 3 \approx 13.7 \text{ minutes}$$

**42.** *Verbal model:*    $\boxed{\begin{array}{c}\text{Work done}\\\text{by Person 1}\end{array}} + \boxed{\begin{array}{c}\text{Work done}\\\text{by Person 2}\end{array}} = \boxed{\begin{array}{c}\text{1 complete}\\\text{job}\end{array}}$

*Equation:*    $\dfrac{1}{x}(6) + \dfrac{1}{x+2}(6) = 1$          $x = \dfrac{10 \pm \sqrt{(-10)^2 - 4(1)(-12)}}{2(1)}$

$\dfrac{6}{x} + \dfrac{6}{x+2} = 1$          $x = \dfrac{10 \pm \sqrt{100 + 48}}{2}$

$x(x+2)\left[\dfrac{6}{x} + \dfrac{6}{x+2}\right] = x(x+2)$          $x = \dfrac{10 \pm \sqrt{148}}{2}$

$6(x+2) + 6x = x^2 + 2x$          $x = \dfrac{10 \pm 2\sqrt{37}}{2} = 5 \pm \sqrt{37} \approx 11.08, -1.08$

$6x + 12 + 6x = x^2 + 2x$

$12x + 12 = x^2 + 2x$

$0 = x^2 - 10x - 12$

Thus, it would take person one 11.08 hours to complete the task alone and it would take person two $11.08 + 2 = 13.08$ hours.

**44.** $h = h_0 - 16t^2$

$0 = 625 - 16t^2$

$16t^2 = 625$

$t^2 = 39.0625$

$t = 6.25$ seconds

**46.** $h = h_0 - 16t^2$

$0 = 984 - 16t^2$

$16t^2 = 984$

$t^2 = 61.5$

$t \approx 7.84$ seconds

**48.**
$$4 = 5 + 25t - 16t^2$$

$16t^2 - 25t - 1 = 0$

$$t = \frac{25 \pm \sqrt{(-25)^2 - 4(16)(-1)}}{2(16)}$$

$$t = \frac{25 \pm \sqrt{625 + 64}}{32}$$

$$t = \frac{25 \pm \sqrt{689}}{32}$$

$$t \approx \frac{25 \pm 26.25}{32}$$

$$t \approx 1.6, -0.04$$

1.6 seconds will pass before you hit the ball.

**50.** $h = -16t^2 + 21t + 5$

(a) $11 = -16t^2 + 21t + 5$

$0 = -16t^2 + 21t - 6$

$$t = \frac{-21 \pm \sqrt{21^2 - 4(-16)(-6)}}{2(-16)}$$

$$t = \frac{-21 \pm \sqrt{441 - 384}}{-32}$$

$$t = \frac{-21 \pm \sqrt{57}}{-32} \approx 0.42 \text{ second and } 0.89 \text{ second}$$

(b) $0 = -16t^2 + 21t + 5$

$$t = \frac{-21 \pm \sqrt{21^2 - 4(-16)(5)}}{2(-16)}$$

$$t = \frac{-21 \pm \sqrt{441 + 320}}{-32}$$

$$t = \frac{-21 \pm \sqrt{761}}{-32}$$

$t \approx -.21$ reject      $t \approx 1.52$ seconds

**52.** *Verbal model:* | Integer | $\cdot$ | Integer | $=$ | Product |

*Equation:* $\quad n \cdot (n + 1) = 1122$

$$n^2 + n + \frac{1}{4} = 1122 + \frac{1}{4}$$

$$\left(n + \frac{1}{2}\right)^2 = \frac{4488 + 1}{4}$$

$$n + \frac{1}{2} = \pm\sqrt{\frac{4489}{4}}$$

$$n = \frac{-1}{2} \pm \frac{67}{2}$$

$n = 33 \qquad n = -34$
$n + 1 = 34 \quad n + 1 = -33$ } reject

The integers are 33 and 34.

**54.** *Verbal model:* | Even integer | $\cdot$ | Even integer | $=$ 528

*Equation:* $\quad 2n(2n + 2) = 528$

$4n^2 + 4n = 528$

$4n^2 + 4n - 528 = 0$

$4(n^2 + n - 132) = 0$

$4(n + 12)(n - 11) = 0$

$n + 12 = 0 \qquad\qquad n - 11 = 0$

$\quad\quad \cancel{n = -12} \qquad\qquad n = 11$

reject $\qquad\qquad\qquad 2n = 22$

$\qquad\qquad\qquad\qquad 2n + 2 = 24$

**56.** *Verbal model:*    | Odd Integer | · | Odd Integer | = | Product |

*Equation:*    $(2n + 1)(2n + 3) = 255$

$$4n^2 + 8n + 3 = 255$$

$$4n^2 + 8n - 252 = 0$$

$$n^2 + 2n - 63 = 0$$

$$(n + 9)(n - 7) = 0$$

$n + 9 = 0 \qquad\qquad n - 7 = 0$

reject $\qquad n = -9 \qquad\qquad n = 7$

$$2n + 1 = 15$$

$$2n + 3 = 17$$

The integers are 15 and 17.

**58.** *Verbal model:*    | Time for part 1 | + | Time for part 2 | = 5

*Equation:*    $\dfrac{100}{r} + \dfrac{135}{r - 5} = 5$

$$r(r - 5)\left[\frac{100}{r} + \frac{135}{r - 5}\right] = 5r(r - 5)$$

$$100(r - 5) + 135r = 5r^2 - 25r$$

$$100r - 500 + 135r = 5r^2 - 25r$$

$$0 = 5r^2 - 260r + 500$$

$$0 = 5(r^2 - 52r + 100)$$

$$0 = 5(r - 50)(r - 2)$$

$r - 50 = 0 \qquad\qquad r - 2 = 0$

$\qquad r = 50 \qquad\qquad r = 2 \text{ reject}$

Thus, the average speed for the first part of the trip was 50 miles per hour.

**60.**    $d_1 = \sqrt{(x_1 - x_2)^2 + (y_1 - y_2)^2}$

$13 = \sqrt{(x_1 - 1)^2 + (14 - 2)^2}$

$169 = (x_1 - 1)^2 + 144$

$25 = (x_1 - 1)^2$

$\pm\sqrt{25} = (x_1 - 1)$

$1 \pm 5 = x_1$

$x_1 = 6 \qquad\qquad x_1 = -4$

$(6, 14) \qquad\qquad (-4, 14)$

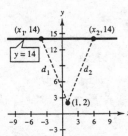

**62.** $s = 12.88t^2 + 43.86t + 300.83, \; 0 \le t \le 6$

(a)

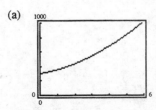

(b)    $400 = 12.88t^2 + 43.86t + 300.83$

$0 = 12.88t^2 + 43.86t - 99.17$

$$t = \frac{-43.86 \pm \sqrt{43.86^2 - 4(12.88)(-99.17)}}{2(12.88)}$$

$$t = \frac{-43.86 \pm \sqrt{7032.938}}{25.76}$$

$t \approx 1.55 \qquad\qquad -4.96 \text{ reject}$

year 1992

**64.** The four strategies that can be used to solve a quadratic equation are factoring, extracting square roots, completing the square and the Quadratic Formula.

**66.** $\dfrac{20 \text{ feet}}{\text{minute}} \cdot \dfrac{1 \text{ minute}}{60 \text{ seconds}} \cdot (45 \text{ seconds}) = 15 \text{ feet}$

**68.** An example of a quadratic equation that has two imaginary solutions is $x^2 + 1 = 0$. Any equation of the form $x^2 + c = 0$, $c$ any positive constant has two imaginary solutions.

## Section 6.5    Quadratic and Rational Inequalities

**2.** $5x(x - 3) = 0$

$5x = 0 \qquad x - 3 = 0$

$\phantom{5}x = 0 \qquad\phantom{x} x = 3$

Critical numbers $= 0, 3$

**4.** $9y^2 - 16 = 0$

$(3y - 4)(3y + 4) = 0$

$3y - 4 = 0 \qquad 3y + 4 = 0$

$\phantom{3}3y = 4 \qquad\phantom{3} 3y = -4$

$\phantom{33}y = \frac{4}{3} \qquad\phantom{33} y = \frac{-4}{3}$

Critical numbers $= \frac{4}{3}, \frac{-4}{3}$

**6.** $y(y - 4) - 3(y - 4) = 0$

$(y - 4)(y - 3) = 0$

$y - 4 = 0 \qquad y - 3 = 0$

$\phantom{y}y = 4 \qquad\phantom{y} y = 3$

Critical numbers: 3, 4

**8.** $3x^2 - 2x - 8 = 0$

$(3x + 4)(x - 2) = 0$

$3x + 4 = 0 \qquad x - 2 = 0$

$\phantom{3}3x = -4 \qquad\phantom{xx} x = 2$

$\phantom{33}x = \dfrac{-4}{3}$

Critical numbers $= -\frac{4}{3}, 2$

**10.** $4x^2 - 4x - 3 = 0$

$(2x + 1)(2x - 3) = 0$

$2x + 1 = 0 \qquad 2x - 3 = 0$

$\phantom{2}2x = -1 \qquad\phantom{2} 2x = 3$

$\phantom{22}x = -\frac{1}{2} \qquad\phantom{22} x = \frac{3}{2}$

Critical numbers: $-\frac{1}{2}, \frac{3}{2}$

**12.** $3 - x$

Positive: $(-\infty, 3)$

Negative: $(3, \infty)$

**14.** $\frac{2}{3}x - 8$

Negative: $(-\infty, 12)$

Positive: $(12, \infty)$

**16.** $7x(3 - x)$

Negative: $(-\infty, 0)$

Positive: $(0, 3)$

Negative: $(3, \infty)$

**18.** $x^2 - 9 = (x - 3)(x + 3)$

Positive: $(-\infty, -3)$

Negative: $(-3, 3)$

Positive: $(3, \infty)$

**20.** $2x^2 - 4x - 3$

$x = \dfrac{4 \pm \sqrt{(-4)^2 - 4(2)(-3)}}{2(2)}$

$x = \dfrac{4 \pm \sqrt{16 + 24}}{4}$

$x = \dfrac{4 \pm \sqrt{40}}{4}$

$x = \dfrac{4 \pm 2\sqrt{10}}{4}$    Positive: $\left(-\infty, \dfrac{2 - \sqrt{10}}{2}\right)$

$x = \dfrac{2(2 \pm \sqrt{10})}{4}$    Negative: $\left(\dfrac{2 - \sqrt{10}}{2}, \dfrac{2 + \sqrt{10}}{2}\right)$

$x = \dfrac{2 \pm \sqrt{10}}{2}$    Positive: $\left(\dfrac{2 + \sqrt{10}}{2}, \infty\right)$

**22.** $5x - 20 < 0$

$5(x - 4) < 0$

Critical number: $x = 4$

Test intervals:

Negative: $(-\infty, 4)$

Positive: $(4, \infty)$

Solution: $(-\infty, 4)$

**24.** $3x - 2 \geq 0$

Critical number: $x = \frac{2}{3}$

Test intervals:

Negative: $\left(-\infty, \frac{2}{3}\right)$

Positive: $\left(\frac{2}{3}, \infty\right)$

Solution: $\left[\frac{2}{3}, \infty\right]$

**26.** $2x(x - 6) > 0$

Critical numbers: $x = 0, 6$

Test intervals:

Positive: $(-\infty, 0)$

Negative: $(0, 6)$

Positive: $(6, \infty)$

Solution: $(-\infty, 0) \cup (6, \infty)$

**28.** $2x(6 - x) > 0$

Critical numbers: $0, 6$

Test intervals:

Negative: $(-\infty, 0)$

Positive: $(0, 6)$

Negative: $(6, \infty)$

Solution: $(0, 6)$

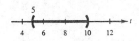

**30.**
$$z^2 \leq 9$$
$$z^2 - 9 \leq 0$$
$$(z - 3)(z + 3) \leq 0$$

Critical numbers: $z = -3, 3$

Test intervals:

Positive: $(-\infty, -3)$

Negative: $(-3, 3)$

Positive: $(3, \infty)$

Solution: $[-3, 3]$

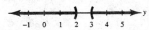

**32.**
$$t^2 - 4t > 12$$
$$t^2 - 4t - 12 > 0$$
$$(t - 6)(t + 2) > 0$$

Critical numbers: $x = 6, -2$

Test intervals:

Positive: $(-\infty, -2)$

Negative: $(-2, 6)$

Positive: $(6, \infty)$

Solution: $(-\infty, -2) \cup (6, \infty)$

**34.** $t^2 - 15t + 50 < 0$

$(t - 10)(t - 5) < 0$

Critical numbers: $5, 10$

Test intervals:

Positive: $(-\infty, 5)$

Negative: $(5, 10)$

Positive: $(10, \infty)$

Solution: $(5, 10)$

**36.** $x^2 + 6x + 10 > 0$

$$x = \frac{-6 \pm \sqrt{36 - 4(10)}}{2}$$

$$x = \frac{-6 \pm \sqrt{-4}}{2}$$

$$x = \frac{-6 \pm 2i}{2}$$

$$x = -3 \pm i$$

No critical numbers

$x^2 + 6x + 10$ is greater than 0 for all values of $x$.

Solution: $(-\infty, \infty)$

**38.** $y^2 - 5y + 6 > 0$

$(y - 3)(y - 2) > 0$

Critical numbers: $y = 2, 3$

Test intervals:

Positive: $(-\infty, 2)$

Negative: $(2, 3)$

Positive: $(3, \infty)$

Solution: $(-\infty, 2) \cup (3, \infty)$

**40.** $-x^2 + 8x - 11 \le 0$

$$0 \le x^2 - 8x + 11$$

$$x = \frac{8 \pm \sqrt{(-8)^2 - 4(1)(11)}}{2(1)}$$

$$x = \frac{8 \pm \sqrt{64 - 44}}{2}$$

$$x = \frac{8 \pm \sqrt{20}}{2}$$

$$x = \frac{8 \pm 2\sqrt{5}}{2}$$

$$x = \frac{2(4 \pm \sqrt{5})}{2}$$

$$x = 4 \pm \sqrt{5}$$

Critical numbers: $x = 4 \pm \sqrt{5}$

Test intervals:

Positive: $\left(-\infty, 4 - \sqrt{5}\right)$

Negative: $\left(4 - \sqrt{5}, 4 + \sqrt{5}\right)$

Positive: $\left(4 + \sqrt{5}, \infty\right)$

Solution: $\left(-\infty, 4 - \sqrt{5}\right] \cup \left[4 + \sqrt{5}, \infty\right)$

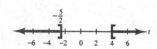

**42.** $x^2 + 8x + 16 < 0$

$$(x + 4)^2 < 0$$

$(x + 4)^2$ is not less than zero for any value of $x$.

Solution: none

**44.** $y^2 + 16y + 64 \le 0$

$$(y + 8)^2 \le 0$$

Critical number: $y = -8$

Test intervals:

Positive: $(-\infty, \infty)$

Solution: $-8$

**46.** $2t^2 - 3t - 20 \ge 0$

$$(2t + 5)(t - 4) \ge 0$$

Critical numbers: $t = -\frac{5}{2}, 4$

Test intervals:

Positive: $\left(-\infty, -\frac{5}{2}\right]$

Negative: $\left[-\frac{5}{2}, 4\right]$

Positive: $[4, \infty)$

Solution: $\left(-\infty, -\frac{5}{2}\right] \cup [4, \infty)$

**48.** $4x^2 - 4x - 63 < 0$

$$(2x + 7)(2x - 9) < 0$$

Critical numbers: $x = \frac{-7}{2}, \frac{9}{2}$

Test intervals:

Positive: $\left(-\infty, \frac{-7}{2}\right)$

Negative: $\left(\frac{-7}{2}, \frac{9}{2}\right)$

Positive: $\left(\frac{9}{2}, \infty\right)$

Solution: $\left(\frac{-7}{2}, \frac{9}{2}\right)$

**50.** $-3x^2 - 4x + 4 \le 0$

$$0 \le 3x^2 + 4x - 4$$

$$0 \le (3x - 2)(x + 2)$$

Critical numbers: $x = -2, \frac{2}{3}$

Test intervals:

Positive: $(-\infty, -2)$

Negative: $\left(-2, \frac{2}{3}\right)$

Positive: $\left(\frac{2}{3}, \infty\right)$

Solution: $(-\infty, -2] \cup \left[\frac{2}{3}, \infty\right)$

**52.** $9x^2 - 24x + 16 \geq 0$

$\qquad (3x - 4)^2 \geq 0$

$(3x - 4)^2 \geq 0$ for all real numbers.

Solution: $(-\infty, \infty)$

**54.** $(y + 3)^2 \geq 0$

$\qquad (y + 3)^2 \geq 0$ for all real numbers.

Solution: $(-\infty, \infty)$

**56.** $\qquad (y + 3)^2 - 6 \geq 0$

$\qquad y^2 + 6y + 9 - 6 \geq 0$

$\qquad\qquad y^2 + 6y + 3 \geq 0$

$\qquad y = \dfrac{-6 \pm \sqrt{6^2 - 4(1)(3)}}{2(1)}$

$\qquad y = \dfrac{-6 \pm \sqrt{36 - 12}}{2}$

$\qquad y = \dfrac{-6 \pm \sqrt{24}}{2}$

$\qquad y = \dfrac{-6 \pm 2\sqrt{6}}{2}$

$\qquad y = \dfrac{2(-3 \pm \sqrt{6})}{2}$

$\qquad y = -3 \pm \sqrt{6}$

Critical numbers: $x = -3 \pm \sqrt{6}$

Test intervals:

Positive: $\left(-\infty, -3 - \sqrt{6}\right)$

Negative: $\left(-3 - \sqrt{6}, -3 + \sqrt{6}\right)$

Positive: $\left(-3 + \sqrt{6}, \infty\right)$

Solution: $\left(-\infty, -3 - \sqrt{6}\right] \cup \left[-3 + \sqrt{6}, \infty\right)$

**58.** $25 \geq (x - 3)^2$

$\qquad 25 \geq x^2 - 6x + 9$

$\qquad 0 \geq x^2 - 6x - 16$

$\qquad 0 \geq (x - 8)(x + 2)$

Critical numbers: $-2, 8$

Test intervals:

Positive: $(-\infty, -2)$

Negative: $(-2, 8)$

Positive: $(8, \infty)$

Solution: $[-2, 8]$

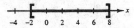

**60.** $x^2(x - 2) \leq 0$

Critical numbers: $0, 2$

Test intervals:

Negative: $(-\infty, 0)$

Negative: $(0, 2)$

Positive: $(2, \infty)$

Solution: $(-\infty, 2]$

**62.** $y = 2x^2 + 5x$, $y > 0$

*Keystrokes:*

$\boxed{Y=}$ 2 $\boxed{X,T,\theta}$ $\boxed{x^2}$ $\boxed{+}$ 5 $\boxed{X,T,\theta}$ $\boxed{GRAPH}$

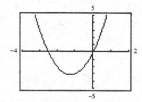

Compare graph to the $x$-axis. Solution is where graph is above the $x$-axis.

Solution: $\left(-\infty, \dfrac{-5}{2}\right) \cup (0, \infty)$

**64.** $y = \dfrac{1}{3}x^2 - 3x$, $y < 0$

*Keystrokes:*

$\boxed{Y=}$ $\boxed{(}$ 1 $\boxed{\div}$ 3 $\boxed{)}$ $\boxed{X,T,\theta}$ $\boxed{x^2}$ $\boxed{-}$ 3 $\boxed{X,T,\theta}$ $\boxed{GRAPH}$

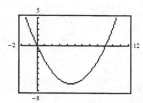

Compare graph to the $x$-axis. Solution is where graph is below the $x$-axis.

Solution: $(0, 9)$

**66.** $y = x^2 - 6x + 9$, $y < 16$

*Keystrokes:*

$y_1$ $\boxed{Y=}$ $\boxed{X,T,\theta}$ $\boxed{x^2}$ $\boxed{-}$ 6 $\boxed{X,T,\theta}$ $\boxed{+}$ 9 $\boxed{ENTER}$

$y_2$ 16 $\boxed{GRAPH}$

Solution is where the graph is below $y_2 = 16$ between the points of intersection: $(-1, 7)$

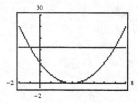

**68.** $y = 8x - x^2$, $y > 12$

*Keystrokes:*

$y_1$ $\boxed{Y=}$ 8 $\boxed{X,T,\theta}$ $\boxed{-}$ $\boxed{X,T,\theta}$ $\boxed{x^2}$ $\boxed{ENTER}$

$y_2$ 12 $\boxed{GRAPH}$

Solution is where the graph is above $y_2 = 12$ between the points of intersection: $(2, 6)$

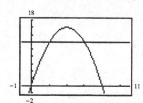

**70.** Critical number: $x = -2$

**72.** Critical numbers: $x = 2, 10$

**74.** $\dfrac{3}{4 - x} > 0$

Critical number: $x = 4$

Test intervals:

Positive: $(-\infty, 4)$

Negative: $(4, \infty)$

Solution: $(-\infty, 4)$

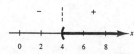

**76.** $\dfrac{-3}{4 - x} > 0$

Critical number: $x = 4$

Test intervals:

Positive: $(-\infty, 4)$

Negative: $(4, \infty)$

Solution: $(4, \infty)$

**78.** $\dfrac{x}{2 - x} < 0$

Critical numbers: $x = 0, 2$

Test intervals:

Negative: $(-\infty, 0)$

Positive: $(0, 2)$

Negative: $(2, \infty)$

Solution: $(-\infty, 0) \cup (2, \infty)$

**80.** $\dfrac{z-1}{z+3} < 0$

Critical numbers: $z = -3, 1$

Test intervals:

Positive: $(-\infty, -3)$

Negative: $(-3, 1)$

Positive: $(1, \infty)$

Solution: $(-3, 1)$

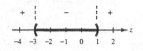

**82.** $\dfrac{u+3}{u+7} \le 0$

Critical numbers: $u = -3, -7$

Test intervals:

Positive: $(-\infty, -7)$

Negative: $(-7, -3]$

Positive: $[-3, \infty)$

Solution: $(-7, -3]$

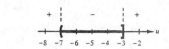

**84.** $\dfrac{x+5}{3x+2} \ge 0$

Critical numbers: $x = -5, -\dfrac{2}{3}$

Test intervals:

Positive: $(-\infty, -5)$

Negative: $\left(-5, -\dfrac{2}{3}\right)$

Positive: $\left(-\dfrac{2}{3}, \infty\right)$

Solution: $(-\infty, -5] \cup \left(-\dfrac{2}{3}, \infty\right)$

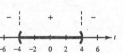

**86.** $\dfrac{u-6}{3u-5} \le 0$

Critical numbers: $u = 6, \dfrac{5}{3}$

Test intervals:

Positive: $\left(-\infty, \dfrac{5}{3}\right)$

Negative: $\left(\dfrac{5}{3}, 6\right]$

Positive: $[6, \infty)$

Solution: $\left(\dfrac{5}{3}, 6\right]$

**88.** $\dfrac{2(4-t)}{4+t} > 0$

Critical numbers: $t = -4, 4$

Test intervals:

Negative: $(-\infty, 4)$

Positive: $(-4, 4)$

Negative: $(4, \infty)$

Solution: $(-4, 4)$

**90.** $\dfrac{1}{x+2} > -3$

$\dfrac{1}{x+2} + 3 > 0$

$\dfrac{1 + 3(x+2)}{x+2} > 0$

$\dfrac{1 + 3x + 6}{x+2} > 0$

$\dfrac{3x+7}{x+2} > 0$

Critical numbers: $x = -\dfrac{7}{3}, -2$

Test intervals:

Positive: $\left(-\infty, -\dfrac{7}{3}\right)$

Negative: $\left(-\dfrac{7}{3}, -2\right)$

Positive: $(-2, \infty)$

Solution: $\left(-\infty, -\dfrac{7}{3}\right) \cup (-2, \infty)$

**92.** $\dfrac{6x}{x-4} < 5$

$\dfrac{6x}{x-4} - 5 < 0$

$\dfrac{6x - 5(x-4)}{x-4} < 0$

$\dfrac{6x - 5x + 20}{x-4} < 0$

$\dfrac{x+20}{x-4} < 0$

Critical numbers: $x = -20, 4$

Test intervals:

Positive: $(-\infty, -20)$

Negative: $(-20, 4)$

Positive: $(4, \infty)$

Solution: $(-20, 4)$

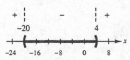

**94.** $\dfrac{x+4}{x-5} \ge 10$

$\dfrac{x+4}{x-5} - 10 \ge 0$

$\dfrac{x + 4 - 10(x-5)}{x-5} \ge 0$

$\dfrac{x + 4 - 10x + 50}{x-5} \ge 0$

$\dfrac{-9x + 54}{x-5} \ge 0$

$\dfrac{-9(x-6)}{x-5} \ge 0$

Critical numbers: $x = 5, 6$

Test intervals:

Negative: $(-\infty, 5)$

Positive: $(5, 6)$

Negative: $(6, \infty)$

Solution: $(5, 6]$

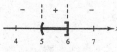

**96.** $\frac{1}{x} - 4 < 0$

*Keystrokes:*

[Y=] 1 [÷] [X,T,θ] [−] 4 [GRAPH]

Solution: $(-\infty, 0) \cup \left(\frac{1}{4}, \infty\right)$

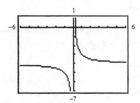

**98.** $\frac{x + 12}{x + 2} - 3 \geq 0$

*Keystrokes:*

[Y=] [(] [X,T,θ] [+] 12 [)] [÷] [(] [X,T,θ] [+] 2 [)] [−] 3 [GRAPH]

Solution: $[-2, 3]$

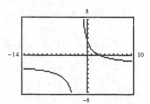

**100.** $\frac{3x - 4}{x - 4} < -5$

*Keystrokes:*

$y_1$ [Y=] [(] 3 [X,T,θ] [−] 4 [)] [÷] [(] [X,T,θ] [−] 4 [)] [ENTER]

$y_2$ [(−)] 5 [GRAPH]

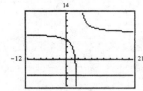

Solution: $(3, 4)$

(Look where graph is below the line $y_2 = -5$.)

**102.** $4 - \frac{1}{x^2} > 1$

*Keystrokes:*

$y_1$ [Y=] 4 [−] 1 [÷] [X,T,θ] [x²] [ENTER]

$y_2$ 1 [GRAPH]

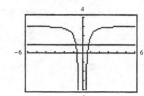

Solution:
$(-\infty, -.58) \cup (.58, \infty)$

(Look where graph is above the line $y_2 = 1$.)

**104.** $y = \frac{2(x - 2)}{x + 1}$   (a) $y \leq 0$   (b) $y \geq 8$

*Keystrokes:*

[Y=] [(] 2 [(] [X,T,θ] [−] 2 [)] [)] [÷]
[(] [X,T,θ] [+] 1 [)] [GRAPH]

Solution: (a) $(-1, 2]$

(Look at $x$-axis and vertical asymptote $x = -1$.)

(b) $[-2, -1)$

(Graph $y = 8$ and find intersection.)

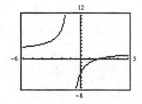

**106.** $y = \frac{5x}{x^2 + 4}$   (a) $y \geq 1$   (b) $y \geq 0$

*Keystrokes:*

[Y=] 5 [X,T,θ] [÷] [(] [X,T,θ] [x²] [+] 4 [)] [GRAPH]

Solution: (a) $[1, 4]$

(Graph $y = 1$ and find intersection.)

(b) $[0, \infty)$

(Look at $x$-axis.)

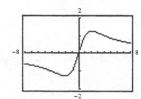

**108.** $h = -16t^2 + 88t$

$$\text{height} > 50$$

$$-16t^2 + 88t > 50$$

$$-16t^2 + 88t - 50 > 0$$

$$8t^2 - 44t + 25 < 0$$

Critical numbers: $t = \dfrac{11 + \sqrt{71}}{4}$   $t = \dfrac{11 - \sqrt{71}}{4}$

Test intervals:

Positive: $\left(-\infty, \dfrac{11 - \sqrt{71}}{4}\right)$

Negative: $\left(\dfrac{11 - \sqrt{71}}{4}, \dfrac{11 + \sqrt{71}}{4}\right)$

Positive: $\left(\dfrac{11 + \sqrt{71}}{4}, \infty\right)$

Solution: $\left(\dfrac{11 - \sqrt{71}}{4}, \dfrac{11 + \sqrt{71}}{4}\right)$

$$(0.64, 4.86)$$

**110.**

$$500(1 + r)^2 > 550$$

$$(1 + r)^2 > \dfrac{550}{500}$$

$$(1 + r)^2 > \dfrac{11}{10}$$

$$1 + 2r + r^2 > \dfrac{11}{10}$$

$$10 + 20r + 10r^2 - 11 > 0$$

$$10r^2 + 20r - 1 > 0$$

Critical numbers: $r = \dfrac{-10 + \sqrt{110}}{10}, \dfrac{-10 - \sqrt{110}}{10}$

$r$ cannot be negative.

Test intervals:

Positive: $\left(\dfrac{-10 + \sqrt{110}}{10}, \infty\right)$

Solution: $\left(\dfrac{-10 + \sqrt{110}}{10}, \infty\right)$

$$(0.0488, \infty) \qquad r > 4.88\%$$

**112.**

$$\text{Area} \geq 500$$

$$l(50 - l) \geq 500$$

$$50l - l^2 \geq 500$$

$$0 \geq l^2 - 50l + 500$$

$$l = \dfrac{50 \pm \sqrt{(-50)^2 - 4(1)(500)}}{2(1)}$$

$$l = \dfrac{50 \pm \sqrt{2500 - 2000}}{2}$$

$$l = \dfrac{50 \pm \sqrt{500}}{2}$$

$$l = \dfrac{50 \pm 10\sqrt{5}}{2}$$

$$l = \dfrac{2\left(25 \pm 5\sqrt{5}\right)}{2}$$

$$l = 25 \pm 5\sqrt{5}$$

Critical numbers: $l = 25 \pm 5\sqrt{5}$

Test intervals:

Positive: $\left[0, 25 - 5\sqrt{5}\right]$

Negative: $\left[25 - 5\sqrt{5}, 25 + 5\sqrt{5}\right]$

Positive: $\left[25 + 5\sqrt{5}, 50\right]$

Solution: $\left[25 - 5\sqrt{5}, 25 + 5\sqrt{5}\right]$

$$[13.28, 36.18]$$

# Review Exercises for Chapter 6

**2.** $u^2 - 18u = 0$

$u(u - 18) = 0$

$u = 0 \qquad u - 18 = 0$

$\qquad\qquad\qquad u = 18$

**4.** $\qquad 2z^2 - 72 = 0$

$2(z^2 - 36) = 0$

$2(z - 6)(z + 6) = 0$

$z - 6 = 0 \qquad z + 6 = 0$

$z = 6 \qquad\quad z = -6$

**6.** $\qquad x^2 + \frac{8}{3}x + \frac{16}{9} = 0$

$9x^2 + 24x + 16 = 0$

$(3x + 4)(3x + 4) = 0$

$3x + 4 = 0 \qquad 3x + 4 = 0$

$x = -\frac{4}{3} \qquad\quad x = -\frac{4}{3}$

**8.** $15x^2 - 30x - 45 = 0$

$15(x^2 - 2x - 3) = 0$

$15(x - 3)(x + 1) = 0$

$x - 3 = 0 \qquad x + 1 = 0$

$x = 3 \qquad\quad x = -1$

**10.** $10x - 8 = 3x^2 - 9x + 12$

$0 = 3x^2 - 19x + 20$

$0 = (3x - 4)(x - 5)$

$3x - 4 = 0 \qquad x - 5 = 0$

$x = \frac{4}{3} \qquad\quad x = 5$

**12.** $2x^2 = 98$

$x^2 = 49$

$x = \pm\sqrt{49}$

$x = \pm 7$

**14.** $y^2 - 8 = 0$

$y^2 = 8$

$y = \pm\sqrt{8}$

$y = \pm 2\sqrt{2}$

**16.** $(x + 3)^2 = 900$

$x + 3 = \pm 30$

$x = -3 \pm 30$

$x = 27, -33$

**18.** $u^2 = -36$

$u = \pm\sqrt{-36}$

$u = \pm 6i$

**20.** $x^2 + 48 = 0$

$x^2 = -48$

$x = \pm\sqrt{-48}$

$x = \pm 4\sqrt{3}i$

**22.** $(x - 2)^2 + 24 = 0$

$(x - 2)^2 = -24$

$x - 2 = \pm\sqrt{-24}$

$x = 2 \pm 2\sqrt{6}i$

**24.** $\qquad x^4 - 10x^2 + 9 = 0$

$(x^2 - 9)(x^2 - 1) = 0$

$x^2 - 9 = 0 \qquad x^2 - 1 = 0$

$x^2 = 9 \qquad\quad x^2 = 1$

$x = \pm\sqrt{9} \qquad x = \pm\sqrt{1}$

$x = \pm 3 \qquad\quad x = \pm 1$

**26.** $\qquad x + 2\sqrt{x} - 3 = 0$

$(\sqrt{x} + 3)(\sqrt{x} - 1) = 0$

$\sqrt{x} + 3 = 0 \qquad \sqrt{x} - 1 = 0$

$\sqrt{x} = -3 \qquad\quad \sqrt{x} = 1$

$(\sqrt{x})^2 = (-3)^2 \qquad (\sqrt{x})^2 = 1^2$

$x = 9 \qquad\qquad x = 1$

not a solution

**Check:**

$9 + 2\sqrt{9} - 3 \overset{?}{=} 0$

$9 + 6 - 3 \overset{?}{=} 0$

$12 \neq 0$

**Check:**

$1 + 2\sqrt{1} - 3 \overset{?}{=} 0$

$1 + 2 - 3 \overset{?}{=} 0$

$0 = 0$

**28.**  $\left(\sqrt{x} - 2\right)^2 + 2\left(\sqrt{x} - 2\right) - 3 = 0$

$\left(\left(\sqrt{x} - 2\right) + 3\right)\left(\left(\sqrt{x} - 2\right) - 1\right) = 0$

$\sqrt{x} - 2 + 3 = 0 \qquad\qquad \sqrt{x} - 2 - 1 = 0$

$\qquad \sqrt{x} + 1 = 0 \qquad\qquad\qquad \sqrt{x} - 3 = 0$

$\qquad\qquad \sqrt{x} = -1 \qquad\qquad\qquad\qquad \sqrt{x} = 3$

$\qquad\qquad \left(\sqrt{x}\right)^2 = (-1)^2 \qquad\qquad \left(\sqrt{x}\right)^2 = 3^2$

$\qquad\qquad\qquad x = 1 \qquad\qquad\qquad\qquad x = 9$

not a solution

**Check:** $\left(\sqrt{1} - 2\right)^2 + 2\left(\sqrt{1} - 2\right) - 3 \overset{?}{=} 0 \quad \left(\sqrt{9} - 2\right)^2 + 2\left(\sqrt{9} - 2\right) - 3 \overset{?}{=} 0$

$\qquad\qquad (1 - 2)^2 + 2(1 - 2) - 3 \overset{?}{=} 0 \qquad (3 - 2)^2 + 2(3 - 2) - 3 \overset{?}{=} 0$

$\qquad\qquad\qquad (-1)^2 + 2(-1) - 3 \overset{?}{=} 0 \qquad\qquad 1^2 + 2(1) - 3 \overset{?}{=} 0$

$\qquad\qquad\qquad\qquad 1 - 2 - 3 \overset{?}{=} 0 \qquad\qquad\qquad 1 + 2 - 3 \overset{?}{=} 0$

$\qquad\qquad\qquad\qquad\qquad -4 \neq 0 \qquad\qquad\qquad\qquad 0 = 0$

**30.**  $x^{2/5} + 4x^{1/5} + 3 = 0$

$\left(x^{1/5} + 3\right)\left(x^{1/5} + 1\right) = 0$

$x^{1/5} + 3 = 0 \qquad\qquad x^{1/5} + 1 = 0$

$\qquad x^{1/5} = -3 \qquad\qquad\qquad x^{1/5} = -1$

$\qquad \sqrt[5]{x} = -3 \qquad\qquad\qquad \sqrt[5]{x} = -1$

$\qquad \left(\sqrt[5]{x}\right)^5 = (-3)^5 \qquad\qquad \left(\sqrt[5]{x}\right)^5 = (-1)^5$

$\qquad\qquad x = -243 \qquad\qquad\qquad x = -1$

**32.**  $x^2 + 12x + 6 = 0$

$x^2 + 12x + 36 = -6 + 36$

$(x + 6)^2 = 30$

$x + 6 = \pm\sqrt{30}$

$x = -6 \pm \sqrt{30}$

**34.**  $u^2 - 5u + 6 = 0$

$\left(u^2 - 5u + \dfrac{25}{4}\right) = -6 + \dfrac{25}{4}$

$\left(u - \dfrac{5}{2}\right)^2 = \dfrac{1}{4}$

$u - \dfrac{5}{2} = \pm\sqrt{\dfrac{1}{4}}$

$u = \dfrac{5}{2} \pm \dfrac{1}{2} = \dfrac{6}{2}, \dfrac{4}{2} = 3, 2$

**36.**  $t^2 + \dfrac{1}{2}t - 1 = 0$

$t^2 + \dfrac{1}{2}t + \dfrac{1}{16} = 1 + \dfrac{1}{16}$

$\left(t + \dfrac{1}{4}\right)^2 = \dfrac{16 + 1}{16}$

$\left(t + \dfrac{1}{4}\right)^2 = \dfrac{17}{16}$

$t + \dfrac{1}{4} = \pm\sqrt{\dfrac{17}{16}}$

$t + \dfrac{1}{4} = \pm\dfrac{\sqrt{17}}{4}$

$t = -\dfrac{1}{4} \pm \dfrac{\sqrt{17}}{4}$

$t = \dfrac{-1 \pm \sqrt{17}}{4}$

**38.** $3x^2 - 2x + 2 = 0$

$$x^2 - \frac{2}{3}x + \frac{2}{3} = 0$$

$$x^2 - \frac{2}{3}x + \frac{1}{9} = -\frac{2}{3} + \frac{1}{9}$$

$$\left(x - \frac{1}{3}\right)^2 = -\frac{6}{9} + \frac{1}{9}$$

$$\left(x - \frac{1}{3}\right)^2 = -\frac{5}{9}$$

$$x - \frac{1}{3} = \pm\sqrt{-\frac{5}{9}}$$

$$x - \frac{1}{3} = \pm\frac{\sqrt{5}}{3}i$$

$$x = \frac{1}{3} \pm \frac{\sqrt{5}}{3}i$$

**40.** $x^2 - x - 72 = 0$

$$x = \frac{1 \pm \sqrt{(-1)^2 - 4(1)(-72)}}{2(1)}$$

$$x = \frac{1 \pm \sqrt{1 + 288}}{2}$$

$$x = \frac{1 \pm \sqrt{289}}{2}$$

$$x = \frac{1 \pm 17}{2}$$

$$x = -8, 9$$

**42.** $2x^2 - 3x - 20 = 0$

$$x = \frac{3 \pm \sqrt{(-3)^2 - 4(2)(-20)}}{2(2)}$$

$$x = \frac{3 \pm \sqrt{9 + 160}}{4}$$

$$x = \frac{3 \pm \sqrt{169}}{4}$$

$$x = \frac{3 \pm 13}{4}$$

$$x = -\frac{5}{2}, 4$$

**44.** $3x^2 + 12x + 4 = 0$

$$x = \frac{-12 \pm \sqrt{12^2 - 4(3)(4)}}{2(3)}$$

$$x = \frac{-12 \pm \sqrt{144 - 48}}{6}$$

$$x = \frac{-12 \pm \sqrt{96}}{6} = \frac{-12 \pm 4\sqrt{6}}{6} = \frac{-6 \pm 2\sqrt{6}}{3}$$

**46.** $-u^2 + 2.5u + 2 = 0$

$$u = \frac{-2.5 \pm \sqrt{(2.5)^2 - 4(-1)(3)}}{2(-1)}$$

$$u = \frac{-2.5 \pm \sqrt{6.25 + 12}}{-2}$$

$$u = \frac{-2.5 \pm \sqrt{18.25}}{-2} = \frac{-2.5 \pm 0.5\sqrt{73}}{-2} = \frac{5 \pm \sqrt{73}}{4}$$

**48.** $y^2 - 26y + 169 = 0$

$$b^2 - 4ac = (-26)^2 - 4(1)(169)$$

$$= 676 - 676$$

$$= 0$$

One repeated rational solution

**50.** $r^2 - 5r - 45 = 0$

$$b^2 - 4ac = (-5)^2 - 4(1)(-45)$$

$$= 25 + 180$$

$$= 205$$

Two distinct irrational solutions

**52.** $7x^2 + 3x - 18 = 0$

$$b^2 - 4ac = 3^2 - 4(7)(-18)$$

$$= 9 + 504$$

$$= 513$$

Two distinct irrational solutions

**54.** $9y^2 + 1 = 0$

$$b^2 - 4ac = 0^2 - 4(9)(1)$$

$$= -36$$

Two distinct imaginary solutions

**56.** $-2x(x - 10) \le 0$

Critical numbers: $x = 0, 10$

Test intervals:

Negative: $(-\infty, 0)$

Positive: $(0, 10)$

Negative: $(10, \infty)$

Solution: $(-\infty, 0] \cup [10, \infty)$

**58.** $(x - 5)^2 - 36 > 0$

$[(x - 5) - 6][(x - 5) + 6] > 0$

$(x - 11)(x + 1) > 0$

Critical numbers: $x = -1, 11$

Test intervals:

Positive: $(-\infty, -1)$

Negative: $(-1, 11)$

Positive: $(11, \infty)$

Solution: $(-\infty, -1) \cup (11, \infty)$

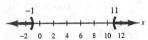

**60.** $3x^2 - 2x - 8 > 0$

$(3x + 4)(x - 2) > 0$

Critical numbers: $x = -\frac{4}{3}, 2$

Test intervals:

Positive: $\left(-\infty, -\frac{4}{3}\right)$

Negative: $\left(-\frac{4}{3}, 2\right)$

Positive: $(2, \infty)$

Solution: $\left(-\infty, -\frac{4}{3}\right) \cup (2, \infty)$

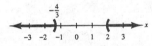

**62.** $\dfrac{2x - 9}{x - 1} \le 0$

Critical numbers: $x = 1, \frac{9}{2}$

Test intervals:

Positive: $(-\infty, 1)$

Negative: $\left(1, \frac{9}{2}\right)$

Positive: $\left(\frac{9}{2}, \infty\right)$

Solution: $\left(1, \frac{9}{2}\right]$

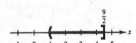

**64.**

$\dfrac{3x + 1}{x - 2} > 4$

$\dfrac{3x + 1}{x - 2} - 4 > 0$

$\dfrac{3x + 1 - 4(x - 2)}{x - 2} > 0$

$\dfrac{3x + 1 - 4x + 8}{x - 2} > 0$

$\dfrac{-x + 9}{x - 2} > 0$

$\dfrac{-(x - 9)}{x - 2} > 0$

$\dfrac{x - 9}{x - 2} < 0$

Critical numbers: $x = 2, 9$

Test intervals:

Positive: $(-\infty, 2)$

Negative: $(2, 9)$

Positive: $(9, \infty)$

Solution: $(2, 9)$

**66.** *Verbal model:*  | Selling price per computer | = | Cost per computer | + | Profit per computer |

*Labels:*  Number computers sold $= x$

Number computers purchased $= x + 5$

*Equation:*

$$\frac{27,000}{x} = \frac{27,000}{x + 5} + 900$$

$$x(x + 5)\left(\frac{2700}{x}\right) = \left(\frac{27,000}{x + 5} + 900\right)x(x + 5)$$

$$27,000(x + 5) = 27,000x + 900\,x(x + 5)$$

$$27,000x + 135,000 = 27,000x + 900x^2 + 4500x$$

$$0 = 900x^2 + 4500x - 135,000$$

$$0 = x^2 + 5x - 150$$

$$0 = (x + 15)(x - 10)$$

$x = -15 \qquad x = 10$ computers
reject

Selling price of each computer $= \dfrac{27,000}{10} = \$2700$

**68.** *Verbal Model:*  | Area | $= \frac{1}{2} \cdot$ | Base | $\cdot$ | Height |

*Labels:*  Base $= x$

Height $= x - 4$

*Equation:*

$$240 = \frac{1}{2} \cdot x \cdot (x - 4)$$

$$480 = x^2 - 4x$$

$$0 = x^2 - 4x - 480$$

$$x = \frac{-(-4) \pm \sqrt{(-4)^2 - 4(1)(-480)}}{2(1)}$$

$$x = \frac{4 \pm \sqrt{16 + 1920}}{2} = \frac{4 \pm \sqrt{1936}}{2} \approx 24, -22 \text{ reject}$$

$x = 24$ centimeters; $x - 4 = 20$ centimeters

Height: 20 centimeters

Base: 24 centimeters

**70.** $A = P(1 + r)^2$

$$38,955.88 = 35,000(1 + r)^2$$

$$1.113025143 = (1 + r)^2$$

$$1.055 \approx 1 + r$$

$$.055 \approx r \approx 5.5\%$$

**72.** *Verbal model:* $\boxed{\begin{array}{c}\text{Cost per person}\\\text{Current group}\end{array}} - \boxed{\begin{array}{c}\text{Cost per person}\\\text{New group}\end{array}} = \boxed{16}$

*Equation:*

$$\frac{360}{x} - \frac{360}{x+6} = 16$$

$$x(x+6)\left[\frac{360}{x} - \frac{360}{x+6}\right] = 16x(x+6)$$

$$360(x+6) - 360x = 16x^2 + 96x$$

$$360x + 2160 - 360x = 16x^2 + 96x$$

$$0 = 16x^2 + 96x - 2160$$

$$0 = 16(x+15)(x-9)$$

$$x + 15 = 0 \qquad x - 9 = 0$$

$$x = -15 \qquad x = 9$$

reject

It is possible to have fewer nonmembers and a greater decrease in the fare if there are fewer, only 9, members going on the trip.

**74.** *Formula:* $c^2 = a^2 + b^2$

*Labels:*
$c = 39 \qquad\qquad a + b = 51$
$a = x \qquad\qquad\quad x + b = 51$
$b = 51 - x \qquad\quad b = 51 - x$

*Equation:* $39^2 = x^2 + (51 - x)^2$

$$1521 = x^2 + 2601 - 102x + x^2$$

$$0 = 2x^2 - 102x + 1080$$

$$0 = x^2 - 51x + 540$$

$$0 = (x - 15)(x - 36)$$

$$x = 15 \qquad\qquad x = 36$$

$$51 - x = 36 \qquad 51 - x = 15$$

15 feet and 36 feet

**76.** *Verbal model:* $\boxed{\begin{array}{c}\text{Work done by}\\\text{Person 1}\end{array}} + \boxed{\begin{array}{c}\text{Work done by}\\\text{Person 2}\end{array}} = \boxed{\text{One complete job}}$

*Labels:*
Time of Person 1 $= x$
Time of Person 2 $= x + 3$

*Equation:*

$$\frac{1}{x}(6) + \frac{1}{x+3}(6) = 1$$

$$x(x+3)\left[(6)\left(\frac{1}{x} + \frac{1}{x+3}\right)\right] = x(x+3)$$

$$6(x+3) + 6x = x(x+3)$$

$$6x + 18 + 6x = x^2 + 3x$$

$$-x^2 + 9x + 18 = 0$$

$$x^2 - 9x - 18 = 0$$

$$x = \frac{9 \pm \sqrt{(-9)^2 - 4(1)(-18)}}{2(1)}$$

$$x = \frac{9 \pm \sqrt{153}}{2}$$

$$x = \frac{9 \pm 12.369317}{2}$$

$$x = 10.684658, \quad \text{reject} -1.6846584 \qquad x \approx 10.7 \text{ hours}$$

$$x + 3 = 13.684658 \qquad\qquad\qquad x + 3 \approx 13.7 \text{ hours}$$

**78.** (a) $h = 200 - 16t^2, t \geq 0$

$164 = 200 - 16t^2$

$-36 = -16t^2$

$\dfrac{9}{4} = t^2$

$\dfrac{3}{2} = t = 1.5$ seconds

(b) $h = 200 - 16t^2, t \geq 0$

$0 = 200 - 16t^2$

$16t^2 = 200$

$t^2 = \dfrac{200}{16}$

$t = +\sqrt{\dfrac{200}{16}}$

$t = \dfrac{10}{4}\sqrt{2} = \dfrac{5}{2}\sqrt{2} \approx 3.54$ sec

**80.** $\overline{C} = \dfrac{C}{x} = \dfrac{100{,}000 + 0.9x}{x} = \dfrac{100{,}000}{x} + 0.9$

$\dfrac{100{,}000}{x} + 0.9 < 2$

$\dfrac{100{,}000}{x} - 1.1 < 0$

$\dfrac{100{,}000 - 1.1x}{x} < 0$

Critical numbers: $x = 0, 90{,}909$

Test intervals:

$x$ must be positive

Positive: $(0, 90{,}909)$

Negative: $(90{,}909, \infty)$

Solution: $(90{,}909, \infty)$

$x > 90{,}909$

**82.** $3000(1 + r)^2 > 3370$

$(1 + r)^2 > 1.12\overline{3}$

$1 + r > 1.05987$

$r > .0599$

$r > 5.99\%$

# CHAPTER 7
# Linear Models and Graphs of Nonlinear Models

# CHAPTER 7
# Linear Models and Graphs of Nonlinear Models

## Section 7.1    Variation
**Solutions to Even-Numbered Exercises**

**2.** $C = kr$

**4.** $s = kt^3$

**6.** $V = k\sqrt[3]{x}$

**8.** $S = \dfrac{k}{v^2}$

**10.** $A = \dfrac{k}{t^4}$

**12.** $V = khr^2$

**14.** $F = \dfrac{km_1m_2}{r^2}$

**16.** The area of a circle varies directly as the square of the radius.

**18.** The surface area of a sphere is proportional to the square of the radius.

**20.** The volume of a sphere varies directly as the cube of the radius.

**22.** The height of a cylinder varies directly as the volume and inversely as the square of the radius.

**24.** $h = kr$

$28 = k(12)$

$\frac{28}{12} = k$

$\frac{7}{3} = k$

$h = \frac{7}{3}r$

**26.** $v = k\sqrt{s}$

$24 = k\sqrt{16}$

$24 = k(4)$

$6 = k$

$v = 6\sqrt{s}$

**28.** $M = kn^3$

$0.012 = k(0.2)^3$

$0.012 = k(0.008)$

$\dfrac{0.012}{0.008} = k$

$1.5 = k$

$M = 1.5n^3$

**30.** $q = \dfrac{k}{P}$

$\dfrac{3}{2} = \dfrac{k}{50}$

$150 = 2k$

$75 = k$

$q = \dfrac{75}{P}$

**32.** $u = \dfrac{k}{v^2}$

$40 = \dfrac{k}{\left(\frac{1}{2}\right)^2}$

$40 = \dfrac{k}{\frac{1}{4}}$

$40 = 4k$

$10 = k$

$u = \dfrac{10}{v^2}$

**34.** $V = khb^2$

$288 = k(6)(12)^2$

$288 = k(6)(144)$

$288 = 864k$

$\dfrac{288}{864} = k$

$\dfrac{1}{3} = k$

$V = \dfrac{1}{3}hb^2$

**36.** $z = \dfrac{kx}{\sqrt{y}}$

$720 = \dfrac{k(48)}{\sqrt{81}}$

$720 = \dfrac{48k}{9}$

$6480 = 48k$

$135 = k$

$z = \dfrac{135x}{\sqrt{y}}$

**38.** (a)  $R = kx$

$300 = k(25)$

$12 = k$

$R = 12(42)$

$R = \$504$

(b) price per unit

**40.** (a)  $d = kF$          $d = \frac{3}{50}F$

  $3 = k(50)$        $d = \frac{3}{50}(20)$

  $\frac{3}{50} = k$          $d = \frac{6}{5}$ inches or 1.2 inches

  (b)         $d = \frac{3}{50}F$

    $1.5 = \frac{3}{50}F$

    $1.5\left(\frac{50}{3}\right) = F$

    25 pounds $= F$

**42.** (a)  $F = kx$          (b)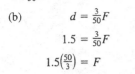

  $50 = k(1.5)$

  $\frac{100}{3} = k$

  $F = \frac{100}{3}x$

The graph is a line with slope $\frac{100}{3}$ and a $y$-intercept at $(0, 0)$.

**44.**  $d = kt^2$        $d = 16t^2$

  $64 = k(2)^2$      $d = 16(6)^2$

  $64 = 4k$        $d = 16(36)$

  $16 = k$        $d = 576$ ft

**46.**    $d = kv^2$          $d = 0.32v^2$

  $0.02 = k\left(\frac{1}{4}\right)^2$      $0.12 = 0.32v^2$

  $0.02 = \frac{1}{16}k$        $0.375 = v^2$

  $0.32 = k$          $\sqrt{0.375} = v$

              $0.61237 \approx v$

              $v \approx 0.61237$ mile per hour

**48.**      $P = kw^3$        $P = 0.048(40)^3$

  $750 = k(25)^3$      $P = 3072$ watts of power

  $\frac{750}{15,625} = k$

  $0.048 = k$

**50.** (a)  $t = \frac{k}{r}$          $t = \frac{k}{80}$

  $3 = \frac{k}{65}$        $t = \frac{195}{80} \approx 2.44$ hr

  $k = 195$ miles

  (b)  Distance

**52.**  $N = \frac{k}{t + 1}$        $N = \frac{500}{t + 1}$

  $500 = \frac{k}{0 + 1}$      $N = \frac{500}{4 + 1}$

  $500 = k$        $N = \frac{500}{5}$

            $N = 100$ prey

**54.**          $F = \frac{k}{r^2}$        $F = \frac{3,040,000,000}{r^2}$

  $190 = \frac{k}{(4000)^2}$    $F = \frac{3,040,000,000}{(5000)^2}$

  $3,040,000,000 = k$      $F = 121.6$ pounds

**56.**  $P = \frac{k}{A}$        $P = \frac{116}{11A}$

  $4 = \frac{k}{29}$        $P = \frac{116}{11(29)}$

  $116 = k$        $P = \frac{116}{319}$

            $P = 0.3\overline{6}$ pound per square inch

Denise weighs 116 pounds.

**58.**  $T = \frac{k}{d}$        $T = \frac{4}{d}$

  $4 = \frac{k}{1}$        $T = \frac{4}{4.385}$

  $4 = k$        $T \approx 0.91°C$

        OR

  $T = \frac{k}{d}$        $T = \frac{4000}{d}$

  $4 = \frac{k}{1000}$      $T = \frac{4000}{4385}$

  $4000 = k$        $T \approx 0.91°C$

**60.** (a) $P = \dfrac{kWD^2}{L}$

(b) Unchanged

(c) Increases by a factor of 8.

(d) Increases by a factor of 4.

(e) Increases by a factor of $\frac{1}{4}$.

(f) $2000 = \dfrac{k(3)8^2}{120}$

$2000 = \dfrac{k(192)}{120}$

$1250 = k$

$L = \dfrac{1250(3)10^2}{120}$

$L = 3125$ pounds

**62.**

| $x$ | 2 | 4 | 6 | 8 | 10 |
|---|---|---|---|---|---|
| $y = kx^2$ | 8 | 32 | 72 | 128 | 200 |

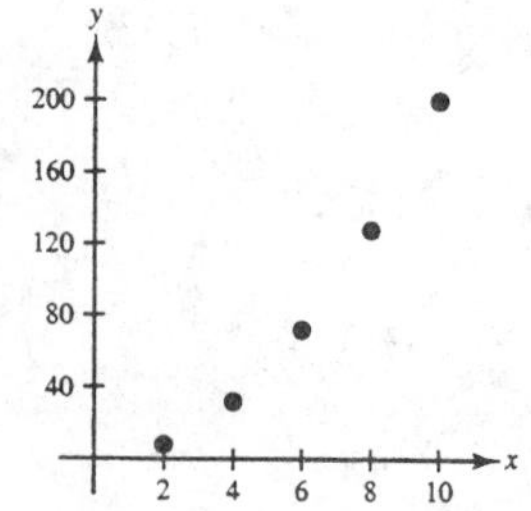


**64.**

| $x$ | 2 | 4 | 6 | 8 | 10 |
|---|---|---|---|---|---|
| $y = kx^2$ | 1 | 4 | 9 | 16 | 25 |

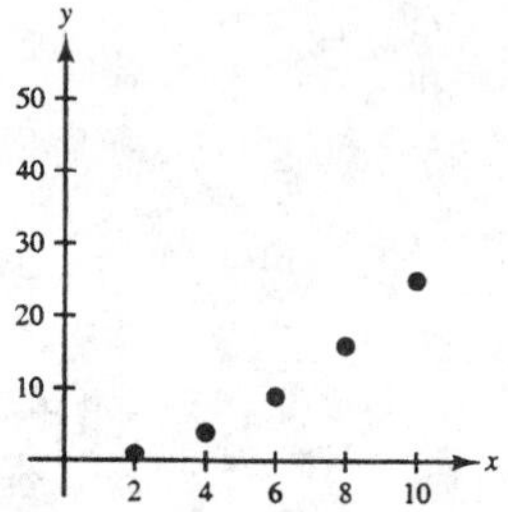


**66.**

| $x$ | 2 | 4 | 6 | 8 | 10 |
|---|---|---|---|---|---|
| $y = \dfrac{k}{x^2}$ | $\dfrac{5}{4}$ | $\dfrac{5}{16}$ | $\dfrac{5}{36}$ | $\dfrac{5}{64}$ | $\dfrac{1}{20}$ |

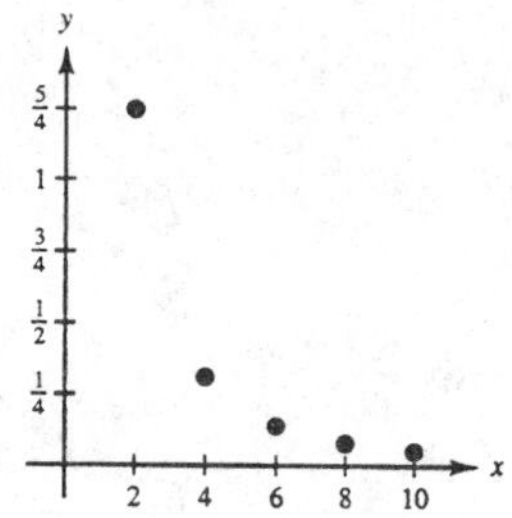


**68.**

| $x$ | 2 | 4 | 6 | 8 | 10 |
|---|---|---|---|---|---|
| $y = \dfrac{k}{x^2}$ | 5 | $\dfrac{5}{4}$ | $\dfrac{5}{9}$ | $\dfrac{5}{16}$ | $\dfrac{1}{5}$ |

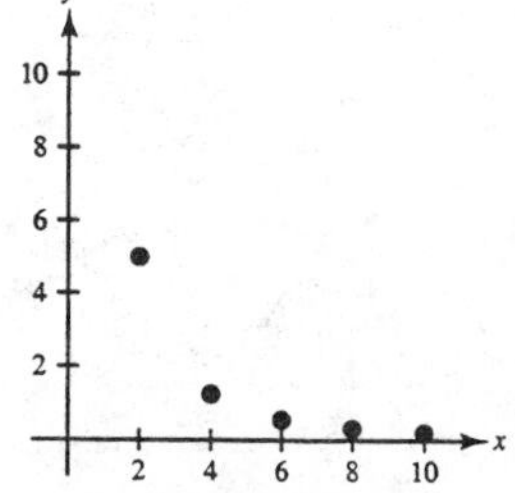


**70.**

| $x$ | 10 | 20 | 30 | 40 | 50 |
|---|---|---|---|---|---|
| $y$ | 2 | 4 | 6 | 8 | 10 |

$y = kx$

$2 = k(10)$

$\frac{1}{5} = k$

$y = \frac{1}{5}x$

**72.**

| $x$ | 10 | 20 | 30 | 40 | 50 |
|---|---|---|---|---|---|
| $y$ | 60 | 30 | 20 | 15 | 12 |

$y = \dfrac{k}{x}$

$60 = \dfrac{k}{10}$

$600 = k$

$y = \dfrac{600}{x}$

**74.** Decrease. $y = k/x$ and $k > 0$ so if one variable increases the other decreases.

**76.** $y = \dfrac{k}{x^2}$

$y = \dfrac{k}{(2x)^2}$

$y = \dfrac{k}{4x^2}$

$y$ will be $\frac{1}{4}$ as great.

## Section 7.2    Graphs of Linear Inequalities

**2.** $x + y < 3$

   (a) $0 + 6 \overset{?}{<} 3$

       $6 \not< 3$

     $(0, 6)$ is *not* a solution.

   (c) $0 + (-2) \overset{?}{<} 3$

       $-2 < 3$

     $(0, -2)$ *is* a solution.

   (b) $4 + 0 \overset{?}{<} 3$

       $4 \not< 3$

     $(4, 0)$ is *not* a solution.

   (d) $1 + 1 \overset{?}{<} 3$

       $2 < 3$

     $(1, 1)$ *is* a solution.

**4.** $-3x + 5y \geq 6$

   (a) $-3(2) + 5(8) \overset{?}{\geq} 6$

       $-6 + 40 \overset{?}{\geq} 6$

          $34 \geq 6$

     $(2, 8)$ *is* a solution.

   (c) $-3(0) + 5(0) \overset{?}{\geq} 6$

       $0 + 0 \overset{?}{\geq} 6$

         $0 \not\geq 6$

     $(0, 0)$ is *not* a solution.

   (b) $-3(-10) + 5(-3) \overset{?}{\geq} 6$

           $30 - 15 \overset{?}{\geq} 6$

              $15 \geq 6$

     $(-10, -3)$ *is* a solution.

   (d) $-3(-3) + 5(3) \overset{?}{\geq} 6$

        $-9 + 15 \overset{?}{\geq} 6$

          $6 \geq 6$

     $(3, 3)$ *is* a solution.

**6.** $y < 3.5x + 7$

   (a) $5 \overset{?}{<} -3.5(1) + 7$

      $5 \overset{?}{<} -3.5 + 7$

      $5 \not< 3.5$

     $(1, 5)$ is *not* a solution.

   (c) $4 \overset{?}{<} -3.5(-1) + 7$

      $4 \overset{?}{<} -3.5 + 7$

      $4 < 10.5$

     $(-1, 4)$ *is* a solution.

   (b) $-1 \overset{?}{<} -3.5(5) + 7$

      $-1 \overset{?}{<} -17.5 + 7$

     $-1 \not< -10.5$

     $(5, -1)$ is *not* a solution.

   (d) $\frac{4}{3} \overset{?}{<} -3.5(0) + 7$

      $\frac{4}{3} \overset{?}{<} 0 + 7$

      $\frac{4}{3} < 7$

     $\left(0, \frac{4}{3}\right)$ *is* a solution.

**8.** $y \geq |x - 3|$

   (a) $0 \overset{?}{\geq} |0 - 3|$

      $0 \overset{?}{\geq} |-3|$

      $0 \not\geq 3$

      $(0, 0)$ is *not* a solution.

   (c) $10 \overset{?}{\geq} |4 - 3|$

      $10 \overset{?}{\geq} |1|$

      $10 \geq 1$

      $(4, 10)$ *is* a solution.

   (b) $2 \overset{?}{\geq} |1 - 3|$

      $2 \overset{?}{\geq} |-2|$

      $2 \geq 2$

      $(1, 2)$ *is* a solution.

   (d) $-1 \overset{?}{\geq} |5 - 3|$

      $-1 \overset{?}{\geq} |2|$

      $-1 \not\geq 2$

      $(5, -1)$ is *not* a solution.

**10.** $x < -2$; (a)

**12.** $3x - 2y > 0$; (e)

**14.** $x + y \leq 4$; (c)

**16.** $x < -3$

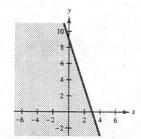

**18.** $y > 2$

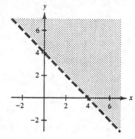

**20.** $y \leq 2x$

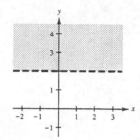

**22.** $y > 4 - x$

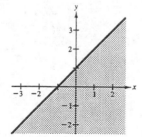

**24.** $y \leq x + 1$

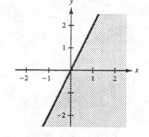

**26.** $x + y \leq 5$

    $y \leq -x + 5$

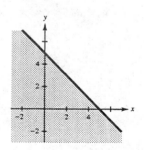

**28.** $3x + y \leq 9$

    $y \leq -3x + 9$

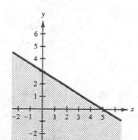

**30.** $3x + 5y \leq 15$

    $5y \leq -3x + 15$

    $y \leq -\frac{3}{5}x + 3$

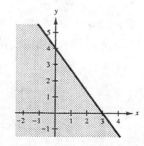

**32.** $4x + 3y \leq 12$

    $3y \leq -4x + 12$

    $y \leq -\frac{4}{3}x + 4$

**34.** $0.25x - 0.75y > 6$

$25x - 75y > 600$

$-75y > -25x + 600$

$y < \frac{1}{3}x - 8$

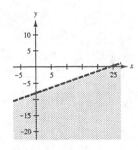

**36.** $y - 2 < -\frac{2}{3}(x - 3)$

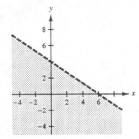

**38.** $\frac{x}{2} + \frac{y}{6} \geq 1$

$3x + y \geq 6$

$y \geq -3x + 6$

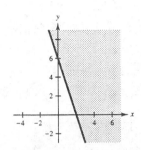

**40.** $y \leq 9 - \frac{3}{2}x$

Keystrokes:

$\boxed{\text{DRAW}}$ 7 $\boxed{(}$ $-10, 9$ $\boxed{-}$ 1.5 $\boxed{\text{X,T,}\theta}$ $\boxed{)}$ $\boxed{\text{ENTER}}$

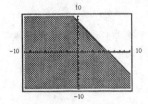

**42.** $y \geq \frac{1}{4}x + 3$

Keystrokes:

$\boxed{\text{Y=}}$ $\boxed{\text{X,T,}\theta}$ $\boxed{\div}$ 4 $\boxed{+}$ 3

$\boxed{\text{DRAW}}$ 7 $\boxed{(}$ $\boxed{\text{Y-VARS}}$ $\boxed{1}$ $\boxed{1}$ $\boxed{,}$ 10 $\boxed{)}$ $\boxed{\text{ENTER}}$

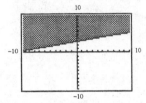

**44.** $2x + 4y - 3 \leq 0$

$4y \leq -2x + 3$

$y \leq -\frac{1}{2}x + \frac{3}{4}$

$\boxed{\text{Y=}}$ $\boxed{(-)}$ $\boxed{\text{X,T,}\theta}$ $\boxed{\div}$ 2 $\boxed{+}$ 3 $\boxed{\div}$ 4

$\boxed{\text{DRAW}}$ 7 $\boxed{(}$ $\boxed{(-)}$ 10 $\boxed{,}$ $\boxed{\text{Y-VARS}}$ $\boxed{1}$ $\boxed{1}$ $\boxed{)}$ $\boxed{\text{ENTER}}$

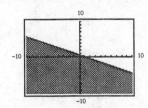

**46.** $x - 3y + 9 \geq 0$

$-3y \geq -x - 9$

$y \leq \frac{1}{3}x + 3$

$\boxed{\text{DRAW}}$ 7 $\boxed{(}$ $-10,$ $\boxed{\text{X,T,}\theta}$ $\boxed{\div}$ 3 $\boxed{+}$ 3 $\boxed{)}$ $\boxed{\text{ENTER}}$

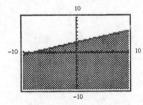

**48.** $m = \dfrac{6 - 2}{4 - 0} = \dfrac{4}{4} = 1$

$y - 2 < 1(x - 0)$

$y - 2 < x$

$y < x + 2$

$-x + y < 2$

**50.** $x < 2$

**52.** $m = \dfrac{1 - 0}{-1 - 0} = \dfrac{1}{-1} = -1$

$y - 0 < -1(x - 0)$

$y < -x$

$x + y < 0$

**54.** $P = 2x + 2y$

$2x + 2y \geq 100$

$x + y \geq 50$

$y \geq -x + 50$

(*Note:* $x$ and $y$ cannot be negative.)

Keystrokes:

$\boxed{Y=}$ $\boxed{(-)}$ $\boxed{X,T,\theta}$ $\boxed{+}$ 50

$\boxed{DRAW}$ 7 $\boxed{(}$ $\boxed{Y\text{-}VARS}$ 1 1 $\boxed{,}$ 100 $\boxed{)}$ $\boxed{ENTER}$

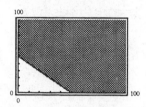

**56.** $15x + 6y \leq 2000$

$6y \leq -15x + 2000$

$y \leq -\frac{15}{6}x + \frac{2000}{6}$

$y \leq -\frac{5}{2}x + \frac{1000}{3}$

(*Note:* $x$ and $y$ cannot be negative.)

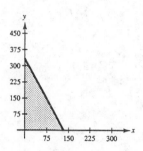

**58.** *Verbal Model:* $\boxed{\begin{array}{c}\text{Cost of}\\\text{cheese pizzas}\end{array}} + \boxed{\begin{array}{c}\text{Cost of}\\\text{extra toppings}\end{array}} + \boxed{\begin{array}{c}\text{Cost for}\\\text{drinks}\end{array}} \leq 26$

*Labels:* Cost of cheese pizzas = 2(8) = \$16

Cost for extra toppings = 0.40x (dollars)

Cost for drinks = 0.80y (dollars)

*Inequality:* $16 + 0.40x + 0.80y \leq 26$

$0.40x + 0.80y \leq 10$

(*Note:* $x$ and $y$ cannot be negative.)

(6, 6)

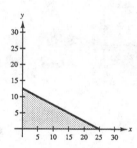

$$0.40(6) + 0.80(6) \overset{?}{\leq} 10$$

$$2.4 + 4.8 \overset{?}{\leq} 10$$

$$7.2 \leq 10$$

Yes

**60.** $30x + 20y \geq 300$

(*Note:* $x$ and $y$ cannot be negative.)

$$20y \geq -30x + 300$$

$$y \geq -\tfrac{3}{2}x + 15$$

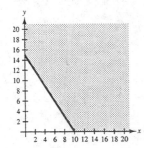

**62.** $(10, 0), (6, 10), (1, 18), (0, 15), (12, 2)$

$6x + 5y \geq 120$

$$5y \geq -6x + 120$$

$$y \geq -\tfrac{6}{5}x + 24$$

(*Note:* $x$ and $y$ cannot be negative.)

Here are some examples of ordered pairs that are solutions. Note that there are other correct answers.

$(15, 10), (22, 0), (12, 10)$

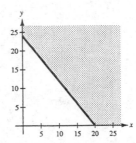

**64.** The four forms of a linear inequality in variables $x$ and $y$:

$ax + by < c, \; ax + by \leq c, \; ax + by > c, \; ax + by \geq c$

**66.** The graph of a line divides the plane into two half-planes. In graphing a linear inequality, graph the corresponding linear equation. The solution to the inequality will be one of the half-planes determined by the line. Example: $x - 2y > 1$

**68.** After graphing the corresponding equation, you determine which half-plane is the solution of a linear inequality by testing a point in one of the half-planes.

# Section 7.3    Graphs of Quadratic Functions

**2.** $f(x) = \tfrac{1}{2}x - 4$    (f)

**4.** $f(x) = -x^2 + 3$    (c)

**6.** $f(x) = 2 - (x - 2)^2$    (a)

**8.** $y = x^2 + 2x = (x^2 + 2x + 1) - 1 = (x + 1)^2 - 1$

vertex $= (-1, -1)$

**10.** $y = x^2 + 6x - 5 = (x^2 + 6x + 9) - 5 - 9 = (x + 3)^2 - 14$

vertex $= (-3, -14)$

**12.** $y = x^2 - 4x + 5$

$y = (x^2 - 4x + 4) + 5 - 4$

$y = (x - 2)^2 + 1$

vertex $= (2, 1)$

**14.** $y = 4 - 8x - x^2$

$y = -(x^2 + 8x - 4)$

$y = -(x^2 + 8x + 16 - 16 - 4)$

$y = -(x^2 + 8x + 16) + 16 + 4$

$y = -(x + 4)^2 + 20$

vertex $= (-4, 20)$

**16.** $y = -x^2 - 10x + 10$

$= -(x^2 + 10x) + 10$

$= -(x^2 + 10x + 25 - 25) + 10$

$= -(x^2 + 10x + 25) + 25 + 10$

$= -(x + 5)^2 + 35$

vertex $= (-5, 35)$

**18.** $y = 3x^2 - 3x - 9$

$y = 3(x^2 - x - 3)$

$y = 3\left(x^2 - x + \dfrac{1}{4} - \dfrac{1}{4} - 3\right)$

$y = 3\left(x^2 - x + \dfrac{1}{4}\right) - \dfrac{3}{4} - 9$

$y = 3\left(x - \dfrac{1}{2}\right)^2 - \dfrac{3}{4} - \dfrac{36}{4}$

$y = 3\left(x - \dfrac{1}{2}\right)^2 - \dfrac{39}{4}$

vertex $= \left(\dfrac{1}{2}, -\dfrac{39}{4}\right)$

**20.** $f(x) = x^2 + 4x + 1$

$a = 1 \quad b = 4$

$x = \dfrac{-b}{2a} = \dfrac{-(4)}{2(1)} = -2$

$f\left(\dfrac{-b}{2a}\right) = (-2)^2 + 4(-2) + 1$

$= 4 - 8 + 1$

$= -3$

vertex $= (-2, -3)$

**22.** $h(x) = -x^2 + 14x - 14$

$a = -1 \quad b = 14$

$x = \dfrac{-b}{2a} = \dfrac{-14}{2(-1)} = 7$

$h\left(\dfrac{-b}{2a}\right) = -7^2 + 14(7) - 14$

$= -49 + 98 - 14 = 35$

vertex $= (7, 35)$

**24.** $y = 9x^2 - 12x$

$a = 9 \quad b = -12$

$x = \dfrac{-b}{2a} = \dfrac{-(-12)}{2(9)} = \dfrac{12}{18} = \dfrac{2}{3}$

$y = 9\left(\dfrac{2}{3}\right)^2 - 12\left(\dfrac{2}{3}\right)$

$= 9\left(\dfrac{4}{9}\right) - 8 = 4 - 8 = -4$

vertex $= \left(\dfrac{2}{3}, -4\right)$

**26.** $y = -3(x + 5)^2 - 3$

$-3 < 0$ opens downward
vertex $= (-5, -3)$

**28.** $y = 2(x - 12)^2 + 3$

$2 > 0$ opens upward
vertex $= (12, 3)$

**30.** $y = -(x + 1)^2$

$-1 < 0$ opens downward
vertex $= (-1, 0)$

**32.** $y = x^2 - 6x$

$1 > 0$ opens upward

$y = x^2 - 6x$

$y = x^2 - 6x + 9 - 9$

$y = (x - 3)^2 - 9$

vertex $= (3, -9)$

**34.** $y = x^2 - 49$ $\qquad\qquad$ $y = x^2 - 49$

$0 = x^2 - 49$ $\qquad\qquad$ $y = 0^2 - 49$

$0 = (x - 7)(x + 7)$ $\qquad$ $y = -49$

$x - 7 = 0 \qquad x + 7 = 0 \qquad (0, -49)$

$x = 7 \qquad\quad x = -7$

$(7, 0)$ and $(-7, 0)$

**36.** $y = x^2 + 4x$ $\qquad\qquad$ $y = x^2 + 4x$

$0 = x^2 + 4x$ $\qquad\qquad$ $y = 0^2 + 4(0)$

$0 = x(x + 4)$ $\qquad\qquad$ $y = 0$

$x = 0 \qquad x + 4 = 0 \qquad (0, 0)$

$\qquad\qquad x = -4$

$(0, 0)$ and $(-4, 0)$

**38.** $y = 10 - x - 2x^2$ $\qquad$ $y = 10 - x - 2x^2$

$0 = 10 - x - 2x^2$ $\qquad$ $y = 10 - 0 - 2(0)^2$

$0 = (5 + 2x)(2 - x)$ $\qquad$ $y = 10$

$5 + 2x = 0 \qquad 2 - x = 0 \qquad (0, 10)$

$2x = -5 \qquad\quad 2 = x$

$x = -\dfrac{5}{2}$

$\left(-\dfrac{5}{2}, 0\right)$ and $(2, 0)$

**40.** $y = x^2 - 3x - 10$      $y = x^2 - 3x - 10$

$0 = x^2 - 3x - 10$      $y = 0^2 - 3(0) - 10$

$0 = (x - 5)(x + 2)$      $y = -10$    $(0, -10)$

$x - 5 = 0$     $x + 2 = 0$

$x = 5$       $x = -2$

$(5, 0)$ and $(-2, 0)$

**42.** $h(x) = x^2 - 9$

x-intercepts               vertex

$0 = x^2 - 9$            $h(x) = x^2 - 9$

$0 = (x - 3)(x + 3)$     $h(x) = (x - 0)^2 - 9$

$x - 3 = 0$     $x + 3 = 0$

$x = 3$       $x = -3$

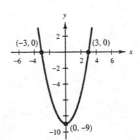

**44.** $f(x) = -x^2 + 9$

x-intercepts       vertex

$0 = -x^2 + 9$     $f(x) = -(x - 0)^2 + 9$

$x^2 = 9$

$x = \pm 3$

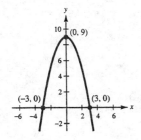

**46.** $g(x) = x^2 - 4x$

x-intercepts       vertex

$0 = x^2 - 4x$       $y = (x^2 - 4x + 4) - 4$

$0 = x(x - 4)$       $= (x - 2)^2 - 4$

$x = 0$     $x - 4 = 0$

$x = 4$

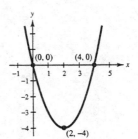

**48.** $y = -x^2 + 4x$

x-intercepts           vertex

$0 = -x^2 + 4x$        $y = -x^2 + 4x$

$0 = -x(x - 4)$       $= -(x^2 - 4x)$

$-x = 0$     $x - 4 = 0$     $= -(x^2 - 4x + 4 - 4)$

$x = 0$       $x = 4$       $= -(x^2 - 4x + 4) + 4$

                     $= -(x - 2)^2 + 4$

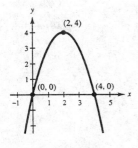

**50.** $y = -(x + 4)^2$

x-intercepts       vertex

$0 = -(x + 4)^2$     $f(x) = -(x + 4)^2$

$x + 4 = 0$

$x = -4$

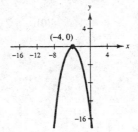

**52.** $y = x^2 + 4x + 2$

*x*-intercepts

$0 = x^2 + 4x + 2$

$x = \dfrac{-4 \pm \sqrt{4^2 - 4(1)(2)}}{2(1)}$

$x = \dfrac{-4 \pm \sqrt{16 - 8}}{2}$

$x = \dfrac{-4 \pm \sqrt{8}}{2}$

$x = \dfrac{-4 \pm 2\sqrt{2}}{2}$

$x = -2 \pm \sqrt{2}$

vertex

$y = x^2 + 4x + 2$

$\quad = (x^2 + 4x + 4) + 2 - 4$

$\quad = (x + 2)^2 - 2$

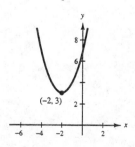

**54.** $y = -x^2 + 2x + 8$

*x*-intercepts

$0 = -x^2 + 2x + 8$

$0 = -(x^2 - 2x - 8)$

$0 = -(x - 4)(x + 2)$

$-1 \neq 0 \quad x - 4 = 0 \quad x + 2 = 0$

$\qquad\qquad\qquad x = 4 \qquad x = -2$

vertex

$y = -(x^2 - 2x) + 8$

$\quad = -(x^2 - 2x + 1 - 1) + 8$

$\quad = -(x^2 - 2x + 1) + 1 + 8$

$\quad = -(x - 1)^2 + 9$

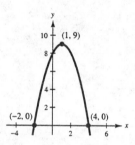

**56.** $f(x) = x^2 + 4x + 7$

vertex

$f(x) = (x^2 + 4x + 4) + 7 - 4$

$f(x) = (x + 2)^2 + 3$

*x*-intercepts

$0 = x^2 + 4x + 7$

$x = \dfrac{-4 \pm \sqrt{4^2 - 4(1)(7)}}{2(1)}$

$x = \dfrac{-4 \pm \sqrt{16 - 28}}{2}$

$x = \dfrac{-4 \pm \sqrt{-12}}{2}$

no *x*-intercepts

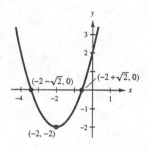

**58.** $y = 3x^2 - 6x + 4$

*x*-intercepts

$0 = 3x^2 - 6x + 4$

$x = \dfrac{6 \pm \sqrt{(-6)^2 - 4(3)(4)}}{2(3)}$

$x = \dfrac{6 \pm \sqrt{36 - 48}}{6}$

$x = \dfrac{6 \pm \sqrt{-12}}{6}$

no *x*-intercepts

vertex

$y = 3x^2 - 6x + 4$

$\quad = 3(x^2 - 2x) + 4$

$\quad = 3(x^2 - 2x + 1 - 1) + 4$

$\quad = 3(x^2 - 2x + 1) - 3 + 4$

$\quad = 3(x - 1)^2 + 1$

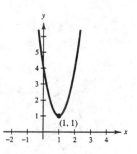

**60.** $y = -\frac{1}{2}(x^2 - 6x + 7)$

$x$-intercepts

$0 = -\frac{1}{2}(x^2 - 6x + 7)$

$x = \frac{6 \pm \sqrt{(-6)^2 - 4(1)(7)}}{2(1)}$

$x = \frac{6 \pm \sqrt{36 - 28}}{2}$

$x = \frac{6 \pm \sqrt{8}}{2}$

$x = \frac{6 \pm 2\sqrt{2}}{2}$

$x = 3 \pm \sqrt{2}$

vertex

$y = -\frac{1}{2}(x^2 - 6x + 7)$

$= -\frac{1}{2}(x^2 - 6x + 9 - 9 + 7)$

$= -\frac{1}{2}(x^2 - 6x + 9) + \frac{9}{2} - \frac{7}{2}$

$= -\frac{1}{2}(x - 3)^2 + 1$

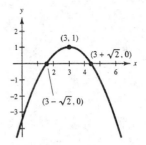

**62.** $y = \frac{1}{5}(2x^2 - 4x + 7)$

$x$-intercepts

$0 = \frac{1}{5}(2x^2 - 4x + 7)$

$x = \frac{4 \pm \sqrt{16 - 4(2)(7)}}{2(2)}$

$x = \frac{4 \pm \sqrt{16 - 56}}{4}$

$x = \frac{4 \pm \sqrt{-40}}{4}$

no $x$-intercepts

vertex

$y = \frac{1}{5}(2x^2 - 4x + 7)$

$= \frac{2}{5}(x^2 - 2x) + \frac{7}{5}$

$= \frac{2}{5}(x^2 - 2x + 1 - 1) + \frac{7}{5}$

$= \frac{2}{5}(x^2 - 2x + 1) - \frac{2}{5} + \frac{7}{5}$

$= \frac{2}{5}(x - 1)^2 + 1$

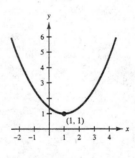

**64.** $f(x) = \frac{1}{3}x^2 - 2$

$x$-intercepts

$0 = \frac{1}{3}x^2 - 2$

$2 = \frac{x^2}{3}$

$6 = x^2$

$\pm\sqrt{6} = x$

vertex

$f(x) = \frac{1}{3}(x^2) - 2$

$f(x) = \frac{1}{3}(x - 0)^2 - 2$

**66.** $h(x) = x^2 - 4$

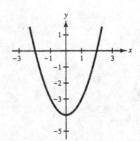

Vertical shift 4 units down

**68.** $h(x) = (x - 4)^2$

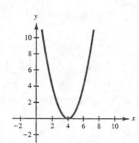

Horizontal shift 4 units right

**70.** $h(x) = (x + 2)^2 - 1$

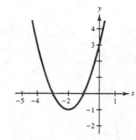

Horizontal shift 2 units left
Vertical shift 1 unit down

**72.** $h(x) = (x - 3)^2 - 2$

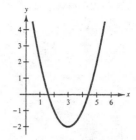

Horizontal shift 3 units right
Vertical shift 2 units down

**74.** $y = -\dfrac{1}{4}(4x^2 - 20x + 13)$

*Keystrokes:*

$\boxed{Y=}$ $\boxed{(}$ $\boxed{(-)}$ 1 $\boxed{\div}$ 4 $\boxed{)}$ $\boxed{(}$ 4 $\boxed{X,T,\theta}$ $\boxed{x^2}$ $\boxed{-}$ 20 $\boxed{X,T,\theta}$ $\boxed{+}$ 13 $\boxed{)}$ $\boxed{GRAPH}$

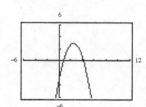

vertex $= \left(\dfrac{5}{2}, 3\right)$

**Check:**

$y = -x^2 + 5x - \dfrac{13}{4}$

$a = -1 \quad b = 5$

$x = \dfrac{-b}{2a} = \dfrac{-5}{2(-1)} = \dfrac{5}{2}$

$y = -\left(\dfrac{5}{2}\right)^2 + 5\left(\dfrac{5}{2}\right) - \dfrac{13}{4}$

$= -\dfrac{25}{4} + \dfrac{25}{2} - \dfrac{13}{4} = \dfrac{-25 + 50 - 13}{4} = \dfrac{12}{4} = 3$

**76.** $y = 0.75x^2 - 7.50x + 23.00$

*Keystrokes:*

$\boxed{Y=}$ .75 $\boxed{X,T,\theta}$ $\boxed{x^2}$ $\boxed{-}$ 7.5 $\boxed{X,T,\theta}$ $\boxed{+}$ 23 $\boxed{GRAPH}$

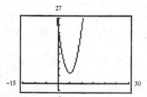

vertex $= (5, 4.25)$

**Check:**

$y = 0.75x^2 - 7.50x + 23$

$a = .75 \quad b = -7.5$

$x = \dfrac{-b}{2a} = \dfrac{-(-7.5)}{2(.75)} = 5$

$y = 0.75(5)^2 - 7.5(5) + 23$

$= 18.75 - 37.5 + 23$

$= 4.25$

**78.** vertex $= (2, 0)$   point $= (0, 4)$

$4 = a(0 - 2)^2 + 0$         $y = 1(x - 2)^2 + 0$

$4 = a(4)$              $y = (x - 2)^2$

$1 = a$               $= x^2 - 4x + 4$

**80.** vertex $= (-2, 4)$   point $= (0, 0)$

$0 = a[0 - (-2)]^2 + 4$    $y = -1[x - (-2)]^2 + 4$

$0 = a(4) + 4$          $y = -(x + 2)^2 + 4$

$-4 = a(4)$            $= -x^2 - 4x - 4 + 4$

$-1 = a$              $= -x^2 - 4x$

**82.** vertex $= (-3, -2)$   point $= (0, 2)$

$2 = a(0 + 3)^2 - 2$       $y = \frac{4}{9}(x + 3)^2 - 2$

$4 = a(9)$              $= \frac{4}{9}(x^2 + 6x + 9) - 2$

$\frac{4}{9} = a$             $= \frac{4}{9}x^2 + \frac{8}{3}x + 4 - 2$

$y = \frac{4}{9}x^2 + \frac{8}{3}x + 2$

**84.** vertex $= (-3, -3)$; $a = 1$

$y = a(x - h)^2 + k$

$y = 1(x - (-3))^2 + (-3)$

$y = (x + 3)^2 - 3$

$y = x^2 + 2x(3) + 3^2 - 3$

$y = x^2 + 6x + 9 - 3$

$y = x^2 + 6x + 6$

**86.** vertex $= (-2, -4)$;   point $= (0, 0)$

$y = a(x - h)^2 + k$

$y = a(x - (-2))^2 + (-4)$

$y = a(x + 2)^2 - 4$

$0 = a(0 + 2)^2 - 4$

$0 = 4a - 4$

$4 = 4a$

$a = 1$

$y = (x + 2)^2 - 4$

$y = x^2 + 4x + 4 - 4$

$y = x^2 + 4x$

**88.** vertex $= (-1, -1)$;   point $= (0, 4)$

$y = a(x - h)^2 + k$

$y = a(x - (-1))^2 + (-1)$

$y = a(x + 1)^2 - 1$

$4 = a(0 + 1)^2 - 1$

$4 = a - 1$

$5 = a$

$y = 5(x + 1)^2 - 1$

$y = 5(x^2 + 2x + 1) - 1$

$y = 5x^2 + 10x + 5 - 1$

$y = 5x^2 + 10x + 4$

**90.** vertex $= (5, 2)$;   point $= (10, 3)$

$y = a(x - h)^2 + k$         $y = \frac{1}{25}(x - 5)^2 + 2$

$y = a(x - 5)^2 + 2$         $y = \frac{1}{25}(x^2 - 10x + 25) + 2$

$3 = a(10 - 5)^2 + 2$        $y = \frac{1}{25}x^2 - \frac{2}{5}x + 1 + 2$

$3 = a(25) + 2$           $y = \frac{1}{25}x^2 - \frac{2}{5}x + 3$

$1 = a(25)$

$\frac{1}{25} = a$

**92.** Vertical shift 1 unit up

**94.** Horizontal shift 4 units left

Vertical shift 1 unit down

**96.** $y = -\dfrac{1}{16}x^2 + 2x + 5$

(a) $y = \dfrac{-1}{16}(0)^2 + 2(0) + 5$

  $= 5$ feet

(b) $y = \dfrac{-1}{16}(x^2 - 32x + 256) + 5 + 16$

  $= \dfrac{-1}{16}(x - 16)^2 + 21$

  Maximum height $= 21$ feet

(c) $0 = \dfrac{-1}{16}x^2 + 2x + 5$

  $0 = x^2 - 32x - 80$

  $x = \dfrac{-(-32) \pm \sqrt{(-32)^2 - 4(1)(-80)}}{2(1)}$

  $x = \dfrac{32 \pm \sqrt{1024 + 320}}{2} = \dfrac{32 \pm \sqrt{1344}}{2}$

  $= \dfrac{32 \pm 8\sqrt{21}}{2} = 16 \pm 4\sqrt{21} \approx 34.3$ feet

**98.** $y = \dfrac{-4}{3}x^2 + \dfrac{10}{3}x + 10$

$y = \dfrac{-4}{3}\left(x^2 - \dfrac{5}{2}x + \dfrac{25}{16}\right) + 10 + \dfrac{25}{12}$

$y = \dfrac{-4}{3}\left(x - \dfrac{5}{4}\right)^2 + \dfrac{145}{12}$

Maximum height $= \dfrac{145}{12} \approx 12.08$ feet

**100.** $P = 230 + 20s - \frac{1}{2}s^2$

*Keystrokes:*

$\boxed{Y=}$ 230 $\boxed{+}$ 20 $\boxed{X,T,\theta}$ $\boxed{-}$ .5 $\boxed{X,T,\theta}$ $\boxed{x^2}$ $\boxed{GRAPH}$

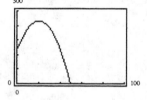

$P = 230 + 20s - \frac{1}{2}s^2$

$P = -\frac{1}{2}s^2 + 20s + 230$

$P = -\frac{1}{2}(s^2 - 40s + 400) + 230 + 200$

$P = -\frac{1}{2}(s - 20)^2 + 430$

Amount of advertising that yields a maximum profit $= \$2000$

Maximum profit $= \$430,000$

**102.** $C = 800 - 10x + \frac{1}{4}x^2, \; 0 < x < 40$

*Keystrokes:*

$\boxed{Y=}$ 800 $\boxed{-}$ 10 $\boxed{X,T,\theta}$ $\boxed{+}$ $\boxed{(}$ 1 $\boxed{\div}$ 4 $\boxed{)}$ $\boxed{X,T,\theta}$ $\boxed{x^2}$ $\boxed{GRAPH}$

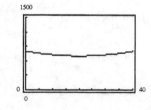

$x = 19.789474$ when $C = 700.01108$

$x \approx 20$ units

**104.** (a)
$$y = 4\left(100 - \frac{x^2}{2500}\right)$$

$$0 = 4\left(100 - \frac{x^2}{2500}\right)$$

$$100 - \frac{x^2}{2500} = 0$$

$$100 = \frac{x^2}{2500}$$

$$250,000 = x^2$$

$$\pm 500 = x$$

Thus, the length of the road across the gorge is 1000 ft.

(c). Length of vertical girders $= 400 - y$

$$100 \text{ ft. from center} = 400 - 4\left(100 - \frac{100^2}{2500}\right)$$

$$= 400 - 4(100 - 4)$$

$$= 400 - 400 + 16$$

$$= 16 \text{ ft.}$$

$$200 \text{ ft. from center} = 400 - 4\left(100 - \frac{200^2}{2500}\right)$$

$$= 400 - 4(100 - 16)$$

$$= 400 - 400 + 64$$

$$= 64 \text{ ft.}$$

$$300 \text{ ft. from center} = 400 - 4\left(100 - \frac{300^2}{2500}\right)$$

$$= 400 - 4(100 - 36)$$

$$= 400 - 400 + 144$$

$$= 144 \text{ ft.}$$

(b) $y = 4\left(100 - \frac{x^2}{2500}\right)$

$$y = 400 - \frac{x^2}{625}$$

$$x = \frac{-0}{2\left(-\frac{1}{625}\right)} = 0$$

$$y = 400 - \frac{0}{625} = 400$$

Thus, the height of the parabolic arch at the center of the span is 400 ft.

| $x$ | $\pm 100$ | $\pm 200$ | $\pm 300$ | $\pm 400$ | $\pm 500$ |
|-----|-----------|-----------|-----------|-----------|-----------|
| $y$ | 16 | 64 | 144 | 256 | 400 |

$$400 \text{ ft. from center} = 400 - 4\left(100 - \frac{400^2}{2500}\right)$$

$$= 400 - 4(100 - 64)$$

$$= 400 - 400 + 256$$

$$= 256 \text{ ft.}$$

$$500 \text{ ft. from center} = 400 - 4\left(100 - \frac{500^2}{2500}\right)$$

$$= 400 - 4(100 - 100)$$

$$= 400 - 400 + 400$$

$$= 400 \text{ ft.}$$

**106.** (a) $y = 1.11x^2 - 1.69x + 50.5$

$1.11 > 0$ so the graph opens upward.

$$x = \frac{-(-1.69)}{2(1.11)} = 0.76 \qquad y = 1.11(0.76)^2 - 1.69(0.76) + 50.5 = 49.86$$

vertex $= (0.76, 49.86)$;   axis: $x = 0.76$

The vertex is a minimum value since the graph opens upward.

(b) *Keystrokes:*

$\boxed{\text{Y=}}$ 1.11 $\boxed{\text{X,T,}\theta}$ $\boxed{x^2}$ $\boxed{-}$ 1.69 $\boxed{\text{X,T,}\theta}$ $\boxed{+}$ 50.5 $\boxed{\text{ENTER}}$

Plot $(1, 50.1)$, $(2, 51.9)$, $(3, 54.8)$, $(4, 59.3)$, $(5, 73.6)$, $(6, 78.7)$

in

$\boxed{\text{STAT}}$ 1 then enter 1, 2, 3, 4, 5, 6, in $L$, and enter 50.1, 51.9, 54.8, 59.3, 73.6, 78.7, in $L_2$.

$\boxed{\text{STAT PLOT}}$ 1 $\boxed{\text{ON}}$ $\boxed{\text{GRAPH}}$

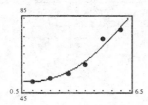

The model appears to be accurate for the restricted domain.

**108.** To find the vertex of the graph of a quadratic function use the method of completing the square to write the function in standard form $f(x) = a(x - h)^2 + k$. The vertex is located at point $(h, k)$.

**110.** The graph of a quadratic function $f(x) = ax^2 + bx + c$ opens upward if $a > 0$ and opens downward if $a < 0$.

**112.** It is not possible for the graph of a quadratic function to have two $y$-intercepts. All functions must have only one $y$ value for each $x$ value.

# Section 7.4    Conic Sections

**2.** $4x^2 + 4y^2 = 9$; (f)

**4.** $\dfrac{x^2}{9} + \dfrac{y^2}{4} = 1$; (b)

**6.** $x^2 - y^2 = -4$; (d)

**8.** Center: $(0, 0)$

Radius: $7$

$x^2 + y^2 = r^2$

$x^2 + y^2 = 7^2$

$x^2 + y^2 = 49$

**10.** Center: $(0, 0)$

Radius: $\frac{5}{2}$

$x^2 + y^2 = r^2$

$x^2 + y^2 = \left(\frac{5}{2}\right)^2$

$x^2 + y^2 = \frac{25}{4}$

**12.** $r = \sqrt{(-2 - 0)^2 + (0 - 0)^2}$

$r = \sqrt{4 + 0}$

$r = \sqrt{4}$

$r = 2$

$x^2 + y^2 = r^2$

$x^2 + y^2 = 2^2$

$x^2 + y^2 = 4$

**14.** $r = \sqrt{(-1 - 0)^2 + (-4 - 0)^2}$

$r = \sqrt{1 + 16}$

$r = \sqrt{17}$

$x^2 + y^2 = r^2$

$x^2 + y^2 = \left(\sqrt{17}\right)^2$

$x^2 + y^2 = 17$

**16.** Center: $(-2, 5)$

Radius: 6

$(x - h)^2 + (y - k)^2 = r^2$

$(x - (-2))^2 + (y - 5)^2 = 6^2$

$(x + 2)^2 + (y - 5)^2 = 36$

**18.** Center: $(-5, -2)$

Radius: $\frac{5}{2}$

$(x - h)^2 + (y - k)^2 = r^2$

$(x - (-5))^2 + (y - (-2))^2 = \left(\frac{5}{2}\right)^2$

$(x + 5)^2 + (y + 2)^2 = \frac{25}{4}$

**20.** Center: $(8, 2)$

Point: $(8, 0)$

$r = \sqrt{(8 - 8)^2 + (0 - 2)^2}$

$r = \sqrt{0 + 4}$

$r = 2$

$(x - h)^2 + (y - k)^2 = r^2$

$(x - 8)^2 + (y - 2)^2 = 2^2$

$(x - 8)^2 + (y - 2)^2 = 4$

**22.** Center: $(-3, -5)$

Point: $(0, 0)$

$r = \sqrt{(0 - (-3))^2 + (0 - (-5))^2}$

$r = \sqrt{9 + 25}$

$r = \sqrt{34}$

$(x - h)^2 + (y - k)^2 = r^2$

$(x - (-3))^2 + (y - (-5))^2 = \left(\sqrt{34}\right)^2$

$(x + 3)^2 + (y + 5)^2 = 34$

**24.** $x^2 + y^2 = 25$

Center: $(0, 0)$

Radius: 5

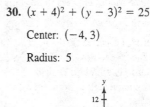

**26.** $x^2 + y^2 = 10$

Center: $(0, 0)$

Radius: $\sqrt{10}$

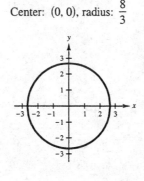

**28.** $9x^2 + 9y^2 = 64$

$\dfrac{9x^2}{9} + \dfrac{9y^2}{9} = \dfrac{64}{9}$

$x^2 + y^2 = \dfrac{64}{9}$

Center: $(0, 0)$, radius: $\dfrac{8}{3}$

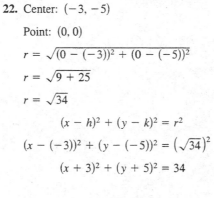

**30.** $(x + 4)^2 + (y - 3)^2 = 25$

Center: $(-4, 3)$

Radius: 5

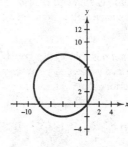

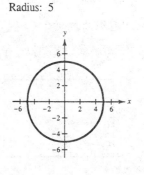

**32.** $(x - 5)^2 + \left(y + \frac{3}{4}\right)^2 = 1$

Center: $\left(5, -\frac{3}{4}\right)$

Radius: 1

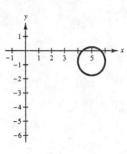

**34.**
$$x^2 + y^2 + 6x - 4y - 3 = 0$$
$$x^2 + 6x + y^2 - 4y = 3$$
$$(x^2 + 6x + 9) + (y^2 - 4y + 4) = 3 + 9 + 4$$
$$(x + 3)^2 + (y - 2)^2 = 16$$

Center: $(-3, 2)$

Radius: 4

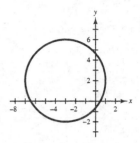

**36.**
$$x^2 + y^2 - 2x + 6y - 15 = 0$$
$$x^2 - 2x + y^2 + 6y = 15$$
$$(x^2 - 2x + 1) + (y^2 + 6y + 9) = 15 + 1 + 9$$
$$(x - 1)^2 + (y + 3)^2 = 25$$

Center: $(1, -3)$

Radius: 5

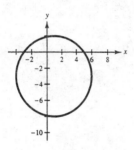

**38.** $4x^2 + 4y^2 = 45$
$$4y^2 = 45 - 4x^2$$
$$y^2 = \frac{45}{4} - x^2$$
$$y = \pm\sqrt{\frac{45}{4} - x^2}$$

Keystrokes:

$y_1$: [Y=] [√] [(] 45 [÷] 4 [−] [X,T,θ] [x²] [)] [ENTER]

$y_2$: [(−)] [√] [(] 45 [÷] 4 [−] [X,T,θ] [x²] [)] [GRAPH]

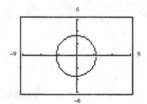

**40.** $(x + 3)^2 + y^2 = 15$
$$y^2 = 15 - (x + 3)^2$$
$$y = \pm\sqrt{15 - (x + 3)^2}$$

Keystrokes:

$y_1$ [Y=] [√] [(] 15 [−] [(] [X,T,θ] [+] 3 [)] [x²] [)] [ENTER]

$y_2$ [(−)] [√] [(] 15 [−] [(] [X,T,θ] [+] 3 [)] [x²] [)] [GRAPH]

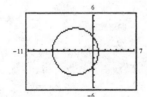

**42.** Center: $(0, 0)$

Vertices: $(\pm 4, 0)$

Co-vertices: $(0, \pm 1)$

$$\frac{x^2}{a^2} + \frac{y^2}{b^2} = 1$$

Major axis is $x$-axis so $a = 4$.

Minor axis is $y$-axis so $b = 1$.

$$\frac{x^2}{4^2} + \frac{y^2}{1^2} = 1$$

$$\frac{x^2}{16} + \frac{y^2}{1} = 1$$

**44.** Center: $(0, 0)$

Vertices: $(\pm 10, 0)$

Co-vertices: $(0, \pm 4)$

$$\frac{x^2}{a^2} + \frac{y^2}{b^2} = 1$$

Major axis is $x$-axis so $a = 10$.

Minor axis is $y$-axis so $b = 4$.

$$\frac{x^2}{10^2} + \frac{y^2}{4^2} = 1$$

$$\frac{x^2}{100} + \frac{y^2}{16} = 1$$

**46.** Center: $(0, 0)$

Vertices: $(0, \pm 5)$

Co-vertices: $(\pm 1, 0)$

$$\frac{x^2}{a^2} + \frac{y^2}{b^2} = 1$$

Major axis is $y$-axis so $a = 5$.

Minor axis is $x$-axis so $b = 1$.

$$\frac{x^2}{1^2} + \frac{y^2}{5^2} = 1$$

$$\frac{x^2}{1} + \frac{y^2}{25} = 1$$

**48.** Center: $(0, 0)$

Vertices: $(0, \pm 8)$

Co-vertices: $(\pm 4, 0)$

$$\frac{x^2}{b^2} + \frac{y^2}{a^2} = 1$$

Major axis is $y$-axis so $a = 8$.

Minor axis is $x$-axis so $b = 4$.

$$\frac{x^2}{4^2} + \frac{y^2}{8^2} = 1$$

$$\frac{x^2}{16} + \frac{y^2}{64} = 1$$

**50.** Center: $(0, 0)$

Major axis (horizontal) 24 units

Minor axis 10 units

$$\frac{x^2}{a^2} + \frac{y^2}{b^2} = 1$$

$a = 12, b = 5$

$$\frac{x^2}{12^2} + \frac{y^2}{5^2} = 1$$

$$\frac{x^2}{144} + \frac{y^2}{25} = 1$$

**52.** Center: $(0, 0)$

Major axis (horizontal) 50 units

Minor axis 30 units

$$\frac{x^2}{a^2} + \frac{y^2}{b^2} = 1$$

$a = 25, b = 15$

$$\frac{x^2}{25^2} + \frac{y^2}{15^2} = 1$$

$$\frac{x^2}{625} + \frac{y^2}{225} = 1$$

**54.** $\dfrac{x^2}{25} + \dfrac{y^2}{9} = 1$

Vertices: $(\pm 5, 0)$

Co-vertices: $(0, \pm 3)$

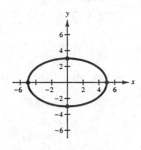

**56.** $\dfrac{x^2}{9} + \dfrac{y^2}{25} = 1$

Vertices: $(0, \pm 5)$

Co-vertices: $(\pm 3, 0)$

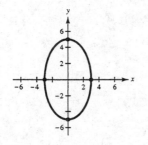

**58.** $\dfrac{x^2}{1} + \dfrac{y^2}{1/4} = 1$

Vertices: $(\pm 1, 0)$

Co-vertices: $\left(0, \pm \frac{1}{2}\right)$

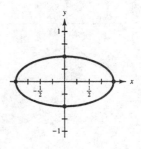

**60.** $4x^2 + 9y^2 - 36 = 0$

$$4x^2 + 9y^2 = 36$$

$$\frac{4x^2}{36} + \frac{9y^2}{36} = \frac{36}{36}$$

$$\frac{x^2}{9} + \frac{y^2}{4} = 1$$

Vertices: $(\pm 3, 0)$

Co-vertices: $(0, \pm 2)$

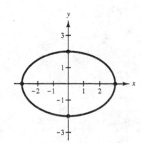

**62.** $16x^2 + 4y^2 - 64 = 0$

$$\frac{16x^2}{64} + \frac{4y^2}{64} = \frac{64}{64}$$

$$\frac{x^2}{4} + \frac{y^2}{16} = 1$$

Vertices: $(0, \pm 4)$

Co-vertices: $(\pm 2, 0)$

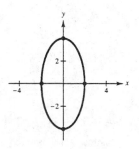

**64.** $9x^2 + y^2 = 64$      Vertices: $(0, \pm 8)$

$$y^2 = 64 - 9x^2$$

$$y = \pm\sqrt{64 - 9x^2}$$

*Keystrokes:*

$y_1$ ⃞Y= ⃞√ ⃞( 64 ⃞− 9 ⃞X,T,θ ⃞x² ⃞) ⃞ENTER

$y_2$ ⃞(-) ⃞√ ⃞( 64 ⃞− 9 ⃞X,T,θ ⃞x² ⃞) ⃞GRAPH

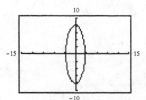

**66.** $5x^2 + 2y^2 - 10 = 0$      Vertices: $\left(0, \pm\sqrt{5}\right)$

$$2y^2 = 10 - 5x^2$$

$$y^2 = \frac{10 - 5x^2}{2}$$

$$y = \pm\sqrt{\frac{10 - 5x^2}{2}}$$

*Keystrokes:*

$y_1$ ⃞Y= ⃞√ ⃞( ⃞( 10 ⃞− 5 ⃞X,T,θ ⃞x² ⃞) ⃞÷ 2 ⃞) ⃞ENTER

$y_2$ ⃞(-) ⃞√ ⃞( ⃞( 10 ⃞− 5 ⃞X,T,θ ⃞x² ⃞) ⃞÷ 2 ⃞) ⃞GRAPH

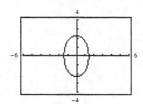

**68.** $y^2 - x^2 = 9$

Vertices: $(0, \pm 3)$

Asymptotes: $y = \pm x$

Equation: $y^2 - x^2 = 9$

$$\frac{y^2}{9} - \frac{x^2}{9} = 1$$

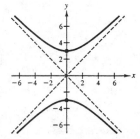

**70.** $x^2 - y^2 = 1$

Vertices: $(\pm 1, 0)$

Asymptotes: $y = \pm x$

Equation: $x^2 - y^2 = 1$

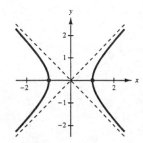

**72.** $\dfrac{y^2}{9} - \dfrac{x^2}{25} = 1$

Vertices: $(0, \pm 3)$

Asymptotes: $y = \pm\dfrac{3}{5}x$

Equation: $\dfrac{y^2}{9} - \dfrac{x^2}{25} = 1$

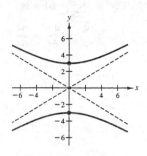

**74.** $\dfrac{y^2}{4} - \dfrac{x^2}{9} = 1$

Vertices: $(\pm 2, 0)$

Asymptotes: $y = \pm\dfrac{3}{2}x$

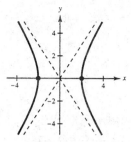

**76.** $\dfrac{y^2}{\frac{1}{4}} - \dfrac{x^2}{\frac{25}{4}} = 1$

Vertices: $\left(0, \pm\dfrac{1}{2}\right)$

Asymptotes:

$y = \pm\dfrac{\frac{1}{2}}{\frac{5}{2}}x$

$y = \pm\dfrac{1}{5}x$

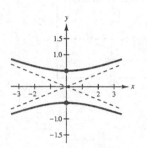

**78.** $4y^2 - 9x^2 - 36 = 0$

$\qquad 4y^2 - 9x^2 = 36$

$\qquad \dfrac{4y^2}{36} - \dfrac{9x^2}{36} = \dfrac{36}{36}$

$\qquad \dfrac{y^2}{9} - \dfrac{x^2}{4} = 1$

Vertices: $(0, \pm 3)$

Asymptotes: $y = \pm\dfrac{3}{2}x$

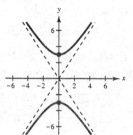

**80.** Vertices: $(\pm 2, 0)$

Asymptotes: $y = \pm\dfrac{1}{3}x$

$\dfrac{x^2}{4} - \dfrac{y^2}{\left(\frac{2}{3}\right)^2} = 1$

$\dfrac{x^2}{4} - \dfrac{9y^2}{4} = 1$

**82.** Vertices: $(0, \pm 2)$

Asymptotes: $y = \pm 3x$

$\dfrac{y^2}{2^2} - \dfrac{x^2}{\left(\frac{2}{3}\right)^2} = 1$

$\dfrac{y^2}{4} - \dfrac{9x^2}{4} = 1$

**84.** Vertices: $(0, \pm 5)$

Asymptotes: $y = \pm x$

$\dfrac{y^2}{5^2} - \dfrac{x^2}{5^2} = 1$

$\dfrac{y^2}{25} - \dfrac{x^2}{25} = 1$

**86.** Vertices: $(\pm 1, 0)$

Asymptotes: $y = \pm\dfrac{1}{2}x$

$\dfrac{x^2}{1^2} - \dfrac{y^2}{\left(\frac{1}{2}\right)^2} = 1$

$\dfrac{x^2}{1} - \dfrac{y^2}{\frac{1}{4}} = 1$

**88.** $\dfrac{y^2}{16} - \dfrac{x^2}{4} = 1$

$y^2 - 4x^2 = 16$

$\qquad y^2 = 16 + 4x^2$

$\qquad\quad y = \pm\sqrt{16 + 4x^2}$

Keystrokes:

$y_1$ [Y=] [√] [(] 16 [+] 4 [X,T,θ] [x²] [)] [ENTER]

$y_2$ [(−)] [√] [(] 16 [+] 4 [X,T,θ] [x²] [)] [GRAPH]

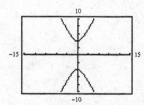

**90.** $x^2 - 2y^2 - 4 = 0$

$$-2y^2 = 4 - x^2$$

$$y^2 = \frac{4 - x^2}{-2}$$

$$y = \pm\sqrt{\frac{x^2 - 4}{2}}$$

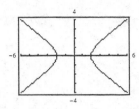

*Keystrokes:*

$y_1$ [Y=] [√] [(] [(] [(] [X,T,θ] [x²] [−] 4 [)] [÷] 2 [)] [ENTER]

$y_2$ [(−)] [√] [(] [(] [(] [X,T,θ] [x²] [−] 4 [)] [÷] 2 [)] [GRAPH]

**92.** Line

$$y = 10 - \tfrac{3}{2}x$$

**94.** Circle

$$4x^2 + 4y^2 = 36$$

**96.** Parabola

$$x^2 - 4y + 2x = 0$$

**98.** Circle

$$2x^2 + 2y^2 = 9$$

**100.** Hyperbola

$$y^2 = x^2 + 2$$

**102.** $x^2 + y^2 = r^2$

$$x^2 + y^2 = 12^2$$

$$x^2 + y^2 = 144$$

$$6^2 + y^2 = 144$$

$$y^2 = 108$$

$$y = \sqrt{108}$$

$$y \approx 10.4 \text{ feet}$$

**104.** $x^2 + y^2 = r^2$

$$x^2 + y^2 = 50^2$$

$$x^2 + y^2 = 2500$$

$$45^2 + y^2 = 2500$$

$$y^2 = 475$$

$$y = 5\sqrt{19}$$

$$y \approx 21.8 \text{ feet}$$

**106.** $\dfrac{x^2}{a^2} + \dfrac{y^2}{b^2} = 1$ or $\dfrac{x^2}{b^2} + \dfrac{y^2}{a^2} = 1$

$a = 4, b = \dfrac{15}{4}$     $b = 4, a = \dfrac{15}{4}$

$\dfrac{x^2}{4^2} + \dfrac{y^2}{\left(\frac{15}{4}\right)^2} = 1$     $\dfrac{x^2}{\left(\frac{15}{4}\right)^2} + \dfrac{y^2}{4^2} = 1$

$\dfrac{x^2}{16} + \dfrac{16y^2}{225} = 1$     $\dfrac{16x^2}{225} + \dfrac{y^2}{16} = 1$

**108.** (a)          (b)

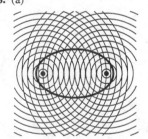
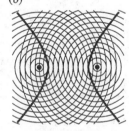

**110.** A circle is the set of all points $(x, y)$ that are a given positive distance $r$ from a fixed point $(h, k)$ called the center.

$$(x - h)^2 + (y - k)^2 = r^2$$

**112.** A hyperbola is the set of all points $(x, y)$ such that the difference of the distances between $(x, y)$ and two distinct fixed points is a constant.

$$\frac{x^2}{a^2} - \frac{y^2}{b^2} = 1 \quad \text{or} \quad \frac{y^2}{a^2} - \frac{x^2}{b^2} = 1$$

**114.** The lengths of the axes of an ellipse are $2a$ for the major axis and $2b$ for the minor axis.

**116.** $x = -\dfrac{2}{3}\sqrt{9 + y^2}$ is the left half of the hyperbola

$$\frac{x^2}{4} - \frac{y^2}{9} = 1.$$

# Section 7.5    Graphs of Rational Functions

**2.** (a)

| $x$ | 0 | 0.5 | 0.9 | 0.99 | 0.999 |
|---|---|---|---|---|---|
| $y$ | 0 | $-2$ | $-18$ | $-198$ | $-1998$ |

| $x$ | 2 | 1.5 | 1.1 | 1.01 | 1.001 |
|---|---|---|---|---|---|
| $y$ | 4 | 6 | 22 | 202 | 2002 |

| $x$ | 2 | 5 | 10 | 100 | 1000 |
|---|---|---|---|---|---|
| $y$ | 4 | 2.5 | 2.222 | 2.0202 | 2.0020 |

(b)

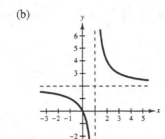

(c) Domain: $x - 1 \neq 0$

$$x \neq 1$$

$$(-\infty, 1) \cup (1, \infty)$$

**4.** (a)

| $x$ | 2 | 2.5 | 2.9 | 2.99 | 2.999 |
|---|---|---|---|---|---|
| $y$ | $-2$ | $-4$ | $-20$ | $-200$ | $-2000$ |

| $x$ | 4 | 3.5 | 3.1 | 3.01 | 3.001 |
|---|---|---|---|---|---|
| $y$ | 2 | 4 | 20 | 200 | 2000 |

| $x$ | 4 | 5 | 10 | 100 | 1000 |
|---|---|---|---|---|---|
| $y$ | 2 | 1 | 0.286 | 0.021 | 0.002 |

(b)

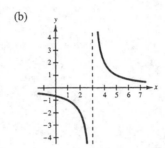

(c) Domain: $x - 3 \neq 0$

$$x \neq 3$$

$$(-\infty, 3) \cup (3, \infty)$$

**6.** (a)

| $x$ | 2 | 2.5 | 2.9 | 2.99 | 2.999 |
|---|---|---|---|---|---|
| $y$ | $-4$ | $-11.364$ | $-71.271$ | $-746.25$ | $-7496.3$ |

| $x$ | 4 | 3.5 | 3.1 | 3.01 | 3.001 |
|---|---|---|---|---|---|
| $y$ | 11.429 | 18.846 | 78.770 | 753.75 | 7503.8 |

| $x$ | 4 | 5 | 10 | 100 | 1000 |
|---|---|---|---|---|---|
| $y$ | 11.429 | 7.813 | 5.495 | 5.005 | 5 |

(b)

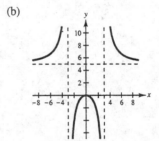

(c) Domain:    $x^2 - 9 \neq 0$

$$(x + 3)(x - 3) \neq 0$$

$$x \neq -3 \quad x \neq 3$$

$$(-\infty, -3) \cup (-3, 3) \cup (3, \infty)$$

**8.** $g(x) = \dfrac{3}{x}$

Domain: $x \neq 0$

$$(-\infty, 0) \cup (0, \infty)$$

Vertical asymptote: $x = 0$

Horizontal asymptote: $y = 0$ since the degree of the numerator is less than the degree of the denominator.

**10.** $f(u) = \dfrac{u^2}{u - 10}$

Domain: $u - 10 \neq 0$

$$u \neq 10$$

$$(-\infty, 10) \cup (10, \infty)$$

Vertical asymptote: $u - 10 = 0$

$$u = 10$$

Horizontal asymptote: none since the degree of the numerator is greater than the degree of the denominator.

**12.** $h(x) = \dfrac{4x - 3}{2x + 5}$

Domain: $2x + 5 \neq 0$

$$x \neq -\frac{5}{2}$$

$$\left(-\infty, -\frac{5}{2}\right) \cup \left(-\frac{5}{2}, \infty\right)$$

Vertical asymptote: $2x + 5 = 0$

$$x = -\frac{5}{2}$$

Horizontal asymptote: $y = 2$ since the degree of the numerator is equal to the degree of the denominator and the leading coefficient of the numerator is 4 and the leading coefficient of the denominator is 2.

**16.** $h(s) = \dfrac{2s + 1}{s(s + 3)}$

Domain: $\qquad s(s + 3) \neq 0$

$$s \neq 0 \qquad s + 3 \neq 0$$

$$s \neq -3$$

$$(-\infty, -3) \cup (-3, 0) \cup (0, \infty)$$

Vertical asymptotes: $s = 0, s = -3$

Horizontal asymptote: $y = 0$ since the degree of the numerator is less than the degree of the denominator.

**20.** $y = \dfrac{x^2 - 9}{x^2 - 2x - 8}$

Domain: $\quad x^2 - 2x - 8 \neq 0$

$$(x - 4)(x + 2) \neq 0$$

$$x \neq 4 \qquad x \neq -2$$

$$(-\infty, -2) \cup (-2, 4) \cup (4, \infty)$$

Vertical asymptotes: $x = 4, x = -2$

Horizontal asymptote: $y = 1$ since the degree of the numerator is equal to the degree of the denominator and the leading coefficients are 1.

**24.** $f(t) = \dfrac{5}{t} - 4t = \dfrac{5}{t} - \dfrac{4t}{1} \cdot \dfrac{t}{t} = \dfrac{5 - 4t^2}{t}$

Domain: $t \neq 0$

$$(-\infty, 0) \cup (0, \infty)$$

Vertical asymptote: $t = 0$

Horizontal asymptote: none since the degree of the numerator is greater than the degree of the denominator.

**14.** $y = \dfrac{3x + 2}{2x - 1}$

Domain: $2x - 1 \neq 0$

$$2x \neq 1$$

$$x \neq \frac{1}{2}$$

$$\left(-\infty, \frac{1}{2}\right) \cup \left(\frac{1}{2}, \infty\right)$$

Vertical asymptote: $x = \dfrac{1}{2}$

Horizontal asymptote: $y = \frac{3}{2}$ since the degree of the numerator is equal to the degree of the denominator and the leading coefficient of the numerator is 3 and the leading coefficient of the denominator is 2.

**18.** $g(t) = \dfrac{3t^3}{t^2 + 1}$

Domain: $t^2 + 1 \neq 0$

$$(-\infty, \infty)$$

No real solution

Vertical asymptote: none

Horizontal asymptote: none since the degree of the numerator is greater than the degree of the denominator.

**22.** $h(v) = \dfrac{3}{v} - 2 = \dfrac{3}{v} - \dfrac{2}{1} \cdot \dfrac{v}{v} = \dfrac{3 - 2v}{v}$

Domain: $v \neq 0$

$$(-\infty, 0) \cup (0, \infty)$$

Vertical asymptote: $v = 0$

Horizontal asymptote: $y = -2$ since the degree of the numerator is equal to the degree of the denominator and the leading coefficient of the numerator is $-2$ and the leading coefficient of the denominator is 1.

**26.** $f(x) = \dfrac{1 - 2x}{x}$ matches graph (a).

Vertical asymptote: $x = 0$

Horizontal asymptote: $y = -2$

**28.** $f(x) = -\dfrac{x+2}{x+1}$ matches graph (c).

Vertical asymptote: $x + 1 = 0$

$$x = -1$$

Horizontal asymptote: $y = -1$

**30.** (b) $f(x) = \dfrac{3x}{x+2}$

**32.** (c) $f(x) = \dfrac{3}{x^2+2}$

**34.** $f(x) = \dfrac{5}{x^2}$

$y$-intercept: $f(0) = \dfrac{5}{0^2} =$ undefined, none

$x$-intercept: none, numerator is never zero.

Vertical asymptote: $x = 0$

Horizontal asymptote: $y = 0$ since the degree of the numerator is less than the degree of the denominator.

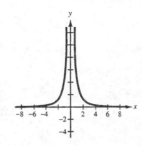

**36.** $f(x) = \dfrac{5}{(x-4)^2}$

$y$-intercept: $f(0) = \dfrac{5}{(0-4)^2} = \dfrac{5}{(-4)^2} = \dfrac{5}{16}$

$x$-intercept: none, numerator is never zero.

Vertical asymptote: $(x-4)^2 = 0$

$$x - 4 = 0$$

$$x = 4$$

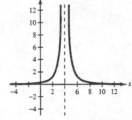

Horizontal asymptote: $y = 0$ since the degree of the numerator is less than the degree of the denominator.

**38.** $f(x) = \dfrac{3}{x+1}$

$y$-intercept: $f(0) = \dfrac{3}{0+1} = \dfrac{3}{1} = 3$

$x$-intercept: none, numerator is never zero.

Vertical asymptote: $x + 1 = 0$

$$x = -1$$

Horizontal asymptote: $y = 0$ since the degree of the numerator is less than the degree of the denominator.

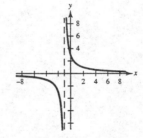

**40.** $g(x) = \dfrac{-3}{x+1}$

$y$-intercept: $g(0) = \dfrac{-3}{0+1} = -\dfrac{3}{1} = -3$

$x$-intercept: none, numerator is never zero.

Vertical asymptote: $x + 1 = 0$

$$x = -1$$

Horizontal asymptote: $y = 0$ since the degree of the numerator is less than the degree of the denominator.

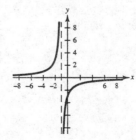

**42.** $y = \dfrac{2x}{x^2 + 4x} = \dfrac{2x}{x(x + 4)} = \dfrac{2}{x + 4}$

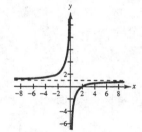

$y$-intercept: $y = \dfrac{2(0)}{0^2 + 4(0)} = $ undefined, none

$x$-intercept: $0 = \dfrac{2x}{x^2 + 4x}$

$\qquad\qquad 0 = \dfrac{2x}{x(x + 4)}$

$\qquad\qquad 0 = \dfrac{2}{x + 4}$, none

Vertical asymptote: $x^2 + 4x = 0$

$\qquad\qquad\qquad x(x + 4) = 0$

$\qquad\qquad\qquad\qquad x = -4$

Horizontal asymptote: $y = 0$ since the degree of the numerator is less than the degree of the denominator.

**44.** $g(v) = \dfrac{2v^2}{v^2 + v} = \dfrac{2v^2}{v(v + 1)} = \dfrac{2v}{v + 1}$

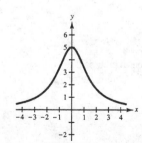

$y$-intercept: $g(0) = \dfrac{2(0)^2}{0^2 + 0} = $ undefined, none

$x$-intercept: $0 = \dfrac{2v}{v + 1}$

$\qquad\qquad 0 = 2v$

$\qquad\qquad 0 = v$

Vertical asymptote: $v(v + 1) = 0$

$\qquad\qquad\qquad v = 0 \qquad v = -1$

Horizontal asymptote: $y = 2$ since the degrees are equal and the leading coefficient of the numerator is 2 and the leading coefficient of the denominator is 1.

**46.** $y = \dfrac{x - 2}{x}$

$y$-intercept: $y = \dfrac{0 - 2}{0}$ is undefined; none

$x$-intercept: $x - 2 = 0$

$\qquad\qquad\qquad x = 2$

Vertical asymptote: $x = 0$

Horizontal asymptote: $y = 1$ since the degree of the numerator is equal to the degree of the denominator and the leading coefficient of the numerator is 1 and the leading coefficient of the denominator is 1.

**48.** $y = \dfrac{10}{x^2 + 2}$

$y$-intercept: $y = \dfrac{10}{0^2 + 2} = 5$

$x$-intercept: $0 = \dfrac{10}{x^2 + 2}$; none

Vertical asymptote: $x^2 + 2 \neq 0$; none

Horizontal asymptote: $y = 0$ since the degree of the numerator is less than the degree of the denominator.

**50.** $y = \dfrac{4x^2}{x^2 + 1}$

$y$-intercept: $y = \dfrac{4(0)^2}{0^2 + 1} = \dfrac{0}{1} = 0$

$x$-intercept: $4x^2 = 0$

$\qquad\qquad x^2 = 0$

$\qquad\qquad\ x = 0$

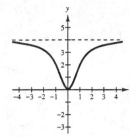

Vertical asymptote: none, $x^2 + 1 \neq 0$

Horizontal asymptote: $y = 4$ since the degree of the numerator is equal to the degree of the denominator and the leading coefficient of the numerator is 4 and the leading coefficient of the denominator is 1.

**52.** $f(x) = \dfrac{4}{x} + 2 = \dfrac{4 + 2x}{x}$

$y$-intercept: $f(0) = \dfrac{4}{0} + 2 =$ undefined; none

$x$-intercept: $\quad 0 = \dfrac{4}{x} + 2$

$\qquad\qquad -2 = \dfrac{4}{x}$

$\qquad\qquad -2x = 4$

$\qquad\qquad\quad x = -2$

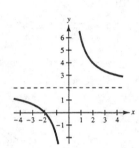

Vertical asymptote: $x = 0$

Horizontal asymptote: $y = 2$ since the degrees are equal and the leading coefficient of the numerator is 2 and the leading coefficient of the denominator is 1.

**54.** $y = \dfrac{4x + 6}{x^2 - 9}$

$y$-intercept: $y = \dfrac{4(0) + 6}{0^2 - 9} = \dfrac{6}{-9} = \dfrac{2}{-3}$

$x$-intercept: $\quad 0 = \dfrac{4x + 6}{x^2 - 9}$

$\qquad\qquad 0 = 4x + 6$

$\qquad\qquad \dfrac{-6}{4} = x$

$\qquad\qquad -\dfrac{3}{2} = x$

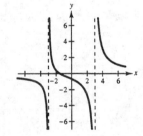

Vertical asymptotes: $x^2 - 9 = 0$

$\qquad\qquad\qquad\quad x = 3, \ x = -3$

Horizontal asymptote: $y = 0$ since the degree of the numerator is less than the degree of the denominator.

**56.** $g(x) = \dfrac{2x^2}{x^2 + 2x - 3}$

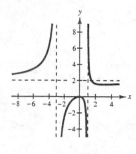

$y$-intercept:  $g(0) = \dfrac{2(0)}{0^2 + 2(0) - 3} = \dfrac{0}{-3} = 0$

$x$-intercept:  $0 = \dfrac{2x^2}{x^2 + 2x - 3}$

$\qquad\qquad\quad 0 = 2x^2$

$\qquad\qquad\quad 0 = x^2$

$\qquad\qquad\quad 0 = x$

Vertical asymptotes:   $x^2 + 2x - 3 = 0$

$\qquad\qquad\qquad\quad (x + 3)(x - 1) = 0$

$\qquad\qquad\qquad\qquad x = -3, \quad x = 1$

Horizontal asymptote: $y = 2$ since the degrees are equal and the leading coefficient of the numerator is 2 and the leading coefficient of the denominator is 1.

**58.** $g(t) = \dfrac{t^2 - 9}{t^2 + 6t + 9}$

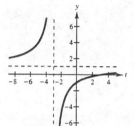

$y$-intercept:  $\dfrac{0^2 - 9}{0^2 + 6(0) + 9} = \dfrac{-9}{9} = -1$

$x$-intercept:  $0 = \dfrac{t^2 - 9}{t^2 + 6t + 9}$

$\qquad\qquad\quad 0 = t^2 - 9$

$\qquad\qquad\quad t = 3, \; \bcancel{-3}$

Vertical asymptote:  $t^2 + 6t + 9 = 0$

$\qquad\qquad\qquad\quad (t + 3)^2 = 0$

$\qquad\qquad\qquad\qquad t = -3$

Horizontal asymptote: $y = 1$ since the degrees are equal and the leading coefficient of the numerator and the denominator is 1.

$$\dfrac{t^2 - 9}{t^2 + 6t + 9} = \dfrac{(t - 3)\cancel{(t + 3)}}{(t + 3)\cancel{(t + 3)}}$$

**60.** $y = \dfrac{3x}{x + 2}$

Domain:  $x + 2 \neq 0$

$\qquad\qquad x \neq -2$

$\qquad (-\infty, -2) \cup (-2, \infty)$

Vertical asymptote:  $x + 2 = 0$

$\qquad\qquad\qquad\quad x = -2$

Horizontal asymptote: $y = 3$ since the degree of the numerator is equal to the degree of the denominator and the leading coefficient of the numerator is 3 and the leading coefficient of the denominator is 1.

Keystrokes:

$y_1$ [Y=] 3 [X,T,θ] [÷] [(] [X,T,θ] [+] 2 [)] [GRAPH]

**62.** $h(x) = \dfrac{x^2}{x-2}$

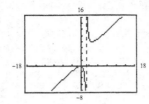

Domain: $x - 2 \neq 0$

$\qquad x \neq 2$

$\qquad (-\infty, 2) \cup (2, \infty)$

Vertical asymptote: $x = 2$

Horizontal asymptote: none since the degree of the numerator is greater than the degree of the denominator.

*Keystrokes:*

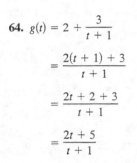

**64.** $g(t) = 2 + \dfrac{3}{t+1}$

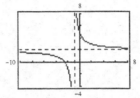

$\qquad = \dfrac{2(t+1)+3}{t+1}$

$\qquad = \dfrac{2t+2+3}{t+1}$

$\qquad = \dfrac{2t+5}{t+1}$

Domain: $t + 1 \neq 0$

$\qquad t \neq -1$

$\qquad (-\infty, -1) \cup (-1, \infty)$

Vertical asymptote: $t = -1$

Horizontal asymptote: $y = 2$ since the degrees are equal and the leading coefficient of the numerator is 2 and the leading coefficient of the denominator is 1.

*Keystrokes:* [Y=] 2 [+] 3 [÷] [(] [X,T,θ] [+] 1 [)] [GRAPH] or

[Y=] [(] 2 [X,T,θ] [+] 5 [)] [÷] [(] [X,T,θ] [+] 1 [)] [GRAPH]

**66.** $y = \dfrac{2(x^2 - 1)}{x^2}$

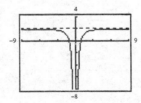

Domain: $x^2 \neq 0$

$\qquad x \neq 0$

$\qquad (-\infty, 0) \cup (0, \infty)$

Vertical asymptote: $x = 0$

Horizontal asymptote: $y = 2$

*Keystrokes:* [Y=] 2 [(] [X,T,θ] [x²] [−] 1 [)] [÷] [X,T,θ] [x²] [GRAPH]

**68.** $y = \dfrac{x}{2} - \dfrac{2}{x} = \dfrac{x^2 - 4}{2x}$

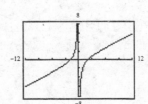

Domain: $2x \neq 0$

$\qquad x \neq 0$

$\qquad (-\infty, 0) \cup (0, \infty)$

Vertical asymptote: $x = 0$

Horizontal asymptote: none since the degree of the numerator is greater than the degree of the denominator.

*Keystrokes:* [Y=] [(] [X,T,θ] [÷] 2 [)] [−] [(] 2 [÷] [X,T,θ] [)] [GRAPH] or

[Y=] [(] [X,T,θ] [x²] [−] 4 [)] [÷] [(] 2 [X,T,θ] [)] [GRAPH]

**70.** Reduce $h(x)$ to lowest terms.

$$h(x) = \frac{x^2 - 9}{x + 3} = \frac{(x + 3)(x - 3)}{x + 3} = x - 3, \quad x \neq -3$$

*Keystrokes:* [Y=] [(] [X,T,θ] [x²] [−] 9 [)] [÷] [(] [X,T,θ] [+] 3 [)] [GRAPH]

There is no vertical asymptote at $x = -3$ like you might expect because $h(x)$ is not written in lowest terms.

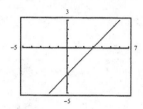

**72.** (a) Average cost $= \dfrac{\text{Cost}}{\text{Number of units}}$

$$\overline{C} = \frac{30{,}000 + 1.25x}{x}, \quad x > 0$$

(b)  $x = 10{,}000$                                    $x = 100{,}000$

$\overline{C} = \dfrac{30{,}000 + 1.25(10{,}000)}{10{,}000}$          $\overline{C} = \dfrac{30{,}000 + 1.25(100{,}000)}{100{,}000}$

$\quad = \dfrac{30{,}000 + 12{,}500}{10{,}000}$          $\quad = \dfrac{30{,}000 + 125{,}000}{100{,}000}$

$\quad = \dfrac{42{,}500}{10{,}000}$          $\quad = \dfrac{155{,}000}{100{,}000}$

$\quad = \$4.25$          $\quad = \$1.55$

(c) *Keystrokes:*  [Y=] [(] 30,000 [+] 1.25 [X,T,θ] [)] [÷] [X,T,θ] [GRAPH]

Horizontal asymptote: $\overline{C} = \$1.25$ since the degree of the numerator is equal to the degree of the denominator and the leading coefficient of the numerator is 1.25 and the leading coefficient of the denominator is 1. As the number of units produced increases, the average cost is approximately $1.25.

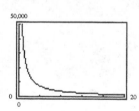

**74.** (a) $x > 0$   and   $x + 5 \neq 0$      and   $x < 20$

$$x \neq -5$$

Domain: $[0, 20]$

(b) *Keystrokes:*  [Y=] [(] 3 [X,T,θ] [+] 5 [)] [÷] [(] [(] 4 [(] [X,T,θ] [+] 5 [)] [)] [GRAPH]

As the container is filled, the concentration of the brine approaches 0.75.

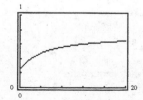

**76.** (a) $t = 1$, $N = \dfrac{150(1)[1 + 4(1)]}{1 + 0.15(1)^2} = \dfrac{140(5)}{1 + 0.15} + \dfrac{750}{1.15} \approx 652.17 \times 1000 = 652{,}170$

$t = 2$, $N = \dfrac{150(2)[1 + 4(2)]}{1 + 0.15(2)^2} = \dfrac{300(9)}{1 + 0.6} = \dfrac{2700}{1.6} = 1687.50 \times 1000 = 1{,}687{,}500$

$t = 4$, $N = \dfrac{150(4)[1 + 4(4)]}{1 + 0.15(4)^2} = \dfrac{600(17)}{1 + 2.4} = \dfrac{10{,}200}{3.4} = 3000.000 \times 1000 = 3{,}000{,}000$

(b) Keystrokes: $\boxed{Y=}$ $\boxed{(}$ 150 $\boxed{X,T,\theta}$ $\boxed{(}$ 1 $\boxed{+}$ 4 $\boxed{X,T,\theta}$ $\boxed{)}$ $\boxed{)}$ $\boxed{\div}$ $\boxed{(}$ 1 $\boxed{+}$ .15 $\boxed{X,T,\theta}$ $\boxed{x^2}$ $\boxed{)}$ $\boxed{GRAPH}$

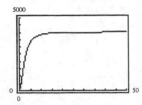

Horizontal asymptote: $y = 4000$ thousand or $4{,}000{,}000$ since the degree of the numerator is equal to the degree of the denominator and the leading coefficient of the numerator is 600 and the leading coefficient of the denominator is 0.15.

(c) The horizontal asymptote indicates an upper limit on the number of units sold.

**78.** Vertical asymptote: $x = -2$
Horizontal asymptote: $y = 0$
Zero of the function: $x = 3$
$$y = \frac{x - 3}{(x + 2)(x^2 + 1)}$$

**80.** Vertical asymptotes: $x = 1, x = -1$
Horizontal asymptote: $y = 1$
Zero of the function: $x = 0$
$$y = \frac{x^2}{(x - 1)(x + 1)} \text{ or } y = \frac{x^2}{x^2 - 1}$$

**82.** The domain of a rational function consists of all values of $x$ for which the denominator is not zero.
$$f(x) = \frac{x}{x - 2}$$

**84.** False. There is a branch of the graph on each side of the asymptote.

# Review Exercises for Chapter 7

**2.** $R$ varies jointly as $u$ and $v$.
$$R = kuv$$

**4.** $F$ varies directly as $g$ and inversely ar $r^2$.
$$F = \frac{kg}{r^2}$$

**6.** $r = \dfrac{k}{s}$

$45 = \dfrac{k}{3/5}$

$45\left(\dfrac{3}{5}\right) = k$

$27 = k$

$r = \dfrac{27}{s}$

**8.** $D = \dfrac{kx^3}{y}$

$810 = \dfrac{k(3)^3}{25}$

$810 = \dfrac{27k}{25}$

$810\left(\dfrac{25}{27}\right) = k$

$750 = k$

$D = \dfrac{750x^3}{y}$

**10.** $x \le 5$

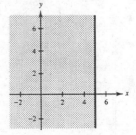

**12.** $y + 3 < 0$

$\qquad y < -3$

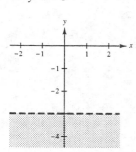

**14.** $3x - 4y > 2$

$\qquad -4y > -3x + 2$

$\qquad y < \frac{3}{4}x - \frac{1}{2}$

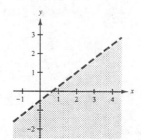

**16.** $(y - 3) \geq 2(x - 5)$

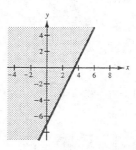

**18.** $y \leq \frac{1}{3}x + 1$

Keystrokes:

Y= ( 1 ÷ 3 X,T,θ + 1

DRAW 7 (−) 10 , Y-VARS 1 1 ) ENTER

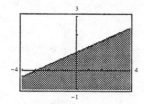

**20.** $4x - 3y \geq 2$

$\qquad -3y \geq -4x + 2$

$\qquad y \leq \frac{4}{3}x - \frac{2}{3}$

Keystrokes:

Y= ( 4 ÷ 3 ) X,T,θ − ( 2 ÷ 3 )

DRAW 7 (−) 10 , Y-VARS 1 1 ) ENTER

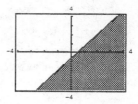

**22.** $g(x) = x^2 + 12x - 9$

$\qquad = (x^2 + 12x + 36) - 9 - 36$

$\qquad = (x + 6)^2 - 45$

Vertex: $(-6, -45)$

**24.** $f(t) = 3t^2 + 2t - 6$

$\qquad = 3\left(t^2 + \frac{2}{3}t\right) - 6$

$\qquad = \left(t^2 + \frac{2}{3}t + \frac{1}{9}\right) - 6 - \frac{1}{3}$

$\qquad = 3\left(t + \frac{1}{3}\right)^2 - \frac{19}{3}$

Vertex: $\left(-\frac{1}{3}, -\frac{19}{3}\right)$

**26.** $y = -x^2 + 3x$

x-intercepts:

$0 = -x^2 + 3x$

$0 = -x(x - 3)$

$-x = 0, \; x - 3 = 0$

$x = 0, \qquad x = 3$

Vertex:

$y = -x^2 + 3x$

$y = -\left(x^2 - 3x + \frac{9}{4}\right) + \frac{9}{4}$

$= -\left(x - \frac{3}{2}\right)^2 + \frac{9}{4}$

$\left(\frac{3}{2}, \frac{9}{4}\right)$

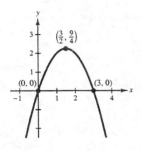

**28.** $y = x^2 + 3x - 10$

x-intercepts:

$0 = x^2 + 3x - 10$

$0 = (x + 5)(x - 2)$

$x = -5, \quad x = 2$

Vertex:

$y = \left(x^2 + 3x + \frac{9}{4}\right) - 10 - \frac{9}{4}$

$= \left(x + \frac{3}{2}\right) - \frac{49}{4}$

$\left(-\frac{3}{2}, -\frac{49}{4}\right)$

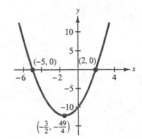

**30.** $h(x) = x^2 - 1$

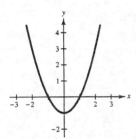

Vertical shift 1 unit down

**32.** $h(x) = (x - 2)^2 + 4$

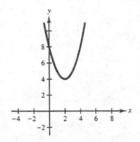

Horizontal shift 2 units right

Vertical shift 4 units up

**34.** Vertex: $(-2, 3)$; $a = 3$

$y = a(x - h)^2 + k$

$y = 3(x + 2)^2 + 3$

**36.** Vertex: $(-4, 0)$

y-intercept: $(0, -6)$

$y = a(x - h)^2 + k$

$y = a(x + 4)^2 + 0$

$-6 = a(0 + 4)^2$

$-\frac{3}{8} = -\frac{6}{16} = a$

$y = -\frac{3}{8}(x + 4)^2$

**38.** Vertex: $(-2, 5)$

Point: $(0, 1)$

$y = a(x - h)^2 + k$

$y = a(x - (-2))^2 + 5$

$y = a(x + 2)^2 + 5$

$1 = a(0 + 2)^2 + 5$

$1 = 4a + 5$

$-4 = 4a$

$-1 = a$

$y = -(x + 2)^2 + 5$

$y = -(x^2 + 4x + 4) + 5$

$y = -x^2 - 4x - 4 + 5$

$y = -x^2 - 4x + 1$

**40.** (d)

**42.** (f)

**44.** (e)

**46.** $x^2 - y^2 = 64$

$\dfrac{x^2}{64} - \dfrac{y^2}{64} = 1$

Hyperbola

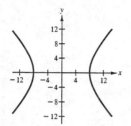

**48.** $x^2 + 4y^2 = 64$

$\dfrac{x^2}{64} + \dfrac{4y^2}{64} = \dfrac{64}{64}$

$\dfrac{x^2}{64} + \dfrac{y^2}{16} = 1$

Ellipse

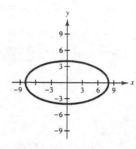

**50.** $y = 9 - (x - 3)^2$

Parabola

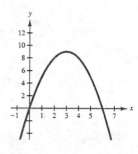

**52.** $\dfrac{x^2}{25} - \dfrac{y^2}{4} = -1$

$\dfrac{y^2}{4} - \dfrac{x^2}{25} = 1$

Hyperbola

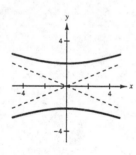

**54.** $x^2 + 9y^2 - 9 = 0$

$x^2 + 9y^2 = 9$

$\dfrac{x^2}{9} + \dfrac{9y^2}{9} = \dfrac{9}{9}$

$\dfrac{x^2}{9} + \dfrac{y^2}{1} = 1$

Ellipse

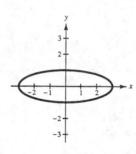

**56.** Vertex: $(-2, 5)$

Point: $(0, 1)$

$y = a(x - h)^2 + k$

$y = a(x - (-2))^2 + 5$

$y = a(x + 2)^2 + 5$

$1 = a(0 + 2)^2 + 5$

$1 = 4a + 5$

$-4 = 4a$

$-1 = a$

$y = -(x + 2)^2 + 5$

$y = -(x^2 + 4x + 4) + 5$

$y = -x^2 - 4x - 4 + 5$

$y = -x^2 - 4x + 1$

**58.** $\dfrac{x^2}{10^2} + \dfrac{y^2}{6^2} = 1$

$\dfrac{x^2}{100} + \dfrac{y^2}{36} = 1$

**60.** $(x - h)^2 + (y - k)^2 = r^2$

$(x - 3)^2 + (y + 2)^2 = 16$

**62.** $\dfrac{y^2}{4^2} - \dfrac{x^2}{2^2} = 1$

$\dfrac{y^2}{16} - \dfrac{x^2}{4} = 1$

**64.** (d)

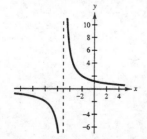

**66.** (c)

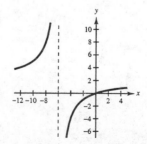

**68.** $f(x) = \dfrac{4}{x}$

$y$-intercept: $f(0) = \dfrac{4}{0} =$ undefined, none

$x$-intercept: $0 = \dfrac{4}{x}$

$\qquad\qquad\quad 0 = 4,$ none

Vertical asymptote: $x = 0$

Horizontal asymptote: $y = 0$ since the degree of the numerator is less than the degree of the denominator.

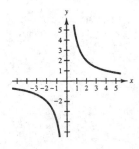

**70.** $s(x) = \dfrac{2x - 6}{x + 4}$

$y$-intercept: $s(0) = \dfrac{2(0) - 6}{0 + 4} = -\dfrac{6}{4} = -\dfrac{3}{2}$

$x$-intercept: $0 = \dfrac{2x - 6}{x + 4}$

$\qquad\qquad\quad 0 = 2x - 6$

$\qquad\qquad\quad 3 = x$

Vertical asymptote: $x = -4$

Horizontal asymptote: $y = 2$ since the degrees are equal and the leading coefficient of the numerator is 2 and the leading coefficient of the denominator is 1.

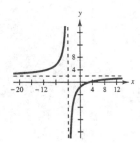

**72.** $h(x) = \dfrac{x - 3}{x - 2}$

$y$-intercept: $h(0) = \dfrac{0 - 3}{0 - 2} = \dfrac{-3}{-2} = \dfrac{3}{2}$

$x$-intercept: $0 = \dfrac{x - 3}{x - 2}$

$\qquad\qquad\quad 0 = x - 3$

$\qquad\qquad\quad 3 = x$

Vertical asymptote: $x = 2$

Horizontal asymptote: $y = 1$ since the degrees are equal and the leading coefficient of the numerator is 1 and the leading coefficient of the denominator is 1.

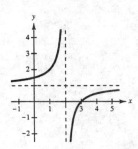

**74.** $f(x) = \dfrac{2x}{x^2 + 4}$

$y$-intercept: $f(0) = \dfrac{2(0)}{0^2 + 4} = 0$

$x$-intercept: $0 = \dfrac{2x}{x^2 + 4}$

$\qquad\qquad\quad 0 = 2x$

$\qquad\qquad\quad 0 = x$

Vertical asymptote: None, $x^2 + 4 \neq 0$

Horizontal asymptote: $y = 0$ since the degree of the numerator is less than the degree of the denominator.

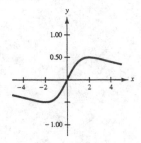

**76.** $g(x) = \dfrac{-2}{(x+3)^2}$

$y$-intercept: $g(0) = \dfrac{-2}{(0+3)^2} = -\dfrac{2}{9}$

$x$-intercept: $0 = -\dfrac{2}{(x+3)^2}$

$\qquad 0 = -2$, none

Vertical asymptote: $x = -3$

Horizontal asymptote: $y = 0$ since the degree of the numerator is less than the degree of the denominator.

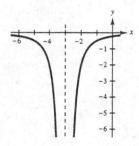

**78.** $y = \dfrac{2x}{x^2 - 4}$

$y$-intercept: $y = \dfrac{2(0)}{0^2 - 4} = 0$

$x$-intercept: $0 = \dfrac{2x}{x^2 - 4}$

$\qquad 0 = 2x$

$\qquad 0 = x$

Vertical asymptote: $x^2 - 4 = 0$

$\qquad\qquad x = 2, \; x = -2$

Horizontal asymptote: $y = 0$ since the degree of the numerator is less than the degree of the denominator.

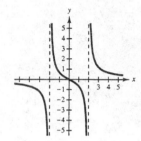

**80.** $y = \dfrac{2x}{x^2 + 3x} = \dfrac{2x}{x(x+3)} = \dfrac{2}{x+3}$

$y$-intercept: $y = \dfrac{2(0)}{0^2 + 3(0)} = \dfrac{0}{0} = $ undefined; none

$x$-intercept: $0 = \dfrac{2}{x+3}$

$\qquad 0 = 2$, none

Vertical asymptote: $x = -3$

Horizontal asymptote: $y = 0$ since the degree of the numerator is less than the degree of the denominator.

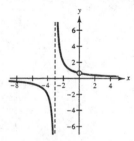

**82.** $y = \dfrac{2x+1}{2x^2 - 5x - 3} = \dfrac{2x+1}{(2x+1)(x-3)} = \dfrac{1}{x-3}$

$y$-intercept: $y = \dfrac{1}{0-3} = \dfrac{1}{-3}$

$x$-intercept: $0 = \dfrac{1}{x-3}$

$\qquad 0 = 1$, none

Vertical asymptote: $x = 3$

Horizontal asymptote: $y = 0$ since the degree of the numerator is less than the degree of the denominator.

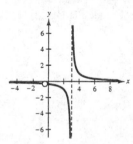

**84.** $y = \dfrac{x-2}{(x+3)^2}$

**86.** $d = ks^2$

$d = k(2s)^2$

$d = 4ks^2 = 4(ks)^2$

The stopping distance increases by a factor of 4.

**88.**
$$F = \frac{k}{r^2}$$

$$200 = \frac{k}{(4000)^2}$$

$$3{,}200{,}000{,}000 = k$$

$$F = \frac{3{,}200{,}000{,}000}{r^2}$$

$$F = \frac{3{,}200{,}000{,}000}{(4500)^2}$$

$$F \approx 158.02 \text{ pounds}$$

**90.** $2x + 2y \le 800$

$x + y \le 400$ (*Note:*  $x$ and $y$ cannot be negative.)

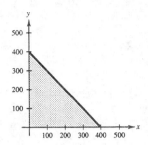

**92.** (a) *Keystrokes:*

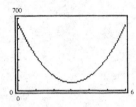

(b) The minimum point occurs at $(3.03, 78.53)$ so $t = 3$ or the year 1993 gives the minimum number of orders.

(c) Year 2000 is $t = 10$ so $N(10) \approx 3130$ orders.

**94.** Horizontal asymptote of $\overline{C} = \dfrac{1}{2}$; as $x$ gets larger, the average cost per unit approaches $0.50.

**96.** (a) $C = \dfrac{528(25)}{100 - 25} = \dfrac{13200}{75} = 176$ million dollars

(b) $C = \dfrac{528(75)}{100 - .75} = \dfrac{39600}{25} = 1584$ million dollars

(c) No. $p = 100$ is a vertical asymptote.

# CHAPTER 8
## Systems of Equations

# CHAPTER 8
# Systems of Equations

## Section 8.1 Systems of Equations
### Solutions to Even-Numbered Exercises

**2.** (a) $(0, 3)$

$5(0) - 4(3) = 34$

$0 - 12 = 34$

$-12 \neq 34$

Not a solution

(b) $(6, -1)$

$5(6) - 4(-1) = 34$

$30 + 4 = 34$

$34 = 34$

$6 - 2(-1) = 8$

$6 + 2 = 8$

$8 = 8$

Solution

**4.** (a) $(-3, -4)$

$-5(-3) - 2(-4) = 23$

$15 + 8 = 23$

$23 = 23$

$-3 + 4(-4) = -19$

$-3 - 16 = -19$

$-19 = -19$

Solution

(b) $(3, 7)$

$-5(3) - 2(7) = 23$

$-15 - 14 = 23$

$-29 \neq 23$

Not a solution

**6.** (a) $\left(0, -\frac{3}{2}\right)$

$2(0) - \left(-\frac{3}{2}\right) = 1.5$

$1.5 = 1.5$

$4(0) - 2\left(-\frac{3}{2}\right) = 3$

$3 = 3$

Solution

(b) $\left(2, \frac{5}{2}\right)$

$2(2) - \frac{5}{2} = 1.5$

$1.5 = 1.5$

$4(2) - 2\left(\frac{5}{2}\right) = 3$

$3 = 3$

Solution

**8.** (a) $(4, 1)$

$4^2 + 1^2 = 17$

$17 = 17$

$4 + 3(1) = 7$

$7 = 7$

Solution

(b) $(2, 1)$

$2^2 + 1^2 = 17$

$5 \neq 17$

Not a solution

**10.** Solve each equation for $y$.

$$x - 2y = 3 \qquad\qquad 2x - 4y = 7$$

$$-2y = -x + 3 \qquad\quad -4y = -2x + 7$$

$$y = \tfrac{1}{2}x - \tfrac{3}{2} \qquad\qquad y = \tfrac{1}{2}x - \tfrac{7}{4}$$

The slopes are equal;
therefore, the system is inconsistent.

**12.** $-5x + 8y = 8 \qquad\qquad 7x - 4y = 14$

$$8y = 5x + 8 \qquad\qquad -4y = -7x + 14$$

$$y = \tfrac{5}{8}x + 1 \qquad\qquad y = \tfrac{7}{4}x - \tfrac{14}{4}$$

The slopes are not equal;
therefore, the system is consistent.

**14.**  $3x + 8y = 28 \qquad\qquad -4x + 9y = 1$

$$8y = -3x + 28 \qquad\qquad 9y = 4x + 1$$

$$y = \frac{-3}{8}x + \frac{28}{8} \qquad\qquad y = \frac{4}{9}x + \frac{1}{9}$$

The slopes are not equal;
therefore, the system is consistent.

**16.**  $9x + 6y = 10 \qquad\qquad -6x - 4y = 3$

$$6y = -9x + 10 \qquad\qquad -4y = 6x + 3$$

$$y = \frac{-9}{6}x + \frac{10}{6} \qquad\qquad y = \frac{-6}{4}x - \frac{3}{4}$$

$$y = \frac{-3}{2}x + \frac{5}{3} \qquad\qquad y = \frac{-3}{2}x - \frac{3}{4}$$

The slopes are equal;
therefore, the system is inconsistent.

**18.** Solve each equation for $y$.

$$x + y = 5 \qquad x - y = 5$$

$$y = 5 - x \qquad x - 5 = y$$

*Keystrokes:*

$y_1$: [Y=] 5 [–] [X,T,θ] [ENTER]

$y_2$: [X,T,θ] [–] 5 [X,T,θ] [GRAPH]

One solution

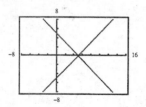

**20.** Solve each equation for $y$.

$$2x - 4y = 9 \qquad x - 2y = 4.5$$

$$-4y = -2x + 9 \qquad -2y = -x + 4.5$$

$$y = \frac{1}{2}x - \frac{9}{4} \qquad y = \frac{1}{2}x - \frac{4.5}{2}$$

$$y = \frac{1}{2}x - \frac{9}{4}$$

*Keystrokes:*

$y_1$: [Y=] [X,T,θ] [÷] 2 [–] 9 [÷] 4 [GRAPH]

Infinitely many solutions.

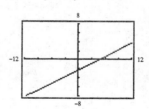

**22.** One solution

$$y = x + 5$$

$$x + 2(x + 5) = 4$$

$$x + 2x + 10 = 4$$

$$3x = -6$$

$$x = -2$$

$$y = -2 + 5$$

$$y = 3$$

$$(-2, 3)$$

**24.** No solutions

**26.** One solution

$$-3y = -2x + 6$$

$$y = \tfrac{2}{3}x - 2$$

$$4x + 3\left(\tfrac{2}{3}x - 2\right) = 12$$

$$4x + 2x - 6 = 12$$

$$6x = 18$$

$$x = 3$$

$$y = \tfrac{2}{3}(3) - 2$$

$$y = 0$$

$$(3, 0)$$

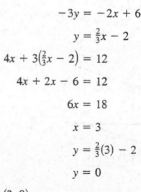

**28.** One solution

$$x - 2\sqrt{x - 2} = 1$$
$$x - 1 = 2\sqrt{x - 2}$$
$$(x - 1)^2 = 4(x - 2)$$
$$x^2 - 2x + 1 = 4x - 8$$
$$x^2 - 6x + 9 = 0$$
$$(x - 3)^2 = 0$$
$$x - 3 = 0$$
$$x = 3$$
$$y = \sqrt{3 - 2}$$
$$y = \sqrt{1}$$
$$y = 1$$

$(3, 1)$

**30.**

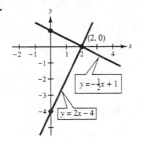

Point of intersection is $(2, 0)$

**32.** Solve each equation for $y$.

$$x - y = 0 \qquad x + y = 4$$
$$-y = -x \qquad y = -x + 4$$
$$y = x$$

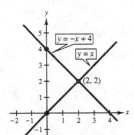

Point of intersection is $(2, 2)$

**34.** Solve each equation for $y$.

$$5x + 2y = 18 \qquad y = 2$$
$$2y = -5x + 18$$
$$y = \frac{-5}{2}x + 9$$

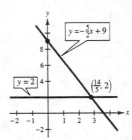

Point of intersection is $\left(\frac{14}{5}, 2\right)$

**36.** Solve each equation for $y$.

$$-x + 3y = 7 \qquad 2x - 6y = 6$$
$$3y = x + 7 \qquad -6y = -2x + 6$$
$$y = \frac{1}{3}x + \frac{7}{3} \qquad y = \frac{1}{3}x - 1$$

No solution

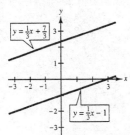

**38.** Solve each equation for $y$.

$$5x + 3y = 24 \qquad x - 2y = 10$$
$$3y = -5x + 24 \qquad -2y = -x + 10$$
$$y = -\frac{5}{3}x + 8 \qquad y = \frac{1}{2}x - 5$$

Point of intersection is $(6, -2)$

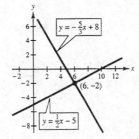

**40.** Solve each equation for *y*.

$$4x - 3y = -3 \qquad 8x - 6y = -6$$

$$-3y = -4x - 3 \qquad -6y = -8x - 6$$

$$y = \frac{4}{3}x + 1 \qquad y = \frac{4}{3}x + 1$$

Same line; infinitely many solutions.

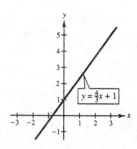

**42.** Solve each equation for *y*.

$$7x + 4y = 6 \qquad 5x - 3y = -25$$

$$4y = -7x + 6 \qquad -3y = -5x - 25$$

$$y = \frac{-7}{4}x + \frac{6}{4} \qquad y = \frac{5}{3}x + \frac{25}{3}$$

Point of intersection is $(-2, 5)$.

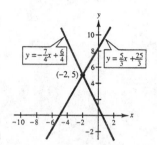

**44.** *Keystrokes:*

$y_1$ [Y=] 8 [−] [X,T,θ] [x²] [ENTER]

$y_2$: 6 [−] [X,T,θ] [GRAPH]

Solutions: $(-1, 7)$ and $(2, 4)$

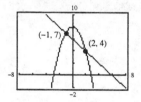

**46.** *Keystrokes:*

$y_1$ [Y=] [X,T,θ] [x²] [−] 2 [X,T,θ] [ENTER]

$y_2$: [Y=] [X,T,θ] [^] 3 [−] 4 [X,T,θ] [GRAPH]

Solutions: $(-1, 3)$, $(0, 0)$, $(2, 0)$

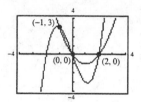

**48.**

$$x - y = 0$$
$$x = y$$
$$5x - 2x = 6$$
$$3x = 6$$
$$x = 2$$
$$y = 2$$
$$(2, 2)$$

**50.** $x - 6(2) = -6$

$$x - 12 = -6$$
$$x = 6$$
$$y = 2$$
$$(6, 2)$$

**52.**

$$y = x + 5$$
$$x - 4(x + 5) = 0$$
$$x - 4x - 20 = 0$$
$$-3x - 20 = 0$$
$$-3x = 20$$
$$x = -\frac{20}{3}$$
$$y = -\frac{20}{3} + 5$$
$$y = -\frac{20}{3} + \frac{15}{3}$$
$$y = -\frac{5}{3}$$
$$\left(-\frac{20}{3}, -\frac{5}{3}\right)$$

**54.**
$$x - 2y = -1$$
$$x = 2y - 1$$
$$2y - 1 - 5y = 2$$
$$-3y - 1 = 2$$
$$-3y = 3$$
$$y = -1$$
$$x = 2(-1) - 1$$
$$x = -2 - 1$$
$$x = -3$$
$$(-3, -1)$$

**56.**
$$x - 5y = -6$$
$$x = 5y - 6$$
$$4(5y - 6) - 3y = 10$$
$$20y - 24 - 3y = 10$$
$$17y = 34$$
$$y = 2$$
$$x = 5(2) - 6$$
$$x = 10 - 6$$
$$x = 4$$
$$(4, 2)$$

**58.**
$$x + 4y = 300$$
$$x = -4y + 300$$
$$(-4y + 300) - 2y = 0$$
$$-6y = -300$$
$$y = 50$$
$$x = -4(50) + 300$$
$$x = -200 + 300$$
$$x = 100$$
$$(100, 50)$$

**60.**
$$2x + 5y = 29$$
$$5x + 2y = 13$$
$$2x = 29 - 5y$$
$$x = \frac{29 - 5y}{2}$$
$$5\left(\frac{29 - 5y}{2}\right) + 2y = 13$$
$$5(29 - 5y) + 4y = 26$$
$$145 - 25y + 4y = 26$$
$$-21y = -119$$
$$y = \frac{119}{21}$$
$$y = \frac{17}{3}$$
$$x = \frac{29 - 5(17/3)}{2} \cdot \frac{3}{3}$$
$$x = \frac{87 - 85}{6}$$
$$x = \frac{1}{3}$$
$$\left(\frac{1}{3}, \frac{17}{3}\right)$$

**62.**
$$5x - 24y = -12$$
$$-24y = -5x - 12$$
$$24y = 5x + 12$$
$$17x - (5x + 12) = 36$$
$$17x - 5x - 12 = 36$$
$$12x - 12 = 36$$
$$12x = 48$$
$$x = 4$$
$$24y = 5(4) + 12$$
$$24y = 20 + 12$$
$$24y = 32$$
$$y = \frac{32}{24}$$
$$y = \frac{4}{3}$$
$$\left(4, \frac{4}{3}\right)$$

**64.**
$$\tfrac{1}{2}x + \tfrac{3}{4}y = 10$$
$$\tfrac{3}{2}x - y = 4$$
$$-y = -\tfrac{3}{2}x + 4$$
$$y = \tfrac{3}{2}x - 4$$
$$\tfrac{1}{2}x + \tfrac{3}{4}\left(\tfrac{3}{2}x - 4\right) = 10$$
$$\tfrac{1}{2}x + \tfrac{9}{8}x - 3 = 10$$
$$\tfrac{13}{8}x = 13$$
$$x = 8$$
$$y = \tfrac{3}{2}(8) - 4$$
$$y = 12 - 4$$
$$y = 8$$
$$(8, 8)$$

**66.**
$$5x^2 = -15x - 10$$
$$5x^2 + 15 + 10 = 0$$
$$5(x^2 + 3x + 2) = 0$$
$$5(x + 2)(x + 1) = 0$$
$$x + 2 = 0 \qquad x + 1 = 0$$
$$x = -2 \qquad x = -1$$
$$y = 5(-2)^2 \qquad y = 5(-1)^2$$
$$y = 20 \qquad y = 5$$
$$(-2, 20) \qquad (-1, 5)$$

**68.**
$$5y^2 + 2y = 16$$
$$5y^2 + 2y - 16 = 0$$
$$(5y - 8)(y + 2) = 0$$
$$y = \tfrac{8}{5} \qquad y = -2$$
$$x = 5\left(\tfrac{8}{5}\right)^2 \qquad x = 5(-2)^2$$
$$x = 5\left(\tfrac{64}{25}\right) \qquad x = 5(4)$$
$$x = \tfrac{64}{5} \qquad x = 20$$
$$\left(\tfrac{64}{5}, \tfrac{8}{5}\right) \text{ and } (20, -2)$$

**70.**
$$x - y = 2$$
$$x = y + 2$$
$$y + 2 - y^2 = 0$$
$$y^2 - y - 2 = 0$$
$$(y - 2)(y + 1) = 0$$
$$y - 2 = 0 \qquad y + 1 = 0$$
$$y = 2 \qquad y = -1$$
$$x = 2 + 2 \qquad x = -1 + 2$$
$$x = 4 \qquad x = 1$$
$$(4, 2) \qquad (1, -1)$$

**72.**
$$x + y = 7$$
$$x = 7 - y$$
$$(7 - y)^2 + y^2 = 169$$
$$49 - 14y + y^2 + y^2 = 169$$
$$2y^2 - 14y - 120 = 0$$
$$2(y^2 - 7y - 60) = 0$$
$$2(y - 12)(y + 5) = 0$$
$$y - 12 = 0 \qquad y + 5 = 0$$
$$y = 12 \qquad y = -5$$
$$x = 7 - 12 \qquad x = 7 - (-5)$$
$$x = -5 \qquad x = 12$$
$$(-5, 12) \qquad (12, -5)$$

**74.**
$$x^2 + 2y = 6$$
$$x - y = -4$$
$$-y = -x - 4$$
$$y = x + 4$$
$$x^2 + 2(x + 4) = 6$$
$$x^2 + 2x + 8 = 6$$
$$x^2 + 2x + 2 = 0$$

No real solution

**76.** $3x - y = 12$
$$-y = -3x + 12$$
$$y = 3x - 12$$
$$x^2 - (3x - 12)^2 = 16$$
$$x^2 - (9x^2 - 72x + 144) = 16$$
$$x^2 - 9x^2 + 72x - 144 = 16$$
$$-8x^2 + 72x - 160 = 0$$
$$-8(x^2 - 9x + 20) = 0$$
$$-8(x - 5)(x - 4) = 0$$

| | |
|---|---|
| $x - 5 = 0$ | $x - 4 = 0$ |
| $x = 5$ | $x = 4$ |
| $y = 3(5) - 12$ | $y = 3(4) - 12$ |
| $y = 15 - 12$ | $y = 12 - 12$ |
| $y = 3$ | $y = 0$ |
| $(5, 3)$ | $(4, 0)$ |

**78.** $35x - 33y = 0$
$$12x - 11y = 92$$
$$35x = 33y$$
$$x = \frac{33}{35}y$$
$$12\left(\frac{33}{35}y\right) - 11y = 92$$
$$\frac{396}{35}y - 11y = 92$$
$$\frac{396}{35}y - \frac{385}{35}y = 92$$
$$\frac{11}{35}y = 92$$
$$y = 92\left(\frac{35}{11}\right)$$
$$y = \frac{3220}{11}$$

$$x = \frac{33}{35}\left(\frac{3220}{11}\right)$$
$$x = \frac{(3)(11)(35)(92)}{(35)(11)}$$
$$x = (3)(92)$$
$$x = 276$$

$$\left(276, \frac{3220}{11}\right)$$

Solve equation for $y$.

$$35x - 33y = 0 \qquad 12x - 11y = 92$$
$$35x = 33y \qquad 12x - 92 = 11y$$
$$\frac{35}{33}x = y \qquad \frac{12}{11}x - \frac{92}{11} = y$$

*Keystrokes:*

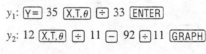

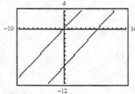

**80.** Answers vary.
$$-x + 2y = 14$$
$$3x + 5y = 24$$

**82.** Answers vary.
$$4x - 5y = -13$$
$$6x - y = 0$$

**84.**    $C = 19.25x + 50,000$

$R = 35.95x$

$R = C$

$35.95x = 19.25x + 50,000$

$16.7x = 50,000$

$x \approx 2994.012$

Thus, 2995 items must be sold to break even.

**86.**    $C = 16.40x + 30,000$

$R = 31.40x$

$R = C$

$31.40x = 16.40x + 30,000$

$15x = 30,000$

$x = 2000$ units

**88.**    $x + y = 12,000$

$0.085x + 0.10y = 1140$

$x = 12,000 - y$

$0.085(12,000 - y) + 0.10y = 1140$

$1020 - 0.085y + 0.10y = 1140$

$0.015 = 120$

$y = 8000$ at $10\%$

$x = 12,000 - 8000$

$x = 4000$ at $8.5\%$

($4000, $8000)

**90.**    $x + y = 80$

$x - y = 18$

$x = 18 + y$

$18 + y + y = 80$

$2y = 62$

$y = 31$

$x = 18 + 31 = 49$

49, 31

**92.**    $x + y = 52$

$x = 2y - 8$

$2y - 8 + y = 52$

$3y = 60$

$y = 20$

$x = 2(20) - 8 = 32$

32, 20

**94.**    $2x + 2y = 50$

$x = 5 + y$

$2(5 + y) + 2y = 50$

$10 + 2y + 2y = 50$

$4y = 40$

$y = 10$

$x = 5 + 10 = 15$

(length $= 15$ feet, width $= 10$ feet)

**96.**    $2x + 2y = 68$

$x = \frac{7}{10}y$

$2\left(\frac{7}{10}y\right) + 2y = 68$

$\frac{7}{5}y + 2y = 68$

$7y + 10y = 340$

$17y = 340$

$y = 20$

$x = \frac{7}{10}(20) = 7(2) = 14$

(width $= 14$ yards, length $= 20$ yards)

# Section 8.2    Linear Systems in Two Variables

**2.**  $x + 3y = 2$

  $\underline{-x + 2y = 3}$

    $5y = 5$

     $y = 1$

  $x + 3(1) = 2$

       $x = -1$

$(-1, 1)$

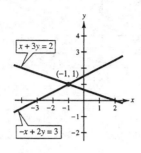

**4.**  $x + \phantom{2}y = \phantom{1}0 \Longrightarrow 2x + 2y = \phantom{10}0$

  $3x - 2y = 10 \Longrightarrow \underline{3x - 2y = \phantom{1}10}$

     $5x \phantom{aaaa} = 10$

      $x \phantom{aaaaa} = \phantom{1}2$

  $2x + y = 0$

       $y = -2$

$(2, -2)$

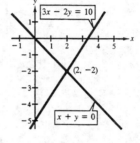

**6.**  $-x + 2y = \phantom{1}2 \Longrightarrow -3x + 6y = \phantom{1}6$

  $-3x + \phantom{2}y = 15 \Longrightarrow \phantom{aa}\underline{3x + y = \phantom{1}15}$

       $7y = 21$

        $y = \phantom{1}3$

  $-x + 2(3) = 2$

    $-x + 6 = 2$

      $-x = -4$

       $x = 4$

$(4, 3)$

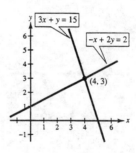

**8.**  $3x + \phantom{0}4y = \phantom{1}2 \Longrightarrow -6x - 8y = -4$

  $0.6x + 0.8y = 1.6 \Longrightarrow \phantom{aa}\underline{6x + 8y = \phantom{1}16}$

        $0 \neq 12$

No solution

Inconsistent

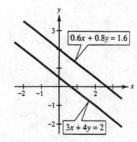

**10.**  $x - 4y = 5$

$\underline{5x + 4y = 7}$

$\qquad 6x = 12$

$\qquad x = 2$

$2 - 4y = 5$

$\quad -4y = 3$

$\qquad y = -\frac{3}{4}$

$\left(2, -\frac{3}{4}\right)$

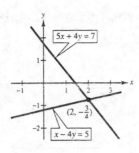

**12.** $3x + 4y = \phantom{0}0 \implies -9x - 12y = \phantom{00}0$

$9x - 5y = 17 \implies \underline{\phantom{-}9x - \phantom{0}5y = \phantom{0}17}$

$\qquad\qquad\qquad\qquad -17y = \phantom{0}17$

$\qquad\qquad\qquad\qquad\quad y = -1$

$3x + 4(-1) = 0$

$\quad 3x - 4 = 0$

$\qquad 3x = 4$

$\qquad x = \frac{4}{3}$

$\left(\frac{4}{3}, -1\right)$

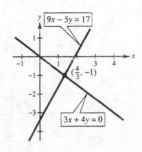

**14.**  $-x + 2y = \phantom{0}9$

$\underline{\phantom{-}x + 3y = 16}$

$\qquad 5y = 25$

$\qquad y = \phantom{0}5$

$-x + 2(5) = \phantom{0}9$

$-x + 10 = \phantom{0}9$

$\quad -x = -1$

$\qquad x = \phantom{0}1$

$(1, 5)$

**16.**  $-3x + 5y = -23$

$\underline{\phantom{-}2x - 5y = \phantom{0}\phantom{0}22}$

$\quad -x = \phantom{0}-1$

$\qquad x = \phantom{00}1$

$-3(1) + 5y = -23$

$-3 + 5y = -23$

$\qquad 5y = -20$

$\qquad y = \phantom{0}-4$

$(1, -4)$

**18.**  $-x + 2y = 12$

$\underline{\phantom{-}x + 6y = 20}$

$\qquad 8y = 32$

$\qquad y = \phantom{0}4$

$-x + 2(4) = \phantom{0}12$

$\quad -x = \phantom{0}4$

$\qquad x = -4$

$(-4, 4)$

**20.** $4x + 3y = \phantom{0}8 \implies \phantom{-}4x + 3y = \phantom{00}8$

$\phantom{4}x - 2y = 13 \implies \underline{-4x + 8y = -52}$

$\qquad\qquad\qquad\qquad 11y = -44$

$\qquad\qquad\qquad\qquad\quad y = \phantom{0}-4$

$x - 2(-4) = 13$

$\quad x + 8 = 13$

$\qquad x = 5$

$(5, -4)$

**22.** $2s - \phantom{0}t = \phantom{00}9 \implies 8s - 4t = \phantom{00}36$

$3s + 4t = -14 \implies \underline{3s + 4t = -14}$

$\qquad\qquad\qquad\qquad 11s = \phantom{00}22$

$\qquad\qquad\qquad\qquad\phantom{1}s = \phantom{000}2$

$2(2) - t = 9$

$\quad 4 - t = 9$

$\quad -t = 5$

$\qquad t = -5$

$(2, -5)$

**24.** $7r - s = -25 \implies 35r - 5s = -125$

$\phantom{7r -}2r + 5s = \phantom{-2}14 \qquad \underline{\phantom{35r -}2r + 5s = \phantom{-12}14}$

$\phantom{7r - s = -25 \implies 35r - 5s} 37r = -111$

$\phantom{7r - s = -25 \implies 35r - 5s = -1}r = \phantom{-1}-3$

$2(-3) + 5s = 14$

$\phantom{2(-3)} -6 + 5s = 14$

$\phantom{2(-3) -6 +} 5s = 20$

$\phantom{2(-3) -6 +} s = 4$

$(-3, 4)$

---

**26.** $\phantom{-}4x - 3y = 25 \implies \phantom{-}12x - 9y = \phantom{-}75$

$-3x + 8y = 10 \implies \underline{-12x + 32y = \phantom{-}40}$

$\phantom{-3x + 8y = 10 \implies -12x +} 23y = 115$

$\phantom{-3x + 8y = 10 \implies -12x + 3}y = \phantom{-1}5$

$4x - 3(5) = 25$

$\phantom{4x} - 15 = 25$

$\phantom{4x - 1} 4x = 40$

$\phantom{4x - 15 =} x = 10$

$(10, 5)$

---

**28.** $-2x + 3y = \phantom{-2}9 \implies -6x + 9y = \phantom{-2}27$

$\phantom{-}6x - 9y = -27 \implies \underline{\phantom{-2}6x - 9y = -27}$

$\phantom{-2x + 3y = 9 \implies -6x + 9} 0 = \phantom{-2}0$

Infinitely many solutions.

---

**30.** $x - \phantom{4}y = \phantom{-}-\frac{1}{2} \implies -4x + \phantom{4}4y = \phantom{-4}2$

$4x - 48y = -35 \implies \underline{\phantom{-}4x - 48y = -35}$

$\phantom{4x - 48y = -35 \implies \phantom{-}4x} -44y = -33$

$\phantom{4x - 48y = -35 \implies \phantom{-}4x -4}y = \phantom{-44}\frac{3}{4}$

$x - \frac{3}{4} = -\frac{1}{2}$

$\phantom{x} x = \frac{3}{4} - \frac{2}{4}$

$\phantom{x} x = \frac{1}{4}$

$\left(\frac{1}{4}, \frac{3}{4}\right)$

---

**32.** $0.02x - 0.05y = -0.19 \implies 2x - 5y = -19 \implies \phantom{-}8x - 20y = \phantom{-2}-76$

$0.03x + 0.04y = \phantom{-}0.52 \implies 3x + 4y = \phantom{-2}52 \implies \underline{15x + 20y = \phantom{-2}260}$

$\phantom{0.03x + 0.04y = 0.52 \implies 3x + 4y = 52 \implies 1} 23x \phantom{+ 20y} = \phantom{-2}184$

$\phantom{0.03x + 0.04y = 0.52 \implies 3x + 4y = 52 \implies 15x} x \phantom{+ 20y} = \phantom{-2}8$

$2(8) - 5y = -19$

$\phantom{2(8)} -5y = -35$

$\phantom{2(8) -5} y = \phantom{-2}7$

$(8, 7)$

---

**34.** $\phantom{-}0.15x - 0.35y = -0.5 \implies \phantom{-}0.6x - 1.4y = \phantom{-2}-2$

$-0.12x + 0.25y = \phantom{-}0.1 \implies \underline{-0.6x + 1.25y = \phantom{-2}0.5}$

$\phantom{-0.12x + 0.25y = 0.1 \implies -0.6x + 1.2} -0.15y = -1.5$

$\phantom{-0.12x + 0.25y = 0.1 \implies -0.6x + 1.25} y = \phantom{-2}10$

$0.15x - 0.35(10) = -0.5$

$\phantom{0.1} 0.15x - 3.5 = -0.5$

$\phantom{0.15x -} 0.15x = 3$

$\phantom{0.15x - 3.5 =} x = 20$

$(20, 10)$

---

**36.** $\phantom{-}12b - 13m = \phantom{-2}2 \implies \phantom{-}12b - 13m = \phantom{-2}2$

$\phantom{-}-6b + 6.5m = -2 \implies \underline{-12b + 13m = -4}$

$\phantom{-6b + 6.5m = -2 \implies -12b + 13m} 0 \neq -2$

Inconsistent

**38.** $12x - 3y = 6 \implies \quad 12x - 3y = \phantom{-}6$

$\phantom{38.}\quad 4x - \phantom{3}y = 2 \implies \underline{-12x + 3y = -6}$

$$0 = \phantom{-}0$$

Infinitely many solutions.
All solutions of the form $4x - y = 2$

**40.** $6x - 6y = 25 \implies 6x - 6y = 25$

$\phantom{40.}\quad 3y = 11 \implies \qquad \underline{6y = 22}$

$$6x = 47$$

$$x = \tfrac{47}{6}$$

$3y = 11$

$y = \tfrac{11}{3}$

$\left(\tfrac{47}{6}, \tfrac{11}{3}\right)$

**42.** $\quad 4x + y = -2 \implies -4x - y = \phantom{-}2$

$\phantom{42.}\quad -6x + y = \phantom{-}18 \implies \underline{-6x + y = \phantom{-}18}$

$$-10x = \phantom{-}20$$

$$x = -2$$

$4(-2) + y = -2$

$\phantom{4(}-8 + y = -2$

$\phantom{4(-8 + }y = 6$

$(-2, 6)$

**44.** $3y = 2x + 21$

$\phantom{44.}\quad x = 50 - 4y$

$\phantom{44.}\quad 3y = 2(50 - 4y) + 21$

$\phantom{44.}\quad 3y = 100 - 8y + 21$

$\phantom{44.}\quad 11y = 121$

$\phantom{44.}\quad y = 11$

$\phantom{44.}\quad x = 50 - 4(11)$

$\phantom{44.}\quad x = 50 - 44$

$\phantom{44.}\quad x = 6$

$(6, 11)$

**46.** $3x - 2y = -20 \implies \quad 9x - 6y = -60$

$\phantom{46.}\quad 5x + 6y = \phantom{-}32 \qquad \underline{5x + 6y = \phantom{-}32}$

$$14x = -28$$

$$x = \phantom{-}-2$$

$3(-2) - 2y = -20$

$\phantom{3(}-6 - 2y = -20$

$\phantom{3(-6 }-2y = -14$

$\phantom{3(-6 -2}y = 7$

$(-2, 7)$

**48.** $\qquad x + 2y = 4 \implies \quad x + 2y = \phantom{-}4$

$\phantom{48.}\quad \tfrac{1}{2}x + \tfrac{1}{3}y = 1 \qquad \underline{-x - \tfrac{2}{3}y = -2}$

$$\tfrac{4}{3}y = \phantom{-}2$$

$$y = 2\left(\tfrac{3}{4}\right)$$

$$y = \phantom{-}\tfrac{3}{2}$$

$x + 2\left(\tfrac{3}{2}\right) = 4$

$\phantom{x + 2}x + 3 = 4$

$\phantom{x + 2(}x = 1$

$\left(1, \tfrac{3}{2}\right)$

**50.** $\quad 4x - \phantom{1}5y = \phantom{1}3 \implies -5y = -4x + \phantom{1}3 \implies y = \tfrac{4}{5}x - \tfrac{3}{5}$

$\phantom{50.}\quad -8x + 10y = 14 \implies \phantom{-}10y = \phantom{-}8x + 14 \implies y = \tfrac{4}{5}x + \tfrac{7}{5}$

No solutions $\implies$ inconsistent

**52.** $\quad x + 10y = 12 \implies 10y = -x + 12 \implies y = -\tfrac{1}{10}x + \tfrac{6}{5}$

$\phantom{52.}\quad -2x + \phantom{1}5y = \phantom{1}2 \implies \phantom{1}5y = 2x + \phantom{1}2 \implies y = \phantom{-}\tfrac{2}{5}x + \tfrac{2}{5}$

One solution $\implies$ consistent

**54.**   $4x - 5y = 28 \implies -5y = -4x + 28 \implies y = \frac{4}{5}x - \frac{28}{5}$

$-2x + 2.5y = -14 \implies \frac{5}{2}y = 2x - 14 \implies y = \frac{4}{5}x - \frac{28}{5}$

Many solutions $\implies$ consistent

**56.**   $12x - 18y = 5 \implies 36x - 54y = 15$

$-18x + ky = 10 \qquad -36x + 2ky = 20$

so $2k = 54$

$k = 27$

**58.** There are many correct answers.

$x + y = 4$

$2x + 3y = 20$

**60.** *Verbal model:* $\boxed{\begin{array}{c}\text{Total}\\\text{cost}\end{array}} = \boxed{\begin{array}{c}\text{Cost}\\\text{per unit}\end{array}} \cdot \boxed{\begin{array}{c}\text{Number}\\\text{of units}\end{array}} + \boxed{\begin{array}{c}\text{Initial}\\\text{cost}\end{array}}$

$\boxed{\begin{array}{c}\text{Total}\\\text{revenue}\end{array}} = \boxed{\begin{array}{c}\text{Price}\\\text{per unit}\end{array}} \cdot \boxed{\begin{array}{c}\text{Number}\\\text{of units}\end{array}}$

*System:*   $C = 1.20x + 8000$

$R = 2.00x$

$R = C$

$2.00x = 1.20x + 8000$

$.80x = 8000$

$x = 10,000$ units

**62.** *Verbal model:* $\boxed{\begin{array}{c}\text{Amount invested}\\\text{in 4\% fund}\end{array}} + \boxed{\begin{array}{c}\text{Amount invested}\\\text{in 5\% fund}\end{array}} = \boxed{\begin{array}{c}\text{Total}\\\text{invested}\end{array}}$

$\boxed{\begin{array}{c}\text{Interest in}\\\text{4\% fund}\end{array}} + \boxed{\begin{array}{c}\text{Interest in}\\\text{5\% fund}\end{array}} = \boxed{\begin{array}{c}\text{Total}\\\text{interest}\end{array}}$

*System:*   $x + y = 4500 \implies -0.04x - 0.04y = -180$

$0.04x + 0.05y = 210 \qquad \underline{0.04x + 0.05y = \phantom{-}210}$

$0.01y = \phantom{-}30$

$y = \$3000$ at 5\%

$x + 3000 = 4500$

$x = \$1500$ at 4\%

**64.** *Verbal model:* $\boxed{\text{Distance}} = \boxed{\text{Rate}} \cdot \boxed{\text{Time}}$

*Labels:* Time at 55 mph $= x$

$D_1 = $ distance at 42 mph for 4 hours $+$ at 55 mph for $x$ hours

$D_2 = $ distance at 50 mph for $(4 + x)$ hours

*System:*   since $D_1 = D_2$

$42(4) + 55x = 50(4 + x)$

$168 + 55x = 200 + 50x$

$5x = 32$

$x = 6.4$ hours

**66.** *Verbal model:*

$$\boxed{\begin{array}{c}\text{Plane speed} \\ \text{(still air)}\end{array}} - \boxed{\begin{array}{c}\text{Speed} \\ \text{of air}\end{array}} = \boxed{\begin{array}{c}\text{Speed into} \\ \text{headwind}\end{array}}$$

$$\boxed{\begin{array}{c}\text{Plane speed} \\ \text{(still air)}\end{array}} + \boxed{\begin{array}{c}\text{Speed} \\ \text{of air}\end{array}} = \boxed{\begin{array}{c}\text{Speed with} \\ \text{headwind}\end{array}}$$

*System:*

$$x - y = \frac{3000}{6.25} \Longrightarrow \quad x - y = 480$$

$$x + y = \frac{3000}{5} \Longrightarrow \quad \begin{array}{r} x + y = 600 \\ \hline 2x = 1080 \end{array}$$

$$x = 540 \text{ miles per hour (plane)}$$

$$540 + y = 600$$

$$y = 60 \text{ miles per hour (wind)}$$

**68.** *Verbal model:*

$$\boxed{\begin{array}{c}\text{Number} \\ \text{adult tickets}\end{array}} \cdot \boxed{\begin{array}{c}\text{Price} \\ \text{adult tickets}\end{array}} + \boxed{\begin{array}{c}\text{Number} \\ \text{children's tickets}\end{array}} \cdot \boxed{\begin{array}{c}\text{Price} \\ \text{children's tickets}\end{array}} = \boxed{\begin{array}{c}\text{Total receipts} \\ \text{(first night)}\end{array}}$$

$$\boxed{\begin{array}{c}\text{Number} \\ \text{adult tickets}\end{array}} \cdot \boxed{\begin{array}{c}\text{Price} \\ \text{adult tickets}\end{array}} + \boxed{\begin{array}{c}\text{Number} \\ \text{children's tickets}\end{array}} \cdot \boxed{\begin{array}{c}\text{Price} \\ \text{children's tickets}\end{array}} = \boxed{\begin{array}{c}\text{Total receipts} \\ \text{(second night)}\end{array}}$$

*Labels:*   Price adult tickets $= x$

Price children's tickets $= y$

*System:*

$$\begin{array}{r} 100x + 175y = 937.50 \\ 200x + 316y = 1790.00 \\ \hline -200x - 350y = -1875.00 \\ 200x + 316y = 1790.00 \\ \hline -34y = -85 \end{array}$$

$$y = \$2.50 \text{ price children's tickets}$$

$$100x + 175(2.50) = 937.50$$

$$100x + 437.50 = 937.50$$

$$100x = 500$$

$$x = \$5.00 \text{ price adult tickets}$$

**70.** *Verbal model:* $8 \left( \boxed{\begin{array}{c}\text{Cost of} \\ \text{regular gasoline}\end{array}} \right) + 12 \left( \boxed{\begin{array}{c}\text{Cost of} \\ \text{premium gasoline}\end{array}} \right) = \boxed{\$27.84}$

$\boxed{\begin{array}{c}\text{Cost of} \\ \text{premium gasoline}\end{array}} = \boxed{\$0.17} + \boxed{\begin{array}{c}\text{Cost of} \\ \text{regular gasoline}\end{array}}$

*Labels:* Cost of regular gasoline $= x$

Cost of premium gasoline $= y$

*System:* $\qquad 8x + 12y = 27.84$

$\qquad\qquad\qquad y = 0.17 + x$

$8x + 12(0.17 + x) = 27.84$

$8x + 2.04 + 12x = 27.84$

$\qquad\qquad 20x = 25.8$

$\qquad\qquad\quad x = \$1.29 \text{ regular gasoline}$

$y = 0.17 + 1.29$

$y = \$1.46 \text{ premium gasoline}$

**72.** *Verbal model:* $\boxed{\begin{array}{c}\text{Amount of} \\ \text{80\% solution}\end{array}} + \boxed{\begin{array}{c}\text{Amount of} \\ \text{50\% solution}\end{array}} = \boxed{50}$

$\boxed{\begin{array}{c}\text{Acid in} \\ \text{80\% solution}\end{array}} + \boxed{\begin{array}{c}\text{Acid in} \\ \text{50\% solution}\end{array}} = \boxed{\begin{array}{c}\text{Acid in} \\ \text{70\% solution}\end{array}}$

*Labels:* Amount of 80% solution $= x$

Amount of 50% solution $= y$

*System:* $\qquad x + y = 50$

$\qquad\quad 0.8x + 0.5y = 35$

$\qquad\qquad\quad y = 50 - x$

$0.8x + 0.5(50 - x) = 35$

$0.8x + 25 - 0.5x = 35$

$\qquad\qquad 0.3x = 10$

$\qquad\qquad\quad 3x = 100$

$\qquad\qquad\quad\; x = \dfrac{100}{3} \text{ gallons of 80\% solution}$

$y = 50 - \dfrac{100}{3} = \dfrac{150 - 100}{3} = \dfrac{50}{3} \text{ gallons of 50\% solution}$

**74.** (a)   $5m + 3b = 7$

$$\underline{3m + 3b = 4}$$

$$2m \phantom{+ 3b} = 3$$

$$m \phantom{+ 3b} = \tfrac{3}{2}$$

$$5\left(\tfrac{3}{2}\right) + 3b = 7$$

$$15 + 6b = 14$$

$$6b = -1$$

$$b = -\tfrac{1}{6}$$

$$y = \tfrac{3}{2}x - \tfrac{1}{6}$$

(b)

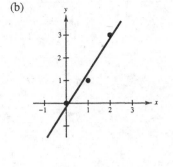

**76.** (a)

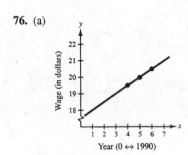

(b)   $3b + 15m = \phantom{0}59.96 \implies -15b - 75m = -299.80$

$\phantom{(b) \;}15b + 77m = 300.79 \implies \underline{\phantom{-}15b + 77m = \phantom{-}300.79}$

$$2m = \phantom{00}.99$$

$$m = \phantom{00}.495$$

$$3b + 15(.495) = 59.96$$

$$3b + 7.425 = 59.96$$

$$3b = 52.535$$

$$b \approx 17.51$$

$$y = 0.495x + 17.51$$

(c) The slope of the line represents the average annual increase in hourly wage.

**80.** There is no solution.

**78.** (a)   $x + \phantom{0}y = \phantom{00}115$

$\phantom{(a) \;}8x + 15y = 1445$

(b)   $x + \phantom{0}y = \phantom{00}115$

$\phantom{(b) \;}8x + 15y = 1445$

$\phantom{(b) \;}y = 115 - x$

$\phantom{(b) \;}8x + 15(115 - x) = \phantom{0}1445$

$\phantom{(b) \;}8x + 1725 - 15x = \phantom{0}1445$

$$-7x = -280$$

$$x = \phantom{00}40 \text{ at } \$8$$

$$115 - x = \phantom{00}75 \text{ at } \$15$$

(c)   $x + \phantom{0}y = \phantom{00}115 \implies -8x - 8y = -920$

$\phantom{(c) \;}8x + 15y = 1445 \implies \underline{\phantom{-}8x + 15y = 1445}$

$\phantom{(c) \;}x = \phantom{00}40 \text{ at } \$8 \phantom{aaaaa} 7y = 525$

$$y = 75 @ \$15$$

$$x = 40 @ \$8$$

**82.** When you add the equations to eliminate one variable, both are eliminated, yielding a contradiction. For example, when you add the equations in the system $x - y = 3$ and $x - y = 8$ after multiplying the first equation by $-1$, you obtain $0 = 5$.

# Section 8.3    Linear Systems in Three Variables

**2.** (a) $3(-2) - (4) + 4(0) \stackrel{?}{=} -10$

$\qquad -6 - 4 + 0 \stackrel{?}{=} -10$

$\qquad\qquad -10 = -10$

$\qquad -(-2) + 4 + 2(0) \stackrel{?}{=} 6$

$\qquad\qquad 2 + 4 \stackrel{?}{=} 6$

$\qquad\qquad\qquad 6 = 6$

$\qquad 2(-2) - (4) + 0 \stackrel{?}{=} -8$

$\qquad\qquad -4 - 4 \stackrel{?}{=} -8$

$\qquad\qquad\qquad -8 = -8$

Solution

(c) $3(-1) - (-1) + 4(5) \stackrel{?}{=} -10$

$\qquad -3 + 1 + 20 \stackrel{?}{=} -10$

$\qquad\qquad 18 \neq -10$

Not a solution

(b) $3(0) - (-3) + 4(10) \stackrel{?}{=} -10$

$\qquad 0 + 3 + 40 \stackrel{?}{=} -10$

$\qquad\qquad 43 \neq -10$

Not a solution

(d) $3(7) - (19) + 4(-3) \stackrel{?}{=} -10$

$\qquad 21 - 19 - 12 \stackrel{?}{=} -10$

$\qquad\qquad -10 = -10$

$\qquad -(7) + (19) + 2(-3) \stackrel{?}{=} 6$

$\qquad\qquad -7 + 19 - 6 \stackrel{?}{=} 6$

$\qquad\qquad\qquad 6 = 6$

$\qquad 2(7) - (19) + (-3) \stackrel{?}{=} -8$

$\qquad\qquad 14 - 19 - 3 \stackrel{?}{=} -8$

$\qquad\qquad\qquad -8 = -8$

Solution

**4.** $10y - 3(3) = 11$

$\qquad 10y - 9 = 11$

$\qquad 10y = 20$

$\qquad y = 2$

$5x + 4(2) - 3 = 0$

$\qquad 5x + 8 - 3 = 0$

$\qquad 5x + 5 = 0$

$\qquad 5x = -5$

$\qquad x = -1$

$(-1, 2, 3)$

**6.** $3(10) + 2y = 2$

$\qquad 30 + 2y = 2$

$\qquad 2y = -28$

$\qquad y = -14$

$10 + (-14) + 2z = 0$

$\qquad -4 + 2z = 0$

$\qquad 2z = 4$

$\qquad z = 2$

$(10, -14, 2)$

**8.** Yes. The first equation was added to the second equation. Then the first equation was multiplied by $-2$ and added to the third equation.

**10.** $2x + 3y = 7 \implies -4x - 6y = -14$

$4x - 2y = -2 \implies \underline{\qquad\qquad -8y = -16}$

Eliminated the $x$-term from the second equation.

**12.** $-2x + 4y - 6z = -10 \qquad x - 2y + 3z = 5$

$\underline{\qquad 2x - 3z = 0} \qquad -x + y + 5z = 4$

$4y - 9z = -10 \qquad 4y - 9z = -10$

Eliminated the $x$-term in Equation 3.

---

**14.**
$$x \qquad\qquad = 3$$
$$-x + 3y \qquad = 3$$
$$y + 2z = 4$$

$$x \qquad\qquad = 3$$
$$3y \qquad = 6$$
$$y + 2z = 4$$

$$x \qquad\qquad = 3$$
$$y \qquad = 2$$
$$y + 2z = 4$$

$$x \qquad\qquad = 3$$
$$y \qquad = 2$$
$$2z = 2$$

$$x \qquad\qquad = 3$$
$$y \qquad = 2$$
$$z = 1$$

$$(3, 2, 1)$$

**16.**
$$x + y + z = 2$$
$$-x + 3y + 2z = 8$$
$$4x + y = 4$$

$$x + y + z = 2$$
$$+4y + 3z = 10$$
$$-3y - 4z = -4$$

$$x + y + z = 2$$
$$y + \tfrac{3}{4}z = \tfrac{5}{2}$$
$$-3y - 4z = -4$$

$$x + y + z = 2$$
$$y + \tfrac{3}{4}z = \tfrac{5}{2}$$
$$-\tfrac{7}{4}z = \tfrac{7}{2}$$

$$x + y + z = 2$$
$$y + \tfrac{3}{4}z = \tfrac{5}{2}$$
$$z = -2$$

$$y + \tfrac{3}{4}(-2) = \tfrac{5}{2}$$
$$y = 4$$
$$x + 4 + (-2) = 2$$
$$x = 0$$

$$(0, 4, -2)$$

**18.**
$$x - y + 2z = -4$$
$$3x + y - 4z = -6$$
$$2x + 3y - 4z = 4$$

$$x - y + 2z = -4$$
$$4y - 10z = 6$$
$$5y - 8z = 12$$

$$x - y + 2z = -4$$
$$4y - 10z = 6$$
$$\tfrac{9}{2}z = \tfrac{9}{2}$$

$$x - y + 2z = -4$$
$$y - \tfrac{5}{2}z = \tfrac{3}{2}$$
$$z = 1$$

$$y - \tfrac{5}{2}(1) = \tfrac{3}{2}$$
$$y - \tfrac{5}{2} = \tfrac{3}{2}$$
$$y = 4$$

$$x - 4 + 2 = -4$$
$$x = -2$$

$$(-2, 4, 1)$$

**20.**

$x + 6y + 2z = 9$

$3x - 2y + 3z = -1$

$5x - 5y + 2z = 7$

$x + 6y + 2z = 9$

$-20y - 3z = -28$

$-35y - 8z = -38$

$x + 6y + 2z = 9$

$-20y - 3z = -28$

$-\frac{11}{4}z = 11$

$x + 6y + 2z = 9$

$y + \frac{3}{20}z = \frac{7}{5}$

$z = -4$

$y + \frac{3}{20}(-4) = \frac{7}{5}$

$y - \frac{3}{5} = \frac{7}{5}$

$y = \frac{10}{5}$

$y = 2$

$x + 6(2) + 2(-4) = 9$

$x + 12 - 8 = 9$

$x = 5$

$(5, 2, -4)$

**22.**

$6y + 4z = -12$

$3x + 3y = 9$

$2x - 3x = 10$

$x + y = 3$

$6y + 4z = -12$

$2x - 3z = 10$

$x + y = 3$

$6y + 4z = -12$

$-2y - 3z = 4$

$x + y = 3$

$y + \frac{2}{3}z = -2$

$-2y - 3z = 4$

$x + y = 3$

$y + \frac{2}{3}z = -2$

$-\frac{5}{3}z = 0$

$x + y = 3$

$y + \frac{2}{3}z = -2$

$z = 0$

$y + \frac{2}{3}(0) = -2$

$y = -2$

$x + (-2) = 3$

$x = 5$

$(5, -2, 0)$

**24.**

$2x - 4y + z = 0$

$3x + 2z = -1$

$-6x + 3y + 2z = -10$

$x - 2y + \frac{1}{2}z = 0$

$3x + 2z = -1$

$-6x + 3y + 2z = -10$

$x - 2y + \frac{1}{2}z = 0$

$6y + \frac{1}{2}z = -1$

$-9y + 5z = -10$

$x - 2y + \frac{1}{2}z = 0$

$y + \frac{1}{12}z = \frac{-1}{6}$

$-9y + 5z = -10$

$x - 2y + \frac{1}{2}z = 0$

$y + \frac{1}{12}z = \frac{-1}{6}$

$\frac{23}{4}z = \frac{-23}{2}$

$x - 2y + \frac{1}{2}z = 0$

$y + \frac{1}{12}z = \frac{-1}{6}$

$z = -2$

$y + \frac{1}{12}(-2) = \frac{-1}{6}$

$y = 0$

$x - 2(0) + \frac{1}{2}(-2) = 0$

$x = 1$

$(1, 0, -2)$

**26.** 
$$3x - y - 2z = 5$$
$$2x + y + 3z = 6$$
$$6x - y - 4z = 9$$

$$x - \tfrac{1}{3}y - \tfrac{2}{3}z = \tfrac{5}{3}$$
$$2x + y + 3z = 6$$
$$6x - y - 4z = 9$$

$$x - \tfrac{1}{3}y - \tfrac{2}{3}z = \tfrac{5}{3}$$
$$\tfrac{5}{3}y + \tfrac{13}{3}z = \tfrac{8}{3}$$
$$y = -1$$

$$x - \tfrac{1}{3}y - \tfrac{2}{3}z = \tfrac{5}{3}$$
$$y = -1$$
$$5y + 13z = 8$$

$$x - \tfrac{1}{3}y - \tfrac{2}{3}z = \tfrac{5}{3}$$
$$y = -1$$
$$13z = 13$$

$$x - \tfrac{1}{3}y - \tfrac{2}{3}z = \tfrac{5}{3}$$
$$y = -1$$
$$z = 1$$

$$x - \tfrac{1}{3}(-1) - \tfrac{2}{3}(1) = \tfrac{5}{3}$$
$$x + \tfrac{1}{3} - \tfrac{2}{3} = \tfrac{5}{3}$$
$$x = 2$$

$$(2, -1, 1)$$

**28.** 
$$5x + 2y = -8$$
$$z = 5$$
$$3x - y + z = 9$$

$$x - \tfrac{1}{3}y + \tfrac{1}{3}z = 3$$
$$5x + 2y = -8$$
$$z = 5$$

$$x - \tfrac{1}{3}y + \tfrac{1}{3}z = 3$$
$$\tfrac{11}{3}y - \tfrac{5}{3}z = -23$$
$$z = 5$$

$$x - \tfrac{1}{3}y + \tfrac{1}{3}z = 3$$
$$y - \tfrac{5}{11}z = -\tfrac{69}{11}$$
$$z = 5$$

$$y - \tfrac{5}{11}(5) = -\tfrac{69}{11}$$
$$y = -\tfrac{44}{11} = -4$$

$$x - \tfrac{1}{3}(-4) + \tfrac{1}{3}(5) = 3$$
$$x + \tfrac{4}{3} + \tfrac{5}{3} = 3$$
$$x = 0$$

$$(0, -4, 5)$$

**30.** 
$$x + 2y - 2z = 4$$
$$2x + 5y - 7z = 5$$
$$3x + 7y - 9z = 10$$

$$x + 2y - 2z = 4$$
$$y - 3z = -3$$
$$y - 3z = -2$$

inconsistent; no solution

**32.** 
$$2x + y - z = 4$$
$$y + 3z = 2$$
$$3x + 2y = 4$$

$$x + \tfrac{1}{2}y - \tfrac{1}{2}z = 2$$
$$y + 3z = 2$$
$$3x + 2y = 4$$

$$x + \tfrac{1}{2}y - \tfrac{1}{2}z = 2$$
$$y + 3z = 2$$
$$\tfrac{1}{2}y + \tfrac{3}{2}z = -2$$

$$x + \tfrac{1}{2}y - \tfrac{1}{2}z = 2$$
$$y + 3z = 2$$
$$y + 3z = -4$$

inconsistent; no solution

**34.** 
$$2x + 3z = 4$$
$$5x + y + z = 2$$
$$11x + 3y - 3z = 0$$

$$x + \tfrac{3}{2}z = 2$$
$$5x + y + z = 2$$
$$11x + 3y - 3z = 0$$

$$x + \tfrac{3}{2}z = 2$$
$$y - \tfrac{13}{2}z = -8$$
$$3y - \tfrac{39}{2}z = -22$$

$$x + \tfrac{3}{2}z = 2$$
$$y - \tfrac{13}{2}z = -8$$
$$0 = 2$$

inconsistent; no solution

**36.** $0.3x - 0.1y + 0.2z = 0.35$

$\qquad 2x + y - 2z = -1$

$\qquad 2x + 4y + 3z = 10.5$

$\qquad 30x - 10y + 20z = 35$

$\qquad 2x + y - 2z = -1$

$\qquad 20x + 40y + 30z = 105$

$\qquad x + \dfrac{1}{2}y - z = \dfrac{-1}{2}$

$\qquad 30x - 10y + 20z = 35$

$\qquad 20x + 40y + 30z = 105$

$\qquad x + \dfrac{1}{2}y - z = -\dfrac{1}{2}$

$\qquad -25y + 50z = 50$

$\qquad 30y + 50z = 115$

$\qquad x + \dfrac{1}{2}y - z = \dfrac{-1}{2}$

$\qquad y - 2z = -2$

$\qquad 30y + 50z = 115$

$\qquad x + \dfrac{1}{2}y - z = \dfrac{-1}{2}$

$\qquad y - 2z = -2$

$\qquad 110z = 175$

$\qquad x + \dfrac{1}{2}y - z = \dfrac{-1}{2}$

$\qquad y - 2z = -2$

$\qquad z = \dfrac{35}{22}$

$\qquad y - 2\left(\dfrac{35}{22}\right) = -2$

$\qquad y = \dfrac{13}{11}$

$\qquad x + \dfrac{1}{2}\left(\dfrac{13}{11}\right) - \dfrac{35}{11} = \dfrac{-1}{2}$

$\qquad x = \dfrac{1}{2}$

$\left(\dfrac{1}{2}, \dfrac{13}{11}, \dfrac{35}{22}\right)$

**38.** $x - 2y - z = 3$

$\qquad 2x + y - 3z = 1$

$\qquad x + 8y - 3z = -7$

$\qquad x - 2y - z = 3$

$\qquad 5y - z = -5$

$\qquad 10y - 2z = -10$

$\qquad x - 2y - z = 3$

$\qquad y - \dfrac{1}{5}z = -1$

$\qquad 10y - 2z = -10$

$\qquad x - 2y - z = 3$

$\qquad y - \dfrac{1}{5}z = -1$

$\qquad 0 = 0$

$y = \dfrac{1}{5}z - 1$

$x - 2\left(\dfrac{1}{5}z - 1\right) - z = 3$

$\qquad x - \dfrac{2}{5}z + 2 - z = 3$

$\qquad\qquad x - \dfrac{7}{5}z = 1$

$\qquad\qquad\qquad x = \dfrac{7}{5}z + 1$

Let $a = z$.

$\left(\dfrac{7}{5}a + 1, \dfrac{1}{5}a - 1, a\right)$

**40.** $\quad x + 6y + 2z = 9$

$\qquad 3x - 2y + 3z = -1$

$\qquad 5x - 5y + 2z = 7$

$\qquad x + 6y + 2z = 9$

$\qquad -20y - 3z = -28$ ✓

$\qquad -35y - 8z = -38$

$\qquad x + 6y + 2z = 9$

$\qquad y + \dfrac{3}{20}z = \dfrac{7}{5}$ ✓

$\qquad -35y - 8z = -38$

$\qquad x + 6y + 2z = 9$

$\qquad y + \dfrac{3}{20}z = \dfrac{7}{5}$

$\qquad \dfrac{-11}{4}z = 11$

$\qquad z = -4$

$\qquad y + \dfrac{3}{20}(-4) = \dfrac{7}{5}$

$\qquad y = \dfrac{7}{5} + \dfrac{3}{5}$

$\qquad y = 2$

$x + 6(2) + 2(-4) = 9$

$\qquad x + 12 - 8 = 9$

$\qquad x = 5$

$(5, 2, -4)$

**42.**   $x + y + z = 2$

$2x - y - 2z = 23$

$x - 2y = -9$

Only one example, many correct answers.

**44.**   $48 = \frac{1}{2}a(1)^2 + v_0(1) + s_0$

$64 = \frac{1}{2}a(2)^2 + v_0(2) + s_0$

$48 = \frac{1}{2}a(3)^2 + v_0(3) + s_0$

$96 = a + 2v_0 + 2s_0$

$64 = 2a + 2v_0 + s_0$

$96 = 9a + 6v_0 + 2s_0$

$96 = a + 2v_0 + 2s_0$

$-128 = -2v_0 - 3s_0$

$-768 = -12v_0 - 16s_0$

$96 = a + 2v_0 + 2s_0$

$64 = v_0 + \frac{3}{2}s_0$

$-768 = -12v_0 - 16s_0$

$96 = a + 2v_0 + 2s_0$

$64 = v_0 + \frac{3}{2}s_0$

$0 = +2s_0$

$0 = s_0$

$64 = v_0 + 0$

$96 = a + 2(64) + 0$

$-32 = a$

$s = -16t^2 + 64t$

**46.**   $10 = \frac{1}{2}a(0)^2 + v_0 + s_0$

$54 = \frac{1}{2}a(1)^2 + v_0(1) + s_0$

$46 = \frac{1}{2}a(3)^2 + v_0(3) + s_0$

$10 = s_0$

$108 = a + 2v_0 + 2s_0$

$92 = 9a + 6v_0 + 2s_0$

$10 = s_0$

$108 = a + 2v_0 + 2s_0$

$-880 = -12v_0 - 16s_0$

$10 = s_0$

$108 = a + 2v_0 + 2s_0$

$\frac{220}{3} = v_0 + \frac{4}{3}s_0$

$\frac{220}{3} = v_0 + \frac{4}{3}(10)$

$60 = v_0$

$108 = a + 2(60) + 2(10)$

$-32 = a$

$s = -16t^2 + 60t + 10$

**48.**   $5 = a(0)^2 + b(0) + c$

$6 = a(1)^2 + b(1) + c$

$5 = a(2)^2 + b(2) + c$

$5 = \qquad\qquad c$

$6 = \quad a + \quad b + c$

$5 = \quad 4a + 2b + c$

$5 = \qquad\qquad c$

$6 = \quad a + \quad b + c$

$-19 = \qquad -2b - 3c$

$-19 = -2b - 3(5)$

$-4 = -2b$

$2 = b$

$6 = a + 2 + 5$

$-1 = a$

$y = -x^2 + 2x + 5$

**50.**    $2 = a(1)^2 + b(1) + c$

$1 = a(2)^2 + b(2) + c$

$-4 = a(3)^2 + b(3) + c$

$2 = a + b + c$

$1 = 4a + 2b + c$

$-4 = 9a + 3b + c$

$2 = a + b + c$

$-7 = -2b - 3c$

$-22 = -6b - 8c$

$2 = a + b + c$

$-7 = -2b - 3c$

$-1 = c$

$c = -1$

$-7 = -2b - 3(-1)$

$-7 = -2b + 3$

$-10 = -2b$

$5 = b$

$2 = a + 5 + (-1)$

$2 = a + 4$

$-2 = a$

$y = -2x^2 + 5x - 1$

**52.**    $-1 = a(-1)^2 + b(-1) + c$

$1 = a(1)^2 + b(1) + c$

$-4 = a(2)^2 + b(2) + c$

$-1 = a - b + c$

$1 = a + b + c$

$-4 = 4a + 2b + c$

$-1 = a - b + c$

$2 = +2b$

$0 = +6b - 3c$

$1 = b$

$0 = 6(1) - 3c$

$-6 = -3c$

$2 = c$

$-1 = a - 1 + 2$

$-2 = a$

$y = -2x^2 + x + 2$

**54.** (a)  Points $(3, 6412)$, $(4, 6336)$, $(5, 6352)$

$a(3)^2 + b(3) + c = 6412 \implies 9a + 3b + c = 6412$

$a(4)^2 + b(4) + c = 6336 \implies 16a + 4b + c = 6336$

$a(5)^2 + b(5) + c = 6352 \implies 25a + 5b + c = 6352$

$9a + 3b + c = 6412 \implies 9a + 3b + c = 6412$

$7a + b = -76 \implies 7a + b = -76$

$16a + 2b = -60 \implies 2a = 92 \implies a = 46$

$7(46) + b = -76 \qquad 9(46) + 3(-398) + c = 6412$

$b = -398 \qquad 414 - 1194 + c = 6412$

$c = 7192$

$y = 46t^2 - 398t + 7192$

(b) *Keystrokes:*

$\boxed{\text{Y=}}$  46  $\boxed{\text{X,T,}\theta}$  $\boxed{x^2}$  $\boxed{-}$  398  $\boxed{\text{X,T,}\theta}$  $\boxed{\div}$  7192  $\boxed{\text{GRAPH}}$

(c) $t = 10$    $y = 46(10)^2 - 398(10) + 7192 = 7812$ thousand metric tons

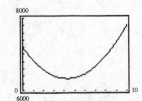

**56.**
$$0^2 + 0^2 + D(0) + E(0) + F = 0$$
$$0^2 + 6^2 + D(0) + E(6) + F = 0$$
$$(-3)^2 + 3^2 + D(-3) + E(3) + F = 0$$

$$F = 0$$
$$6E + F = -36$$
$$-3D + 3E + F = -18$$

$$6E + 0 = -36$$
$$E = -6$$
$$-3D + 3(-6) + 0 = -18$$
$$-3D = 0$$
$$D = 0$$

$$x^2 + y^2 - 6y = 0$$

**58.**
$$0^2 + 0^2 + D(0) + E(0) + F = 0$$
$$0^2 + 2^2 + D(0) + E(2) + F = 0$$
$$3^2 + 0^2 + D(3) + E(0) + F = 0$$

$$F = 0$$
$$2E + F = 0$$
$$3D + F = 0$$

$$2E + 0 = -4$$
$$2E = -4$$
$$E = -2$$

$$3D + 0 = -9$$
$$3D = -9$$
$$D = -3$$

$$x^2 + y^2 - 3x - 2y = 0$$

**60.**
$$5^2 + 13^2 + D(5) + E(13) + F = 0$$
$$17^2 + 5^2 + D(17) + E(5) + F = 0$$
$$10^2 + 12^2 + D(10) + E(12) + F = 0$$

$$5D + 13E + F = -194$$
$$17D + 5E + F = -314$$
$$10D + 12E + F = -244$$

$$F + 5D + 13E = -194$$
$$F + 17D + 5E = -314$$
$$F + 10D + 12E = -244$$

$$F + 5D + 13E = -194$$
$$12D - 8E = -120$$
$$5D - E = -50$$

$$F + 5D + 13E = -194$$
$$3D - 2E = -30$$
$$5D - E = -50$$

$$F + 5D + 13E = -194$$
$$3D - 2E = -30$$
$$\tfrac{7}{3}E = 0$$

$$F + 5D + 13E = -194$$
$$D - \tfrac{2}{3}E = -10$$
$$E = 0$$

$$D - \tfrac{2}{3}(0) = -10$$
$$D = -10$$

$$F + 5(-10) + 13(0) = -194$$
$$F - 50 = -194$$
$$F = -144$$

$$x^2 + y^2 - 10x - 144 = 0$$

**62.**

$$x + y + z = 10$$

$$0.1x + 0.2y + 0.5z = 2.5$$

$$\begin{cases} x + y + z = 10 \\ 0.1x + 0.2y + 0.5z = 2.5 \end{cases} \implies \begin{cases} x + y + z = 10 \\ x + 2y + 5z = 25 \end{cases}$$

$$\implies \begin{cases} x + y + z = 10 \\ y + 4z = 15 \end{cases}$$

$$y = -4z + 15$$

$$x + (-4z + 15) + z = 10$$

$$x - 4z + 15 + z = 10$$

$$x - 3z = -5$$

$$z = a \quad y = -4a + 15 \quad x = 3a - 5$$

$$-4a + 15 \geq 0 \quad 3a - 5 \geq 0 \quad a \geq 0$$

$$-4a \geq -15 \quad 3a \geq 5$$

$$a \leq \frac{15}{4} \qquad a \geq \frac{5}{3}$$

$z = a$ must be in the interval $\left[\frac{5}{3}, \frac{15}{4}\right]$

(a) $z = a = 2 \quad y = -4(2) + 15 \quad x = 3(2) - 5$    10% solution: $6\frac{1}{4}$ liters

$\qquad\qquad\qquad y = -8 + 15 \qquad x = 6 - 5$    20% solution: 0 liters

$\qquad\qquad\qquad y = 7 \qquad\qquad x = 1$    50% solution: $3\frac{3}{4}$ liters

(b) $z = \frac{5}{3} \quad y = -4\left(\frac{5}{3}\right) + 15 \quad x = 3\left(\frac{5}{3}\right) - 5$    10% solution: 1 liters

$\qquad\qquad y = -\frac{20}{3} + \frac{45}{3} \qquad x = 5 - 5$    20% solution: 7 liters

$\qquad\qquad y = \frac{25}{3} \qquad\qquad x = 0$    50% solution: 2 liters

(c) $z = a = \frac{15}{4} \quad y = -4\left(\frac{15}{4}\right) + 15 \quad x = 3\left(\frac{15}{4}\right) - 5$    10% solution: 0 liters

$\qquad\qquad y = -15 + 15 \qquad x = \frac{45}{4} - \frac{20}{4}$    20% solution: $8\frac{1}{3}$ liters

$\qquad\qquad y = 0 \qquad\qquad x = \frac{25}{4}$    50% solution: $1\frac{2}{3}$ liters

**64.**

$$A + B + C = 0$$
$$-B + C = 0$$
$$-A = 1$$

$$A + B + C = 0$$
$$A + 2C = 0$$
$$-A = 1$$
$$A = -1$$

$$-1 + 2C = 0$$
$$2C = 1$$
$$C = \frac{1}{2}$$

$$-1 + B + \frac{1}{2} = 0$$
$$B = \frac{1}{2}$$

$$\frac{1}{x^3 - x} = \frac{-1}{x} + \frac{\frac{1}{2}}{x + 1} + \frac{\frac{1}{2}}{x - 1}$$

$$= \frac{-1}{x} + \frac{1}{2(x + 1)} + \frac{1}{2(x - 1)}$$

$$= \frac{-2(x^2 - 1) + x(x - 1) + x(x + 1)}{2x(x + 1)(x - 1)}$$

$$= \frac{-2x^2 + 2 + x^2 - x + x^2 + x}{2x(x + 1)(x - 1)}$$

$$= \frac{2}{2x(x + 1)(x - 1)}$$

$$= \frac{1}{x(x^2 - 1)}$$

$$= \frac{1}{x^3 - x}$$

**66.**
$$x + 2y = 2$$
$$y = 3$$

**68.** (a) Interchange two equations.

(b) Multiply one of the equations by a nonzero constant.

(c) Add a multiple of one of the equations to another equation to replace the latter equation.

## Section 8.4   Matrices and Linear Systems

**2.** $3 \times 3$

**4.** $2 \times 2$

**6.** $1 \times 4$

**8.** $\begin{bmatrix} 8 & 3 & \vdots & 25 \\ 3 & -9 & \vdots & 12 \end{bmatrix}$

**10.** $\begin{bmatrix} 9 & -3 & 1 & \vdots & 13 \\ 12 & 0 & -8 & \vdots & 5 \\ 3 & 4 & -1 & \vdots & 6 \end{bmatrix}$

**12.** $\begin{bmatrix} 10 & 6 & -8 & \vdots & -4 \\ -4 & -7 & 0 & \vdots & 9 \end{bmatrix}$

**14.**
$$9x - 4y = 0$$
$$6x + y = -4$$

**16.**
$$4x - y + 3z = 5$$
$$2x \quad\quad - 2z = -1$$
$$-x + 6y \quad\quad = 3$$

**18.**
$$7x + 3y - 2z + 4w = 2$$
$$-x \quad\quad + 4z - w = 6$$
$$8x + 3y \quad\quad = -4$$
$$2y - 4z + 3w = 12$$

**20.**
$$\begin{bmatrix} 3 & 6 & 8 \\ 4 & -3 & 6 \end{bmatrix}$$

$-R_1 + R_2 \begin{bmatrix} 3 & 6 & 8 \\ 1 & -9 & -2 \end{bmatrix}$

**22.**
$$\begin{bmatrix} 2 & 3 & -5 & 6 \\ 5 & -7 & 12 & 9 \\ -4 & 6 & 9 & 5 \end{bmatrix}$$

$2R_1 + R_3 \begin{bmatrix} 2 & 3 & -5 & 6 \\ 5 & -7 & 12 & 9 \\ 0 & 12 & -1 & 17 \end{bmatrix}$

**24.**
$$\begin{bmatrix} 2 & 4 & 8 & 3 \\ 1 & -1 & -3 & 2 \\ 2 & 6 & 4 & 9 \end{bmatrix}$$

$\frac{1}{2}R_1 \begin{bmatrix} 1 & 2 & 4 & \frac{3}{2} \\ 1 & -1 & -3 & 2 \\ 2 & 6 & 4 & 9 \end{bmatrix}$

$\begin{matrix} \\ -R_1 + R_2 \\ 2R_1 + R_3 \end{matrix} \begin{bmatrix} 1 & 2 & 4 & \frac{3}{2} \\ 0 & -3 & -7 & \frac{1}{2} \\ 0 & 2 & -4 & 6 \end{bmatrix}$

**26.**
$$\begin{bmatrix} 1 & 3 & 6 \\ -4 & -9 & 3 \end{bmatrix}$$

$4R_1 + R_2 \begin{bmatrix} 1 & 3 & 6 \\ 0 & 3 & 27 \end{bmatrix}$

$\frac{1}{3}R_2 \begin{bmatrix} 1 & 3 & 6 \\ 0 & 1 & 9 \end{bmatrix}$

**28.**
$$\begin{bmatrix} 3 & 2 & 6 \\ 2 & 3 & -3 \end{bmatrix}$$

$\begin{matrix} \frac{1}{3}R_1 \\ \frac{1}{2}R_2 \end{matrix} \begin{bmatrix} 1 & \frac{2}{3} & 2 \\ 1 & \frac{3}{2} & -\frac{3}{2} \end{bmatrix}$

$-R_1 + R_2 \begin{bmatrix} 1 & \frac{2}{3} & 2 \\ 0 & \frac{5}{6} & -\frac{7}{2} \end{bmatrix}$

$\frac{6}{5}R_2 \begin{bmatrix} 1 & \frac{2}{3} & 2 \\ 0 & 1 & -\frac{21}{5} \end{bmatrix}$

**30.**
$$\begin{bmatrix} 1 & 2 & -1 & 3 \\ 3 & 7 & -5 & 14 \\ -2 & -1 & -3 & 8 \end{bmatrix}$$

$\begin{matrix} \\ -3R_1 + R_2 \\ 2R_1 + R_3 \end{matrix} \begin{bmatrix} 1 & 2 & -1 & 3 \\ 0 & 1 & -2 & 5 \\ 0 & 3 & -5 & 14 \end{bmatrix}$

$\begin{matrix} \\ \\ -3R_2 + R_3 \end{matrix} \begin{bmatrix} 1 & 2 & -1 & 3 \\ 0 & 1 & -2 & 5 \\ 0 & 0 & 1 & -1 \end{bmatrix}$

**32.**
$$\begin{bmatrix} 1 & -3 & 0 & -7 \\ -3 & 10 & 1 & 23 \\ 4 & -10 & 2 & -24 \end{bmatrix}$$

$\begin{matrix} 3R_1 + R_2 \\ -4R_1 + R_3 \end{matrix} \begin{bmatrix} 1 & -3 & 0 & -7 \\ 0 & 1 & 1 & 2 \\ 0 & 2 & 2 & 4 \end{bmatrix}$

$\begin{matrix} \\ \\ -2R_2 + R_3 \end{matrix} \begin{bmatrix} 1 & -3 & 0 & -7 \\ 0 & 1 & 1 & 2 \\ 0 & 0 & 0 & 0 \end{bmatrix}$

**34.**
$$\begin{bmatrix} 1 & -3 & -2 & -8 \\ 1 & 3 & -2 & 17 \\ 1 & 2 & -2 & -5 \end{bmatrix}$$

$\begin{matrix} -R_1 + R_2 \\ -R_1 + R_3 \end{matrix} \begin{bmatrix} 1 & -3 & -2 & -8 \\ 0 & 6 & 0 & 25 \\ 0 & 5 & 0 & 3 \end{bmatrix}$

$\frac{1}{6}R_2 \begin{bmatrix} 1 & -3 & -2 & -8 \\ 0 & 1 & 0 & \frac{25}{6} \\ 0 & 5 & 0 & 3 \end{bmatrix}$

$-5R_2 + R_3 \begin{bmatrix} 1 & -3 & -2 & -8 \\ 0 & 1 & 0 & \frac{25}{6} \\ 0 & 0 & 0 & \frac{107}{6} \end{bmatrix}$

$\frac{6}{107}R_3 \begin{bmatrix} 1 & -3 & -2 & -8 \\ 0 & 1 & 0 & \frac{25}{6} \\ 0 & 0 & 0 & 1 \end{bmatrix}$

**36.** $x + 5y = 0$

$\qquad y = -1$

$x + 5(-1) = 0$

$\qquad x - 5 = 0$

$\qquad x = 5$

$(5, -1)$

**38.** $x + 5y - 3z = 0$

$\qquad y = 6$

$\qquad z = -5$

$x + 5(6) - 3(-5) = 0$

$\qquad x + 30 + 15 = 0$

$\qquad x + 45 = 0$

$\qquad\qquad x = -45$

$(-45, 6, -5)$

**40.** $x + 2y - 2z = -1$

$\qquad y + z = 9$

$\qquad z = -3$

$y + (-3) = 9 \qquad x + 2(12) - 2(-3) = -1$

$\qquad y = 12 \qquad\qquad x + 24 + 6 = -1$

$\qquad\qquad\qquad\qquad x + 30 = -1$

$\qquad\qquad\qquad\qquad\qquad x = -31$

$(-31, 12, -3)$

**42.** $2x + 6y = 16$

$2x + 3y = 7$

$$\begin{bmatrix} 2 & 6 & \vdots & 16 \\ 2 & 3 & \vdots & 7 \end{bmatrix}$$

$\tfrac{1}{2}R_1 \begin{bmatrix} 1 & 3 & \vdots & 8 \\ 2 & 3 & \vdots & 7 \end{bmatrix}$

$-2R_1 + R_2 \begin{bmatrix} 1 & 3 & \vdots & 8 \\ 0 & -3 & \vdots & -9 \end{bmatrix}$

$-\tfrac{1}{3}R_2 \begin{bmatrix} 1 & 3 & \vdots & 8 \\ 0 & 1 & \vdots & 3 \end{bmatrix}$

$y = 3 \quad x + 3(3) = 8$

$\qquad\qquad x = -1$

$(-1, 3)$

**44.** $x - 3y = 5$

$-2x + 6y = -10$

$$\begin{bmatrix} 1 & -3 & \vdots & 5 \\ -2 & 6 & \vdots & -10 \end{bmatrix}$$

$2R_1 + R_2 \begin{bmatrix} 1 & -3 & \vdots & 5 \\ 0 & 0 & \vdots & 0 \end{bmatrix}$

$x - 3y = 5 \qquad$ Let $y = a$. ($a$ is any real number)

$x - 3a = 5$

$\quad x = 3a + 5$

$(3a + 5, a)$

**46.** $2x - y = -0.1$

$3x + 2y = 1.6$

$$\begin{bmatrix} 2 & -1 & & -0.1 \\ 3 & 2 & & 1.6 \end{bmatrix}$$

$\tfrac{1}{2}R_1 \begin{bmatrix} 1 & -0.5 & \vdots & -0.05 \\ 3 & 2 & \vdots & 1.6 \end{bmatrix}$

$-3R_1 + R_2 \begin{bmatrix} 1 & -0.5 & \vdots & -0.05 \\ 3 & 3.5 & \vdots & 1.75 \end{bmatrix}$

$\tfrac{1}{3.5}R_2 \begin{bmatrix} 1 & -0.5 & \vdots & -0.05 \\ 3 & 1 & \vdots & 0.5 \end{bmatrix}$

$y = 0.5 \quad x - 0.5(0.5) = -0.05$

$\qquad\qquad x - 0.25 = -0.05$

$\qquad\qquad\qquad x = 0.2$

$(0.2, 0.5)$

**48.** $x - 3z = -2$

$3x + y - 2z = 5$

$2x + 2y + z = 4$

$$\begin{bmatrix} 1 & 0 & -3 & \vdots & -2 \\ 3 & 1 & -2 & \vdots & 5 \\ 2 & 2 & 1 & \vdots & 4 \end{bmatrix}$$

$\begin{matrix} -3R_1 + R_2 \\ -2R_1 + R_3 \end{matrix} \begin{bmatrix} 1 & 0 & -3 & \vdots & -2 \\ 0 & 1 & 7 & \vdots & 11 \\ 0 & 2 & 7 & \vdots & 8 \end{bmatrix}$

$-2R_2 + R_3 \begin{bmatrix} 1 & 0 & -3 & \vdots & -2 \\ 0 & 1 & 7 & \vdots & 11 \\ 0 & 0 & -7 & \vdots & -14 \end{bmatrix}$

$-\tfrac{1}{7}R_3 \begin{bmatrix} 1 & 0 & -3 & \vdots & -2 \\ 0 & 1 & 7 & \vdots & 11 \\ 0 & 0 & 1 & \vdots & 2 \end{bmatrix}$

$z = 2 \quad y + 7(2) = 11 \quad x - 3(2) = -2$

$\qquad\quad y + 14 = 11 \qquad x - 6 = -2$

$\qquad\qquad\quad y = -3 \qquad\qquad x = 4$

$(4, -3, 2)$

**50.**
$$2y + \phantom{4}z = 3$$
$$-4y - 2z = 0$$
$$x + \phantom{4}y + \phantom{2}z = 2$$

$$\begin{bmatrix} 0 & 2 & 1 & \vdots & 3 \\ 0 & -4 & -2 & \vdots & 0 \\ 1 & 1 & 1 & \vdots & 2 \end{bmatrix}$$

$R_1 \leftrightarrow R_3 \begin{bmatrix} 1 & 1 & 1 & \vdots & 2 \\ 0 & -4 & -2 & \vdots & 0 \\ 0 & 2 & 1 & \vdots & 3 \end{bmatrix}$

$R_2 \leftrightarrow R_3 \begin{bmatrix} 1 & 1 & 1 & \vdots & 2 \\ 0 & 2 & 1 & \vdots & 3 \\ 0 & -4 & -2 & \vdots & 0 \end{bmatrix}$

$2R_2 + R_3 \begin{bmatrix} 1 & 1 & 1 & \vdots & 2 \\ 0 & 2 & 1 & \vdots & 3 \\ 0 & 0 & 0 & \vdots & 6 \end{bmatrix}$

Inconsistent; no solution

**52.**
$$2x - \phantom{3}y + 3z = 24$$
$$2y - \phantom{3}z = 14$$
$$7x - 5y \phantom{+ 0z} = \phantom{0}6$$

$$\begin{bmatrix} 2 & -1 & 3 & \vdots & 24 \\ 0 & 2 & -1 & \vdots & 14 \\ 7 & -5 & 0 & \vdots & 6 \end{bmatrix}$$

$\begin{matrix} \frac{1}{2}R_1 \\ \\ \frac{1}{2}R_2 \end{matrix} \begin{bmatrix} 1 & -\frac{1}{2} & \frac{3}{2} & \vdots & 12 \\ 0 & 1 & -\frac{1}{2} & \vdots & 7 \\ 7 & -5 & 0 & \vdots & 6 \end{bmatrix}$

$-7R_1 + R_3 \begin{bmatrix} 1 & -\frac{1}{2} & \frac{3}{2} & \vdots & 12 \\ 0 & 1 & -\frac{1}{2} & \vdots & 7 \\ 0 & -\frac{3}{2} & -\frac{21}{2} & \vdots & -78 \end{bmatrix}$

$\begin{matrix} \frac{1}{2}R_2 + R_1 \\ \\ \frac{3}{2}R_2 + R_3 \end{matrix} \begin{bmatrix} 1 & 0 & \frac{5}{4} & \vdots & \frac{31}{2} \\ 0 & 1 & -\frac{1}{2} & \vdots & 7 \\ 0 & 0 & -\frac{45}{4} & \vdots & -\frac{135}{2} \end{bmatrix}$

$\dfrac{-4}{45}R_3 \begin{bmatrix} 1 & 0 & \frac{5}{4} & \vdots & \frac{31}{2} \\ 0 & 1 & -\frac{1}{2} & \vdots & 7 \\ 0 & 0 & 1 & \vdots & 6 \end{bmatrix} \checkmark$

$\begin{matrix} -\frac{5}{4}R_3 + R_1 \\ \frac{1}{2}R_3 + R_2 \end{matrix} \begin{bmatrix} 1 & 0 & 0 & \vdots & 8 \\ 0 & 1 & 0 & \vdots & 10 \\ 0 & 0 & 1 & \vdots & 6 \end{bmatrix}$

$(8, 10, 6)$

**54.**
$$2x \phantom{- 0y} + 3z = 3$$
$$4x - 3y + \phantom{1}7z = 5$$
$$8x - 9y + 15z = 9$$

$$\begin{bmatrix} 2 & 0 & 3 & \vdots & 3 \\ 4 & -3 & 7 & \vdots & 5 \\ 8 & -9 & 15 & \vdots & 9 \end{bmatrix}$$

$\frac{1}{2}R_1 \begin{bmatrix} 1 & 0 & \frac{3}{2} & \vdots & \frac{3}{2} \\ 4 & -3 & 7 & \vdots & 5 \\ 8 & -9 & 15 & \vdots & 9 \end{bmatrix}$

$\begin{matrix} -4R_1 + R_2 \\ -8R_1 + R_3 \end{matrix} \begin{bmatrix} 1 & 0 & \frac{3}{2} & \vdots & \frac{3}{2} \\ 0 & -3 & 1 & \vdots & -1 \\ 0 & -9 & 3 & \vdots & -3 \end{bmatrix}$

$-\frac{1}{3}R_2 \begin{bmatrix} 1 & 0 & \frac{3}{2} & \vdots & \frac{3}{2} \\ 0 & 1 & -\frac{1}{3} & \vdots & \frac{1}{3} \\ 0 & -9 & 3 & \vdots & -3 \end{bmatrix}$

$9R_2 + R_3 \begin{bmatrix} 1 & 0 & \frac{3}{2} & \vdots & \frac{3}{2} \\ 0 & 1 & -\frac{1}{3} & \vdots & \frac{1}{3} \\ 0 & 0 & 0 & \vdots & 0 \end{bmatrix}$

$$y - \tfrac{1}{3}z = \tfrac{1}{3} \qquad x + \tfrac{3}{2}z = \tfrac{3}{2}$$
$$y = \tfrac{1}{3}z + \tfrac{1}{3} \qquad x = \tfrac{3}{2} - \tfrac{3}{2}z$$

Let $a = z$.   ($a$ is any real number)

$\left( \tfrac{3}{2} - \tfrac{3}{2}a, \tfrac{1}{3}a + \tfrac{1}{3}, a \right)$

**56.** $2x + 4y + 5z = 5$

$\quad x + 3y + 3z = 2$

$\quad 2x + 4y + 4z = 2$

$$\begin{bmatrix} 2 & 4 & 5 & \vdots & 5 \\ 1 & 3 & 3 & \vdots & 2 \\ 2 & 4 & 4 & \vdots & 2 \end{bmatrix}$$

$-\frac{1}{2}R_3 \leftrightarrow R_1 \begin{bmatrix} 1 & 2 & 2 & \vdots & 1 \\ 1 & 3 & 3 & \vdots & 2 \\ 2 & 4 & 5 & \vdots & 5 \end{bmatrix}$

$\begin{matrix} \\ -R_1 + R_2 \\ -2R_1 + R_3 \end{matrix} \begin{bmatrix} 1 & 2 & 2 & \vdots & 1 \\ 0 & 1 & 1 & \vdots & 1 \\ 0 & -2 & -1 & \vdots & 1 \end{bmatrix}$

$\begin{matrix} -2R_2 + R_1 \\ \\ 2R_2 + R_3 \end{matrix} \begin{bmatrix} 1 & 0 & 0 & \vdots & -1 \\ 0 & 1 & 1 & \vdots & 1 \\ 0 & 0 & 1 & \vdots & 3 \end{bmatrix}$

$\begin{matrix} -R_3 + R_1 \\ \\ \\ \end{matrix} \begin{bmatrix} 1 & 0 & 0 & \vdots & -1 \\ 0 & 1 & 0 & \vdots & -2 \\ 0 & 0 & 1 & \vdots & 3 \end{bmatrix}$

$(-1, -2, 3)$

**60.** $2x + 2y + z = 8$

$\quad 2x + 3y + z = 7$

$\quad 6x + 8y + 3z = 22$

$$\begin{bmatrix} 2 & 2 & 1 & \vdots & 8 \\ 2 & 3 & 1 & \vdots & 7 \\ 6 & 8 & 3 & \vdots & 22 \end{bmatrix}$$

$\begin{matrix} \\ -R_1 + R_2 \\ -3R_1 + R_3 \end{matrix} \begin{bmatrix} 2 & 2 & 1 & \vdots & 8 \\ 0 & 1 & 0 & \vdots & -1 \\ 0 & 2 & 0 & \vdots & -2 \end{bmatrix}$

$\begin{matrix} \\ \\ -2R_2 + R_3 \end{matrix} \begin{bmatrix} 2 & 2 & 1 & \vdots & 8 \\ 0 & 1 & 0 & \vdots & -1 \\ 0 & 0 & 0 & \vdots & 0 \end{bmatrix}$

$\begin{matrix} \frac{1}{2}R_1 \\ \\ \\ \end{matrix} \begin{bmatrix} 1 & 1 & \frac{1}{2} & & 4 \\ 0 & 1 & 0 & & -1 \\ 0 & 0 & 0 & & 0 \end{bmatrix}$

$y = -1 \quad x + (-1) + \frac{1}{2}z = 4 \qquad$ Let $z = a$.

$\qquad\qquad x + \frac{1}{2}z = 5 \qquad$ ($a$ is any real number.)

$\qquad\qquad\qquad x = 5 - \frac{1}{2}z$

$\left(5 - \frac{1}{2}a, -1, a\right)$

**58.** $3x + y - 2z = 2$

$\quad 6x + 2y - 4z = 1$

$\quad -3x - y + 2z = 1$

$$\begin{bmatrix} 3 & 1 & -2 & \vdots & 2 \\ 6 & 2 & -4 & \vdots & 1 \\ -3 & -1 & 2 & \vdots & 1 \end{bmatrix}$$

$\frac{1}{3}R_1 \begin{bmatrix} 1 & \frac{1}{3} & -\frac{2}{3} & \vdots & \frac{2}{3} \\ 6 & 2 & -4 & \vdots & 1 \\ -3 & -1 & 2 & \vdots & 1 \end{bmatrix}$

$\begin{matrix} \\ -6R_1 + R_2 \\ +3R_1 + R_3 \end{matrix} \begin{bmatrix} 1 & \frac{1}{3} & -\frac{2}{3} & \vdots & \frac{2}{3} \\ 0 & 0 & 0 & \vdots & -3 \\ 0 & 0 & 0 & \vdots & 3 \end{bmatrix}$

Inconsistent; no solution

**62.** $3x + 3y + z = 4$

$\quad 2x + 6y + z = 5$

$\quad -x - 3y + 2z = -5$

$$\begin{bmatrix} 3 & 3 & 1 & \vdots & 4 \\ 2 & 6 & 1 & \vdots & 5 \\ -1 & -3 & 2 & \vdots & -5 \end{bmatrix}$$

$-R_3 \leftrightarrow R_1 \begin{bmatrix} 1 & 3 & -2 & \vdots & 5 \\ 2 & 6 & 1 & \vdots & 5 \\ 3 & 3 & 1 & \vdots & 4 \end{bmatrix}$

$\begin{matrix} \\ -2R_1 + R_2 \\ -3R_1 + R_3 \end{matrix} \begin{bmatrix} 1 & 3 & -2 & \vdots & 5 \\ 0 & 0 & 5 & \vdots & -5 \\ 0 & -6 & 7 & \vdots & -11 \end{bmatrix}$

$\frac{1}{5}R_2 \leftrightarrow R_3 \begin{bmatrix} 1 & 3 & -2 & \vdots & 5 \\ 0 & -6 & 7 & \vdots & -11 \\ 0 & 0 & 1 & \vdots & -1 \end{bmatrix}$

$-\frac{1}{6}R_2 \begin{bmatrix} 1 & 3 & -2 & \vdots & 5 \\ 0 & 1 & -\frac{7}{6} & \vdots & \frac{11}{6} \\ 0 & 0 & 1 & \vdots & -1 \end{bmatrix}$

$-3R_2 + R_1 \begin{bmatrix} 1 & 0 & \frac{3}{2} & \vdots & -\frac{1}{2} \\ 0 & 1 & -\frac{7}{6} & \vdots & \frac{11}{6} \\ 0 & 0 & 1 & \vdots & -1 \end{bmatrix}$

$\begin{matrix} -\frac{3}{2}R_3 + R_1 \\ \frac{7}{6}R_3 + R_2 \\ \\ \end{matrix} \begin{bmatrix} 1 & 0 & 0 & \vdots & 1 \\ 0 & 1 & 0 & \vdots & \frac{2}{3} \\ 0 & 0 & 1 & \vdots & -1 \end{bmatrix}$

$\left(1, \frac{2}{3}, -1\right)$

**64.** *Verbal model:*   Investment 1 + Investment 2 + Investment 3 = 16,000

$0.05 \cdot$ Investment 1 $+ 0.06 \cdot$ Investment 2 $+ 0.07 \cdot$ Investment 3 $= 990$

Investment 1 $-$ Investment 3 $= -3000$

Investment 2 $-$ Investment 3 $= -2000$

*Labels:*   Investment 1 $= x$

Investment 2 $= y$

Investment 3 $= z$

*System:*
$$x + y + z = 16,000$$
$$0.05x + 0.06y + 0.07z = 990$$
$$x - z = -3000$$
$$y - z = -2000$$

$$100R_2 \begin{bmatrix} 1 & 1 & 1 & : & 16,000 \\ 5 & 6 & 7 & : & 99,000 \\ 1 & 0 & -1 & : & -3000 \\ 0 & 1 & -1 & : & -2000 \end{bmatrix}$$

$$\begin{matrix} \\ -5R_1 + R_2 \\ -R_1 + R_3 \\ \\ \end{matrix} \begin{bmatrix} 1 & 1 & 1 & : & 16,000 \\ 0 & 1 & 2 & : & 19,000 \\ 0 & -1 & -2 & : & -19,000 \\ 0 & 1 & -1 & : & -2000 \end{bmatrix}$$

$$\begin{matrix} \\ \\ R_2 + R_3 \\ -R_2 + R_4 \end{matrix} \begin{bmatrix} 1 & 1 & 1 & : & 16,000 \\ 0 & 1 & 2 & : & 19,000 \\ 0 & 0 & 0 & : & 0 \\ 0 & 0 & -3 & : & -21,000 \end{bmatrix}$$

$$R_4 \Leftrightarrow R_3 \begin{bmatrix} 1 & 1 & 1 & : & 16,000 \\ 0 & 1 & 2 & : & 19,000 \\ 0 & 0 & -3 & : & -21,000 \\ 0 & 0 & 0 & : & 0 \end{bmatrix}$$

| | | |
|---|---|---|
| $z = 7000$ | $y + 2(7000) = 19,000$ | $x + 5000 + 7000 = 16,000$ |
| | $y + 14,000 = 19,000$ | $x + 12,000 = 16,000$ |
| | $y = 5000$ | $x = 4000$ |

5% account: $4000     6% account: $5,000     7% account: $7,000

**66.** *Verbal model:*   $.09$ CDs $+ .05$ Bonds $+ .12$ BC Stocks $+ .14$ G Stocks $= 50,000$

BC Stocks $+$ G Stocks $= 125,000$

CDs $+$ Bonds $= 375,000$

*Labels:*   $x =$ certificates of deposit

$y =$ municipal bonds

$z =$ blue-chip stocks

$w =$ growth stocks

**—CONTINUED—**

**66. —CONTINUED—**

$$\text{System:} \begin{cases} 0.09x + 0.05y + 0.12z + 0.14w = 50,000 \\ z + w = 125,000 \\ x + y = 375,000 \end{cases}$$

$$100R_1 \begin{bmatrix} 9 & 5 & 12 & 14 & \vdots & 5,000,000 \\ 0 & 0 & 1 & 1 & \vdots & 125,000 \\ 1 & 1 & 0 & 0 & \vdots & 375,000 \end{bmatrix}$$

$$R_1 \leftrightarrow R_3 \begin{bmatrix} 1 & 1 & 0 & 0 & \vdots & 375,000 \\ 0 & 0 & 1 & 1 & \vdots & 125,000 \\ 9 & 5 & 12 & 14 & \vdots & 5,000,000 \end{bmatrix}$$

$$\begin{matrix} \\ \\ -9R_1 + R_3 \end{matrix} \begin{bmatrix} 1 & 1 & 0 & 0 & \vdots & 375,000 \\ 0 & 0 & 1 & 1 & \vdots & 125,000 \\ 0 & -4 & 12 & 14 & \vdots & 1,625,000 \end{bmatrix}$$

$$R_2 \leftrightarrow R_3 \begin{bmatrix} 1 & 1 & 0 & 0 & \vdots & 375,000 \\ 0 & -4 & 12 & 14 & \vdots & 1,625,000 \\ 0 & 0 & 1 & 1 & \vdots & 125,000 \end{bmatrix}$$

$$-\tfrac{1}{4}R_2 \begin{bmatrix} 1 & 1 & 0 & 0 & \vdots & 375,000 \\ 0 & 1 & -3 & -\tfrac{7}{2} & \vdots & -406,250 \\ 0 & 0 & 1 & 1 & \vdots & 125,000 \end{bmatrix}$$

so let $w = s$,

then $z + s = 125,000$

$z = 125,000 - s$.

$y - 3(125,000 - s) - \tfrac{7}{2}s = -406,250$

$y - 375,000 + 3s - \tfrac{7}{2}s = -406,250$

$y - 375,000 - \tfrac{1}{2}s = -406,250$

$y = -31,250 + \tfrac{1}{2}s$

$x + \left(-31,250 + \tfrac{1}{2}s\right) = 375,000$

$x - 31,250 + \tfrac{1}{2}s = 375,000$

$x = 406,250 - \tfrac{1}{2}s$

Certificates of deposit: $406,250 - \tfrac{1}{2}s$

Municipal bonds: $-31,250 + \tfrac{1}{2}s$

Blue-chip stocks: $125,000 - s$

Growth stocks: $s$, where $s \geq 62,500$

If $s = \$100,000$:

CD = $\$356,250$

Bonds = $\$18,750$

BC Stocks = $\$25,000$

G Stocks = $\$100,000$

**68.** *Verbal model:*

$$\boxed{\begin{matrix}\text{Pounds}\\\text{Nut 1}\end{matrix}} + \boxed{\begin{matrix}\text{Pounds}\\\text{Nut 2}\end{matrix}} + \boxed{\begin{matrix}\text{Pounds}\\\text{Nut 3}\end{matrix}} = \boxed{\begin{matrix}50\\\text{Pounds}\end{matrix}}$$

$$\boxed{3.00(\text{Nut 1})} + \boxed{4.00(\text{Nut 2})} + \boxed{6.00(\text{Nut 3})} = 50(4.10)$$

$$\boxed{\begin{matrix}\text{Pounds}\\\text{Nut 1}\end{matrix}} + \boxed{\begin{matrix}\text{Pounds}\\\text{Nut 2}\end{matrix}} \qquad\qquad = 37.50$$

*Labels:* Pounds Nut 1 = $x$

Pounds Nut 2 = $y$

Pounds Nut 3 = $z$

*System:*

$$\begin{aligned} x + y + z &= 50 \\ 3x + 4y + 6z &= 205 \\ x + y &= 37.50 \end{aligned}$$

$$\begin{bmatrix} 1 & 1 & 1 & \vdots & 50 \\ 3 & 4 & 6 & \vdots & 205 \\ 1 & 1 & 0 & \vdots & 37.50 \end{bmatrix}$$

$$\begin{matrix} \\ -3R_1 + R_2 \\ -R_1 + R_3 \end{matrix} \begin{bmatrix} 1 & 1 & 1 & \vdots & 50 \\ 0 & 1 & 3 & \vdots & 55 \\ 0 & 0 & -1 & \vdots & -12.5 \end{bmatrix}$$

$$\begin{matrix} \\ \\ -R_3 \end{matrix} \begin{bmatrix} 1 & 1 & 1 & \vdots & 50 \\ 0 & 1 & 3 & \vdots & 55 \\ 0 & 0 & 1 & \vdots & 12.5 \end{bmatrix}$$

$z = 12.5$ pounds at $6

$y + 3(12.5) = 55$

$y = 17.5$ pounds at $4

$x + 17.5 + 12.5 = 50$

$x = 20$ pounds at $3

**70.** *Verbal model:*    $\boxed{\text{Number} \atop 1}$ + $\boxed{\text{Number} \atop 2}$ + $\boxed{\text{Number} \atop 3}$ = $\boxed{24}$

$\boxed{\text{Number} \atop 2}$ = $\boxed{\text{Number} \atop 1}$ + 4

$\boxed{\text{Number} \atop 3}$ = 3 · $\boxed{\text{Number} \atop 1}$

*Labels:*  Number 1 = $x$
Number 2 = $y$
Number 3 = $z$

*System:*  $x + y + z = 24$
$y \phantom{+ z} = x + 4$
$z = 3x$

$x + y + z = 24$
$-x + y \phantom{+ z} = 4$
$-3x \phantom{+ y} + z = 0$

$z = 12$    $y + .5(12) = 14$    $x + 8 + 12 = 24$
$y = 8$              $x = 4$

$$\begin{bmatrix} 1 & 1 & 1 & : & 24 \\ -1 & 1 & 0 & : & 4 \\ -3 & 0 & 1 & : & 0 \end{bmatrix}$$

$$\begin{array}{c} R_1 + R_2 \\ 3R_1 + R_3 \end{array} \begin{bmatrix} 1 & 1 & 1 & : & 24 \\ 0 & 2 & 1 & : & 28 \\ 0 & 3 & 4 & : & 72 \end{bmatrix}$$

$$\tfrac{1}{2}R_2 \begin{bmatrix} 1 & 1 & 1 & : & 24 \\ 0 & 1 & .5 & : & 14 \\ 0 & 3 & 4 & : & 72 \end{bmatrix}$$

$$-3R_2 + R_3 \begin{bmatrix} 1 & 1 & 1 & : & 24 \\ 0 & 1 & .5 & : & 14 \\ 0 & 0 & 2.5 & : & 30 \end{bmatrix}$$

$$\tfrac{2}{5}R_3 \begin{bmatrix} 1 & 1 & 1 & : & 24 \\ 0 & 1 & .5 & : & 14 \\ 0 & 0 & 1 & : & 12 \end{bmatrix}$$

**72.** $(1, 11), (2, 10), (3, 7)$

$11 = a(1)^2 + b(1) + c \implies 11 = a + b + c$
$10 = a(2)^2 + b(2) + c \implies 10 = 4a + 2b + c$
$7 = a(3)^2 + b(3) + c \implies 7 = 9a + 3b + c$

$$\begin{bmatrix} 1 & 1 & 1 & : & 11 \\ 4 & 2 & 1 & : & 10 \\ 9 & 3 & 1 & : & 7 \end{bmatrix}$$

$$\begin{array}{c} -4R_1 + R_2 \\ -9R_1 + R_3 \end{array} \begin{bmatrix} 1 & 1 & 1 & : & 11 \\ 0 & -2 & -3 & : & -34 \\ 0 & -6 & -8 & : & -92 \end{bmatrix}$$

$$-3R_2 + R_3 \begin{bmatrix} 1 & 1 & 1 & : & 11 \\ 0 & -2 & -3 & : & -34 \\ 0 & 0 & 1 & : & 10 \end{bmatrix}$$

$$-\tfrac{1}{2}R_2 \begin{bmatrix} 1 & 1 & 1 & : & 11 \\ 0 & 1 & \tfrac{3}{2} & : & 17 \\ 0 & 0 & 1 & : & 10 \end{bmatrix}$$

$z = 10$    $y + \tfrac{3}{2}(10) = 17$    $x + 2 + 10 = 11$    $y = -x^2 + 2x + 10$
$y + 15 = 17$        $x + 12 = 11$
$y = 2$              $x = -1$

**74.**

$$1 = a(1)^2 + b(1) + c \implies 1 = a + b + c$$
$$17 = a(-3)^2 + b(-3) + c \implies 17 = 9a - 3b + c$$
$$-\tfrac{1}{2} = a(2)^2 + b(2) + c \implies -1 = 8a + 4b + 2c$$

$$\begin{bmatrix} 1 & 1 & 1 & \vdots & 1 \\ 9 & -3 & 1 & \vdots & 17 \\ 8 & 4 & 2 & \vdots & -1 \end{bmatrix}$$

$$\begin{matrix} \\ -9R_1 + R_2 \\ -8R_1 + R_3 \end{matrix} \begin{bmatrix} 1 & 1 & 1 & \vdots & 1 \\ 0 & -12 & -8 & \vdots & 8 \\ 0 & -4 & -6 & \vdots & -9 \end{bmatrix}$$

$$-\tfrac{1}{12}R_2 \begin{bmatrix} 1 & 1 & 1 & \vdots & 1 \\ 0 & 1 & \tfrac{2}{3} & \vdots & -\tfrac{2}{3} \\ 0 & -4 & -6 & \vdots & -9 \end{bmatrix}$$

$$4R_2 + R_3 \begin{bmatrix} 1 & 1 & 1 & \vdots & 1 \\ 0 & 1 & \tfrac{2}{3} & \vdots & -\tfrac{2}{3} \\ 0 & 0 & -\tfrac{10}{3} & \vdots & -\tfrac{35}{3} \end{bmatrix}$$

$$-\tfrac{3}{10}R_3 \begin{bmatrix} 1 & 1 & 1 & \vdots & 1 \\ 0 & 1 & \tfrac{2}{3} & \vdots & -\tfrac{2}{3} \\ 0 & 0 & 1 & \vdots & \tfrac{7}{2} \end{bmatrix}$$

$$c = \tfrac{7}{2} \qquad b + \tfrac{2}{3}\left(\tfrac{7}{2}\right) = -\tfrac{2}{3} \qquad a + -3 + \tfrac{7}{2} = 1 \qquad y = \tfrac{1}{2}x^2 - 3x + \tfrac{7}{2}$$
$$b + \tfrac{7}{3} = -\tfrac{9}{3} \qquad a + -\tfrac{6}{2} + \tfrac{7}{2} = 1$$
$$b = -3 \qquad a = 1 - \tfrac{1}{2}$$
$$a = \tfrac{1}{2}$$

**76.** $(-1, 2), (2, 3), (3, 2)$

$$(-1)^2 + 2^2 + D(-1) + E(2) + F = 0 \implies -D + 2E + F = -5$$
$$2^2 + 3^2 + D(2) + E(3) + F = 0 \implies 2D + 3E + F = -13$$
$$3^2 + 2^2 + D(3) + E(2) + F = 0 \implies 3D + 2E + F = -13$$

$$z = -3 \qquad y + \tfrac{3}{7}(-3) = \frac{-23}{7} \qquad\qquad x - 2(-2) - (-3) = 5$$
$$\qquad\qquad\qquad x = -2$$
$$y = \frac{-23}{7} + \frac{9}{7} \qquad x^2 + y^2 - 2x - 2y - 3 = 0$$
$$y = -2$$

$$\begin{bmatrix} -1 & 2 & 1 & \vdots & -5 \\ 2 & 3 & 1 & \vdots & -13 \\ 3 & 2 & 1 & \vdots & -13 \end{bmatrix}$$

$$-R_1 \begin{bmatrix} 1 & -2 & -1 & \vdots & 5 \\ 2 & 3 & 1 & \vdots & -13 \\ 3 & 2 & 1 & \vdots & -13 \end{bmatrix}$$

$$\begin{matrix} \\ -2R_1 + R_2 \\ -3R_1 + R_3 \end{matrix} \begin{bmatrix} 1 & -2 & -1 & \vdots & 5 \\ 0 & 7 & 3 & \vdots & -23 \\ 0 & 8 & 4 & \vdots & -28 \end{bmatrix}$$

$$\begin{matrix} \\ \tfrac{1}{7}R_2 \\ \tfrac{1}{4}R_3 \end{matrix} \begin{bmatrix} 1 & -2 & -1 & \vdots & 5 \\ 0 & 1 & \tfrac{3}{7} & \vdots & -\tfrac{23}{7} \\ 0 & 2 & 1 & \vdots & -7 \end{bmatrix}$$

$$-2R_2 + R_3 \begin{bmatrix} 1 & -2 & -1 & \vdots & 5 \\ 0 & 1 & \tfrac{3}{7} & \vdots & -\tfrac{23}{7} \\ 0 & 0 & \tfrac{1}{7} & \vdots & -\tfrac{3}{7} \end{bmatrix}$$

$$7R_3 \begin{bmatrix} 1 & -2 & -1 & \vdots & 5 \\ 0 & 1 & \tfrac{3}{7} & \vdots & -\tfrac{23}{7} \\ 0 & 0 & 1 & \vdots & -3 \end{bmatrix}$$

**78.** (a)

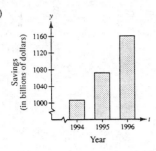

(b) $(4, 1006.3), (5, 1072.3), (6, 1161.0)$

$a(4)^2 + b(4) + c = 1006.3$

$a(5)^2 + b(5) + c = 1072.3$

$a(6)^2 + b(6) + c = 1161.0$

$$\begin{bmatrix} 16 & 4 & 1 & \vdots & 1006.3 \\ 25 & 5 & 1 & \vdots & 1072.3 \\ 36 & 6 & 1 & \vdots & 1161.0 \end{bmatrix}$$

$\dfrac{-1}{16}R_1 \begin{bmatrix} 1 & .25 & .0625 & \vdots & 62.89375 \\ 25 & 5 & 1 & \vdots & 1072.3 \\ 36 & 6 & 1 & \vdots & 1161.0 \end{bmatrix}$

$\begin{matrix} \\ -25R_1 + R_2 \\ 36R_1 + R_3 \end{matrix} \begin{bmatrix} 1 & .25 & .0625 & \vdots & 62.89375 \\ 0 & -1.25 & -.56 & \vdots & -500.044 \\ 0 & -3 & -1.25 & \vdots & -1103.175 \end{bmatrix}$

$\dfrac{1}{-1.25}R_2 \begin{bmatrix} 1 & .25 & .0625 & \vdots & 62.89375 \\ 0 & 1 & .45 & \vdots & 400.035 \\ 0 & -3 & -1.25 & \vdots & -1103.175 \end{bmatrix}$

$\begin{matrix} \\ \\ 3R_2 + R_3 \end{matrix} \begin{bmatrix} 1 & .25 & .0625 & \vdots & 62.89375 \\ 0 & 1 & .45 & \vdots & 400.035 \\ 0 & 0 & .1 & \vdots & 96.93 \end{bmatrix}$

$\begin{matrix} \\ \\ 10R_3 \end{matrix} \begin{bmatrix} 1 & .25 & .0625 & \vdots & 62.89375 \\ 0 & 1 & .45 & \vdots & 400.035 \\ 0 & 0 & 1 & \vdots & 969.3 \end{bmatrix}$

$z = 969.3 \qquad y + .45(969.3) = 400.035 \qquad\qquad x + .25(-36.15) + .0625(969.3) = 62.89375$

$$y = -36.15 \qquad\qquad\qquad x = 11.35$$

$$y = 11.35t^2 - 36.15t + 969.3$$

(c)

(d) $y = 11.35(10)^2 - 36.15(10) + 969.3$

$\quad = 1742.8$

**80.** $2A + B + C = 2$

$2A + B - C = 0$

$5A + 4B + 2C = 3$

$$\begin{bmatrix} 2 & 1 & 1 & \vdots & 2 \\ 2 & 1 & -1 & \vdots & 0 \\ 5 & 4 & 2 & \vdots & 3 \end{bmatrix}$$

$$\begin{matrix} \\ -R_1 + R_2 \\ -\frac{5}{2}R_1 + R_3 \end{matrix} \begin{bmatrix} 2 & 1 & 1 & \vdots & 2 \\ 0 & 0 & -2 & \vdots & -2 \\ 0 & \frac{3}{2} & -\frac{1}{2} & \vdots & -2 \end{bmatrix}$$

$$\begin{matrix} \frac{1}{2}R_1 \\ \\ R_2 \leftrightarrow R_3 \end{matrix} \begin{bmatrix} 1 & \frac{1}{2} & \frac{1}{2} & \vdots & 2 \\ 0 & \frac{3}{2} & -\frac{1}{2} & \vdots & -2 \\ 0 & 0 & -2 & \vdots & -2 \end{bmatrix}$$

$$\begin{matrix} \\ \frac{2}{3}R_2 \\ -\frac{1}{2}R_3 \end{matrix} \begin{bmatrix} 1 & \frac{1}{2} & \frac{1}{2} & \vdots & 1 \\ 0 & 1 & -\frac{1}{3} & \vdots & -\frac{4}{3} \\ 0 & 0 & 1 & \vdots & 1 \end{bmatrix}$$

$C = 1 \qquad B - \dfrac{1}{3}(1) = -\dfrac{4}{3}$

$$B = -1$$

$$A + \frac{1}{2}(-1) + \frac{1}{2}(1) = 1$$

$$A - \frac{1}{2} + \frac{1}{2} = 1$$

$$A = 1$$

$$\frac{x+1}{x(x^2+1)} = \frac{1}{x} + \frac{-x+1}{x^2+1}$$

$$= \frac{x^2 + 1 + (-x+1)x}{x(x^2+1)}$$

$$= \frac{x^2 + 1 - x^2 + x}{x(x^2+1)}$$

$$= \frac{x+1}{x(x^2+1)}$$

**82.** They are the same.

**84.** $\begin{bmatrix} 1 & -2 & 6 \\ 0 & 1 & 5 \\ 0 & 0 & 0 \end{bmatrix}$

Answers vary.

**86.** The row-echelon form of the matrix will have fewer rows with nonzero entries than there are variables in the system.

# Section 8.5 Determinants and Linear Systems

**2.** $\det(A) = \begin{vmatrix} -3 & 1 \\ 5 & 2 \end{vmatrix} = -3(2) - 5(1) = -6 - 5 = -11$

**4.** $\det(A) = \begin{vmatrix} 2 & -2 \\ 4 & 3 \end{vmatrix} = 6 - (-8) = 14$

**6.** $\det(A) = \begin{vmatrix} 4 & -3 \\ 0 & 0 \end{vmatrix} = 0 - 0 = 0$

**8.** $\det(A) = \begin{vmatrix} -2 & 3 \\ 6 & -9 \end{vmatrix} = 18 - 18 = 0$

**10.** $\det(A) = \begin{vmatrix} \frac{2}{3} & \frac{5}{6} \\ 14 & -2 \end{vmatrix} = \frac{2}{3}(-2) - 14\left(\frac{5}{6}\right) = -\frac{4}{3} - \frac{35}{3} = -\frac{39}{3} = -13$

**12.** $\det(A) = \begin{vmatrix} -1.2 & 4.5 \\ 0.4 & -0.9 \end{vmatrix} = -1.2(-0.9) - 0.4(4.5) = 1.08 - 1.8 = -0.72$

**14.** $\det(A) = \begin{vmatrix} 10 & 2 & -4 \\ 8 & 0 & -2 \\ 4 & 0 & 2 \end{vmatrix}$

second column $= -2\begin{vmatrix} 8 & -2 \\ 4 & 2 \end{vmatrix} + 0\begin{vmatrix} 10 & -4 \\ 4 & 2 \end{vmatrix} - 0\begin{vmatrix} 10 & -4 \\ 8 & -2 \end{vmatrix} = -2(16 + 8) + 0 - 0$

$$= -2(24)$$

$$= -48$$

**16.** $\begin{bmatrix} 2 & 1 & 3 \\ 1 & 4 & 4 \\ 1 & 0 & 2 \end{bmatrix}$ $\det(A) = \begin{vmatrix} 2 & 1 & 3 \\ 1 & 4 & 4 \\ 1 & 0 & 2 \end{vmatrix}$

$$\text{third row} = (1)\begin{vmatrix} 1 & 3 \\ 4 & 4 \end{vmatrix} - (0)\begin{vmatrix} 2 & 3 \\ 1 & 4 \end{vmatrix} + (2)\begin{vmatrix} 2 & 1 \\ 1 & 4 \end{vmatrix}$$

$$= (1)(-8) - 0 + (2)(7)$$

$$= -8 + 14$$

$$= 6$$

**18.** $\begin{bmatrix} 2 & 3 & 1 \\ 0 & 5 & -2 \\ 0 & 0 & -2 \end{bmatrix}$ $\det(A) = \begin{vmatrix} 2 & 3 & 1 \\ 0 & 5 & -2 \\ 0 & 0 & -2 \end{vmatrix}$

$$\text{third row} = 0 - 0 + (-2)\begin{vmatrix} 2 & 3 \\ 0 & 5 \end{vmatrix}$$

$$= (-2)(10)$$

$$= -20$$

**20.** $\begin{bmatrix} 3 & 2 & 2 \\ 2 & 2 & 2 \\ -4 & 4 & 3 \end{bmatrix}$ $\det(A) = \begin{vmatrix} 3 & 2 & 2 \\ 2 & 2 & 2 \\ -4 & 4 & 3 \end{vmatrix}$

$$\text{second row} = -(2)\begin{vmatrix} 2 & 2 \\ 4 & 3 \end{vmatrix} + (2)\begin{vmatrix} 3 & 2 \\ -4 & 3 \end{vmatrix} - (2)\begin{vmatrix} 3 & 2 \\ -4 & 4 \end{vmatrix}$$

$$= (-2)(-2) + (2)(17) - (2)(20)$$

$$= 4 + 34 - 40$$

$$= -2$$

**22.** $\begin{bmatrix} 2 & -1 & 0 \\ 4 & 2 & 1 \\ 4 & 2 & 1 \end{bmatrix}$ $\det(A) = \begin{vmatrix} 2 & -1 & 0 \\ 4 & 2 & 1 \\ 4 & 2 & 1 \end{vmatrix}$

$$\text{third column} = 0 - (1)\begin{vmatrix} 2 & -1 \\ 4 & 2 \end{vmatrix} + (1)\begin{vmatrix} 2 & -1 \\ 4 & 2 \end{vmatrix}$$

$$= 0 - (1)(8) + (1)(8)$$

$$= 0$$

**24.** $\begin{bmatrix} -3 & 4 & 2 \\ 6 & 3 & 1 \\ 4 & -7 & -8 \end{bmatrix}$ $\det(A) = \begin{vmatrix} -3 & 4 & 2 \\ 6 & 3 & 1 \\ 4 & -7 & -8 \end{vmatrix}$

$$\text{second row} = -(6)\begin{vmatrix} 4 & 2 \\ -7 & -8 \end{vmatrix} + 3\begin{vmatrix} -3 & 2 \\ 4 & -8 \end{vmatrix} - (1)\begin{vmatrix} -3 & 4 \\ 4 & -7 \end{vmatrix}$$

$$= (-6)(-18) + (3)(16) - (1)(5)$$

$$= 108 + 48 - 5$$

$$= 151$$

**26.** $\begin{bmatrix} 6 & 8 & -7 \\ 0 & 0 & 0 \\ 4 & -6 & 22 \end{bmatrix} \det(A) = \begin{vmatrix} 6 & 8 & -7 \\ 0 & 0 & 0 \\ 4 & -6 & 22 \end{vmatrix}$

$$\text{second row} = -0 + 0 - 0$$
$$= 0$$

**28.** $\begin{bmatrix} 8 & 7 & 6 \\ -4 & 0 & 0 \\ 5 & 1 & 4 \end{bmatrix} \det(A) = \begin{vmatrix} 8 & 7 & 6 \\ -4 & 0 & 0 \\ 5 & 1 & 4 \end{vmatrix}$

$$\text{second row} = -(-4)\begin{vmatrix} 7 & 6 \\ 1 & 4 \end{vmatrix} + 0\begin{vmatrix} 8 & 6 \\ 5 & 4 \end{vmatrix} - 0\begin{vmatrix} 8 & 7 \\ 5 & 1 \end{vmatrix}$$
$$= 4(22) + 0 - 0$$
$$= 88$$

**30.** $\begin{bmatrix} -0.4 & 0.4 & 0.3 \\ 0.2 & 0.2 & 0.2 \\ 0.3 & 0.2 & 0.2 \end{bmatrix} \det(A) = \begin{vmatrix} -0.4 & 0.4 & 0.3 \\ 0.2 & 0.2 & 0.2 \\ 0.3 & 0.2 & 0.2 \end{vmatrix}$

$$\text{second row} = -(0.2)\begin{vmatrix} 0.4 & 0.3 \\ 0.2 & 0.2 \end{vmatrix} + (0.2)\begin{vmatrix} -0.4 & 0.3 \\ 0.3 & 0.2 \end{vmatrix} - (0.2)\begin{vmatrix} -0.4 & 0.4 \\ 0.3 & 0.2 \end{vmatrix}$$
$$= (-0.2)(0.02) + (0.2)(-0.17) + (-0.2)(-0.2)$$
$$= -0.004 - 0.034 + 0.04$$
$$= 0.002$$

**32.** $\begin{bmatrix} x & y & 1 \\ -2 & -2 & 1 \\ 1 & 5 & 1 \end{bmatrix} \det(A) = \begin{vmatrix} x & y & 1 \\ -2 & -2 & 1 \\ 1 & 5 & 1 \end{vmatrix}$

$$\text{third column} = (1)\begin{vmatrix} -2 & -2 \\ 1 & 5 \end{vmatrix} - (1)\begin{vmatrix} x & y \\ 1 & 5 \end{vmatrix} + (1)\begin{vmatrix} x & y \\ -2 & -2 \end{vmatrix}$$
$$= (1)(-8) - (1)(5x - y) + (1)(-2x + 2y)$$
$$= -8 - 5x + y - 2x + 2y$$
$$= -7x + 3y - 8$$

**34.** *Keystrokes:*

$\boxed{\text{MATRX}}$ $\boxed{\text{EDIT}}$ $\boxed{1}$ 3 $\boxed{\text{ENTER}}$ 3 $\boxed{\text{ENTER}}$ $\boxed{(-)}$1$\boxed{\div}$ 2 $\boxed{\text{ENTER}}$ $\boxed{(-)}$1 $\boxed{\text{ENTER}}$

6 $\boxed{\text{ENTER}}$ 8 $\boxed{\text{ENTER}}$ $\boxed{(-)}$ 1$\boxed{\div}$ 4 $\boxed{\text{ENTER}}$ $\boxed{(-)}$4 $\boxed{\text{ENTER}}$ 1 $\boxed{\text{ENTER}}$

2 $\boxed{\text{ENTER}}$ 1 $\boxed{\text{ENTER}}$ $\boxed{\text{QUIT}}$

$\boxed{\text{MATRX}}$ $\boxed{\text{MATH}}$ $\boxed{1}$ $\boxed{\text{MATRX}}$ $\boxed{1}$ $\boxed{\text{ENTER}}$

Solution is 105.625.

**36.** *Keystrokes:*

MATRX  EDIT  1  3  ENTER  3  ENTER  1  ÷  2  ENTER  3  ÷  2  ENTER

1  ÷  2  ENTER  4  ENTER  8  ENTER  10  ENTER  (−) 2  ENTER

(−) 6  ENTER  12  ENTER  QUIT

MATRX  MATH  1  MATRX  1  ENTER

Solution is $-28$.

**38.** *Keystrokes:*

MATRX  EDIT  1  3  ENTER  3  ENTER  .4  ENTER  .3  ENTER

.3  ENTER  (−) .2  ENTER  .6  ENTER  .6  ENTER  3  ENTER

1  ENTER  1  ENTER  QUIT

MATRX  MATH  1  MATRX  1  ENTER

Solution is $0$.

**40.** $\begin{bmatrix} 2 & -1 & \vdots & -10 \\ 3 & 2 & \vdots & -1 \end{bmatrix}$

$D = \begin{vmatrix} 2 & -1 \\ 3 & 2 \end{vmatrix} = 4 - (-3) = 7$

$x = \dfrac{D_x}{D} = \dfrac{\begin{vmatrix} -10 & -1 \\ -1 & 2 \end{vmatrix}}{7} = \dfrac{-20 - 1}{7} = \dfrac{-21}{7} = -3$

$y = \dfrac{D_y}{D} = \dfrac{\begin{vmatrix} 2 & -10 \\ 3 & -1 \end{vmatrix}}{7} = \dfrac{-2 + 30}{7} = \dfrac{28}{7} = 4$

$(-3, 4)$

**42.** $18x + 12y = 13$

$30x + 24y = 23$

$\begin{bmatrix} 18 & 12 & \vdots & 13 \\ 30 & 24 & \vdots & 23 \end{bmatrix}$

$D = \begin{vmatrix} 18 & 12 \\ 30 & 24 \end{vmatrix} = 432 - 360 = 72$

$x = \dfrac{D_x}{D} = \dfrac{\begin{vmatrix} 13 & 12 \\ 23 & 24 \end{vmatrix}}{72} = \dfrac{312 - 276}{72} = \dfrac{36}{72} = \dfrac{1}{2}$

$y = \dfrac{D_y}{D} = \dfrac{\begin{vmatrix} 18 & 13 \\ 30 & 23 \end{vmatrix}}{72} = \dfrac{14 - 390}{72} = \dfrac{24}{72} = \dfrac{1}{3}$

$\left( \dfrac{1}{2}, \dfrac{1}{3} \right)$

**44.** $\begin{bmatrix} 13 & -6 & \vdots & 17 \\ 26 & -12 & \vdots & 8 \end{bmatrix}$

$D = \begin{vmatrix} 13 & -6 \\ 26 & -12 \end{vmatrix} = -156 + 156 = 0$

Cannot be solved by Cramer's Rule because $D = 0$.

**46.** $-0.4x + 0.8y = 1.6$

$0.2x + 0.3y = 2.2$

$$\begin{bmatrix} -0.4 & 0.8 & \vdots & 1.6 \\ 0.2 & 0.3 & \vdots & 2.2 \end{bmatrix}$$

$$D = \begin{vmatrix} -0.4 & 0.8 \\ 0.2 & 0.3 \end{vmatrix} = -0.12 - 0.16 = -0.28$$

$$x = \frac{D_x}{D} = \frac{\begin{vmatrix} 1.6 & 0.8 \\ 2.2 & 0.3 \end{vmatrix}}{-0.28} = \frac{0.48 - 1.76}{-0.28} = \frac{-1.28}{-0.28} = \frac{128}{28} = \frac{32}{7}$$

$$y = \frac{D_y}{D} = \frac{\begin{vmatrix} -0.4 & 1.6 \\ 0.2 & 2.2 \end{vmatrix}}{-0.28} = \frac{-0.88 - 0.32}{-0.28} = \frac{-1.2}{-0.28} = \frac{120}{28} = \frac{30}{7}$$

$$\left( \frac{32}{7}, \frac{30}{7} \right)$$

**48.** $3x_1 + 2x_2 = 1$

$2x_1 + 10x_2 = 6$

$$\begin{bmatrix} 3 & 2 & \vdots & 1 \\ 2 & 10 & \vdots & 6 \end{bmatrix}$$

$$D = \begin{vmatrix} 3 & 2 \\ 2 & 10 \end{vmatrix} = 30 - 4 = 26$$

$$x = \frac{D_x}{D} = \frac{\begin{vmatrix} 1 & 2 \\ 6 & 10 \end{vmatrix}}{26} = \frac{10 - 12}{26} = \frac{-2}{26} = -\frac{1}{13}$$

$$y = \frac{D_y}{D} = \frac{\begin{vmatrix} 3 & 1 \\ 2 & 6 \end{vmatrix}}{26} = \frac{18 - 2}{26} = \frac{16}{28} = \frac{8}{13}$$

$$\left( \frac{-1}{13}, \frac{8}{13} \right)$$

**50.** $4x - 2y + 3z = -2$

$2x + 2y + 5z = 16$

$8x - 5y - 2z = 4$

$$\begin{bmatrix} 4 & -2 & 3 & \vdots & -2 \\ 2 & 2 & 5 & \vdots & 16 \\ 8 & -5 & -2 & \vdots & 4 \end{bmatrix}$$

$$D = \begin{vmatrix} 4 & -2 & 3 \\ 2 & 2 & 5 \\ 8 & -5 & -2 \end{vmatrix} = -(2)\begin{vmatrix} -2 & 3 \\ -5 & -2 \end{vmatrix} + (2)\begin{vmatrix} 4 & 3 \\ 8 & -2 \end{vmatrix} - (5)\begin{vmatrix} 4 & -2 \\ 8 & -5 \end{vmatrix}$$

$$= (-2)(19) + (2)(-32) - (5)(-4)$$

$$= -38 - 64 + 20$$

$$= -82$$

$$x = \frac{\begin{vmatrix} -2 & -2 & 3 \\ 16 & 2 & 5 \\ 4 & -5 & -2 \end{vmatrix}}{-82} = \frac{(-2)\begin{vmatrix} 2 & 5 \\ -5 & -2 \end{vmatrix} - (-2)\begin{vmatrix} 16 & 5 \\ 4 & -2 \end{vmatrix} + (3)\begin{vmatrix} 16 & 2 \\ 4 & -5 \end{vmatrix}}{-82}$$

$$= \frac{(-2)(21)(2)(-52) + (3)(-88)}{-82}$$

$$= \frac{-42 - 104 - 264}{-82} = \frac{-410}{-82} = 5$$

$$y = \frac{\begin{vmatrix} 4 & -2 & 3 \\ 2 & 16 & 5 \\ 8 & 4 & -2 \end{vmatrix}}{-82} = \frac{(4)\begin{vmatrix} 16 & 5 \\ 4 & -2 \end{vmatrix} - (2)\begin{vmatrix} -2 & 3 \\ 4 & -2 \end{vmatrix} + (8)\begin{vmatrix} -2 & 3 \\ 16 & 5 \end{vmatrix}}{-82}$$

$$= \frac{(4)(-52) - (2)(-8) + (8)(-58)}{-82}$$

$$= \frac{-208 + 16 - 464}{-82} = \frac{-656}{-82} = 8$$

$$z = \frac{\begin{vmatrix} 4 & -2 & -2 \\ 2 & 2 & 16 \\ 8 & -5 & 4 \end{vmatrix}}{-82} = \frac{(4)\begin{vmatrix} 2 & 16 \\ -5 & 4 \end{vmatrix} - (2)\begin{vmatrix} -2 & -2 \\ -5 & 4 \end{vmatrix} + (8)\begin{vmatrix} -2 & -2 \\ 2 & 16 \end{vmatrix}}{-82}$$

$$= \frac{(4)(88) - (2)(-18) + (8)(-28)}{-82}$$

$$= \frac{352 + 36 - 224}{-82} = \frac{164}{-82} = -2$$

$(5, 8, -2)$

**52.** $14x_1 - 21x_2 - 7x_3 = 10$

$-4x_1 + 2x_2 - 2x_3 = 4$

$56x_1 - 21x_2 + 7x_3 = 5$

$$\begin{bmatrix} 14 & -21 & -7 & \vdots & 10 \\ -4 & 2 & -2 & \vdots & 4 \\ 56 & -21 & 7 & \vdots & 5 \end{bmatrix}$$

$$D = \begin{vmatrix} 14 & -21 & -7 \\ -4 & 2 & -2 \\ 56 & -21 & 7 \end{vmatrix} = -(-4)\begin{vmatrix} -21 & -7 \\ -21 & 7 \end{vmatrix} + (2)\begin{vmatrix} 14 & -7 \\ 56 & 7 \end{vmatrix} - (-2)\begin{vmatrix} 14 & -21 \\ 56 & -21 \end{vmatrix}$$

$$= 4(-294) + (2)(490) + (2)(882)$$

$$= (-1176) + 980 + 1764$$

$$= 1568$$

$$x = \frac{\begin{vmatrix} 10 & -21 & -7 \\ 4 & 2 & -2 \\ 5 & -21 & 7 \end{vmatrix}}{1568} = \frac{(10)\begin{vmatrix} 2 & -2 \\ -21 & 7 \end{vmatrix} - (4)\begin{vmatrix} -21 & -7 \\ -21 & 7 \end{vmatrix} + (5)\begin{vmatrix} -21 & -7 \\ 2 & -2 \end{vmatrix}}{1568}$$

$$= \frac{(10)(-28) - (4)(-294) + (5)(56)}{1568}$$

$$= \frac{-280 + 1176 + 280}{1568} = \frac{1176}{1568} = \frac{21}{28} = \frac{3}{4}$$

$$y = \frac{\begin{vmatrix} 14 & 10 & -7 \\ -4 & 4 & -2 \\ 56 & 5 & 7 \end{vmatrix}}{1568} = \frac{-(-4)\begin{vmatrix} 10 & -7 \\ 5 & 7 \end{vmatrix} + (4)\begin{vmatrix} 14 & -7 \\ 56 & 7 \end{vmatrix} - (-2)\begin{vmatrix} 14 & 10 \\ 56 & 5 \end{vmatrix}}{1568}$$

$$= \frac{(4)(105) + (4)(490) + (2)(-490)}{1568}$$

$$= \frac{420 + 1960 - 980}{1568} = \frac{1400}{1568} = \frac{25}{28}$$

$$z = \frac{\begin{vmatrix} 14 & -21 & 10 \\ -4 & 2 & 4 \\ 56 & -21 & 5 \end{vmatrix}}{1568} = \frac{(10)\begin{vmatrix} -4 & 2 \\ 56 & -21 \end{vmatrix} - (4)\begin{vmatrix} 14 & -21 \\ 56 & -21 \end{vmatrix} + (5)\begin{vmatrix} 14 & -21 \\ -4 & 2 \end{vmatrix}}{1568}$$

$$= \frac{(10)(-28) - (4)(882) + (5)(-56)}{1568}$$

$$= \frac{-280 - 3528 - 280}{1568} = \frac{-4088}{1568} = \frac{-73}{28}$$

$$\left(\frac{3}{4}, \frac{25}{28}, \frac{-73}{28}\right)$$

**54.** $2x + 3y + \phantom{0}5z = \phantom{0}4$

$3x + 5y + \phantom{0}9z = \phantom{0}7$

$5x + 9y + 17z = 13$

$$\begin{bmatrix} 2 & 3 & 5 & \vdots & 4 \\ 3 & 5 & 9 & \vdots & 7 \\ 5 & 9 & 17 & \vdots & 13 \end{bmatrix}$$

$$D = \begin{vmatrix} 2 & 3 & 5 \\ 3 & 5 & 9 \\ 5 & 9 & 17 \end{vmatrix} = (2)\begin{vmatrix} 5 & 9 \\ 9 & 17 \end{vmatrix} - (3)\begin{vmatrix} 3 & 9 \\ 5 & 17 \end{vmatrix} + 5\begin{vmatrix} 3 & 5 \\ 5 & 9 \end{vmatrix}$$

$$= (2)(4) - (3)(6) + 5(2)$$

$$= 8 - 18 + 10$$

$$= 0$$

Cannot be solved by Cramer's Rule because $D = 0$.

**56.** $\phantom{0}3x + 2y + 5z = 4$

$\phantom{0}4x - 3y - 4z = 1$

$-8x + 2y + 3z = 0$

$$\begin{bmatrix} 3 & 2 & 5 & \vdots & 4 \\ 4 & -3 & -4 & \vdots & 1 \\ -8 & 2 & 3 & \vdots & 0 \end{bmatrix}$$

$$D = \begin{vmatrix} 3 & 2 & 5 \\ 4 & -3 & -4 \\ -8 & 2 & 3 \end{vmatrix} = (3)\begin{vmatrix} -3 & -4 \\ 2 & 3 \end{vmatrix} - (2)\begin{vmatrix} 4 & -4 \\ -8 & 3 \end{vmatrix} + (5)\begin{vmatrix} 4 & -3 \\ -8 & 2 \end{vmatrix}$$

$$= (3)(-1) - (2)(-20) + (5)(-16)$$

$$= -3 + 40 - 80$$

$$= -43$$

$$x = \frac{\begin{vmatrix} 4 & 2 & 5 \\ 1 & -3 & -4 \\ 0 & 2 & 3 \end{vmatrix}}{-43} = \frac{(4)\begin{vmatrix} -3 & -4 \\ 2 & 3 \end{vmatrix} - (1)\begin{vmatrix} 2 & 5 \\ 2 & 3 \end{vmatrix} + 0}{-43}$$

$$= \frac{(4)(-1) - (1)(-4)}{-43}$$

$$= \frac{-4 + 4}{-43} = 0$$

$$y = \frac{\begin{vmatrix} 3 & 4 & 5 \\ 4 & 1 & -4 \\ -8 & 0 & 3 \end{vmatrix}}{-43} = \frac{-(4)\begin{vmatrix} 4 & -4 \\ -8 & 3 \end{vmatrix} + (1)\begin{vmatrix} 3 & 5 \\ -8 & 3 \end{vmatrix} - 0}{-43}$$

$$= \frac{-(4)(-20) + (1)(49)}{-43}$$

$$= \frac{+80 + 49}{-43} = \frac{+129}{-43} = -3$$

**—CONTINUED—**

**56.** **—CONTINUED—**

$$z = \frac{\begin{vmatrix} 3 & 2 & 4 \\ 4 & 123 & 1 \\ -8 & 2 & 0 \end{vmatrix}}{-43} = \frac{(43)\begin{vmatrix} 4 & -3 \\ -8 & 2 \end{vmatrix} - (1)\begin{vmatrix} 3 & 2 \\ -8 & 2 \end{vmatrix} + 0}{-43}$$

$$= \frac{(4)(-16) - (1)(22)}{-43}$$

$$= \frac{-64 - 22}{-43} = \frac{-86}{-43} = 2$$

$(0, -3, 2)$

**58.** $\begin{bmatrix} 3 & 7 & \vdots & 3 \\ 7 & 25 & \vdots & 11 \end{bmatrix}$

$$D = \begin{vmatrix} 3 & 7 \\ 7 & 25 \end{vmatrix} = 75 - 49 = 26$$

$$x = \frac{D_x}{D} = \frac{\begin{vmatrix} 3 & 7 \\ 11 & 25 \end{vmatrix}}{26} = \frac{75 - 77}{26} = \frac{-2}{26} = \frac{-1}{13}$$

$$y = \frac{D_y}{D} = \frac{\begin{vmatrix} 3 & 3 \\ 7 & 11 \end{vmatrix}}{26} = \frac{33 - 21}{26} = \frac{12}{26} = \frac{6}{13}$$

*Keystrokes:*

det $D$ [MATRX] [EDIT] [1] 2 [ENTER] 2 [ENTER] 3 [ENTER] 7 [ENTER]

7 [ENTER] 25 [ENTER] [QUIT]

[MATRX] [MATH] [1] [MATRX] [1] [ENTER]

det $Dx$ [MATRX] [EDIT] [2] 2 [ENTER] 2 [ENTER] 3 [ENTER] 7 [ENTER]

11 [ENTER] 25 [ENTER] [QUIT]

[MATRX] [MATH] [1] [MATRX] [2] [ENTER]

det $Dy$ [MATRX] [EDIT] [3] 2 [ENTER] 2 [ENTER] 3 [ENTER] 3 [ENTER]

7 [ENTER] 11 [ENTER] [QUIT]

[MATRX] [MATH] [1] [MATRX] [3] [ENTER]

**60.** $\begin{bmatrix} 6 & 4 & -8 & \vdots & -22 \\ -2 & 2 & 3 & \vdots & 13 \\ -2 & 2 & -1 & \vdots & 5 \end{bmatrix}$

$$D = \begin{vmatrix} 6 & 4 & -8 \\ -2 & 2 & 3 \\ -2 & 2 & -1 \end{vmatrix} = -80$$

$$x = \frac{D_x}{D} = \frac{\begin{vmatrix} -22 & 4 & -9 \\ 13 & 2 & 3 \\ 5 & 2 & -1 \end{vmatrix}}{-80} = \frac{160}{-80} = -2$$

$$y = \frac{D_y}{D} = \frac{\begin{vmatrix} 6 & -22 & -8 \\ -2 & 13 & 3 \\ -2 & 5 & -1 \end{vmatrix}}{-80} = \frac{-120}{-80} = \frac{3}{2}$$

$$z = \frac{D_z}{D} = \frac{\begin{vmatrix} 6 & 4 & -22 \\ -2 & 2 & 13 \\ -2 & 2 & 5 \end{vmatrix}}{-80} = \frac{-160}{-80} = 2$$

*Keystrokes:*

det $D$   [MATRX] [EDIT] [1] 3 [ENTER] 3 [ENTER]

Enter each number in matrix followed by [ENTER]

[QUIT]

[MATRX] [MATH] [1] [MATRX] [ENTER]

det $Dx$   [MATRX] [EDIT] [2] 3 [ENTER] 3 [ENTER]

Enter each number in matrix followed by [ENTER]

[QUIT]

[MATRX] [MATH] [1] [MATRX] [2] [ENTER]

det $Dy$   [MATRX] [EDIT] [3] 3 [ENTER] 3 [ENTER]

Enter each number in matrix followed by [ENTER]

[MATRX] [MATH] [1] [MATRX] [3] [ENTER]

det $Dz$   [MATRX] [EDIT] [4] 3 [ENTER] 3 [ENTER]

Enter each number in matrix followed by [ENTER]

[MATRX] [MATH] [1] [MATRX] [4] [ENTER]

**62.** $(4 - x)(1 - x) - (-2) = 0$

$$4 - 4x - x + x^2 + 2 = 0$$

$$x^2 - 5x + 6 = 0$$

$$(x - 3)(x - 2) = 0$$

$x - 3 = 0 \quad x - 2 = 0$

$\quad x = 3 \qquad x = 2$

**64.** $(x_1, y_1) = (2, 0), (x_2, y_2) = (0, 5), (x_3, y_3) = (6, 3)$

$$\begin{vmatrix} x_1 & y_1 & 1 \\ x_2 & y_2 & 1 \\ x_3 & y_3 & 1 \end{vmatrix} = \begin{vmatrix} 2 & 0 & 1 \\ 0 & 5 & 1 \\ 6 & 3 & 1 \end{vmatrix} = 2\begin{vmatrix} 5 & 1 \\ 3 & 1 \end{vmatrix} - 0 + 6\begin{vmatrix} 0 & 1 \\ 5 & 1 \end{vmatrix}$$

$$= 2(2) + 6(-5)$$

$$= -26$$

Area $= -\frac{1}{2}(-26) = 13$

**66.** $(x_1, y_1) = (-2, -3), (x_2, y_2) = (2, -3), (x_3, y_3) = (0, 4)$

$$\begin{vmatrix} x_1 & y_1 & 1 \\ x_2 & y_2 & 1 \\ x_3 & y_3 & 1 \end{vmatrix} = \begin{vmatrix} -2 & -3 & 1 \\ 2 & -3 & 1 \\ 0 & 4 & 1 \end{vmatrix} = -2\begin{vmatrix} -3 & 1 \\ 4 & 1 \end{vmatrix} - 2\begin{vmatrix} -3 & 1 \\ 4 & 1 \end{vmatrix} + 0\begin{vmatrix} -3 & 1 \\ -3 & 1 \end{vmatrix}$$

$$= -2(-7) - 2(-7) + 0$$

$$= 14 + 14$$

$$= 28$$

Area $= \frac{1}{2}(28) = 14$

**68.** $(x_1, y_1) = (-4, 2), (x_2, y_2) = (1, 5), (x_3, y_3) = (4, -4)$

$$\begin{vmatrix} x_1 & y_1 & 1 \\ x_2 & y_2 & 1 \\ x_3 & y_3 & 1 \end{vmatrix} = \begin{vmatrix} -4 & 2 & 1 \\ 1 & 5 & 1 \\ 4 & -4 & 1 \end{vmatrix} = -4\begin{vmatrix} 5 & 1 \\ -4 & 1 \end{vmatrix} - 2\begin{vmatrix} 1 & 1 \\ 4 & 1 \end{vmatrix} + 1\begin{vmatrix} 1 & 5 \\ 4 & 4 \end{vmatrix}$$

$$= -4(5 - -4) - 2(1 - 4) + 1(-4 - 20)$$

$$= -4(9) - 2(-3) + 1(-24)$$

$$= -36 + 6 - 24$$

$$= -54$$

Area $= -\frac{1}{2}(-54) = 27$

**70.** $(x_1, y_1) = \left(\frac{1}{4}, 0\right), (x_2, y_2) = \left(0, \frac{3}{4}\right), (x_3, y_3) = (8, -2)$

$$\begin{vmatrix} x_1 & y_1 & 1 \\ x_2 & y_2 & 1 \\ x_3 & y_3 & 1 \end{vmatrix} = \begin{vmatrix} \frac{1}{4} & 0 & 1 \\ 0 & \frac{3}{4} & 1 \\ 8 & -2 & 1 \end{vmatrix} = \frac{1}{4}\begin{vmatrix} \frac{3}{4} & 1 \\ -2 & 1 \end{vmatrix} - 0 + 1\begin{vmatrix} 0 & \frac{3}{4} \\ 8 & -2 \end{vmatrix}$$

$$= \frac{1}{4}\left(\frac{11}{4}\right) + 1(-6) = \frac{-85}{16}$$

Area $= -\frac{1}{2}\left(\frac{-85}{16}\right) = \frac{85}{32}$

**72.** *Verbal model:* 
$$\boxed{\begin{array}{c}\text{Area of}\\\text{Triangle 1}\end{array}} + \boxed{\begin{array}{c}\text{Area of}\\\text{Triangle 2}\end{array}} = \boxed{\begin{array}{c}\text{Area of}\\\text{shaded region}\end{array}}$$

*Equation:* $A = (6) + (9)$

$\qquad\qquad A = 15$ square units

Let $(x_1, y_1) = (-1, 2)$, $(x_2, y_2) = (5, 2)$, $(x_3, y_3) = (0, 0)$.

$$\begin{vmatrix} x_1 & y_1 & 1 \\ x_2 & y_2 & 1 \\ x_3 & y_3 & 1 \end{vmatrix} = \begin{vmatrix} -1 & 2 & 1 \\ 5 & 2 & 1 \\ 0 & 0 & 1 \end{vmatrix} = 0\begin{vmatrix} 2 & 1 \\ 2 & 1 \end{vmatrix} - 0\begin{vmatrix} -1 & 1 \\ 5 & 1 \end{vmatrix} + 1\begin{vmatrix} -1 & 2 \\ 5 & 2 \end{vmatrix} = 0 - 0 + (-2 - 10) = -12$$

Area $= \dfrac{-1}{2}(-12) = 6$

Let $(x_1, y_1) = (0, 0)$, $(x_2, y_2) = (5, 2)$, $(x_3, y_3) = (4, -2)$.

$$\begin{vmatrix} x_1 & y_1 & 1 \\ x_2 & y_2 & 1 \\ x_3 & y_3 & 1 \end{vmatrix} = \begin{vmatrix} 0 & 0 & 1 \\ 5 & 2 & 1 \\ 4 & -2 & 1 \end{vmatrix} = 0\begin{vmatrix} 2 & 1 \\ -2 & 1 \end{vmatrix} - 0\begin{vmatrix} 5 & 1 \\ 4 & 1 \end{vmatrix} + 1\begin{vmatrix} 5 & 2 \\ 4 & -2 \end{vmatrix} = 0 - 0 + (-10 - 8) = -18$$

Area $= -\dfrac{1}{2}(-18) = 9$

**74.** *Verbal model:* 
$$\boxed{\begin{array}{c}\text{Area of}\\\text{Rectangle}\end{array}} - \boxed{\begin{array}{c}\text{Area of}\\\text{Triangle}\end{array}} = \boxed{\begin{array}{c}\text{Area of}\\\text{shaded region}\end{array}}$$

*Equation:* $35 - 12 = 23$

Let $(x_1, y_1) = (-3, 0)$, $(x_2, y_2) = (1, 4)$, $(x_3, y_3) = (2, -1)$.

$$\begin{vmatrix} x_1 & y_1 & 1 \\ x_2 & y_2 & 1 \\ x_3 & y_3 & 1 \end{vmatrix} = \begin{vmatrix} -3 & 0 & 1 \\ 1 & 4 & 1 \\ 2 & -1 & 1 \end{vmatrix} = -3\begin{vmatrix} 4 & 1 \\ -1 & 1 \end{vmatrix} - 0 + 1\begin{vmatrix} 1 & 4 \\ 2 & -1 \end{vmatrix}$$

$$= (-3)(5) + (1)(-9) = -24$$

Area $= \dfrac{-1}{2}(-24) = 12$

Area of rectangle $=$ length $\cdot$ width

$$= |4 - (-3)| \cdot |(-1) - (4)|$$

$$= 7 \cdot 5$$

$$= 35$$

**76.** From the diagram, the coordinates of $A$, $B$, and $C$ are determined to be $A(-85, 80)$, $B(-65, 0)$, and $C(0, 50)$.

$$\begin{vmatrix} x_1 & y_1 & 1 \\ x_2 & y_2 & 1 \\ x_3 & y_3 & 1 \end{vmatrix} = \begin{vmatrix} -85 & 80 & 1 \\ -65 & 0 & 1 \\ 0 & 50 & 1 \end{vmatrix} = -85\begin{vmatrix} 0 & 1 \\ 50 & 1 \end{vmatrix} - (-65)\begin{vmatrix} 80 & 1 \\ 50 & 1 \end{vmatrix} + 0\begin{vmatrix} 8 & 1 \\ 0 & 1 \end{vmatrix}$$

$$= -85(0 - 50) + 65(80 - 50) + 0$$

$$= 4250 + 1950 + 0$$

$$= 6200$$

Area $= \frac{1}{2}(6200) = 3100$ square feet

**78.** $(x_1, y_1) = (-1, -1), (x_2, y_2) = (1, 9), (x_3, y_3) = (2, 13)$

$$\begin{vmatrix} x_1 & y_1 & 1 \\ x_2 & y_2 & 1 \\ x_3 & y_3 & 1 \end{vmatrix} = \begin{vmatrix} -1 & -1 & 1 \\ 1 & 9 & 1 \\ 2 & 13 & 1 \end{vmatrix} = (1)\begin{vmatrix} 1 & 9 \\ 2 & 13 \end{vmatrix} - (1)\begin{vmatrix} -1 & -1 \\ 2 & 13 \end{vmatrix} + (1)\begin{vmatrix} -1 & -1 \\ 1 & 9 \end{vmatrix}$$

$$= (1)(-5) - (1)(-11) + (1)(-8)$$

$$= -5 + 11 - 8$$

$$= -2$$

The three points are *not* collinear.

**80.** $(x_1, y_1) = (-1, 8), (x_2, y_2) = (1, 2), (x_3, y_3) = (2, 0)$

$$\begin{vmatrix} x_1 & y_1 & 1 \\ x_2 & y_2 & 1 \\ x_3 & y_3 & 1 \end{vmatrix} = \begin{vmatrix} -1 & 8 & 1 \\ 1 & 2 & 1 \\ 2 & 0 & 1 \end{vmatrix} = -1\begin{vmatrix} 2 & 1 \\ 0 & 1 \end{vmatrix} - 8\begin{vmatrix} 1 & 1 \\ 2 & 1 \end{vmatrix} + 1\begin{vmatrix} 1 & 2 \\ 2 & 0 \end{vmatrix}$$

$$= -1(2 - 0) - 8(1 - 2) + 1(0 - 4)$$

$$= -1(2) - 8(-1) + 1(-4)$$

$$= -2 + 8 - 4$$

$$= 2$$

The three points are *not* collinear.

**82.** Let $(x_1, y_1) = \left(0, \frac{1}{2}\right), (x_2, y_2) = \left(1, \frac{7}{6}\right), (x_3, y_3) = \left(9, \frac{13}{2}\right)$.

$$\begin{vmatrix} x_1 & y_1 & 1 \\ x_2 & y_2 & 1 \\ x_3 & y_3 & 1 \end{vmatrix} = \begin{vmatrix} 0 & \frac{1}{2} & 1 \\ 1 & \frac{7}{6} & 1 \\ 9 & \frac{13}{2} & 1 \end{vmatrix} = 0\begin{vmatrix} \frac{7}{6} & 1 \\ \frac{13}{2} & 1 \end{vmatrix} - \frac{1}{2}\begin{vmatrix} 1 & 1 \\ 9 & 1 \end{vmatrix} + 1\begin{vmatrix} 1 & \frac{7}{6} \\ 9 & \frac{13}{2} \end{vmatrix}$$

$$= 0 - \frac{1}{2}(-8) + 1(-4)$$

$$= 4 - 4$$

$$= 0$$

The three points are collinear.

**84.** $(x_1, y_1) = (-4, 3), (x_2, y_2) = (2, 1)$

$$\begin{vmatrix} x & y & 1 \\ -4 & 3 & 1 \\ 2 & 1 & 1 \end{vmatrix} = 0$$

$$(1)\begin{vmatrix} -4 & 3 \\ 2 & 1 \end{vmatrix} - (1)\begin{vmatrix} x & y \\ 2 & 1 \end{vmatrix} + (1)\begin{vmatrix} x & y \\ -4 & 3 \end{vmatrix} = 0$$

$$(1)(-10) - (1)(x - 2y) + (1)(3x + 4y) = 0$$

$$-10 - x + 2y + 3x + 4y = 0$$

$$2x + 6y - 10 = 0$$

**86.** $(x_1, y_1) = (-8, 3), (x_2, y_2) = (4, 6)$

$$\begin{vmatrix} x & y & 1 \\ -8 & 3 & 1 \\ 4 & 6 & 1 \end{vmatrix} = 0$$

$$x\begin{vmatrix} 3 & 1 \\ 6 & 1 \end{vmatrix} - y\begin{vmatrix} -8 & 1 \\ 4 & 1 \end{vmatrix} + 1\begin{vmatrix} -8 & 3 \\ 4 & 6 \end{vmatrix} = 0$$

$$x(-3) - y(-12) + 1(-60) = 0$$

$$-3x + 12y - 60 = 0$$

$$x - 4y + 20 = 0$$

**88.** $(x_1, y_1) = \left(-\frac{1}{2}, 3\right), (x_2, y_2) = \left(\frac{5}{2}, 1\right)$

$$\begin{vmatrix} x & y & 1 \\ -\frac{1}{2} & 3 & 1 \\ \frac{5}{2} & 1 & 1 \end{vmatrix} = 0$$

$$(1)\begin{vmatrix} -\frac{1}{2} & 3 \\ \frac{5}{2} & 1 \end{vmatrix} - (1)\begin{vmatrix} x & y \\ \frac{5}{2} & 1 \end{vmatrix} + (1)\begin{vmatrix} x & y \\ -\frac{1}{2} & 3 \end{vmatrix} = 0$$

$$(1)(-8) - (1)\left(x - \frac{5}{2}y\right) + (1)\left(3x + \frac{1}{2}y\right) = 0$$

$$-8 - x + \frac{5}{2}y + 3x + \frac{1}{2}y = 0$$

$$2x + 3y - 8 = 0$$

**90.** $(x_1, y_1) = (3, 1.6), (x_2, y_2) = (5, -2.2)$

$$\begin{vmatrix} x & y & 1 \\ 3 & 1.6 & 1 \\ 5 & -2.2 & 1 \end{vmatrix} = 0$$

$$x\begin{vmatrix} 1.6 & 1 \\ -2.2 & 1 \end{vmatrix} - y\begin{vmatrix} 3 & 1 \\ 5 & 1 \end{vmatrix} + 1\begin{vmatrix} 3 & 1.6 \\ 5 & -2.2 \end{vmatrix} = 0$$

$$x(3.8) - y(-2) + 1(-14.6) = 0$$

$$3.8x + 2y - 14.6 = 0$$

$$19x + 10y - 73 = 0$$

**92.**

$$0 = a(-1)^2 + b(-1) + c \implies 0 = a - b + c$$

$$4 = a(1)^2 + b(1) + c \implies 4 = a + b + c$$

$$-5 = a(4)^2 + b(4) + c \implies -5 = 16a - 4b + c$$

$$\begin{bmatrix} 1 & -1 & 1 & \vdots & 0 \\ 1 & 1 & 1 & \vdots & 4 \\ 16 & 4 & 1 & \vdots & -5 \end{bmatrix} D = \begin{vmatrix} 1 & -1 & 1 \\ 1 & 1 & 1 \\ 16 & 4 & 1 \end{vmatrix} = -30$$

$$a = \frac{D_a}{D} = \frac{\begin{vmatrix} 0 & -1 & 1 \\ 4 & 1 & 1 \\ -5 & 4 & 1 \end{vmatrix}}{-30} = \frac{30}{-30} = -1$$

$$b = \frac{D_b}{D} = \frac{\begin{vmatrix} 1 & 0 & 1 \\ 1 & 4 & 1 \\ 16 & -5 & 1 \end{vmatrix}}{-30} = \frac{-60}{-30} = 2$$

$$c = \frac{D_c}{D} = \frac{\begin{vmatrix} 1 & -1 & 0 \\ 1 & 1 & 4 \\ 16 & 4 & -5 \end{vmatrix}}{-30} = \frac{-90}{-30} = 3$$

$$y = -x^2 + 2x + 3$$

**94.** $y = ax^2 + bx + c$

$6 = a(-2)^2 + b(-2) + c \implies 6 = 4a - 2b + c$

$9 = a(1)^2 + b(1) + c \implies 9 = a + b + c$

$1 = a(3)^2 + b(3) + c \implies 1 = 9a + 3b + c$

$$\begin{bmatrix} 4 & -2 & 1 & \vdots & 6 \\ 1 & 1 & 1 & \vdots & 9 \\ 9 & 3 & 1 & \vdots & 1 \end{bmatrix}$$

$$d = \begin{vmatrix} 4 & -2 & 1 \\ 1 & 1 & 1 \\ 9 & 3 & 1 \end{vmatrix} = (1)\begin{vmatrix} 1 & 1 \\ 9 & 3 \end{vmatrix} - (1)\begin{vmatrix} 4 & -2 \\ 9 & 3 \end{vmatrix} + (1)\begin{vmatrix} 4 & -2 \\ 1 & 1 \end{vmatrix}$$

$$= (1)(-6) - (1)(30) + (1)(6)$$

$$= -6 - 30 + 6$$

$$= -30$$

$$a = \frac{\begin{vmatrix} 6 & -2 & 1 \\ 9 & 1 & 1 \\ 1 & 3 & 1 \end{vmatrix}}{-30} = \frac{(1)\begin{vmatrix} 9 & 1 \\ 1 & 3 \end{vmatrix} - (1)\begin{vmatrix} 6 & -2 \\ 1 & 3 \end{vmatrix} + (1)\begin{vmatrix} 6 & -2 \\ 9 & 1 \end{vmatrix}}{-30}$$

$$= \frac{(1)(26) - (1)(20) + (1)(24)}{-30}$$

$$= \frac{26 - 20 + 24}{-30} = \frac{30}{-30} = -1$$

$$b = \frac{\begin{vmatrix} 4 & 6 & 1 \\ 1 & 9 & 1 \\ 9 & 1 & 1 \end{vmatrix}}{-30} = \frac{(1)\begin{vmatrix} 1 & 9 \\ 9 & 1 \end{vmatrix} - (1)\begin{vmatrix} 4 & 6 \\ 9 & 1 \end{vmatrix} + (1)\begin{vmatrix} 4 & 6 \\ 1 & 9 \end{vmatrix}}{-30}$$

$$= \frac{(1)(-80) - (1)(-50) + (1)(30)}{-30}$$

$$= \frac{-80 + 50 + 30}{-30} = 0$$

$$c = \frac{\begin{vmatrix} 4 & -2 & 6 \\ 1 & 1 & 9 \\ 9 & 3 & 1 \end{vmatrix}}{-30} = \frac{-(1)\begin{vmatrix} -2 & 6 \\ 3 & 1 \end{vmatrix} + (1)\begin{vmatrix} 4 & 6 \\ 9 & 1 \end{vmatrix} - (9)\begin{vmatrix} 4 & -2 \\ 9 & 3 \end{vmatrix}}{-30}$$

$$= \frac{-(1)(-20) + (1)(-50) - (9)(30)}{-30}$$

$$= \frac{20 - 50 - 270}{-30} = \frac{-300}{-30} = 10$$

$y = -x^2 + 10$

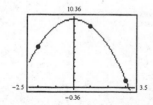

**96.** $y = ax^2 + bx + c$

$$3 = a(2)^2 + b(2) + c \implies 3 = 4a + 2b + c$$

$$\frac{9}{2} = a(-1)^2 + b(-1) + c \implies \frac{9}{2} = a - b + c$$

$$9 = a(-2)^2 + b(-2) + c \implies 9 = 4a - 2b + c$$

$$\begin{bmatrix} 4 & 2 & 1 & \vdots & 3 \\ 1 & -1 & 1 & \vdots & \frac{9}{2} \\ 4 & -2 & 1 & \vdots & 9 \end{bmatrix}$$

$$D = \begin{vmatrix} 4 & 2 & 1 \\ 1 & -1 & 1 \\ 4 & -2 & 1 \end{vmatrix} = (1)\begin{vmatrix} 1 & -1 \\ 4 & -2 \end{vmatrix} - (1)\begin{vmatrix} 4 & 2 \\ 4 & -2 \end{vmatrix} + (1)\begin{vmatrix} 4 & 2 \\ 1 & -1 \end{vmatrix}$$

$$= (1)(2) - (1)(-16) + (1)(-6)$$

$$= 2 + 16 - 6$$

$$= 12$$

$$a = \frac{\begin{vmatrix} 3 & 2 & 1 \\ \frac{9}{2} & -1 & 1 \\ 9 & -2 & 1 \end{vmatrix}}{12} = \frac{(1)\begin{vmatrix} \frac{9}{2} & -1 \\ 9 & -2 \end{vmatrix} - (1)\begin{vmatrix} 3 & 2 \\ 9 & -2 \end{vmatrix} + (1)\begin{vmatrix} 3 & 2 \\ \frac{9}{2} & -1 \end{vmatrix}}{12}$$

$$= \frac{(1)(0) - (1)(-24) + (1)(-12)}{12}$$

$$= \frac{24 - 12}{12} = \frac{12}{12} = 1$$

$$b = \frac{\begin{vmatrix} 4 & 3 & 1 \\ 1 & \frac{9}{2} & 1 \\ 4 & 9 & 1 \end{vmatrix}}{12} = \frac{(1)\begin{vmatrix} 1 & \frac{9}{2} \\ 4 & 9 \end{vmatrix} - (1)\begin{vmatrix} 4 & 3 \\ 4 & 9 \end{vmatrix} + (1)\begin{vmatrix} 4 & 3 \\ 1 & \frac{9}{2} \end{vmatrix}}{12}$$

$$= \frac{(1)(-9) - (1)(24) + (1)(15)}{12}$$

$$= \frac{-9 - 24 + 15}{12} = \frac{-8}{12} = -\frac{3}{2}$$

$$c = \frac{\begin{vmatrix} 4 & 2 & 3 \\ 1 & -1 & \frac{9}{2} \\ 4 & -2 & 9 \end{vmatrix}}{12} = \frac{(4)\begin{vmatrix} -1 & \frac{9}{2} \\ -2 & 9 \end{vmatrix} - (1)\begin{vmatrix} 2 & 3 \\ -2 & 9 \end{vmatrix} + (4)\begin{vmatrix} 2 & 3 \\ -1 & \frac{9}{2} \end{vmatrix}}{12}$$

$$= \frac{(4)(0) - (1)(24) + (4)(12)}{12}$$

$$= \frac{-24 + 48}{12} = \frac{24}{12} = 2$$

$$y = 1x^2 - \frac{3}{2}x + 2$$

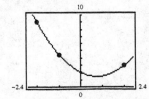

**98.** (a)  $(5, 56.0)$ $(6, 60.6)$ $(7, 57.1)$

$$a(5)^2 + b(5) + c = 56.0 \implies 25a + 5b + c = 56.0$$

$$a(6)^2 + b(6) + c = 60.6 \implies 36a + 6b + c = 60.6$$

$$a(7)^2 + b(7) + c = 57.1 \implies 49a + 7b + c = 57.1$$

$$\det A = \begin{vmatrix} 25 & 5 & 1 \\ 36 & 6 & 1 \\ 49 & 7 & 1 \end{vmatrix} = -2$$

$$a = \frac{\begin{vmatrix} 56.0 & 5 & 1 \\ 60.6 & 6 & 1 \\ 57.1 & 7 & 1 \end{vmatrix}}{-2} = \frac{8.1}{-2} = -4.05$$

$$b = \frac{\begin{vmatrix} 25 & 56.0 & 1 \\ 36 & 60.6 & 1 \\ 49 & 57.1 & 1 \end{vmatrix}}{-2} = \frac{-98.3}{-2} = 49.15$$

$$c = \frac{\begin{vmatrix} 25 & 5 & 56.0 \\ 36 & 6 & 60.6 \\ 49 & 7 & 57.1 \end{vmatrix}}{-2} = \frac{177}{-2} = -88.5$$

$$y_1 = -4.05t^2 + 49.15t - 88.5$$

(b)  $(5, 29.3)$, $(6, 32.6)$, $(7, 35.2)$

$$a(5)^2 + b(5) + c = 29.3 \implies 25a + 5b + c = 29.3$$

$$a(6)^2 + b(6) + c = 32.6 \implies 36a + 6b + c = 32.6$$

$$a(7)^2 + b(7) + c = 35.2 \implies 49a + 7b + c = 35.2$$

$$\det A = \begin{vmatrix} 25 & 5 & 1 \\ 36 & 6 & 1 \\ 49 & 7 & 1 \end{vmatrix} = -2$$

$$a = \frac{\begin{vmatrix} 29.3 & 5 & 1 \\ 32.6 & 6 & 1 \\ 35.2 & 7 & 1 \end{vmatrix}}{-2} = \frac{0.7}{-2} = -0.35$$

$$b = \frac{\begin{vmatrix} 25 & 29.3 & 1 \\ 36 & 32.6 & 1 \\ 49 & 35.2 & 1 \end{vmatrix}}{-2} = \frac{-14.3}{-2} = 7.15$$

$$c = \frac{\begin{vmatrix} 25 & 5 & 29.3 \\ 36 & 6 & 32.6 \\ 49 & 7 & 35.2 \end{vmatrix}}{-2} = \frac{-4.6}{-2} = 2.3$$

$$y_2 = -0.35t^2 + 7.15t + 2.3$$

(c)

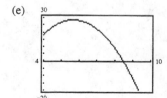

(d)  $y_1 - y_2 = (-4.05t^2 + 49.15t - 88.5) - (-0.35t^2 + 7.15t + 2.3)$

$$= -3.7t^2 + 42.0t - 90.8$$

(e)

The model forecasts that imports will exceed exports starting in 1998.

**100.** (g) 
$$\begin{aligned} x + y + z &= 200 \\ 8x + 15y + 100z &= 4995 \\ x \quad\quad - 4z &= 0 \end{aligned} \Longrightarrow \left[\begin{array}{ccc:c} 1 & 1 & 1 & 200 \\ 8 & 15 & 100 & 4995 \\ 1 & 0 & -4 & 0 \end{array}\right] \Longrightarrow$$

$$-8R_1 + R_2 \left[\begin{array}{ccc:c} 1 & 1 & 1 & 200 \\ 0 & 7 & 92 & 3395 \\ 1 & 0 & -4 & 0 \end{array}\right] \Longrightarrow \begin{array}{c} -R_1 + R_3 \end{array}\left[\begin{array}{ccc:c} 1 & 1 & 1 & 200 \\ 0 & 7 & 92 & 3395 \\ 0 & -1 & -5 & -200 \end{array}\right] \Longrightarrow$$

$$-R_3 \leftrightarrow R_2 \left[\begin{array}{ccc:c} 1 & 1 & 1 & 200 \\ 0 & 1 & 5 & 200 \\ 0 & 7 & 92 & 3395 \end{array}\right] \Longrightarrow \begin{array}{c} -7R_2 + R_3 \end{array}\left[\begin{array}{ccc:c} 1 & 1 & 1 & 200 \\ 0 & 1 & 5 & 200 \\ 0 & 0 & 57 & 1995 \end{array}\right] \begin{array}{c} \frac{1}{57}R_3 \end{array}\Longrightarrow \left[\begin{array}{ccc:c} 1 & 1 & 1 & 200 \\ 0 & 1 & 5 & 200 \\ 0 & 0 & 1 & 35 \end{array}\right]$$

$$z = 35 \qquad y + 5(35) = 200 \qquad x + 25 + 35 = 200$$
$$\qquad\qquad\qquad\quad y = 25 \qquad\qquad\qquad x = 140$$

140 students, 25 nonstudents, 35 contributors

(h) 
$$\begin{aligned} x + y + z &= 200 \\ 8x + 15y + 100z &= 4995 \\ x \quad\quad - 4z &= 0 \end{aligned}$$

$$\det A = \begin{vmatrix} 1 & 1 & 1 \\ 8 & 15 & 100 \\ 1 & 0 & -4 \end{vmatrix} = 57 \qquad\qquad x = \dfrac{\begin{vmatrix} 200 & 1 & 1 \\ 4995 & 15 & 100 \\ 0 & 0 & -4 \end{vmatrix}}{57} = \dfrac{7980}{57} = 140$$

$$y = \dfrac{\begin{vmatrix} 1 & 200 & 1 \\ 8 & 4995 & 100 \\ 1 & 0 & -4 \end{vmatrix}}{57} = \dfrac{1425}{57} = 25 \qquad\qquad z = \dfrac{\begin{vmatrix} 1 & 1 & 200 \\ 8 & 15 & 4995 \\ 1 & 0 & 0 \end{vmatrix}}{57} = \dfrac{1995}{57} = 35$$

140 students, 25 nonstudents, 35 contributors

**102.** No. The matrix must be square.

**104.** The determinant of the coefficient matrix must be a nonzero real number.

# Review Exercises for Chapter 8

**2.** 
$$\begin{aligned} -2x + 5y &= 21 \\ 9x - y &= 13 \end{aligned}$$

(a) $(2, 5)$

$$-2(2) + 5(5) \overset{?}{=} 21 \qquad 9(2) - 5 \overset{?}{=} 13$$
$$21 = 21 \qquad\qquad\qquad 13 = 13$$

Solution

(b) $(-2, 4)$

$$-2(-2) + 5(4) \overset{?}{=} 21$$
$$24 \neq 21$$

Not a solution

**4.** 
$$\begin{aligned} x^2 + 2y &= 1 \\ 2x - 5y &= 26 \end{aligned}$$

(a) $(3, -4)$

$$3^2 + 2(-4) \overset{?}{=} 1 \qquad 2(3) - 5(-4) \overset{?}{=} 26$$
$$1 = 1 \qquad\qquad\qquad\quad 26 = 26$$

Solution

(b) $(2, 8)$

$$2^2 + 2(8) \overset{?}{=} 1$$
$$20 \neq 1$$

Not a solution

**6.** $2x = 3(y - 1)$

$y = x$

Solve each equation for $y$.

$2x = 3(y - 1)$    $y = x$

$2x = 3y - 3$

$2x + 3 = 3y$

$\frac{2}{3}x + 1 = y$

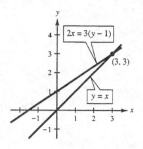

Point of intersection is $(3, 3)$.

**8.** $x + y = -1$

$3x + 2y = 0$

Solve each equation for $y$.

$x + y = -1$    $3x + 2y = 0$

$y = -x - 1$    $2y = -3x$

$y = -\frac{3}{2}x$

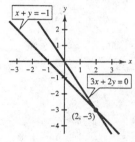

Point of intersection is $(2, -3)$.

**10.** $x = y + 3$

$x = y + 1$

Solve each equation for $y$.

$x = y + 3$    $x = y + 1$

$x - 3 = y$    $x - 1 = y$

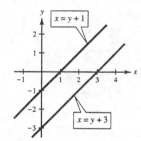

Inconsistent; no solution

**12.**  $3x - 2y = 6$

$-6x + 4y = 12$

Solve each equation for $y$.

$3x - 2y = 6$        $-6x + 4y = 12$

$-2y = -3x + 6$        $4y = 6x + 12$

$y = \frac{3}{2}x - 3$        $y = \frac{3}{2}x + 3$

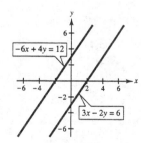

Inconsistent; no solution

**14.**  $2x - y = 4$

$-3x + 4y = -11$

Solve each equation for $y$.

$2x - y = 4$            $-3x + 4y = -11$

$-y = -2x + 4$            $4y = 3x - 11$

$y = 2x - 4$            $y = \frac{3}{4}x - \frac{11}{4}$

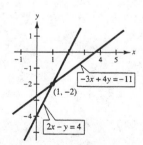

Point of intersection is $(1, -2)$.

**16.** $8x + 5y = 1$

$3x - 4y = 18$

Solve each equation for $y$.

$8x + 5y = 1 \qquad\qquad 3x - 4y = 18$

$5y = -8x + 1 \qquad -4y = -3x + 18$

$y = -\frac{8}{5}x + \frac{1}{5} \qquad y = \frac{3}{4}x - \frac{9}{2}$

*Keystrokes:*

$y_1$: [Y=] [(−)] 8 [X,T,θ] 5 [÷] 1 [+] 5 [÷] [ENTER]

$y_2$: 3 [X,T,θ] [÷] 4 [−] 9 [÷] 2 [GRAPH]

Solution is $(2, -3)$.

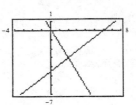

**18.** $\quad y = 9 - x^2$

$2x + y = 6$

Solve each equation for $y$.

$y = 9 - x^2 \quad 2x + y = 6$

$\qquad\qquad y = -2x + 6$

*Keystrokes:*

$y_1$: [Y=] 9 [−] [X,T,θ] [x²] [ENTER]

$y_2$: [−] 2 [X,T,θ] [+] 6 [GRAPH]

Solutions are $(-1, 8)$ and $(3, 0)$.

**20.** $3x - 7y = 10$

$-2x + y = -14$

Solve for $y$ and substitute into first equation.

$y = 2x - 14$

$3x - 7(2x - 14) = 10$

$3x - 14x + 98 = 10$

$-11x = -88$

$x = 8$

$y = 2(8) - 14$

$y = 16 - 14$

$y = 2$

$(8, 2)$

**22.** $5x + 2y = 3$

$2x + 3y = 10$

Solve for $y$ and substitute into second equation.

$5x + 2y = 3$

$2y = -5x + 3$

$y = -\frac{5}{2}x + \frac{3}{2}$

$2x + 3\left(-\frac{5}{2}x + \frac{3}{2}\right) = 10$

$2x - \frac{15}{2}x + \frac{9}{2} = 10$

$-\frac{11}{2}x = \frac{11}{2}$

$x = -1$

$y = -\frac{5}{2}(-1) + \frac{3}{2}$

$y = \frac{5}{2} + \frac{3}{2}$

$y = \frac{8}{2}$

$y = 4$

$(-1, 4)$

**24.** $24x - 4y = 20$

$6x - y = 5$

Solve for $y$ and substitute into first equation.

$y = 6x - 5$

$24x - 4(6x - 5) = 20$

$24x - 24x + 20 = 20$

$20 = 20$

Infinitely many solutions.

**26.**   $y^2 = 16x$

$x - y = -4$

Solve for $y$ and substitute into first equation.

$$-y = -x - 4$$
$$y = x + 4$$
$$(x + 4)^2 = 16x$$
$$x^2 + 8x + 16 = 16x$$
$$x^2 - 8x + 16 = 0$$
$$(x - 4)^2 = 0$$
$$x = 4 \qquad y = 4 + 4 = 8$$

$(4, 8)$

**28.**   $x^2 + y^2 = 32$

$x + y = 0$

Solve for $y$ and substitute into first equation.

$$y = -x$$
$$x^2 + (-x)^2 = 32$$
$$x^2 + x^2 = 32$$
$$2x^2 = 32$$
$$x^2 = 16$$
$$x = \pm 4$$
$$y = -(4) \qquad y = -(-4)$$
$$y = -4 \qquad\ y = 4$$

$(4, -4)$ and $(-4, 4)$

**30.**
$$
\begin{array}{rl}
4x + y = & 1 \\
\underline{x - y = } & \underline{4} \\
5x = & 5 \\
x = & 1 \\
1 - y = & 4 \\
-y = & 3 \\
y = & -3
\end{array}
$$

$(1, -3)$

**32.**
$$
\begin{array}{rlrl}
3x + 2y = & 11 & \Rightarrow \quad 3x + 2y = & 11 \\
x - 3y = & -11 & \Rightarrow \quad \underline{-3x + 9y = } & \underline{+33} \\
& & 11y = & 44 \\
& & y = & 4
\end{array}
$$

$$x - 3(4) = -11$$
$$x = 1$$

$(1, 4)$

**34.**
$$
\begin{array}{rlrl}
0.1x + 0.5y = & -0.17 & \Rightarrow \quad 0.3x + 1.5y = & -0.51 \\
-0.3x - 0.2y = & -0.01 & \Rightarrow \quad \underline{-0.3x - 0.2y = } & \underline{-0.01} \\
& & 1.3y = & -0.52 \\
& & y = & -0.4
\end{array}
$$

$$0.1x + 0.5(-0.4) = -0.17$$
$$0.1x - 0.20 = -0.17$$
$$0.1x = 0.03$$
$$x = 0.3$$

$(0.3, -0.4)$

**36.** 
$$2x + 3y + z = 10$$
$$2x - 3y - 3z = 22$$
$$4x - 2y + 3z = -2$$

$$2x + 3y + z = 10$$
$$-6y - 4z = 12$$
$$-8y + z = -22$$

$$x + \tfrac{3}{2}y + \tfrac{1}{2}z = 5$$
$$y + \tfrac{2}{3}z = -2$$
$$-8y + z = -22$$

$$x + \tfrac{3}{2}y + \tfrac{1}{2}z = 5$$
$$y + \tfrac{2}{3}z = -2$$
$$\tfrac{19}{3}z = -38$$

$$x + \tfrac{3}{2}y + \tfrac{1}{2}z = 5$$
$$y + \tfrac{2}{3}z = -2$$
$$z = -6$$

$$y + \tfrac{2}{3}(-6) = -2$$
$$y - 4 = -2$$
$$y = 2$$

$$x + \tfrac{3}{2}(2) + \tfrac{1}{2}(-6) = 5$$
$$x + 3 - 3 = 5$$
$$x = 5$$

$$(5, 2, -6)$$

**38.** 
$$-3x + y + 2z = -13$$
$$-x - y + z = 0$$
$$2x + 2y - 3z = -1$$

$$x + y - z = 0$$
$$-3x + y + 2z = -13$$
$$2x + 2y - 3z = -1$$

$$x + y - z = 0$$
$$4y - z = -13$$
$$-z = -1$$

$$z = 1$$
$$4y - 1 = -13$$
$$4y = -12$$
$$y = -3$$

$$x + (-3) - 1 = 0$$
$$x = 4$$

$$(4, -3, 1)$$

**40.** 
$$2x + 3y - 5z = 3$$
$$-x + 2y = 3$$
$$3x + 5y + 2z = 15$$

$$7y - 5z = 9$$
$$-x + 2y = 3$$
$$11y + 2z = 24$$

$$-x + 2y = 3$$
$$y - \tfrac{5}{7}z = \tfrac{9}{7}$$
$$11y + 2z = 24$$

$$-x + 2y = 3$$
$$y - \tfrac{5}{7}z = \tfrac{9}{7}$$
$$\tfrac{69}{7}z = \tfrac{69}{7}$$
$$z = 1$$

$$y - \tfrac{5}{7}(1) = \tfrac{9}{7}$$
$$y = \tfrac{14}{7}$$
$$y = 2$$

$$-x + 2(2) = 3$$
$$-x = -1$$
$$x = 1$$

$$(1, 2, 1)$$

**42.** 
$$2x - 5y = 2$$
$$3x - 7y = 1$$

$$\begin{bmatrix} 2 & -5 & \vdots & 2 \\ 3 & -7 & \vdots & 1 \end{bmatrix}$$

$$\tfrac{1}{2}R_1 \begin{bmatrix} 1 & -\tfrac{5}{2} & \vdots & 1 \\ 3 & -7 & \vdots & 1 \end{bmatrix}$$

$$-3R_1 + R_2 \begin{bmatrix} 1 & -\tfrac{5}{2} & \vdots & 1 \\ 0 & \tfrac{1}{2} & \vdots & -2 \end{bmatrix}$$

$$2R_2 \begin{bmatrix} 1 & -\tfrac{5}{2} & \vdots & 1 \\ 0 & 1 & \vdots & -4 \end{bmatrix}$$

$$y = -4$$
$$x - \tfrac{5}{2}(-4) = 1$$
$$x + 10 = 1$$
$$x = -9$$

$$(-9, -4)$$

**44.** $2x + y = 0.3$

$3x - y = -1.3$

$$\begin{bmatrix} 2 & 1 & \vdots & 0.3 \\ 3 & -1 & \vdots & -1.3 \end{bmatrix}$$

$$\tfrac{1}{2}R_1 \begin{bmatrix} 1 & .5 & \vdots & 1.5 \\ 3 & -1 & \vdots & -1.3 \end{bmatrix}$$

$$-3R_1 + R_2 \begin{bmatrix} 1 & .5 & \vdots & .15 \\ 0 & -2.5 & \vdots & -1.75 \end{bmatrix}$$

$$-\tfrac{2}{5}R_2 \begin{bmatrix} 1 & .5 & \vdots & .15 \\ 0 & 1 & \vdots & 0.7 \end{bmatrix}$$

$y = 0.7$

$x + .5(0.7) = .15$

$x + .35 = .15$

$x = -0.2$

$(-0.2, 0.7)$

**46.**

$-x + 3y - z = -4$

$2x \qquad + 6z = 14$

$-3x - y + z = 10$

$$\begin{bmatrix} -1 & 3 & -1 & \vdots & -4 \\ 2 & 0 & 6 & \vdots & 14 \\ -3 & -1 & 1 & \vdots & 10 \end{bmatrix}$$

$$\begin{matrix} -R_1 \\ \\ \tfrac{1}{2}R_2 \end{matrix} \begin{bmatrix} 1 & -3 & 1 & \vdots & 4 \\ 1 & 0 & 3 & \vdots & 7 \\ -3 & -1 & 1 & \vdots & 10 \end{bmatrix}$$

$$\begin{matrix} -R_1 + R_2 \\ 3R_1 + R_3 \\ \\ \end{matrix} \begin{bmatrix} 1 & -3 & 1 & \vdots & 4 \\ 0 & 3 & 2 & \vdots & 3 \\ 0 & -10 & 4 & \vdots & 22 \end{bmatrix}$$

$$\tfrac{1}{3}R_2 \begin{bmatrix} 1 & -3 & 1 & \vdots & 4 \\ 0 & 1 & \tfrac{2}{3} & \vdots & 1 \\ 0 & -10 & 4 & \vdots & 22 \end{bmatrix}$$

$$10R_2 + R_3 \begin{bmatrix} 1 & -3 & 1 & \vdots & 4 \\ 0 & 1 & \tfrac{2}{3} & \vdots & 1 \\ 0 & 0 & \tfrac{32}{3} & \vdots & 32 \end{bmatrix}$$

$$\tfrac{3}{32}R_3 \begin{bmatrix} 1 & -3 & 1 & \vdots & 4 \\ 0 & 1 & \tfrac{2}{3} & \vdots & 1 \\ 0 & 0 & 1 & \vdots & 3 \end{bmatrix}$$

$z = 3$

$y + \tfrac{2}{3}(3) = 1$

$y + 2 = 1$

$y = -1$

$x - 3(-1) + 3 = 4$

$x + 6 = 4$

$x = -2$

$(-2, -1, 3)$

**48.**

$-x_1 + 2x_2 + 3x_3 = 4$

$2x_1 - 4x_2 - x_3 = -13$

$3x_1 + 2x_2 - 4x_3 = -1$

$$\begin{bmatrix} -1 & 2 & 3 & \vdots & 4 \\ 2 & -4 & -1 & \vdots & -13 \\ 3 & 2 & -4 & \vdots & -1 \end{bmatrix}$$

$$-R_1 \begin{bmatrix} 1 & -2 & -3 & \vdots & -4 \\ 2 & -4 & -1 & \vdots & -13 \\ 3 & 2 & -4 & \vdots & -1 \end{bmatrix}$$

$$\begin{matrix} \\ -2R_1 + R_2 \\ -3R_1 + R_3 \end{matrix} \begin{bmatrix} 1 & -2 & -3 & \vdots & -4 \\ 0 & 0 & 5 & \vdots & -5 \\ 0 & 8 & 5 & \vdots & 11 \end{bmatrix}$$

$$\tfrac{1}{5}R_2 \leftrightarrow R_3 \begin{bmatrix} 1 & -2 & -3 & \vdots & -4 \\ 0 & 8 & 5 & \vdots & 11 \\ 0 & 0 & 1 & \vdots & -1 \end{bmatrix}$$

$x_3 = -1$

$8x_2 + 5(-1) = 11$

$8x_2 = 16$

$x_2 = 2$

$x_1 - 2(2) - 3(-1) = -4$

$x_1 - 1 = -4$

$x_1 = -3$

$(-3, 2, -1)$

**50.** $\det(A) = \begin{vmatrix} -3.4 & 1.2 \\ -5 & 2.5 \end{vmatrix} = (3.4)(2.5) - (-5)(1.2) = -8.5 + 6 = -2.5$

**52.** $\det(A) = \begin{vmatrix} 7 & -1 & 10 \\ -3 & 0 & -2 \\ 12 & 1 & 1 \end{vmatrix}$

(using the second row)

$= -(-3)\begin{vmatrix} -1 & 10 \\ 1 & 1 \end{vmatrix} + 0\begin{vmatrix} 7 & 10 \\ 12 & 1 \end{vmatrix} - (-2)\begin{vmatrix} 7 & -1 \\ 12 & 1 \end{vmatrix}$

$= 3(-11) + 0 + 2(19)$

$= -33 + 38$

$= 5$

**54.** $\det(A) = \begin{vmatrix} 4 & 0 & 10 \\ 0 & 10 & 0 \\ 10 & 0 & 34 \end{vmatrix}$

(using the second row)

$= -0\begin{vmatrix} 0 & 10 \\ 0 & 34 \end{vmatrix} + 10\begin{vmatrix} 4 & 10 \\ 10 & 34 \end{vmatrix} - 0\begin{vmatrix} 4 & 0 \\ 10 & 0 \end{vmatrix}$

$= 0 + 10(136 - 100) - 0$

$= 10(36)$

$= 360$

**56.** $12x + 42y = -17$

$30x - 18y = 19$

$\begin{bmatrix} 12 & 42 & \vdots & -17 \\ 30 & -18 & \vdots & 19 \end{bmatrix}$

$D = \begin{vmatrix} 12 & 42 \\ 30 & -18 \end{vmatrix} = -216 - 1260 = -1476$

$x = \dfrac{D_x}{D} = \dfrac{\begin{vmatrix} -17 & 42 \\ 19 & -18 \end{vmatrix}}{-1476} = \dfrac{306 - 798}{-1476} = \dfrac{-492}{-1476} = \dfrac{1}{3}$

$y = \dfrac{D_y}{D} - \dfrac{\begin{vmatrix} 12 & -17 \\ 30 & 19 \end{vmatrix}}{-1476} = \dfrac{228 + 510}{-1476} = \dfrac{738}{-1476} = -\dfrac{1}{2}$

$\left(\dfrac{1}{3}, -\dfrac{1}{2}\right)$

**58.** $4x + 24y = 20$

$-3x + 12y = -5$

$\begin{bmatrix} 4 & 24 & \vdots & 20 \\ -3 & 12 & \vdots & -5 \end{bmatrix}$

$D = \begin{vmatrix} 4 & 24 \\ -3 & 12 \end{vmatrix} = 48 + 72 = 120$

$x = \dfrac{D_x}{D} = \dfrac{\begin{vmatrix} 20 & 24 \\ -5 & 12 \end{vmatrix}}{120} = \dfrac{240 + 120}{120} = \dfrac{360}{120} = 3$

$y = \dfrac{D_y}{D} = \dfrac{\begin{vmatrix} 4 & 20 \\ -3 & -5 \end{vmatrix}}{120} = \dfrac{-20 + 60}{120} = \dfrac{40}{120} = \dfrac{1}{3}$

$\left(3, \dfrac{1}{3}\right)$

**60.** $2x_1 + x_2 + 2x_3 = 4$

$\qquad 2x_1 + 2x_2 = 5$

$2x_1 - x_2 + 6x_3 = 2$

$\begin{bmatrix} 2 & 1 & 2 & \vdots & 4 \\ 2 & 2 & 0 & \vdots & 5 \\ 2 & -1 & 6 & \vdots & 2 \end{bmatrix}$

$D = \begin{vmatrix} 2 & 1 & 2 \\ 2 & 2 & 0 \\ 2 & -1 & 6 \end{vmatrix} = -(2)\begin{vmatrix} 1 & 2 \\ -1 & 6 \end{vmatrix} + (2)\begin{vmatrix} 2 & 2 \\ 2 & 6 \end{vmatrix} - 0$

$= (-2)(8) + (2)(8)$

$= -16 + 16 = 0$

**—CONTINUED—**

**60. —CONTINUED—**

Cannot be solved by Cramer's Rule because $D = 0$.

Solve by elimination:

$$\begin{bmatrix} 2 & 1 & 2 & \vdots & 4 \\ 2 & 2 & 0 & \vdots & 5 \\ 2 & -1 & 6 & \vdots & 2 \end{bmatrix}$$

$$\tfrac{1}{2}R_1 \begin{bmatrix} 1 & \tfrac{1}{2} & 1 & \vdots & 2 \\ 2 & 2 & 0 & \vdots & 5 \\ 2 & -1 & 6 & \vdots & 2 \end{bmatrix}$$

$$\begin{matrix} \\ -2R_1 + R_2 \\ -2R_1 + R_3 \end{matrix} \begin{bmatrix} 1 & \tfrac{1}{2} & 1 & \vdots & 2 \\ 0 & 1 & -2 & \vdots & 1 \\ 0 & -2 & 4 & \vdots & -2 \end{bmatrix}$$

$$\begin{matrix} \\ \\ 2R_2 + R_3 \end{matrix} \begin{bmatrix} 1 & \tfrac{1}{2} & 1 & \vdots & 2 \\ 0 & 1 & -2 & \vdots & 1 \\ 0 & 0 & 0 & \vdots & 0 \end{bmatrix}$$

$$y = 2z + 1$$
$$x + \tfrac{1}{2}y + z = 2$$
$$x = 2 - \tfrac{1}{2}y - z$$
$$\phantom{x} = 2 - \tfrac{1}{2}(2z + 1) - z$$
$$\phantom{x} = 2 - z - \tfrac{1}{2} - z$$
$$\phantom{x} = \tfrac{3}{2} - 2z$$

Let $z = a$. ($a$ is any real number)

$$\left(\tfrac{3}{2} - 2a, \; 2a + 1, \; a\right)$$

**62.** $x + y = 2$

$2x - y = -32$

There are many other correct solutions.

**64.** *Verbal model:*

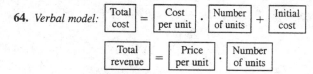

*Labels:* Total cost $= C$

Cost per unit $= \$1.70$

Number of units $= x$

Initial cost $= \$33,000$

Total revenue $= R$

Price per unit $= \$5.00$

*System:* $\quad C = 1.70x + 33,000$

$\qquad\quad R = 5.00x$

$\qquad\quad R = C$

$5.00x = 1.70x + 33,000$

$3.30x = 33,000$

$\quad x = 10,000$ units

**66.** *Verbal model:*

| Gallons Solution 1 | + | Gallons Solution 2 | = 50 |

| Value Solution 1 | + | Value Solution 2 | = .90(50) |

*Labels:* Gallons Solution 1 $= x$

Gallons Solution 2 $= y$

*System:* $\qquad x + y = 50$

$\qquad 1.00x + .75y = .90(50)$

$y = 50 - x$

$1.00x + .75(50 - x) = .90(50)$

$1.00x + 37.5 - .75x = 45$

$\qquad\qquad .25x = 7.5$

$\qquad\qquad\quad x = 30$ gallons at 100%

$\qquad\qquad\quad y = 50 - 30$

$\qquad\qquad\quad\phantom{y} = 20$ gallons at 75%

**68.** *Verbal model:*  $\boxed{\begin{array}{c}\text{Length}\\\text{Piece 1}\end{array}} + \boxed{\begin{array}{c}\text{Length}\\\text{Piece 2}\end{array}} = 128$

$\qquad\qquad\quad \boxed{\begin{array}{c}\text{Length}\\\text{Piece 1}\end{array}} = 3 \cdot \boxed{\begin{array}{c}\text{Length}\\\text{Piece 2}\end{array}}$

*Labels:*  Length Piece 1 $= x$

$\qquad\quad$ Length Piece 2 $= y$

*System:*   $x + y = 128$

$\qquad\qquad x = 3y$

$\qquad\quad 3y + y = 128$

$\qquad\qquad 4y = 128$

$\qquad\qquad y = 32$ inches

$\qquad\quad x = 3(32) = 96$ inches

**70.** *Verbal model:*  $\boxed{\begin{array}{c}\text{Speed}\\\text{Plane 1}\end{array}} \cdot \boxed{\text{Time}} + \boxed{\begin{array}{c}\text{Speed}\\\text{Plane 2}\end{array}} \cdot \boxed{\text{Time}} = \boxed{\text{Distance}}$

$\qquad\qquad\quad \boxed{\begin{array}{c}\text{Speed}\\\text{Plane 2}\end{array}} = \boxed{\begin{array}{c}\text{Speed}\\\text{Plane 1}\end{array}} + 25$

*Labels:*  Speed Plane 1 $= x$

$\qquad\quad$ Speed Plane 2 $= y$

$\qquad\quad$ Time $= \frac{2}{3}$ hour

$\qquad\quad$ Distance $= 275$

*System:*   $x \cdot \frac{2}{3} + y \cdot \frac{2}{3} = 275$

$\qquad\qquad\qquad y = x + 25$

$\qquad \frac{2}{3}x + \frac{2}{3}(x + 25) = 275$

$\qquad\quad \frac{2}{3}(2x + 25) = 275$

$\qquad\qquad 2x + 25 = 412.5$

$\qquad\qquad\quad 2x = 387.5$

$\qquad\qquad\quad x = 193.75$ mph

$\qquad\quad x + 25 = 218.75$ mph

**72.** Verbal model:

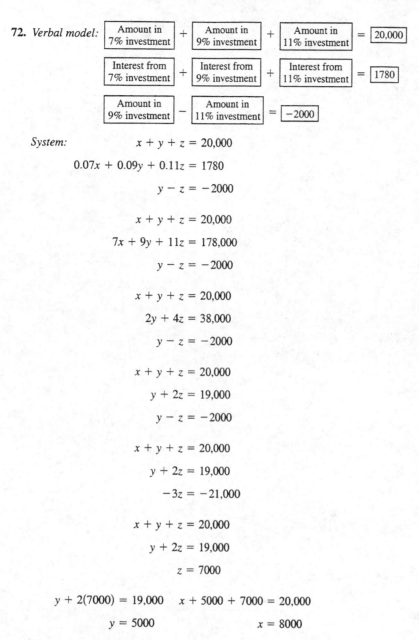

$$\boxed{\begin{array}{c}\text{Amount in}\\7\%\text{ investment}\end{array}} + \boxed{\begin{array}{c}\text{Amount in}\\9\%\text{ investment}\end{array}} + \boxed{\begin{array}{c}\text{Amount in}\\11\%\text{ investment}\end{array}} = \boxed{20{,}000}$$

$$\boxed{\begin{array}{c}\text{Interest from}\\7\%\text{ investment}\end{array}} + \boxed{\begin{array}{c}\text{Interest from}\\9\%\text{ investment}\end{array}} + \boxed{\begin{array}{c}\text{Interest from}\\11\%\text{ investment}\end{array}} = \boxed{1780}$$

$$\boxed{\begin{array}{c}\text{Amount in}\\9\%\text{ investment}\end{array}} - \boxed{\begin{array}{c}\text{Amount in}\\11\%\text{ investment}\end{array}} = \boxed{-2000}$$

System:
$$x + y + z = 20{,}000$$
$$0.07x + 0.09y + 0.11z = 1780$$
$$y - z = -2000$$

$$x + y + z = 20{,}000$$
$$7x + 9y + 11z = 178{,}000$$
$$y - z = -2000$$

$$x + y + z = 20{,}000$$
$$2y + 4z = 38{,}000$$
$$y - z = -2000$$

$$x + y + z = 20{,}000$$
$$y + 2z = 19{,}000$$
$$y - z = -2000$$

$$x + y + z = 20{,}000$$
$$y + 2z = 19{,}000$$
$$-3z = -21{,}000$$

$$x + y + z = 20{,}000$$
$$y + 2z = 19{,}000$$
$$z = 7000$$

$$y + 2(7000) = 19{,}000 \qquad x + 5000 + 7000 = 20{,}000$$
$$y = 5000 \qquad\qquad x = 8000$$

7% investment: $8000

9% investment: $5000

11% investment $7000

**74.** $-6 = a(0)^2 + b(0) + c \Rightarrow -6 = \qquad c$

$\qquad -3 = a(1)^2 + b(1) + c \qquad -3 = a + b + c$

$\qquad 4 = a(2)^2 + b(2) + c \qquad 4 = 4a + 2b + c$

$$\begin{bmatrix} 0 & 0 & 1 & \vdots & -6 \\ 1 & 1 & 1 & \vdots & -3 \\ 4 & 2 & 1 & \vdots & 4 \end{bmatrix}$$

$$D = \begin{vmatrix} 0 & 0 & 1 \\ 1 & 1 & 1 \\ 4 & 2 & 1 \end{vmatrix} = -2$$

$$a = \frac{D_a}{D} = \frac{\begin{vmatrix} -6 & 0 & 1 \\ -3 & 1 & 1 \\ 4 & 2 & 1 \end{vmatrix}}{-2} = \frac{-4}{-2} = 2$$

$$b = \frac{D_b}{D} = \frac{\begin{vmatrix} 0 & -6 & 1 \\ 1 & -3 & 1 \\ 4 & 4 & 1 \end{vmatrix}}{-2} = \frac{-2}{-2} = 1$$

$$y = 2x^2 + 1x - 6$$

**76.** (a) Points: $(0, 11), (15, 16), (30, 11)$

$11 = a(0)^2 + b(0) + c \Rightarrow 11 = \qquad c$

$16 = a(15)^2 + b(15) + c \Rightarrow 16 = 225a + 15b + c$

$11 = a(30)^2 + b(30) + c \Rightarrow 11 = 900a + 30b + c$

$$\det(A) = \begin{vmatrix} 0 & 0 & 1 \\ 225 & 15 & 1 \\ 900 & 30 & 1 \end{vmatrix} = -6750$$

$$a = \frac{\begin{vmatrix} 11 & 0 & 1 \\ 16 & 15 & 1 \\ 11 & 30 & 1 \end{vmatrix}}{-6750} = \frac{150}{-6750} = -\frac{1}{45} \qquad c = 11$$

$$b = \frac{\begin{vmatrix} 0 & 11 & 1 \\ 225 & 16 & 1 \\ 900 & 11 & 1 \end{vmatrix}}{-6750} = \frac{-4500}{-6750} = \frac{2}{3}$$

$$y = -\frac{1}{45}x^2 + \frac{2}{3}x + 11$$

(b) *Keystrokes:* [Y=] [(] [−] 1 [÷] 45 [)] [X,T,θ] [x²] [+] [(] 2 [÷] 3 [)] [X,T,θ] [÷] 11 [GRAPH]

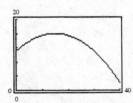

(c) Trace until the *y*-coordinate is approximately 5 feet. The *x*-coordinate is approximately $-7.25$ so the child is $7\frac{1}{4}$ feet from the edge of the garage.

**78.** $(x_1, y_1) = (-4, 0), (x_2, y_2) = (4, 0), (x_3, y_3) = (0, 6)$

$$\begin{vmatrix} x_1 & y_1 & 1 \\ x_2 & y_2 & 1 \\ x_3 & y_3 & 1 \end{vmatrix} = \begin{vmatrix} -4 & 0 & 1 \\ 4 & 0 & 1 \\ 0 & 6 & 1 \end{vmatrix} = -0 + 0 - (6)\begin{vmatrix} -4 & 1 \\ 4 & 1 \end{vmatrix} = (-6)(-8) = 48$$

Area $= +\frac{1}{2}(48) = 24$

**80.** $(x_1, y_1) = \left(\frac{3}{2}, 1\right), (x_2, y_2) = \left(4, -\frac{1}{2}\right), (x_3, y_3) = (4, 2)$

$$\begin{vmatrix} x_1 & y_1 & 1 \\ x_2 & y_2 & 1 \\ x_3 & y_3 & 1 \end{vmatrix} = \begin{vmatrix} \frac{3}{2} & 1 & 1 \\ 4 & -\frac{1}{2} & 1 \\ 4 & 2 & 1 \end{vmatrix} = (1)\begin{vmatrix} 4 & -\frac{1}{2} \\ 4 & 2 \end{vmatrix} - (1)\begin{vmatrix} \frac{3}{2} & 1 \\ 4 & 2 \end{vmatrix} + (1)\begin{vmatrix} \frac{3}{2} & 1 \\ 4 & -\frac{1}{2} \end{vmatrix}$$

$$= (1)(10) - (1)(-1) + (1)\left(-\frac{19}{4}\right)$$

$$= 10 + 1 - \frac{19}{4}$$

$$= \frac{25}{4}$$

Area $= +\frac{1}{2}\left(\frac{25}{4}\right) = \frac{25}{8}$ or $3\frac{1}{8}$

**82.** $\begin{vmatrix} 2 & y & 1 \\ 2 & 5 & 1 \\ 6 & -1 & 1 \end{vmatrix} = 0$

$$(1)\begin{vmatrix} 2 & 5 \\ 6 & -1 \end{vmatrix} - (1)\begin{vmatrix} x & y \\ 6 & -1 \end{vmatrix} + (1)\begin{vmatrix} x & y \\ 2 & 5 \end{vmatrix} = 0$$

$$(1)(-32) - (1)(-x - 6y) + (1)(5x - 2y) = 0$$

$$-32 + x + 6y + 5x - 2y = 0$$

$$6x + 4y - 32 = 0 \text{ or}$$

$$3x + 2y - 16 = 0$$

**84.** $\begin{vmatrix} x & y & 1 \\ -0.8 & 0.2 & 1 \\ 0.7 & 3.2 & 1 \end{vmatrix} = 0$

$$(1)\begin{vmatrix} -0.8 & 0.2 \\ 0.7 & 3.2 \end{vmatrix} - (1)\begin{vmatrix} x & y \\ 0.7 & 3.2 \end{vmatrix} + (1)\begin{vmatrix} x & y \\ -0.8 & 0.2 \end{vmatrix} = 0$$

$$(1)(-2.7) - (1)(3.2x - 0.7y) + (1)(0.2x + 0.8y) = 0$$

$$-2.7 - 3.2x + 0.7y + 0.2x + 0.8y = 0$$

$$-3x + 1.5y - 2.7 = 0 \text{ or } 10x - 5y + 9 = 0$$

# CHAPTER 9
# Exponential and Logarithmic Functions

# CHAPTER 9
# Exponential and Logarithmic Functions

## Section 9.1  Exponential Functions

Solutions to Even-Numbered Exercises

**2.** $10e^{2x} \cdot e^{-x} = 10e^{2x+(-x)} = 10e^{x}$

**4.** $\dfrac{3^{2x+3}}{3^{x+1}} = 3^{2x+3-(x+1)} = 3^{2x+3-x-1} = 3^{x+2}$

**6.** $-4e^{-2x} = \dfrac{-4}{e^{2x}}$

**8.** $\sqrt{4e^{6x}} = \sqrt{4}\sqrt{(e^{3x})^2} = 2e^{3x}$

**10.** $6^{-\pi} \approx 0.004$

*Keystrokes:*

6 $\boxed{y^x}$ $\pi$ $\boxed{+/-}$ $\boxed{=}$ Scientific

6 $\boxed{\wedge}$ $\boxed{(-)}$ $\pi$ $\boxed{\text{ENTER}}$ Graphing

**12.** $e^{-1/3} \approx 0.717$

*Keystrokes:*

$\boxed{(}$ 1 $\div$ 3 $\boxed{+/-}$ $\boxed{)}$ $\boxed{\text{INV}}$ $\boxed{\ln x}$ Scientific

$\boxed{e^x}$ $\boxed{(}$ $\boxed{(-)}$ 1 $\div$ 3 $\boxed{)}$ $\boxed{\text{ENTER}}$ Graphing

**14.** $(9e^2)^{3/2} = 9^{3/2} \cdot e^{2 \cdot 3/2} = \left(\sqrt{9}\right)^3 e^3 = 27e^3 \approx 542.309$

*Keystrokes:*

27 $\boxed{\times}$ 3 $\boxed{\text{INV}}$ $\boxed{\ln x}$ $\boxed{=}$ Scientific

27 $\boxed{e^x}$ 3 $\boxed{\text{ENTER}}$ Graphing

**16.** $\dfrac{6e^5}{10e^7} = \dfrac{3e^{5-7}}{5} = \dfrac{3e^{-2}}{5} = \dfrac{3}{5e^2} \approx 0.081$

*Keystrokes:*

3 $\boxed{\div}$ $\boxed{(}$ 5 $\boxed{\times}$ 2 $\boxed{\text{INV}}$ $\boxed{\ln x}$ $\boxed{)}$ $\boxed{=}$ Scientific

3 $\boxed{\div}$ $\boxed{(}$ 5 $\boxed{e^x}$ 2 $\boxed{)}$ $\boxed{\text{ENTER}}$ Graphing

**18.** $F(x) = 3^{-x}$

(a) $F(-2) = 3^{-(-2)} = 3^2 = 9$

(b) $F(0) = 3^{-0} = 3^0 = 1$

(c) $F(1) = 3^{-1} = \frac{1}{3}$

**20.** $G(x) = 2.04^{-x}$

(a) $G(-1) = 2.04^{-(-1)} \approx 2.040$

(b) $G(1) = 2.04^{-1} \approx 0.490$

(c) $G\left(\sqrt{3}\right) = 2.04^{-\sqrt{3}} \approx 0.291$

**22.** $g(s) = 1200\left(\frac{2}{3}\right)^s$

(a) $g(0) = 1200\left(\frac{2}{3}\right)^0 = 1200(1) = 1200$

(b) $g(2) = 1200\left(\frac{2}{3}\right)^2 = 1200\left(\frac{4}{9}\right) \approx 533.333$

(c) $g\left(\sqrt{2}\right) = 1200\left(\frac{2}{3}\right)^{\sqrt{2}} = 1200(0.5635978) \approx 676.317$

**24.** $g(t) = 10{,}000(1.03)^{4t}$

(a) $g(1) = 10{,}000(1.03)^{4(1)} \approx 11{,}255.088$

(b) $g(3) = 10{,}000(1.03)^{4(3)} \approx 14{,}257.609$

(c) $g(5.5) = 10{,}000(1.03)^{4(5.5)} \approx 19{,}161.034$

**26.** $P(t) = \dfrac{10{,}000}{(1.01)^{12t}}$

(a) $P(2) = \dfrac{10{,}000}{(1.01)^{12(2)}} = \dfrac{10{,}000}{(1.01)^{24}} \approx 7875.661$

(b) $P(10) = \dfrac{10{,}000}{(1.01)^{12(10)}} = \dfrac{10{,}000}{(1.01)^{120}} \approx 3029.948$

(c) $P(20) = \dfrac{10{,}000}{(1.01)^{12(20)}} = \dfrac{10{,}000}{(1.01)^{240}} \approx 918.058$

**28.** $A(t) = 200e^{0.1t}$

(a) $A(10) = 200e^{0.1(10)} = 200e^1 \approx 543.656$

(b) $A(20) = 200e^{0.1(20)} = 200e^2 \approx 1477.811$

(c) $A(40) = 200e^{0.1(40)} = 200e^4 \approx 10{,}919.630$

**30.** $f(z) = \dfrac{100}{1 + e^{-0.05z}}$

(a) $f(0) = \dfrac{100}{1 + e^{-0.05(0)}} = \dfrac{100}{1 + e^0} = \dfrac{100}{1 + 1} = \dfrac{100}{2} = 50$

(b) $f(10) = \dfrac{100}{1 + e^{-0.05(10)}} = \dfrac{100}{1 + e^{-0.5}} \approx 62.246$

(c) $f(20) = \dfrac{100}{1 + e^{-0.05(20)}} = \dfrac{100}{1 + e^{-1}} \approx 73.106$

**32.** $f(x) = 3^{-x} = \left(\dfrac{1}{3}\right)^x$

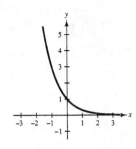

**34.** $h(x) = \frac{1}{2}(3^{-x})$

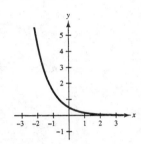

**36.** $g(x) = 3^x + 1$

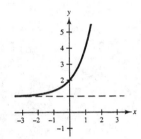

**38.** $f(x) = 4^{x+1}$

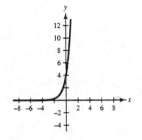

**40.** $g(x) = 4^x + 1$

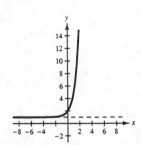

**42.** $f(t) = 2^{t^2}$

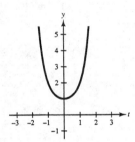

**44.** $h(t) = -2^{-0.5t}$

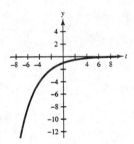

**46.** $g(x) = 2^{-0.5x}$

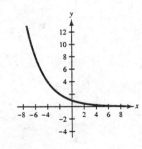

**48.** $f(x) = \left(\dfrac{3}{4}\right)^x + 1$

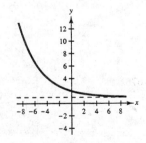

**50.** $h(y) = 27\left(\dfrac{2}{3}\right)^y$

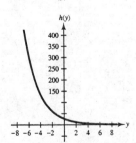

**52.** $f(x) = -2^x$   (a)

**54.** $f(x) = 2^x - 1$   (d)

**56.** $f(x) = 2^{x+1}$   (c)

**58.** $f(x) = e^{-x^2}$   (g)

**60.** $y = 5^{-x/3}$

*Keystrokes:*

Y= 5 ^ ( (–) X,T,θ ÷ 3 ) GRAPH

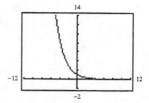

**62.** $y = 5^{-x/3} + 2$

*Keystrokes:*

Y= 5 ^ ( (–) X,T,θ ÷ 3 ) + 2 GRAPH

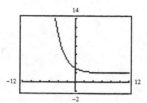

**64.** $y = 100(1.06)^{-t}$

*Keystrokes:*

Y= 100 ( 1.06 ) ^ (–) X,T,θ GRAPH

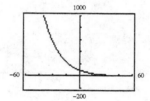

**66.** $y = 50e^{-0.05x}$

*Keystrokes:*

Y= 50 eˣ (–) .05 X,T,θ GRAPH

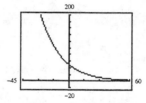

**68.** $A(t) = 1000e^{0.08t}$

*Keystrokes:*

Y= 1000 eˣ .08 X,T,θ GRAPH

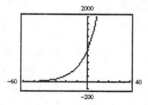

**70.** $g(t) = 7e^{(x+1)/2}$

*Keystrokes:*

Y= 7 eˣ ( X,T,θ + 1 ) ÷ 2 Graph

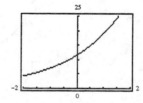

**72.** $h(x) = 4^x + 2$

Vertical shift 2 units up

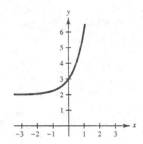

**74.** $h(x) = 4^{x-4}$

Horizontal shift 4 units right

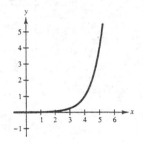

**76.** $h(x) = -4^x + 1$

Reflection in the x-axis
Vertical shift 1 unit up

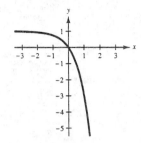

**78.** $y_1 \leftrightarrow a$

$y_2 \leftrightarrow b$

$y_3 \leftrightarrow c$

When $k > 0$ increases, the graph increases at a greater rate.

**80.** $y = 23\left(\frac{1}{2}\right)^{t/45}, t \geq 0$

$y = 23\left(\frac{1}{2}\right)^{150/45} \approx 2.282$ grams

**82.**

| $n$ | 1 | 4 | 12 | 365 | Continuous |
|---|---|---|---|---|---|
| $A$ | \$18,760.65 | \$20,993.96 | \$21,551.27 | \$21,829.69 | \$21,839.26 |

(a) $A = 400\left(1 + \dfrac{0.08}{1}\right)^{1(50)}$

$\approx \$18,760.65$

(b) $A = 400\left(1 + \dfrac{0.08}{4}\right)^{4(50)}$

$\approx \$20,993.96$

(c) $A = 400\left(1 + \dfrac{0.08}{12}\right)^{12(50)}$

$\approx \$21,551.27$

(d) $A = 400\left(1 + \dfrac{0.08}{365}\right)^{365(50)}$

$\approx \$21,829.69$

(e) $A = Pe^{rt}$

$= 400e^{0.08(50)}$

$\approx \$21,839.26$

**84.**

| $n$ | 1 | 4 | 12 | 365 | Continuous |
|---|---|---|---|---|---|
| $A$ | \$1717.35 | \$1723.32 | \$1724.71 | \$1725.39 | \$1725.41 |

(a) $A = 1500\left(1 + \dfrac{0.07}{1}\right)^{1(2)}$

$\approx \$1717.35$

(b) $A = 1500\left(1 + \dfrac{0.07}{4}\right)^{4(2)}$

$\approx \$1723.32$

(c) $A = 1500\left(1 + \dfrac{0.07}{12}\right)^{12(2)}$

$\approx \$1724.71$

(d) $A = 1500\left(1 + \dfrac{0.07}{365}\right)^{365(2)}$

$\approx \$1725.39$

(e) $A = 1500e^{(0.07)(2)}$

$\approx \$1725.41$

**86.**

| $n$ | 1 | 4 | 12 | 365 | Continuous |
|---|---|---|---|---|---|
| $A$ | \$152,203.13 | \$167,212.90 | \$170,948.62 | \$172,813.72 | \$172,877.82 |

(a) $A = 10,000\left(1 + \dfrac{0.095}{1}\right)^{1(30)}$

$\approx \$152,203.13$

(b) $A = 10,000\left(1 + \dfrac{0.095}{4}\right)^{4(30)}$

$\approx \$167,212.90$

(c) $A = 10,000\left(1 + \dfrac{0.095}{12}\right)^{12(30)}$

$\approx \$170,948.62$

(d) $A = 10,000\left(1 + \dfrac{0.095}{365}\right)^{365(30)}$

$\approx \$172,813.72$

(e) $A = 10,000e^{0.095(30)}$

$\approx \$172,877.82$

**88.**

| $n$ | 1 | 4 | 12 | 365 | Continuous |
|---|---|---|---|---|---|
| $P$ | \$17,843.09 | \$16,862.99 | \$16,641.28 | \$16,533.56 | \$16,529.89 |

(a) $100,000 = P\left(1 + \dfrac{0.09}{1}\right)^{1(20)}$

$\dfrac{100,000}{(1.09)^{20}} = P$

$\$17,843.09 \approx P$

(b) $100,000 = P\left(1 + \dfrac{0.09}{4}\right)^{4(20)}$

$\dfrac{100,000}{\left(1 + \dfrac{0.09}{4}\right)^{80}} = P$

$\$16,862.99 \approx P$

(c) $100,000 = P\left(1 + \dfrac{0.09}{12}\right)^{(12)(20)}$

$\dfrac{100,000}{\left(1 + \dfrac{0.09}{12}\right)^{240}} = P$

$\$16,641.28 \approx P$

(d) $100,000 = P\left(1 + \dfrac{0.09}{365}\right)^{(365)(20)}$

$\dfrac{100,000}{\left(1 + \dfrac{0.09}{365}\right)^{7300}} = P$

$\$16,533.5 \approx P$

(e) $100,000 = Pe^{(0.09)(20)}$

$\dfrac{100,000}{e^{1.8}} = P$

$\$16,529.8 \approx P$

**90.**

| $n$ | 1 | 4 | 12 | 365 | Continuous |
|---|---|---|---|---|---|
| $A$ | \$2163.33 | \$2154.76 | \$2152.77 | \$2151.80 | \$2151.77 |

(a) $2500 = P\left(1 + \dfrac{0.075}{1}\right)^{1(2)}$

$\dfrac{2500}{\left(1 + \dfrac{0.075}{1}\right)^{1(2)}} = P$

$\$2163.33 \approx P$

(b) $2500 = P\left(1 + \dfrac{0.075}{4}\right)^{4(2)}$

$\dfrac{2500}{\left(1 + \dfrac{0.075}{4}\right)^{4(2)}} = P$

$\$2154.76 \approx P$

—CONTINUED—

**90.** —CONTINUED—

(c)
$$2500 = P\left(1 + \frac{0.075}{12}\right)^{12(2)}$$

$$\frac{2500}{\left(1 + \frac{0.075}{12}\right)^{12(2)}} = P$$

$$\$2152.77 \approx P$$

(e)
$$2500 = Pe^{0.075(2)}$$

$$\frac{2500}{e^{0.075(2)}} = P$$

$$\$2151.77 \approx P$$

(d)
$$2500 = P\left(1 + \frac{0.075}{365}\right)^{365(2)}$$

$$\frac{2500}{\left(1 + \frac{0.075}{365}\right)^{365(2)}} = P$$

$$\$2152.80 \approx P$$

**92.** $P(t) = 205.7(1.0098)^t$

(a) $P(30) = 205.7(1.0098)^{30} \approx 275.6$ million

(b) $P(40) = 205.7(1.0098)^{40} \approx 303.8$ million

**94.** $C(t) = P(1.05)^t, 0 \le t \le 10$

$C(10) = 24.95(1.05)^{10} \approx \$40.64$

**96.** (a) $V(t) = 16,000 - 3000t$

(b)

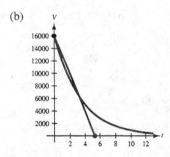

(c) If you are selling the car after owning it for 2 years, you would prefer the straight-line model because it indicates that the value of the car is greater than that of the exponential model.

(d) After 4 years, you would prefer the exponential model over the straight-line model because the value of the car is greater.

**98.** $f(t) = 2^{t-1}$

$f(30) = 2^{29} = 536,870,912$

**100.** $h = 2940 + 60e^{-1.7t} - 22t$

(a) *Keystrokes:*

$\boxed{Y=}$ 2940 $\boxed{+}$ 60 $\boxed{e^x}$ $\boxed{(-)}$ 1.7 $\boxed{X,T,\theta}$ $\boxed{-}$ 22 $\boxed{X,T,\theta}$ $\boxed{GRAPH}$

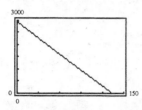

(b) $h = 2940 + 60e^{-1.7(0)} - 22(0) = 3000$ feet

$h = 2940 + 60e^{-1.7(50)} - 22(50) = 1840$ feet

$h = 2940 + 60e^{-1.7(100)} - 22(100) = 740$ feet

| $t$ | 0 | 50 | 100 |
|---|---|---|---|
| $h$ | 3000 ft | 1840 ft | 740 ft |

(c) $0 = 2940 + 60e^{-1.7t} - 22t$

Approximate time $\approx 133.6$ seconds

Use calculator to find the *x*-intercept of the graph.

**102.** (a) $y = 97{,}107e^{0.0317t}$

$y = 97{,}107e^{0.0317(1)} \approx \$100{,}235$

$y = 97{,}107e^{0.0317(2)} \approx \$103{,}463$

$y = 97{,}107e^{0.0317(3)} \approx \$106{,}795$

$y = 97{,}107e^{0.0317(4)} \approx \$110{,}235$

$y = 97{,}107e^{0.0317(5)} \approx \$113{,}785$

$y = 97{,}107e^{0.0317(6)} \approx \$117{,}450$

(b) Keystrokes:

$\boxed{Y=}$ 97107 $\boxed{e^x}$ .0317 $\boxed{X,T,\theta}$ $\boxed{GRAPH}$

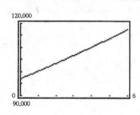

(c) If the model were used to predict home prices in the years ahead, the predictions would be increasing at a faster rate with increasing *t*. Home prices probably will not increase at a faster rate indefinitely.

(d) The coefficient *t* represents the continuous rate of growth.

**104.** (a) $A = P\left(1 + \dfrac{r}{n}\right)^{nt}$

$A = 5000\left(1 + \dfrac{0.06}{1}\right)^{1(3)} \approx \$5955.08$ annually

$A = 5000\left(1 + \dfrac{0.06}{4}\right)^{4(3)} \approx \$5978.09$ quarterly

$A = 5000\left(1 + \dfrac{0.06}{12}\right)^{12(3)} \approx \$5983.40$ monthly

$A = 5000\left(1 + \dfrac{0.06}{365}\right)^{365(3)} \approx \$5986.00$ daily

$A = 5000\left(1 + \dfrac{0.06}{8760}\right)^{8760(3)} \approx \$5986.08$ hourly

$A = 5000e^{.06(3)} \approx \$5986.09$ continuously

| Compounding | Amount |
|---|---|
| annually | 5955.08 |
| quarterly | 5978.09 |
| monthly | 5983.40 |
| daily | 5986.00 |
| hourly | 5986.08 |

(b) First investment:    Second investment:

$A = 5000e^{.07(3)}$     $A = 5000\left(1 + \dfrac{0.08}{4}\right)^{4(3)}$

$\approx \$6168.39$      $\approx \$6341.21$

**106.** By definition, the base is positive and not equal to 1. If the base were 1, the function would simplify to the constant function $y = 1$.

**108.** The function $f$ is applied to growth problems such as those involving compound interest and population growth. The function $g$ is applied to decay problems such as those involving radioactive decay and depreciation.

**110.** Because $1 < \sqrt{2} < 2$ and $2 > 0$, $2^1 < 2^{\sqrt{2}} < 2^2$.

# Section 9.2    Inverse Functions

**2.** $f(x) = x + 1$, $g(x) = 5 - 3x$

  (a) $(f \circ g)(x) = (5 - 3x) + 1 = 6 - 3x$

  (b) $(g \circ f)(x) = 5 - 3(x + 1) = 5 - 3x - 3 = 2 - 3x$

  (c) $(f \circ g)(3) = 6 - 3(3) = 6 - 9 = -3$

  (d) $(g \circ f)(3) = 2 - 3(3) = 2 - 9 = -7$

**4.** $f(x) = x^2 - 3x$, $g(x) = 3 - 2x$

  (a) $(f \circ g)(x) = (3 - 2x)^2 - 3(3 - 2x)$

  $\qquad = 9 - 12x + 4x^2 - 9 + 6x$

  $\qquad = 4x^2 - 6x$

  (b) $(g \circ f)(x) = 3 - 2(x^2 - 3x)$

  $\qquad = 3 - 2x^2 + 6x$

  (c) $(f \circ g)(-1) = 4(-1)^2 - 6(-1)$

  $\qquad = 4 + 6 = 10$

  (d) $(g \circ f)(3) = 3 - 2(3)^2 + 6(3)$

  $\qquad = 3 - 18 + 18 = 3$

**6.** $f(x) = |x|$, $g(x) = 2x + 5$

  (a) $(f \circ g)(x) = |2x + 5|$

  (b) $(g \circ f)(x) = 2|x| + 5$

  (c) $(f \circ g)(-2) = |2(-2) + 5| = |-4 + 5| = |1| = 1$

  (d) $(g \circ f)(-4) = 2|-4| + 5 = 2(4) + 5 = 8 + 5 = 13$

**8.** $f(x) = \sqrt{x + 6}$, $g(x) = 2x - 3$

  (a) $(fg)(x) = f[g(x)] = f(2x - 3) = \sqrt{2x - 3 + 6} = \sqrt{2x + 3}$

  (b) $(g \circ f)(x) = g[f(x)] = g(\sqrt{x + 6}) = 2\sqrt{x + 6} - 3$

  (c) $(f \circ g)(3) = f[g(3)] = f(3) = \sqrt{3 + 6} = \sqrt{9} = 3$

  (d) $(g \circ f)(-2) = g[f(-2)] = g(2) = 2(2) - 3 = 4 - 3 = 1$

**10.** $f(x) = \dfrac{4}{x^2 - 4}$, $g(x) = \dfrac{1}{x}$

  (a) $(f \circ g)(x) = \dfrac{4}{\left(\dfrac{1}{x}\right)^2 - 4} = \dfrac{4}{\dfrac{1}{x^2} - 4} = \dfrac{4}{\dfrac{1}{x^2} - \dfrac{4x^2}{x^2}} = \dfrac{4}{\dfrac{1 - 4x^2}{x^2}} = \dfrac{4x^2}{1 - 4x^2}$

  (b) $(g \circ f)(x) = \dfrac{1}{\dfrac{4}{x^2 - 4}} = \dfrac{x^2 - 4}{4}$

  (c) $(f \circ g)(-2) = \dfrac{4}{\left(\dfrac{1}{-2}\right)^2 - 4} = \dfrac{4}{\dfrac{1}{4} - 16} = \dfrac{4}{-\dfrac{15}{4}} = \dfrac{-16}{15}$

  (d) $(g \circ f)(1) = \dfrac{1}{\dfrac{4}{1^2 - 4}} = \dfrac{1}{\dfrac{4}{-3}} = -\dfrac{3}{4}$

**12.** (a) $g(0) = 2$

(b) $f(2) = -3$

(c) $(f \circ g)(0) = f(g(0)) = f(2) = -3$

**14.** $(f \circ g)(2) = f[g(2)] = f(2) = -3$

$(g \circ f)(2) = g[f(2)] = g(-3) = 1$

**16.** (a) $g(2) = 3$

(b) $f(0) = 1$

(c) $(f \circ g)(10) = f[g(10)] = f(1) = 2$

**18.** $(f \circ g)(1) = f[g(1)] = f(2) = 5$

$(g \circ f)(0) = g[f(0)] = g(1) = 2$

**20.** $f(x) = 2 - 3x, g(x) = 5x + 3$

(a) $f \circ g = 2 - 3(5x + 3) = 2 - 15x - 9 = -7 - 15x$    Domain: $(-\infty, \infty)$

(b) $g \circ f = 5(2 - 3x) + 3 = 10 - 15x + 3 = 13 - 15x$    Domain: $(-\infty, \infty)$

**22.** $f(x) = \sqrt{x - 5}, g(x) = x + 3$

(a) $f \circ g = \sqrt{(x + 3) - 5} = \sqrt{x - 2}$    Domain: $[2, \infty)$

(b) $g \circ f = \sqrt{x - 5} + 3$    Domain: $[5, \infty)$

**24.** $f(x) = \sqrt{2x - 1}, g(x) = x^2 + 1$

(a) $f \circ g = \sqrt{2(x^2 + 1) - 1} = \sqrt{2x^2 + 2 - 1} = \sqrt{2x^2 + 1}$    Domain: $(-\infty, \infty)$

(b) $g \circ f = \left(\sqrt{2x - 1}\right)^2 + 1 = 2x - 1 + 1 = 2x$    Domain: $(-\infty, \infty)$

**26.** (a) $f \circ g = \dfrac{\sqrt{x}}{\sqrt{x} - 4}$    Domain: $[0, 16) \cup (16, \infty)$

(b) $g \circ f = \sqrt{\dfrac{x}{x - 4}}$    Domain: $(-\infty, 0] \cup (4, \infty)$

**28.** $f(x) = \dfrac{1}{5}x$

Yes, it has an inverse, because no horizontal line intersects the graph of $f$ at more than one point.

**30.** $f(x) = \sqrt{-x}$

Yes, it has an inverse because no horizontal line intersects the graph of $f$ at more than one point.

**32.** $g(x) = |x - 4|$

No it does not have an inverse, because it is possible to find a horizontal line that intersects the graph of $g$ at more than one point.

**34.** $f(x) = (2 - x)^3$

*Keystrokes:*

$\boxed{Y=}$ $\boxed{(}$ $\boxed{2}$ $\boxed{-}$ $\boxed{X,T,\theta}$ $\boxed{)}$ $\boxed{\wedge}$ $3$ $\boxed{GRAPH}$

one-to-one

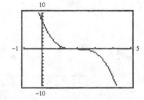

**36.** $h(t) = 4 - \sqrt[3]{t}$

*Keystrokes:*

$\boxed{Y=}$ $4$ $\boxed{-}$ $\boxed{MATH}$ $\boxed{4}$ $\boxed{X,T,\theta}$ $\boxed{GRAPH}$

one-to-one

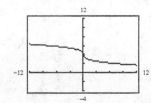

**38.** $f(x) = (x + 2)^5$

*Keystrokes:*

$\boxed{\text{Y=}}$ $\boxed{(}$ $\boxed{\text{X,T,}\theta}$ $\boxed{+}$ 2 $\boxed{)}$ $\boxed{\wedge}$ 5 $\boxed{\text{GRAPH}}$

one-to-one

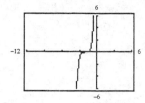

**40.** $g(t) = \dfrac{5}{t^2}$

*Keystrokes:*

$\boxed{\text{Y=}}$ 5 $\boxed{\div}$ $\boxed{\text{X,T,}\theta}$ $\boxed{x^2}$ $\boxed{\text{GRAPH}}$

not one-to-one

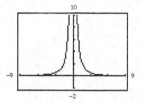

**42.** $f(x) = \dfrac{1}{x^2 - 2}$

*Keystrokes:*

$\boxed{\text{Y=}}$ 1 $\boxed{\div}$ $\boxed{(}$ $\boxed{\text{X,T,}\theta}$ $\boxed{x^2}$ $\boxed{-}$ 2 $\boxed{)}$ $\boxed{\text{GRAPH}}$

not one-to-one

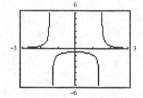

**44.** $f(x) = \frac{2}{3}x, g(x) = \frac{3}{2}x$

$f(g(x)) = f\left(\frac{3}{2}x\right) = \frac{2}{3}\left(\frac{3}{2}x\right) = x$

$g(f(x)) = g\left(\frac{2}{3}x\right) = \frac{3}{2}\left(\frac{2}{3}x\right) = x$

**46.** $f(x) = 3 - x, g(x) = 3 - x$

$f(g(x)) = f(3 - x) = 3 - (3 - x) = 3 - 3 + x = x$

$g(f(x)) = g(3 - x) = 3 - (3 - x) = 3 - 3 + x = x$

**48.** $f(x) = 2x - 1, g(x) = \frac{1}{2}(x + 1)$

$f(g(x)) = f\left(\frac{1}{2}(x + 1)\right) = 2\left(\frac{1}{2}(x + 1)\right) - 1 = x + 1 - 1 = x$

$g(f(x)) = g(2x - 1) = \frac{1}{2}(2x - 1 + 1) = \frac{1}{2}(2x) = x$

**50.** $f(x) = -\frac{1}{4}x + 3, g(x) = -4(x - 3)$

$f(g(x)) = f(-4(x - 3)) = -\frac{1}{4}(-4(x - 3)) + 3 = x - 3 + 3 = x$

$g(f(x)) = g\left(-\frac{1}{4}x + 3\right) = -4\left(-\frac{1}{4}x + 3 - 3\right) = -4\left(-\frac{1}{4}x\right) = x$

**52.** $f(x) = x^7, g(x) = \sqrt[7]{x}$

$f(g(x)) = f(\sqrt[7]{x}) = (\sqrt[7]{x})^7 = x$

$g(f(x)) = g(x^7) = \sqrt[7]{x^7} = x$

**54.** $f(x) = \dfrac{1}{x - 3}, g(x) = 3 + \dfrac{1}{x}$

$f(g(x)) = f\left(3 + \dfrac{1}{x}\right) = \dfrac{1}{\left(3 + \dfrac{1}{x}\right) - 3} = \dfrac{1}{\dfrac{1}{x}} = x$

$g(f(x)) = g\left(\dfrac{1}{x - 3}\right) = 3 + \dfrac{1}{\dfrac{1}{x - 3}} = 3 + x - 3 = x$

**56.** $f(x) = 2x$

$f^{-1}(x) = \dfrac{x}{2}$

$f(f^{-1}(x)) = f\left(\dfrac{x}{2}\right) = 2\left(\dfrac{x}{2}\right) = x$

$f^{-1}(f(x)) = f^{-1}(2x) = \dfrac{2x}{2} = x$

**58.** $f(x) = \dfrac{1}{3}x$

$f^{-1}(x) = 3x$

$f(f^{-1}(x)) = f(3x) = \dfrac{1}{3}(3x) = x$

$f^{-1}(f(x)) = f^{-1}\left(\dfrac{1}{3}x\right) = 3\left(\dfrac{1}{3}x\right) = x$

**60.** $f(x) = x - 5$

$f^{-1}(x) = x + 5$

$f(f^{-1}(x)) = f(x + 5) = (x + 5) - 5 = x$

$f^{-1}(f(x)) = f^{-1}(x - 5) = (x - 5) + 5 = x$

**62.** $f(x) = 8 - x$

$f^{-1}(x) = 8 - x$

$f(f^{-1}(x)) = f(8 - x) = 8 - (8 - x)$

$\qquad\qquad = 8 - 8 + x$

$\qquad\qquad = x$

$f^{-1}(f(x)) = f^{-1}(8 - x) = 8 - (8 - x)$

$\qquad\qquad = 8 - 8 + x$

$\qquad\qquad = x$

**64.** $f(x) = x^5$

$f^{-1}(x) = \sqrt[5]{x}$

$f(f^{-1}(x)) = f(\sqrt[5]{x}) = (\sqrt[5]{x})^5 = x$

$f^{-1}(f(x)) = f^{-1}(x^5) = \sqrt[5]{x^5} = x$

**66.** $f(x) = x^{1/5}$

$f^{-1}(x) = x^5$

$f(f^{-1}(x)) = f(x^5) = (x^5)^{1/5} = x$

$f^{-1}(f(x)) = f^{-1}(x^{1/5}) = (x^{1/5})^5 = x$

**68.**   $f(x) = \dfrac{x}{10}$

$y = \dfrac{x}{10}$

$x = \dfrac{y}{10}$

$10x = y$

$f^{-1}(x) = 10x$

**70.**   $f(x) = 7 - x$

$y = 7 - x$

$x = 7 - y$

$x - 7 = -y$

$-(x - 7) = y$

$7 - x = y$

$f^{-1}(x) = 7 - x$

**72.**   $g(t) = 6t + 1$

$y = 6t + 1$

$t = 6y + 1$

$t - 1 = 6y$

$\dfrac{1}{6}(t - 1) = y$

$g^{-1}(t) = \dfrac{1}{6}(t - 1)$

**74.**   $h(s) = 5 - \dfrac{3}{2}s$

$y = 5 - \dfrac{3}{2}s$

$s = 5 - \dfrac{3}{2}y$

$\dfrac{3}{2}y = 5 - s$

$y = \dfrac{2}{3}(5 - s)$

$h^{-1}(s) = \dfrac{2}{3}(5 - s)$

**76.**   $h(x) = \sqrt{x + 5}$

$y = \sqrt{x + 5}$

$x = \sqrt{y + 5}$

$x^2 = (y + 5)^2$

$x^2 = y + 5$

$x^2 - 5 = y$

$h^{-1}(x) = x^2 - 5, x \geq 0$

**78.**   $h(t) = t^5 + 8$

$y = t^5 + 8$

$t = y^5 + 8$

$t - 8 = y^5$

$\sqrt[5]{t - 8} = y$

$h^{-1}(t) = \sqrt[5]{t - 8}$

**80.** $f(s) = \dfrac{2}{3 - s}$

$y = \dfrac{2}{3 - s}$

$s = \dfrac{2}{3 - y}$

$s(3 - y) = 2$

$3 - y = \dfrac{2}{s}$

$-y = -3 + \dfrac{2}{s}$

$y = 3 - \dfrac{2}{s}, s \neq 0$

$f^{-1}(s) = 3 - \dfrac{2}{s}, s \neq 0$

**82.** $f(x) = \sqrt{x^2 - 4}, x \geq 2$

$y = \sqrt{x^2 - 4}$

$x = \sqrt{y^2 - 4}$

$x^2 = y^2 - 4$

$x^2 + 4 = y^2$

$\sqrt{x^2 + 4} = y$

$f^{-1}(x) = \sqrt{x^2 + 4}, x \geq 0$

**84.** $f(x) = x - 7,\quad f^{-1}(x) = x + 7$

(0, −7)  (−7, 0)

(7, 0)  (0, 7)

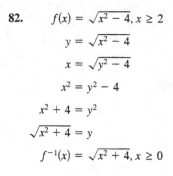

**86.** $f(x) = 5 - 4x,\quad f^{-1}(x) = -\frac{1}{4}(x - 5)$

(0, 5)  (5, 0)

$\left(\frac{5}{4}, 0\right)$  $\left(0, \frac{5}{4}\right)$

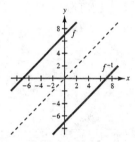

**88.** $f(x) = (x + 2)^2, x \geq -2,\quad f^{-1}(x) = \sqrt{x} - 2$

(0, 4)  (4, 0)

(−2, 0)  (0, −2)

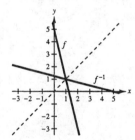

**90.** (c)

**92.** (a)

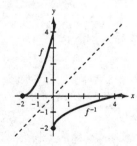

**94.** $f(x) = \frac{1}{5}x - 1$, $g(x) = 5x + 5$

*Keystrokes:*

$y_1$: [Y=] [X,T,θ] [÷] 5 [−] 1 [ENTER]

$y_2$: 5 [X,T,θ] [+] 5 [GRAPH]

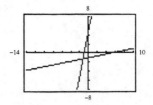

**96.** $f(x) = \sqrt{4 - x}$, $g(x) = 4 - x^2$, $x \geq 0$

*Keystrokes:*

$y_1$: [Y=] [√] [(] 4 [−] [X,T,θ] [)] [ENTER]

$y_2$: 4 [−] [X,T,θ] [$x^2$] [÷] [(] [X,T,θ] [TEST] [4] [0] [)] [GRAPH]

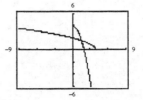

**98.** $f(x) = \sqrt[3]{x + 2}$, $g(x) = x^3 - 2$

*Keystrokes:*

$y_1$: [Y=] [MATH 4] [(] [X,T,θ] [+] 2 [)] [ENTER]

$y_2$: [X,T,θ] [MATH 3] [−] 2 [GRAPH]

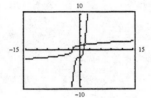

**100.** $f(x) = |x - 2|$, $x \geq 2$, $g(x) = x + 2$, $x \geq 0$

*Keystrokes:*

$y_1$: [Y=] [ABS] [(] [X,T,θ] [−] 2 [)] [÷] [(] [X,T,θ] [TEST] [4] 2 [)] [ENTER]

$y_2$: [X,T,θ] [+] 2 [÷] [(] [X,T,θ] [TEST] [4] 0 [)] [GRAPH]

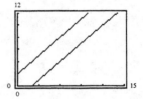

**102.**
$$f(x) = 9 - x^2, \ x \geq 0$$
$$y = 9 - x^2$$
$$x = 9 - y^2$$
$$x - 9 = -y^2$$
$$9 - x = y^2$$
$$\sqrt{9 - x} = y$$
$$\sqrt{9 - x} = f^{-1}(x), \ x \leq 9$$

**104.**
$$f(x) = |x - 2|, \ x \geq 2$$
$$y = x - 2$$
$$x = y - 2$$
$$x + 2 = y$$
$$f^{-1}(x) = x + 2, \ x \geq 0$$

**106.**

| $x$ | $-1$ | 0 | 1 | 5 |
|-----|------|---|---|---|
| $f^{-1}$ | $-2$ | 0 | 3 | 4 |

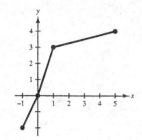

**108.**

| $x$ | $-3$ | $-2$ | 0 | 6 |
|-----|------|------|---|---|
| $f^{-1}$ | 4 | 3 | $-1$ | $-2$ |

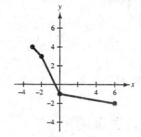

**110.** (a) $y = 9 + 0.65x$

$x = 9 + 0.65y$

$$\frac{x - 9}{0.65} = y$$

$$\frac{100}{65}(x - 9) = y$$

$$\frac{20}{13}(x - 9) = y$$

(b) $x$: hourly wage

$y$: number of units produced

(c) $14.20 = 9 + 0.65x$

$5.20 = 0.65x$

$8 = x$

**112.** $A(r(t)) = A(0.6t) = \pi(0.6t)^2 = 0.36\pi t^2$

This function gives the area of a ripple as a function of time.
Input : time    Output : area

**114.** $(C \circ x)(t) = C[x(t)] = 8.5(12t) + 300$

$= 102t + 300$

Production cost after $t$ hours of operation

**116.** $f(x) = 4x$, $g(x) = x + 6$

(a) $(f \circ g)(x) = f[g(x)] = f(x + 6) = 4(x + 6) = 4x + 24$

(b)  $(f \circ g)^{-1}(x) = 4x + 24$

$y = 4x + 24$

$x = 4y + 24$

$x - 24 = 4y$

$$\frac{x - 24}{4} = y$$

$$(f \circ g)^{-1}(x) = \frac{x - 24}{4}$$

(c)  $f(x) = 4x$          $g(x) = x + 6$

$y = 4x$          $y = x + 6$

$x = 4y$          $x = y + 6$

$\dfrac{x}{4} = y$          $x - 6 = y$

$f^{-1}(x) = \dfrac{x}{4}$          $g^{-1}(x) = x - 6$

—CONTINUED—

**116.** —CONTINUED—

(d) $(g^{-1} \circ f^{-1})(x) = g^{-1}[f^{-1}(x)] = g^{-1}\left(\dfrac{x}{4}\right) = \dfrac{x}{4} - 6 = \dfrac{x}{4} - \dfrac{24}{4} = \dfrac{x-24}{4}$

$(g^{-1} \circ f^{-1})(x) = (f \circ g)^{-1}(x)$

(e) $f(x) = x^3 + 1, g(x) = 2x$

$(f \circ g)(x) = f(g(x)) = f(2x) = (2x)^3 + 1 = 8x^3 + 1$

$(f \circ g)^{-1}(x) = \dfrac{\sqrt[3]{x-1}}{2}$

$y = 8x^3 + 1$

$x = 8y^3 + 1$

$x - 1 = 8y^3$

$\dfrac{x-1}{8} = y^3$

$\sqrt[3]{\dfrac{x-1}{8}} = y$

$\dfrac{\sqrt[3]{x-1}}{2} = (f \circ g)^{-1}(x)$

$f^{-1}(x) = \sqrt[3]{x-1} \quad g^{-1}(x) = \dfrac{x}{2}$

$(g^{-1} \circ f^{-1})(x) = g^{-1}(f^{-1}(x)) = g^{-1}\left(\sqrt[3]{x-1}\right) = \dfrac{\sqrt[3]{x-1}}{2} = (f \circ g)^{-1}(x)$

(f) Answers will vary.

(g) $(f \circ g)^{-1}(x) = (g^{-1} \circ f^{-1})(x)$

**118.** False. $f(x) = \dfrac{1}{x}$

**120.** True. Since $f$ and $f^{-1}$ are inverses,

$(f \circ f^{-1})(x) = (f^{-1} \circ f)(x) = x.$

In particular, $(f \circ f^{-1})(2) = (f^{-1} \circ f)(2) = 2.$

$(f \circ f^{-1})(2) = f(f^{-1}(2)) = f(2) = 2$

$(f^{-1} \circ f)(2) = f^{-1}(f(2)) = f^{-1}(2) = 2$

**122.** Interchange the coordinates of each ordered pair. The inverse of the function defined by $\{(3, 6), (5, -2)\}$ is $\{(6, 3), (-2, 5)\}$.

**124.** $f(x) = x^4$

Answers will vary.
Any function that is even.
Any function whose graph does not pass the Horizontal Line Test.

**126.** They are reflections in the line $y = x$.

# Section 9.3   Logarithmic Functions

**2.** $\log_6 36 = 2$

$6^2 = 36$

**4.** $\log_8 \frac{1}{8} = -1$

$8^{-1} = \frac{1}{8}$

**6.** $\log_{10} 10{,}000 = 4$

$10^4 = 10{,}000$

**8.** $\log_{32} 4 = \frac{2}{5}$

$32^{2/5} = 4$

**10.** $\log_{16} 8 = \frac{3}{4}$

$16^{3/4} = 8$

**12.** $\log_3 1.179 \approx 0.15$

$3^{0.15} \approx 1.179$

**14.**    $6^4 = 1296$

$\log_6 1296 = 4$

**16.**    $5^{-4} = \frac{1}{625}$

$\log_5 \frac{1}{625} = -4$

**18.**    $81^{3/4} = 27$

$\log_{81} 27 = \frac{3}{4}$

**20.**    $6^{-3} = \frac{1}{216}$

$\log_6 \frac{1}{216} = -3$

**22.**    $6^1 = 6$

$\log_6 6 = 1$

because $6^1 = 6$

**24.**    $10^{0.12} \approx 1.318$

$\log_{10} 1.318 \approx 0.12$

**26.** $\log_3 27 = 3$

because $3^3 = 27$

**28.** $\log_8 8 = 1$

because $8^1 = 8$

**30.** $\log_{10} 0.00001 = -5$

because $10^{-5} = 0.00001$

**32.** $\log_3 \frac{1}{9} = -2$

because $3^{-2} = \frac{1}{9}$

**34.** $\log_5 \frac{1}{125} = -3$

because $5^{-3} = \frac{1}{125}$

**36.** $\log_{10} \frac{1}{100} = -2$

because $10^{-2} = \frac{1}{100}$

**38.** $\log_4(-4) = $ not possible

There is no power to which
4 can be raised to obtain $-4$.

**40.** $\log_3 1 = 0$

because $3^0 = 1$

**42.** $\log_2 0 = $ not possible

There is no power to which
2 can be raised to obtain 0.

**44.** $\log_{25} 125 = \frac{3}{2}$

because $25^{3/2} = 125$

**46.** $\log_{144} 12 = \frac{1}{2}$

because $144^{1/2} = 12$

**48.** $\log_5 5^3 = 3$

because $5^3 = 5^3$

**50.** $\log_{10} 5310 \approx 3.7251$

**52.** $\log_{10} 0.345 \approx -0.4622$

**54.** $\log_{10} \frac{\sqrt{3}}{2} \approx -0.0625$

**56.** $f(x) = \log_4 x, g(x) = 4^x$

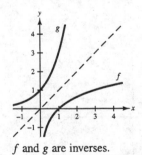

$f$ and $g$ are inverses.

**58.** $f(x) = \log_{1/2} x, g(x) = \left(\frac{1}{2}\right)^x$

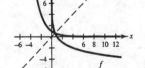

$f$ and $g$ are inverse functions.

**60.** $f(x) = 5^x, g(x) = \log_5 x$

$f$ and $g$ are inverse functions.

**62.** $f(x) = 10^x, g(x) = \log_{10} x$

$f$ and $g$ are inverse functions.

**64.** $h(x) = -4 + \log_2 x$

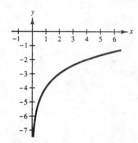

Vertical shift 4 units down

**66.** $h(x) = \log_2(x + 4)$

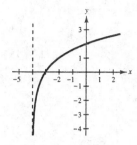

Horizontal shift 4 units left

**68.** $h(x) = -\log_2 x$

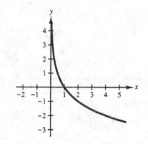

Reflection in the $x$-axis

**70.** $f(x) = -2 + \log_3 x$;  (b)

**72.** $f(x) = \log_3(-x)$;  (c)

**74.** $f(x) = \log_3(x + 2)$;  (f)

**76.** $y(x) = \log_8 x$

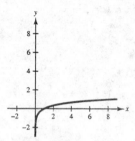

Table of values:

| $x$ | 1 | 8 |
|---|---|---|
| $y$ | 0 | 1 |

**78.** $h(s) = -2 \log_3 s$

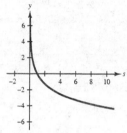

Table of values:

| $x$ | 1 | 3 |
|---|---|---|
| $y$ | 0 | -2 |

**80.** $f(x) = -2 + \log_3 x$

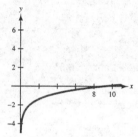

Table of values:

| $x$ | 1 | 3 |
|---|---|---|
| $y$ | -2 | -1 |

**82.** $h(x) = \log_3(x + 1)$

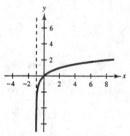

Table of values:

| $x$ | 0 | 2 |
|---|---|---|
| $y$ | 0 | 1 |

**84.** $g(x) = \log_4(4x)$

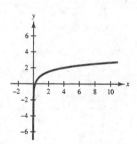

Table of values:

| $x$ | 1 | 4 |
|---|---|---|
| $y$ | 1 | 2 |

**86.** $g(x) = \log_6 x$

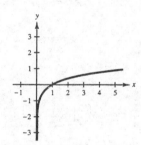

Domain: $(0, \infty)$

Vertical asymptote: $x = 0$

**88.** $f(x) = -\log_6(x + 2)$

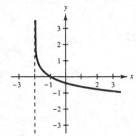

Domain: $(-2, \infty)$

Vertical asymptote: $x = -2$

**90.** $y = \log_5(x - 1) + 4$

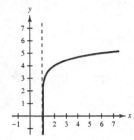

Domain: $(1, \infty)$

Vertical asymptote: $x = 1$

**92.** $y = 5 \log_{10}(x - 3)$

*Keystrokes:*

$\boxed{Y=}$ 5 $\boxed{LOG}$ $\boxed{(}$ $\boxed{X,T,\theta}$ $\boxed{-}$ 3 $\boxed{)}$ $\boxed{GRAPH}$

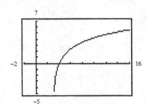

Domain: $(3, \infty)$

Vertical asymptote: $x = 3$

**94.** $y = 5 \log_{10}(3x)$

*Keystrokes:*

$\boxed{Y=}$ 5 $\boxed{LOG}$ $\boxed{(}$ 3 $\boxed{X,T,\theta}$ $\boxed{)}$ $\boxed{GRAPH}$

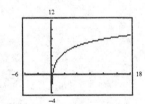

Domain: $(0, \infty)$

Vertical asymptote: $x = 0$

**96.** $y = \log_{10}(-x)$

*Keystrokes:*

$\boxed{Y=}$ log $\boxed{(}$ $\boxed{(-)}$ $\boxed{X,T,\theta}$ $\boxed{)}$ $\boxed{GRAPH}$

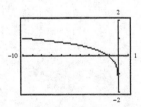

Domain: $(-\infty, 0)$

Vertical asymptote: $x = 0$

**98.** $\ln 6.57 \approx 1.8825$

**100.** $\ln(\sqrt{3} + 1) \approx 1.0051$

**102.** $\ln\left(1 + \dfrac{0.10}{12}\right) \approx 0.0083$

**104.** $f(x) = 4 - \ln(x + 4)$

(e) Basic graph shifted 4 left and 4 up and reflected in the *x*-axis

**106.** $f(x) = -\frac{3}{2} \ln x$

(c) Basic graph reflected in the *x*-axis and multiplied by $\frac{3}{2}$

**108.** $f(x) = \ln(-x)$

(a) Basic graph reflected in the *y*-axis

**110.** $f(x) = -2 \ln x$

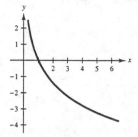

Table of values:

| $x$ | 1 | $e$ |
|---|---|---|
| $y$ | 0 | $-2$ |

**112.** $h(t) = 4 \ln t$

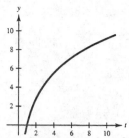

Table of values:

| $x$ | 1 | $e$ |
|---|---|---|
| $y$ | 0 | 4 |

**114.** $h(x) = 2 + \ln x$

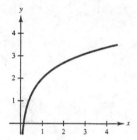

Table of values:

| $x$ | 1 | $e$ |
|---|---|---|
| $y$ | 2 | 3 |

**116.** $g(x) = -3 \ln(x + 3)$

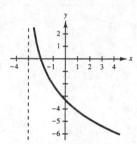

Table of values:

| $x$ | $-2$ | 0 |
|---|---|---|
| $y$ | 0 | $-3.3$ |

**118.** $h(x) = -\ln(x - 2)$

*Keystrokes:*

Y= (−) LN (( X,T,θ − 2 )) GRAPH

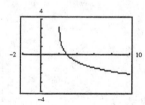

Domain: $(2, \infty)$

Vertical asymptote: $x = 2$

**120.** $g(t) = \ln(3 - t)$

*Keystrokes:*

Y= LN (( 3 − X,T,θ )) GRAPH

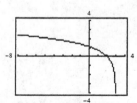

Domain: $(-\infty, 3)$

Vertical asymptote: $t = 3$

**122.** $\log_5 510 = \dfrac{\log 510}{\log 5} \approx 3.8737$

$\qquad = \dfrac{\ln 510}{\ln 5} \approx 3.8737$

**124.** $\log_7 4 = \dfrac{\log 4}{\log 7} \approx 0.7124$

$\qquad = \dfrac{\ln 4}{\ln 7} \approx 0.7124$

**126.** $\log_{12} 0.6 = \dfrac{\log 0.6}{\log 12} \approx -0.2056$

$\qquad = \dfrac{\ln 0.6}{\ln 12} \approx -0.2056$

**128.** $\log_{20} 125 = \dfrac{\log 125}{\log 20} \approx 1.6117$

$\qquad = \dfrac{\ln 125}{\ln 20} \approx 1.6117$

**130.** $\log_{(1/3)}(18) = \dfrac{\log(18)}{\log\left(\frac{1}{3}\right)} \approx -2.6309$

$\qquad = \dfrac{\ln(18)}{\ln\left(\frac{1}{3}\right)} \approx -2.6309$

**132.** $\log_3 \sqrt{26} = \dfrac{\log \sqrt{26}}{\log 3} \approx 1.4828$

$\qquad = \dfrac{\ln \sqrt{26}}{\ln 3} \approx 1.4828$

**134.** $\log_4(2 + e^3) = \dfrac{\log(2 + e^3)}{\log 4} \approx 2.2325$

$\qquad = \dfrac{\ln(2 + e^3)}{\ln 4} \approx 2.2325$

**136.** $B = 10 \log_{10}\left(\dfrac{I}{10^{-16}}\right)$

$\quad B = 10 \log_{10}\left(\dfrac{10^{-4}}{10^{-16}}\right)$

$\quad = 10 \log_{10}(10^{12})$

$\quad = 10(12)\log_{10}(10)$

$\quad = 10(12)(1)$

$\quad = 120$ decibels

**138.** $S = 93 \log_{10} d + 65$

$\quad S = 93 \log_{10} 220 + 65$

$\quad S \approx 282.8$ miles per hour

**140.** $t = 10.042 \ln\left(\dfrac{x}{x - 1250}\right), \; 1250 < x$

(a) *Keystrokes:*

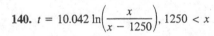 $\boxed{\text{Y=}}$ 10.042 $\boxed{\text{LN}}$ $\boxed{(}$ $\boxed{\text{X,T,}\theta}$ $\boxed{\div}$ $\boxed{(}$ $\boxed{\text{X,T,}\theta}$ $\boxed{-}$ 1250 $\boxed{)}$ $\boxed{)}$ $\boxed{\text{GRAPH}}$

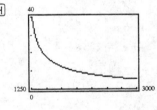

The length of mortgage decreases as the monthly payment increases.

(b) Trace to $x = 1316.35 \cdot t \approx 30$ years

(c) *Verbal model:* $\boxed{\text{Total amount paid}} = \boxed{\text{Number of months}} \cdot \boxed{\text{Monthly payment}}$

$\quad$ *Labels:* $\qquad$ Total amount paid $= x$

$\qquad\qquad\qquad$ Number of months $= 12 \times 30$

$\qquad\qquad\qquad$ Monthly payment $= 1316.35$

$\quad$ *Equation:* $\qquad x = 360 \cdot 1316.35$

$\qquad\qquad\qquad x = \$473,886$

$\qquad\qquad\qquad \boxed{\text{Interest}} = \boxed{\text{Total amount paid}} - \boxed{\text{Mortgage cost}}$

$\qquad\qquad\qquad i = \$473,886 - 150,000$

$\qquad\qquad\qquad i = \$323,886$

**142.** "Logarithm of $x$ with base 5" is $\log_5 x$.

**144.** $\log_a a = 1$ because $a^1 = a$.

**146.** Common logarithms are logarithms with base 10. Natural logarithms are logarithms with base e.

**148.** Domain = positive real numbers    $(0, \infty)$

**150.** If $1000 \le x \le 10,000$, then $f(x) = \log_{10} x$ lies $3 \le f(x) \le 4$.

**152.** When $f(x)$ increases by 1 unit, $x$ increases by a factor of 10.

## Section 9.4    Properties of Logarithms

**2.** $\log_3 9 = 2$ because $3^2 = 9$.

**4.** $\log_8 \left(\frac{1}{64}\right)^5 = \log_8 (8^{-2})^5 = \log_8 8^{-10} = -10$

**6.** $\ln \sqrt[3]{e} = \ln e^{1/3} = \frac{1}{3} \ln e = \frac{1}{3}$

**8.** $\log_4 4^2 = 2$

**10.** $\ln e^{-4} = -4 \ln e = -4(1) = -4$

**12.** $\log_6 2 + \log_6 3 = \log_6 (2 \cdot 3) = \log_6 6 = 1$

**14.** $\log_{10} 5 + \log_{10} 20 = \log_{10}(5 \cdot 20) = \log_{10} 100 = 2$

**16.** $\log_5 50 - \log_5 2 = \log_5 \left(\frac{50}{2}\right) = \log_5 25 = 2$ because $5^2 = 25$.

**18.** $\log_3 324 - \log_3 4 = \log_3 \frac{324}{4} = \log_3 81 = 4$

**20.** $\log_3 \left(\frac{2}{3}\right) + \log_3 \left(\frac{1}{2}\right) = \log_3 \left(\frac{2}{3} \cdot \frac{1}{2}\right) = \log_3 \left(\frac{1}{3}\right) = \log_3 3^{-1} = -1$

**22.** $\ln e^5 - \ln e^2 = \ln \frac{e^5}{e^2} = \ln e^3 = 3 \ln e = 3$

**24.** $\ln(e^2 \cdot e^4) = \ln e^2 + \ln e^4 = 2 \ln e + 4 \ln e = 2(1) + 4(1) = 2 + 4 = 6$

**26.** $\log_4 8 = \log_4 2^3 = 3 \log_4 2 \approx 3(0.500) = 1.5000$

**28.** $\begin{aligned} \log_4 24 &= \log_4 (3 \cdot 8) = \log_4 3 + \log_4 8 \\ &= \log_4 3 + \log_4 2^3 \\ &= \log_4 3 + 3 \log_4 2 \\ &\approx 0.7925 + 3(0.500) \\ &\approx 2.2925 \end{aligned}$

**30.** $\begin{aligned} \log_4 \frac{9}{2} &= \log_4 9 - \log_4 2 \\ &= \log_4 3^2 - \log_4 2 \\ &= 2 \log_4 3 - \log_4 2 \\ &\approx 2(0.7925) - 0.500 \\ &\approx 1.0850 \end{aligned}$

**32.** $\log_4 \sqrt[3]{9} = \log_4 (9)^{1/3}$

$\qquad = \log_4 (3^2)^{1/3}$

$\qquad = \log_4 3^{2/3}$

$\qquad = \frac{2}{3} \log_4 3$

$\qquad \approx \frac{2}{3}(0.7925)$

$\qquad \approx 0.5283$

**34.** $\log_4 \sqrt{3 \cdot 2^5} = \log_4 (3 \cdot 2^5)^{1/2}$

$\qquad = \frac{1}{2}[\log_4 (3 \cdot 2^5)]$

$\qquad = \frac{1}{2}[\log_4 3 + \log_4 2^5]$

$\qquad = \frac{1}{2}[\log_4 3 + 5 \log_4 2]$

$\qquad \approx \frac{1}{2}[0.7925 + 5(0.5000)]$

$\qquad \approx 1.6463$

**36.** $\log_4 4^3 = 3$

**38.** $\log_{10} \frac{1}{4} = \log_{10} 1 - \log_{10} 4$

$\qquad = \log_{10} 1 - \log_{10}\left(\frac{12}{3}\right)$

$\qquad = \log_{10} 1 - (\log_{10} 12 - \log_{10} 3)$

$\qquad = \log_{10} 1 - \log_{10} 12 + \log_{10} 3$

$\qquad \approx 0 - 1.079 + 0.477$

$\qquad \approx -0.602$

**40.** $\log_{10} 144 = \log_{10} 12^2$

$\qquad = 2 \log_{10} 12$

$\qquad \approx 2(1.079)$

$\qquad \approx 2.158$

**42.** $\log_{10} 5^0 = \log_{10} 1$

$\qquad = 0$

**44.** $\log_2 3x = \log_2 3 + \log_2 x$

**46.** $\log_3 x^3 = 3 \log_3 x$

**48.** $\log_2 s^{-4} = -4 \log_2 s$

**50.** $\log_3 \sqrt[3]{5y} = \log_3 (5y)^{1/3} = \frac{1}{3} \log_3 (5y)$

$\qquad\qquad\qquad = \frac{1}{3}(\log_3 5 + \log_3 y)$

**52.** $\ln 5x = \ln 5 + \ln x$

**54.** $\log_{10} \frac{7}{y} = \log_{10} 7 - \log_{10} y$

**56.** $\log_4 \frac{1}{\sqrt{t}} = \log_4 1 - \log_4 \sqrt{t}$

$\qquad\qquad = 0 - \log_4 t^{1/2}$

$\qquad\qquad = -\frac{1}{2} \log_4 t$

**58.** $\ln y(y-1)^2 = \ln y + \ln(y-1)^2$

$\qquad\qquad = \ln y + 2 \ln(y-1)$

**60.** $\log_8[(x-y)^4 z^6] = \log_8(x-y)^4 + \log_8 z^6$

$\qquad\qquad = 4 \log_8(x-y) + 6 \log_8 z$

**62.** $\log_5 \sqrt{xy} = \log_5 (xy)^{1/2} = \frac{1}{2} \log_5 (xy) = \frac{1}{2}(\log_5 x + \log_5 y)$

**64.** $\ln \sqrt[3]{x(x+5)} = \frac{1}{3}[\ln x + \ln(x+5)]$    or    $\frac{1}{3} \ln x + \frac{1}{3} \ln(x+5)$

**66.** $\log_2\left(\frac{x^2}{x-3}\right)^3 = 3 \log_2\left(\frac{x^2}{x-3}\right) = 3(\log_2 x^2 - \log_2(x-3))$

$\qquad\qquad\qquad = 3(2 \log_2 x - \log_2(x-3))$

**68.** $\ln\sqrt{\dfrac{3x}{x-5}} = \ln\left(\dfrac{3x}{x-5}\right)^{1/2} = \dfrac{1}{2}\left(\ln\dfrac{3x}{x-5}\right)$

$$= \dfrac{1}{2}(\ln 3x - \ln(x-5))$$

$$= \dfrac{1}{2}[\ln 3 + \ln x - \ln(x-5)]$$

**70.** $\log_3 \dfrac{x^2 y}{z^7} = \log_3 x^2 + \log_3 y - \log_3 z^7$

$$= 2\log_3 x + \log_3 y - 7\log_3 z$$

**72.** $\log_4 \dfrac{\sqrt[3]{a+1}}{(ab)^4} = \log_4 \sqrt[3]{a+1} - \log_4(ab)^4$

$$= \log_4(a+1)^{1/3} - 4\log_4(ab)$$

$$= \dfrac{1}{3}\log_4(a+1) - 4(\log_4 a + \log_4 b)$$

**74.** $\ln[(xy)^2(x+3)^4] = \ln(xy)^2 + \ln(x+3)^4$

$$= 2\ln(xy) + 4\ln(x+3)$$

$$= 2(\ln x + \ln y) + 4\ln(x+3)$$

**76.** $\ln\left[(u-v)\dfrac{\sqrt[3]{u-4}}{3v}\right] = \ln(u-v) + \ln\sqrt[3]{u-4} - \ln(3v)$

$$= \ln(u-v) + \ln(u-4)^{1/3} - (\ln 3 + \ln v)$$

$$= \ln(u-v) + \dfrac{1}{3}\ln(u-4) - (\ln 3 + \ln v)$$

**78.** $\log_6 12 + \log_6 y = \log_6(12 \cdot y)$

$$= \log_6(12y)$$

**80.** $\log_5 2x + \log_5 3y = \log_5(2x \cdot 3y) = \log_5(6xy)$

**82.** $\ln 10x - \ln z = \ln\left(\dfrac{10x}{z}\right)$

**84.** $10\log_4 z = \log_4 z^{10}, \; z > 0$

**86.** $-5\ln(x+3) = \ln(x+3)^{-5}$ or $\ln\left(\dfrac{1}{(x+3)^5}\right)$

**88.** $-\dfrac{1}{2}\log_3 5y = \log_3(5y)^{-1/2} = \log_3 \dfrac{1}{\sqrt{5y}}$

**90.** $\ln 6 - 3\ln z = \ln 6 - \ln z^3 = \ln\dfrac{6}{z^3}$

**92.** $4\ln 3 - 2\ln x - \ln y = \ln 3^4 - \ln x^2 - \ln y$

$$= \ln 81 - \ln x^2 - \ln y$$

$$= \ln\left(\dfrac{81}{x^2}\right) - \ln y$$

$$= \ln\left(\dfrac{81}{x^2 y}\right), \; x > 0$$

**94.** $4\ln 2 + 2\ln x - \dfrac{1}{2}\ln y = \ln 2^4 + \ln x^2 - \ln y^{1/2}$

$$= \ln 16 + \ln x^2 - \ln\sqrt{y}$$

$$= \ln(16x^2) - \ln\sqrt{y}$$

$$= \ln\left(\dfrac{16x^2}{\sqrt{y}}\right), \; x > 0$$

**96.** $\dfrac{1}{2}(\ln 8 + \ln 2x) = \dfrac{1}{2}[\ln(8 \cdot 2x)]$

$$= \dfrac{1}{2}\ln(16x)$$

$$= \ln\sqrt{16x}$$

$$= \ln 4\sqrt{x}$$

**98.** $5\left[\ln x - \dfrac{1}{2}\ln(x + 4)\right] = 5[\ln x - \ln(x + 4)^{1/2}]$

$$= 5[\ln x - \ln\sqrt{x + 4}]$$

$$= 5\ln\left(\dfrac{x}{\sqrt{x + 4}}\right)$$

$$= \ln\left(\dfrac{x}{\sqrt{x + 4}}\right)^5$$

$$= \ln\left(\dfrac{x^5}{(x + 4)^2\sqrt{x + 4}}\right)$$

$$= \ln\left(\dfrac{x^5}{(x + 4)^{5/2}}\right)$$

**100.** $5\log_3 x + \log_3(x - 6) = \log_3 x^5 + \log_3(x - 6) = \log_3 x^5(x - 6), \, x > 6$

**102.** $\dfrac{1}{4}\log_6(x + 1) - 5\log_6(x - 4) = \log_6(x + 1)^{1/4} - \log_6(x - 4)^5$

$$= \log_6\dfrac{\sqrt[4]{x + 1}}{(x - 4)^5}$$

**104.** $2\log_5(x + y) + 3\log_5 w = \log_5(x + y)^2 + \log_5 w^3 = \log_5(x + y)^2 w^3, \, x + y > 0$

**106.** $\dfrac{1}{3}[\ln(x - 6) - 4\ln y - 2\ln z] = \dfrac{1}{3}[\ln(x - 6) - \ln y^4 - \ln z^2]$

$$= \dfrac{1}{3}\left[\ln\dfrac{(x - 6)}{y^4 z^2}\right]$$

$$= \ln\sqrt[3]{\dfrac{(x - 6)}{y^4 z^2}}, \, y > 0, z > 0$$

**108.** $3\left[\dfrac{1}{2}\log_9(a + 6) - 2\log_9(a - 1)\right] = 3[\log_9(a + 6)^{1/2} - \log_9(a - 1)^2]$

$$= 3\log_9\dfrac{\sqrt{a + 6}}{(a - 1)^2}$$

$$= \log_9\left(\dfrac{\sqrt{a + 6}}{(a - 1)^2}\right)^3, \, a > 1$$

**110.** $\log_3(3^2 \cdot 4) = \log_3 3^2 + \log_3 4$

$$= 2 + \log_3 4$$

**112.** $\log_2\sqrt{22} = \log_2(22)^{1/2}$

$$= \dfrac{1}{2}\log_2 22$$

$$= \dfrac{1}{2}[\log_2 2 + \log_2 11]$$

$$= \dfrac{1}{2}[1 + \log_2 11]$$

$$= \dfrac{1}{2} + \dfrac{1}{2}\log_2 11$$

**114.** $\ln\dfrac{6}{e^5} = \ln 6 - \ln e^5$

$$= \ln 6 - 5\ln e$$

$$= \ln 6 - 5(1)$$

$$= \ln 6 - 5$$

**116.** *Keystrokes:*

$y_1$: [Y=] [LN] [√] [(] [X,T,θ] [(] [X,T,θ] [+] 1 [)] [)] [)] [ENTER]

$y_2$: [(] [LN] [X,T,θ] [+] [LN] [(] [X,T,θ] [+] 1 [)] [)] [÷] 2 [GRAPH]

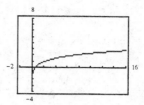

**118.** *Keystrokes:*

$y_1$: [Y=] [LN] [(] [√] [X,T,θ] [÷] [(] [X,T,θ] [−] 3 [)] [)] [)] [ENTER]

$y_2$: [LN] [X,T,θ] [÷] 2 [−] [LN] [(] [X,T,θ] [−] 3 [)] [GRAPH]

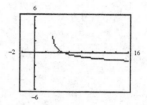

**120.**  $\ln 1 = 0$

$\ln 2 \approx 0.6931$

$\ln 3 \approx 1.0986$

$\ln 4 = \ln 2^2 = 2 \ln 2 \approx 2(0.6931) \approx 1.3862$

$\ln 5 \approx 1.6094$

$\ln 6 = \ln(2 \cdot 3) = \ln 2 + \ln 3 \approx 0.6931 + 1.0986 \approx 1.7917$

$\ln 7 \approx 1.9459$

$\ln 8 = \ln 2^3 = 3 \ln 2 \approx 3(0.6931) \approx 2.0793$

$\ln 9 = \ln 3^2 = 2 \ln 3 \approx 2(1.0986) \approx 2.1972$

$\ln 10 = \ln(2 \cdot 5) = \ln 2 + \ln 5 \approx 0.6931 + 1.6094 \approx 2.3025$

$\ln 12 = \ln(3 \cdot 4) = \ln 3 + \ln 4 = \ln 3 + \ln 2^2 = \ln 3 + 2 \ln 2$

$$\approx 1.0986 + 2(0.6931)$$

$$\approx 2.4848$$

$\ln 14 = \ln(7 \cdot 2) = \ln 7 + \ln 2 \approx 1.9459 + 0.6931 \approx 2.6390$

$\ln 15 = \ln(3 \cdot 5) = \ln 3 + \ln 5 \approx 1.0986 + 1.6094 \approx 2.708$

$\ln 16 = \ln 2^4 = 4 \ln 2 \approx 4(0.6931) \approx 2.7724$

$\ln 18 = \ln(2 \cdot 9) = \ln 2 + \ln 9 = \ln 2 + \ln 3^2 = \ln 2 + 2 \ln 3$

$$\approx 0.6931 + 2(1.0986)$$

$$\approx 2.8903$$

$\ln 20 = \ln(4 \cdot 5) = \ln 4 + \ln 5 = \ln 2^2 + \ln 5 = 2 \ln 2 + \ln 5$

$$\approx 2(0.6931) + 1.6094$$

$$\approx 2.9956$$

Any differences are due to round off errors.

**122.** (a) $f(2) = 80 - \log_{10}(2 + 1)^{12}$

$\qquad = 80 - \log_{10} 3^{12}$

$\qquad = 80 - 12 \log_{10} 3$

$\qquad \approx 74.27$

$\quad f(8) = 80 - \log_{10}(8 + 1)^{12}$

$\qquad = 80 - \log_{10} 9^{12}$

$\qquad = 80 - 12 \log_{10} 9$

$\qquad \approx 68.55$

(b) *Keystrokes:*

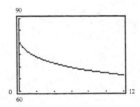

**124.** $E = 1.4(\log_{10} C_2 - \log_{10} C_1)$

$\quad E = 1.4(\log_{10}(2C_1) - \log_{10} C_1)$

$\qquad = 1.4(\log_{10} 2 + \log_{10} C_1 - \log_{10} C_1)$

$\qquad = 1.4(\log_{10} 2)$

$\qquad \approx 0.4214$ kilocalories per gram molecule

**126.** True

**128.** False

$\quad \log_3 u + \log_3 v = \log_3(uv)$

**130.** False

$\quad \log_6 10 - \log_6 3 = \log_6\left(\frac{10}{3}\right)$

**132.** True

**134.** True

**136.** False

$\quad \frac{1}{2} \ln x = \ln x^{1/2} = \ln \sqrt{x} = f\left(\sqrt{x}\right)$

**138.** True

## Section 9.5    Solving Exponential and Logarithmic Equations

**2.** (a) $4^{-1+3} \overset{?}{=} 16$

$\qquad 4^2 \overset{?}{=} 16$

$\qquad 16 = 16$

$\qquad$ Solution

(b) $4^{0+3} \overset{?}{=} 16$

$\qquad 4^3 \overset{?}{=} 16$

$\qquad 64 \neq 16$

$\qquad$ Not a solution

**4.** (a) $2^{3(3.1133)-1} \overset{?}{=} 324$

$\qquad 2^{9.3399-1} \overset{?}{=} 324$

$\qquad 2^{8.3399} \overset{?}{=} 324$

$\qquad 324 = 324$

$\qquad$ Solution

(b) $2^{3(2.4327)-1} \overset{?}{=} 324$

$\qquad 2^{7.2981-1} \overset{?}{=} 324$

$\qquad 2^{6.2981} \overset{?}{=} 324$

$\qquad 78.7 \neq 324$

$\qquad$ Not a solution

**6.** (a) $\ln(-3 + e^{2.5} + 3) \overset{?}{=} 2.5$

$\qquad \ln(e^{2.5}) \overset{?}{=} 2.5$

$\qquad 2.5 = 2.5$

$\quad$ Solution

(b) $\ln(9.1825 + 3) \overset{?}{=} 2.5$

$\qquad \ln(12.1825) \overset{?}{=} 2.5$

$\qquad 2.5 = 2.5$

$\quad$ Solution

**8.** $5^x = 5^3$

so $x = 3$

**10.** $10^{1-x} = 10^4$

so $1 - x = 4$

$1 = x + 4$

$-3 = x$

**12.** $4^{x+4} = 4^3$

so $x + 4 = 3$

$x = -1$

**14.** $3^{2x} = 81$

$3^{2x} = 3^4$

so $2x = 4$

$x = 2$

**16.** $5^{3-2x} = 625$

$5^{3-2x} = 5^4$

so $3 - 2x = 4$

$-2x = 1$

$x = -\frac{1}{2}$

**18.** $3^x = \frac{1}{243}$

$3^x = 3^{-5}$

so $x = -5$

**20.** $3^{2-x} = 9$

$3^{2-x} = 3^2$

so $2 - x = 2$

$-x = 0$

$x = 0$

**22.** $9^{x-2} = 243^{x+1}$

$(3^2)^{x-2} = (3^5)^{x+1}$

$2(x - 2) = 5(x + 1)$

$2x - 4 = 5x + 5$

$-9 = 3x$

$-3 = x$

**24.** $\ln 3x = \ln 24$

so $3x = 24$

$x = 8$

**26.** $\log_5 2x = \log_5 36$

so $2x = 36$

$x = 18$

**28.** $\ln(2x - 3) = \ln 17$

so $2x - 3 = 17$

$2x = 20$

$x = 10$

**30.** $\log_4(x - 4) = \log_4 12$

so $x - 4 = 12$

$x = 16$

**32.** $\log_3(4 - 3x) = \log_3(2x + 9)$

so $4 - 3x = 2x + 9$

$-5 = 5x$

$-1 = x$

**34.** $\log_2(3x - 1) = 5$

so $3x - 1 = 2^5$

$3x - 1 = 32$

$3x = 33$

$x = 11$

**36.** $\log_3 3^{x^2} = x^2$

**38.** $e^{\ln(x+1)} = x + 1,\ x > -1$

**40.** $5^x = 21$

$\log_5 5^x = \log_5 21$

$x = \log_5 21$

$x = \dfrac{\log 21}{\log 5}$

$x \approx 1.89$

**42.** $2^x = 1.5$

$\log_2 2^x = \log_2 1.5$

$x = \log_2 1.5$

$= \dfrac{\log 1.5}{\log 2}$

$\approx 0.58$

**44.** $8^{4x} = 20$

$\log_8 8^{4x} = \log_8 20$

$4x = \log_8 20$

$4x = \dfrac{\log 20}{\log 8}$

$x = \dfrac{\log 20}{4 \log 8}$

$x \approx 0.36$

**46.** $5^{5y} = 305$

$\log_5 5^{5y} = \log_5 305$

$5y = \log_5 305$

$5y = \dfrac{\log 305}{\log 5}$

$y = \dfrac{\log 305}{5 \log 5}$

$y \approx 0.71$

**48.** $5^{3-x} = 15$

$\log_5 5^{3-x} = \log_5 15$

$3 - x = \log_5 15$

$3 - x = \dfrac{\log 15}{\log 5}$

$3 - \dfrac{\log 15}{\log 5} = x$

$1.32 \approx x$

**50.**    $12^{x-1} = 324$

$$\log_{12} 12^{x-1} = \log_{12} 324$$

$$x - 1 = \log_{12} 324$$

$$x = 1 + \log_{12} 324$$

$$x = 1 + \frac{\log 324}{\log 12}$$

$$x \approx 3.33$$

**52.**    $6e^{-x} = 3$

$$e^{-x} = \frac{3}{6}$$

$$e^{-x} = \frac{1}{2}$$

$$\ln e^{-x} = \ln\left(\frac{1}{2}\right)$$

$$-x = \ln\left(\frac{1}{2}\right)$$

$$x = -\ln\left(\frac{1}{2}\right)$$

$$x \approx 0.69$$

**54.**    $\frac{2}{3}e^x = 1$

$$e^x = \frac{3}{2}$$

$$\ln e^x = \ln \frac{3}{2}$$

$$x = \ln \frac{3}{2}$$

$$x \approx 0.41$$

**56.**    $4e^{-3x} = 6$

$$e^{-3x} = \frac{6}{4}$$

$$\ln e^{-3x} = \ln \frac{3}{2}$$

$$-3x = \ln \frac{3}{2}$$

$$x = \frac{\ln \frac{3}{2}}{-3}$$

$$x \approx -0.14$$

**58.**    $32(1.5)^x = 640$

$$1.5^x = 20$$

$$\log_{1.5} 1.5^x = \log_{1.5} 20$$

$$x = \frac{\log 20}{\log 1.5}$$

$$x \approx 7.39$$

**60.**    $6000e^{-2t} = 1200$

$$e^{-2t} = 0.2$$

$$\ln e^{-2t} = \ln 0.2$$

$$-2t = \ln 0.2$$

$$t = \frac{\ln 0.2}{-2}$$

$$t \approx 0.80$$

**62.**    $10{,}000e^{-0.1t} = 4000$

$$e^{-0.1t} = \frac{4000}{10{,}000}$$

$$e^{-0.1t} = \frac{2}{5}$$

$$\ln e^{-0.1t} = \ln \frac{2}{5}$$

$$-0.1t = \ln \frac{2}{5}$$

$$t = \frac{\ln \frac{2}{5}}{-0.1} = -10 \ln \frac{2}{5}$$

$$t \approx 9.16$$

**64.**    $3(2^{t+4}) = 350$

$$2^{t+4} = \frac{350}{3}$$

$$\log_2 2^{t+4} = \log_2\left(\frac{350}{3}\right)$$

$$t + 4 = \log_2\left(\frac{350}{3}\right)$$

$$t = \log_2\left(\frac{350}{3}\right) - 4$$

$$t = \frac{\log\left(\frac{350}{3}\right)}{\log 2} - 4$$

$$t \approx 2.87$$

**66.**    $5^{x+6} - 4 = 12$

$$5^{x+6} = 16$$

$$\log_5 5^{x+6} = \log_5 16$$

$$x + 6 = \log_5 16$$

$$x + 6 = \frac{\log 16}{\log 5}$$

$$x = \frac{\log 16}{\log 5} - 6$$

$$x \approx -4.28$$

**68.**    $9 + e^{5-x} = 32$

$$e^{5-x} = 23$$

$$\ln e^{5-x} = \ln 23$$

$$5 - x = \ln 23$$

$$5 - \ln 23 = x$$

$$1.86 \approx x$$

**70.**    $4 - 2e^x = -23$

$$-2e^x = -27$$

$$e^x = \frac{27}{2}$$

$$\ln e^x = \ln\left(\frac{27}{2}\right)$$

$$x = \ln\left(\frac{27}{2}\right) \approx 2.60$$

**72.**    $10 + e^{4x} = 18$

$$e^{4x} = 8$$

$$\ln e^{4x} = \ln 8$$

$$4x = \ln 8$$

$$x = \frac{\ln 8}{4}$$

$$x \approx 0.52$$

**74.** $50 - e^{x/2} = 35$

$-e^{x/2} = -15$

$e^{x/2} = 15$

$\ln e^{x/2} = \ln 15$

$\dfrac{x}{2} = \ln 15$

$x = 2 \ln 15$

$x \approx 5.42$

**76.** $2e^x + 5 = 115$

$2e^x = 110$

$e^x = 55$

$\ln e^x = \ln 55$

$x = \ln 55$

$x \approx 4.01$

**78.** $50(3 - e^{2x}) = 125$

$3 - e^{2x} = \dfrac{125}{50}$

$3 - e^{2x} = \dfrac{5}{2}$

$-e^{2x} = \dfrac{5}{2} - 3$

$-e^{2x} = -\dfrac{1}{2}$

$e^{2x} = \dfrac{1}{2}$

$\ln e^{2x} = \ln \dfrac{1}{2}$

$2x = \ln \dfrac{1}{2}$

$x = \dfrac{\ln \frac{1}{2}}{2}$

$x \approx -0.35$

**80.** $\dfrac{5000}{(1.05)^x} = 250$

$5000 = 250(1.05)^x$

$20 = (1.05)^x$

$\log_{1.05} 20 = \log_{1.05}(1.05)^x$

$\log_{1.05} 20 = x$

$\dfrac{\log 20}{\log 1.05} = x$

$61.40 \approx x$

**82.** $\dfrac{500}{1 + e^{-0.1t}} = 400$

$500 = 400(1 + e^{-0.1t})$

$\dfrac{500}{400} = 1 + e^{-0.1t}$

$\dfrac{5}{4} = 1 + e^{-0.1t}$

$\dfrac{1}{4} = e^{-0.1t}$

$\ln \dfrac{1}{4} = \ln e^{-0.1t}$

$\ln \dfrac{1}{4} = -0.1t$

$\dfrac{\ln .25}{-0.1} = t$

$13.86 \approx t$

**84.** $\log_{10} = -2$

$10^{\log 10x} = 10^{-2}$

$x = 10^{-2}$

$x = \dfrac{1}{100}$

$x = 0.01$

**86.** $\log_4 x = 2.1$

$4^{\log_4 x} = 4^{2.1}$

$x = 4^{2.1}$

$x \approx 18.38$

**88.** $6 \log_2 x = 18$

$\log_2 x = 3$

$2^{\log_2 x} = 2^3$

$x = 8.00$

**90.** $12 \ln x = 20$

$\ln x = \dfrac{20}{12}$

$e^{\ln x} = e^{5/3}$

$x = e^{5/3}$

$x \approx 5.29$

**92.** $\log_3 6x = 4$

$3^{\log_3 6x} = 3^4$

$6x = 81$

$x = 13.50$

**94.** $\ln(0.5t) = \dfrac{1}{4}$

$e^{\ln(0.5t)} = e^{0.25}$

$0.5t = e^{0.25}$

$t = \dfrac{e^{0.25}}{0.5}$

$t \approx 2.57$

**96.** $\ln\sqrt{x} = 6.5$

$e^{\ln\sqrt{x}} = e^{6.5}$

$\sqrt{x} = e^{6.5}$

$x = (e^{6.5})^2$

$x = e^{13}$

$x \approx 442{,}413.39$

**98.** $5\log_{10}(x + 2) = 15$

$\log_{10}(x + 2) = 3$

$10^{\log_{10}(x+2)} = 10^3$

$x + 2 = 1000$

$x = 998.00$

**100.** $\dfrac{2}{3}\ln(x + 1) = -1$

$\ln(x + 1) = -1.5$

$e^{\ln(x+1)} = e^{-1.5}$

$x + 1 = e^{-1.5}$

$x = e^{-1.5} - 1$

$x \approx -0.78$

**102.** $5 - 4\log_2 x = 2$

$-4\log_2 x = -3$

$\log_2 x = \dfrac{3}{4}$

$2^{\log_2 x} = 2^{0.75}$

$x = 2^{0.75}$

$x \approx 1.68$

**104.** $-5 + 2\ln 3x = 5$

$2\ln 3x = 10$

$\ln 3x = 5$

$3x = e^5$

$x = \dfrac{1}{3}e^5 \text{ or } \dfrac{e^5}{3}$

$x \approx 49.47$

**106.** $\log_5 x - \log_5 4 = 2$

$\log_5 \dfrac{x}{4} = 2$

$5^{\log_5 x/4} = 5^2$

$\dfrac{x}{4} = 25$

$x = 100$

**108.** $\log_7(x - 1) - \log_7 4 = 1$

$\log_7 \dfrac{x - 1}{4} = 1$

$7^{\log_7 (x - 1/4)} = 7^1$

$\dfrac{x - 1}{4} = 7$

$x - 1 = 28$

$x = 29.00$

**110.** $\log_3(x - 2) + \log_3 5 = 3$

$\log_3(x - 2)5 = 3$

$3^{\log_3(5x - 10)} = 3^3$

$5x - 10 = 27$

$5x = 37$

$x = 7.40$

**112.** $\log_{10} x + \log_{10}(x + 1) = 0$

$\log_{10} x(x + 1) = 0$

$10^{\log_{10} x(x+1)} = 10^0$

$x(x + 1) = 1$

$x^2 + x = 1$

$x^2 + x - 1 = 0$

$x = \dfrac{-1 \pm \sqrt{1^2 - 4(1)(-1)}}{2(1)}$

$= \dfrac{-1 \pm \sqrt{1 + 4}}{2}$

$= \dfrac{-1 \pm \sqrt{5}}{2}$

$x \approx 0.62$

$x \approx -1.62 \quad \text{Extraneous}$

**114.** $\log_6(x-5) + \log_6 x = 2$

$\log_6(x-5)x = 2$

$6^{\log_6(x^2-5x)} = 6^2$

$x^2 - 5x = 36$

$x^2 - 5x - 36 = 0$

$(x-9)(x+4) = 0$

$x = 9.00 \quad x = -4 \quad$ Extraneous

**116.** $\log_{10}(25x) - \log_{10}(x-1) = 2$

$\log_{10}\left(\frac{25x}{x-1}\right) = 2$

$\frac{25x}{x-1} = 10^2$

$\frac{25x}{x-1} = 100$

$25x = 100(x-1)$

$x = 4(x-1)$

$x = 4x - 4$

$0 = 3x - 4$

$4 = 3x$

$\frac{4}{3} = x \approx 1.33$

**118.** $\log_3 2x + \log_3(x-1) - \log_3 4 = 1$

$\log_3 \frac{2x(x-1)}{4} = 1$

$\log_3\left(\frac{x^2-x}{2}\right) = 1$

$3^{\log_3(x^2-x/2)} = 3^1$

$\frac{x^2-x}{2} = 3$

$x^2 - x = 6$

$x^2 - x - 6 = 0$

$(x-3)(x+2) = 0$

$x = 3.00 \quad x = -2 \quad$ Extraneous

**120.** *Keystrokes:*

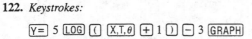

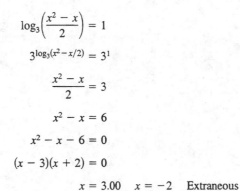

*x*-intercept: (2.35, 0)

**122.** *Keystrokes:*

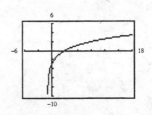

*x*-intercept: (2.98, 0)

**124.** *Keystrokes:*

$y_1$: [Y=] 2 [ENTER]

$y_2$: [LN] [X,T,$\theta$] [GRAPH]

Point of intersection = (7.39, 2)

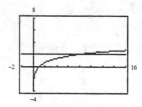

**126.** *Keystrokes:*

$y_1$: [Y=] 200 [ENTER]

$y_2$: 1000 [$e^x$] [(] [(−)] [X,T,$\theta$] [÷] 2 [)] [GRAPH]

Point of intersection = (3.22, 200)

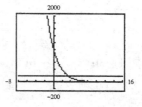

**128.** *Verbal model:*  $A = Pe^{rt}$

*Labels:*    $A$ = amount of balance = \$2847.07

$P$ = principal = \$2500

$r$ = annual interest rate

$t$ = time = 2 years

*Equation:*    $2847.07 = 2500e^{r(2)}$

$1.138828 = e^{2r}$

$\ln 1.138828 = \ln e^{2r}$

$\ln 1.138828 = 2r$

$\dfrac{\ln 1.138828}{2} = r \approx 6.5\%$

**130.**  $10{,}000 = 5000e^{10r}$

$2 = e^{10r}$

$\ln 2 = \ln e^{10r}$

$\ln 2 = 10r$

$\dfrac{\ln 2}{10} = r$

$r \approx 0.069$

$r \approx 6.9\%$

**132.**

$$B = 10 \log_{10}\left(\frac{I}{10^{-16}}\right)$$

$$90 = 10 \log_{10}\left(\frac{I}{10^{-16}}\right)$$

$$9 = \log_{10}\left(\frac{I}{10^{-16}}\right)$$

$$10^9 = 10^{\log_{10}(I/10^{-16})}$$

$$10^9 = \frac{I}{10^{-16}}$$

$$10^9 \cdot 10^{-16} = I$$

$$10^{-7} \text{ watts per square centimeter} = I$$

**134.**

$$F = 200e^{-0.5\pi\theta/180}$$

$$80 = 200e^{-0.5\pi\theta/180}$$

$$0.4 = e^{-0.5\pi\theta/180}$$

$$\ln 0.4 = \ln e^{-0.5\pi\theta/180}$$

$$\ln 0.4 = \frac{-0.5\pi\theta}{180}$$

$$(\ln 0.4)\left(\frac{180}{-0.5\pi}\right) = \theta$$

$$\theta \approx 105°$$

**136.** $N = 273.1 + 355.8e^{-t}, 0 \le t \le 6$

    (a) *Keystrokes:*

        $\boxed{Y=}$ 273.1 $\boxed{+}$ 355.8 $\boxed{e^x}$ $\boxed{(-)}$ $\boxed{X,T,\theta}$ $\boxed{GRAPH}$

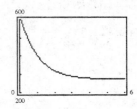

    (b) Let $y_2 = 500$ and find the intersection of the graphs.

        $t \approx 0.45$

**138.** (a)
$$8 = 7.9 \ln(1.0245 - d) + 61.84$$
$$-53.84 = 7.9 \ln(1.0245 - d)$$
$$-6.815189873 = \ln(1.0245 - d)$$
$$e^{-6.815189873} = e^{\ln(1.0245 - d)}$$
$$e^{-6.815189873} = 1.0245 - d$$
$$d = 1.0245 - e^{-6.815189873}$$
$$d \approx 1.0234 \text{ grams per cubic centimeter}$$

(b)
$$0 = 7.9 \ln(1.0245 - d) + 61.84$$
$$-61.84 = 7.9 \ln(1.0245 - d)$$
$$-7.827848101 = \ln(1.0245 - d)$$
$$e^{-7.827848101} = e^{\ln(1.0245 - d)}$$
$$e^{-7.827848101} = 1.0245 - d$$
$$d = 1.0245 - d^{-7.827848101}$$
$$d \approx 1.0241 \text{ grams per cubic centimeter}$$

**140.** Three basic properties of logarithms:
$$\log_a(uv) = \log_a u + \log_a v$$
$$\log_a\left(\frac{u}{v}\right) = \log_a u - \log_a v$$
$$\log_a u^n = n \log_a u$$

**142.** To solve $10^{2x-1} = 5316$ take the common logarithm of both sides of the equation.
$$\log_{10} 10^{2x-1} = \log_{10} 5316$$
$$2x - 1 = \log_{10} 5316$$
The linear equation can now be solved.

## Section 9.6  Applications

**2.**
$$21,628.70 = 3000\left(1 + \frac{r}{4}\right)^{4(20)}$$
$$7.2095667 = \left(1 + \frac{r}{4}\right)^{80}$$
$$(7.2095667)^{1/80} = 1 + \frac{r}{4}$$
$$1.025 = 1 + \frac{r}{4}$$
$$0.025 = \frac{r}{4}$$
$$0.1 = r$$
$$10\% = r$$

**4.**
$$314.85 = 200\left(1 + \frac{r}{1}\right)^{1(5)}$$
$$1.57425 = (1 + r)^5$$
$$(1.57425)^{1/5} = 1 + r$$
$$1.0950016 = 1 + r$$
$$0.0950016 = r$$
$$9.5\% = r$$

**6.**    $4234 = 2000e^{r(10)}$

$2.117 = e^{10r}$

$\ln 2.117 = \ln e^{10r}$

$\ln 2.117 = 10r$

$\dfrac{\ln 2.117}{10} = r$

$0.075 = r$

$7.5\% = r$

**8.**    $110{,}202.78 = 10{,}000\left(1 + \dfrac{r}{365}\right)^{365(30)}$

$11.020278 = \left(1 + \dfrac{r}{365}\right)^{10950}$

$(11.020278)^{1/10950} = 1 + \dfrac{r}{365}$

$1.000219178 = 1 + \dfrac{r}{365}$

$0.000219178 = \dfrac{r}{365}$

$0.08 = r$

$r = 8\%$

**10.**    $1000 = 500\left(\dfrac{0.0525}{12}\right)^{12t}$

$2 = (1.004375)^{12t}$

$\log_{1.004375} 2 = \log_{1.004375} 1.004375^{12t}$

$\dfrac{\log 2}{\log 1.004375} = 12t$

$\dfrac{\log 2}{\log 1.004375} \div 12 = t$

$13.23 \text{ years} \approx t$

**12.**    $20{,}000 = 10{,}000\left(1 + \dfrac{0.095}{1}\right)^{1(t)}$

$2 = (1.095)^t$

$\log_{1.095} 2 = \log_{1.095}(1.095)^t$

$7.64 \text{ years} \approx t$

**14.**    $200 = 100e^{0.06t}$

$2 = e^{0.06t}$

$\ln 2 = \ln e^{0.06t}$

$\ln 2 = 0.06t$

$\dfrac{\ln 2}{0.96} = t$

$11.55 \text{ years} \approx t$

**16.**    $24{,}000 = 12{,}000e^{0.04t}$

$2 = e^{0.04t}$

$\ln 2 = \ln e^{0.04t}$

$\ln 2 = 0.04t$

$\dfrac{\ln 2}{0.04} = t$

$17.33 \text{ years} \approx t$

**18.** $73{,}890.56 = 10{,}000\left(1 + \dfrac{0.10}{n}\right)^{n(20)}$

$73{,}890.56 = 10{,}000e^{0.10(20)}$

$73{,}890.56 = 73{,}890.56$

continuous compounding

**20.** $4788.76 = 4000\left(1 + \dfrac{0.09}{n}\right)^{n(2)}$

$4788.76 = 4000\left(1 + \dfrac{0.09}{365}\right)^{365(2)}$

$4788.76 = 4788.76$

daily compounding

22. $A = P\left(1 + \dfrac{r}{n}\right)^{nt}$

$A = 1000\left(1 + \dfrac{0.095}{365}\right)^{365(1)}$

$A = \$1099.65$

effective yield $= \dfrac{99.64}{1000} = 0.09964$

$\approx 9.96\%$

24. $A = 1000\left(1 + \dfrac{0.08}{1}\right)^{1(1)}$

$A = \$1080.00$

effective yield $= \dfrac{80.00}{1000} = 0.08$

$= 8\%$

26. $A = 1000\left(1 + \dfrac{0.09}{4}\right)^{4(1)}$

$A = \$1093.08$

effective yield $= \dfrac{93.08}{1000} = 0.09308$

$\approx 9.31\%$

28. $A = 1000\left(1 + \dfrac{0.0525}{365}\right)^{365(1)}$

$A = \$1053.90$

effective yield $= \dfrac{53.90}{1000} = 0.0539$

$= 5.39\%$

30. (a) No. The effective yield is the ratio of the year's interest to the amount invested. This *ratio* will be the same regardless of the amount invested.

(b) When interest is compounded more frequently, the difference between the effective yield and the annual interest rate becomes greater.

32. $5000 = Pe^{0.08(5)}$

$\dfrac{5000}{e^{0.4}} = P$

$\$3351.60 \approx P$

34. $3000 = P\left(1 + \dfrac{0.07}{12}\right)^{12(10)}$

$\dfrac{3000}{(1.0058333)^{120}} = P$

$\$1492.79 \approx P$

36. $8000 = P\left(1 + \dfrac{0.06}{12}\right)^{12(2)}$

$\dfrac{8000}{(1.005)^{24}} = P$

$\$7097.49 \approx P$

38. $100{,}000 = P\left(1 + \dfrac{0.09}{365}\right)^{365(40)}$

$\dfrac{100{,}000}{(1.000246575)^{14600}} = P$

$\$2733.59 \approx P$

40. $A = \dfrac{P(e^{rt} - 1)}{e^{r/12} - 1}$

$A = \dfrac{100(e^{0.09(30)} - 1)}{e^{0.09/12} - 1}$

$A \approx \$184{,}369.97$

42. $A = \dfrac{P(e^{rt} - 1)}{e^{r/12} - 1}$

$A = \dfrac{20(e^{0.07(20)} - 1)}{e^{0.07/12} - 1}$

$A \approx \$10{,}444.45$

44. $A = \dfrac{P(e^{rt} - 1)}{e^{r/12} - 1}$

$A = \dfrac{30(e^{0.08(40)} - 1)}{e^{0.08/12} - 1}$

$A \approx \$105{,}543.80$

Total interest $= \$105{,}543.80 - 14{,}400 = \$91{,}143.80$

Total deposits $= \$30 \cdot 12 \cdot 40 = \$14{,}400.00$

**46.** (a)    $y = Ce^{kt}$

$100 = Ce^{k(0)}$

$100 = C$

(b)    $300 = 100e^{k(5)}$

$3 = e^{5k}$

$\ln 3 = \ln e^{5k}$

$\ln 3 = 5k$

$\dfrac{\ln 3}{5} = k$

$0.2197 \approx k$

**48.** (a)    $y = Ce^{kt}$

$1000 = Ce^{k(0)}$

$1000 = C$

(b)    $500 = 1000e^{k(7)}$

$\dfrac{1}{2} = e^{7k}$

$\ln \dfrac{1}{2} = \ln e^{7k}$

$\ln \dfrac{1}{2} = 7k$

$\dfrac{\ln \frac{1}{2}}{7} = k$

$-0.0990 \approx k$

**50.** (a)  $y = Ce^{kt}$

$16.3 = Ce^{k(0)}$

$16.3 = C$

(b)    $17.6 = 16.3e^{k(21)}$

$\dfrac{17.6}{16.3} = e^{21k}$

$\ln \dfrac{17.6}{16.3} = \ln e^{21k}$

$\ln \dfrac{17.6}{16.3} = 21k$

$\dfrac{1}{21} \ln \dfrac{17.6}{16.3} = k$

$0.0037 \approx k$

$y = 16.3e^{0.0037t}$

(c) $y = 16.3e^{0.0037(26)}$

$y \approx 17.9$ million

**52.** (a)  $y = Ce^{kt}$

$11.0 = Ce^{k(0)}$

$11.0 = C$

(b)    $21.2 = 11.0e^{k(21)}$

$\dfrac{21.2}{11.0} = e^{21k}$

$\ln \dfrac{21.2}{11.0} = \ln e^{21k}$

$\ln \dfrac{21.2}{11.0} = 21k$

$\dfrac{1}{21} \ln \dfrac{21.2}{11.0} = k$

$0.0312 \approx k$

$y = 11.0e^{0.0312t}$

(c) $y = 11.0e^{0.0312(26)}$

$y \approx 24.8$ million

**54.** (a) $y = Ce^{kt}$

$11.5 = Ce^{k(0)}$

$11.5 = C$

(b) $\quad 13.1 = 11.5e^{k(21)}$

$\dfrac{13.1}{11.5} = e^{21k}$

$\ln \dfrac{13.1}{11.5} = \ln e^{21k}$

$\ln \dfrac{13.1}{11.5} = 21k$

$\dfrac{1}{21} \ln \dfrac{13.1}{11.5} = k$

$0.0062 \approx k$

$y = 11.5e^{0.0062t}$

(c) $y = 11.5e^{0.0062(26)}$

$y \approx 13.5$ million

**56.** (a) $y = Ce^{kt}$

$16.1 = Ce^{k(0)}$

$16.1 = C$

(b) $\quad 20.8 = 16.1e^{k(21)}$

$\dfrac{20.8}{16.1} = e^{21k}$

$\ln \dfrac{20.8}{16.1} = \ln e^{21k}$

$\ln \dfrac{20.8}{16.1} = 21k$

$\dfrac{1}{21} \ln \dfrac{20.8}{16.1} = k$

$0.0122 \approx k$

$y = 16.1e^{0.0122t}$

(c) $y = 16.1e^{0.0122(26)}$

$y \approx 22.1$ million

**58.** (a) $P = \dfrac{11.14}{1 + 1.101e^{-0.051(30)}}$

$= \dfrac{11.14}{1 + 1.101e^{-1.53}}$

$\approx 8.995$

Thus, the population will be approximately 9 billion in 2020.

(b) *Keystrokes:*

$\boxed{Y=}$ 11.14 $\boxed{\div}$ $\boxed{(}$ 1 $\boxed{+}$ 1.101 $\boxed{e^x}$ $\boxed{(-)}$ .051 $\boxed{X,T,\theta}$ $\boxed{)}$ $\boxed{\text{GRAPH}}$

The population will be twice what it was in 1990 in approximately 2050.

**60.** (a) $\quad y = Ce^{kt}$

$\dfrac{1}{2}C = Ce^{k(1620)}$

$\dfrac{1}{2} = e^{1620k}$

$\ln \dfrac{1}{2} = \ln e^{1620k}$

$\ln \dfrac{1}{2} = 1620k$

$\dfrac{\ln \frac{1}{2}}{1620} = k$

$-0.00043 \approx k$

(b) $\quad 0.25 = Ce^{-0.00043(1000)}$

$0.25 = Ce^{-0.43}$

$\dfrac{0.25}{e^{-0.43}} = C$

$0.38$ grams $\approx C$

**62.** (a) $y = Ce^{kt}$

$10 = Ce^{k(0)}$

$10 = C$

(b) $\quad 5 = 10e^{k(5730)}$

$\dfrac{1}{2} = e^{5730k}$

$\ln \dfrac{1}{2} = \ln e^{5730k}$

$\ln \dfrac{1}{2} = 5730k$

$\dfrac{\ln \frac{1}{2}}{5730} = k$

$-0.00012 \approx k$

(c) $y = 10e^{-0.00012(1000)}$

$y \approx 8.86$ grams

**64.** (a) $\quad y = Ce^{kt}$

$\dfrac{1}{2}C = Ce^{k(24,360)}$

$\dfrac{1}{2} = e^{24,360k}$

$\ln \dfrac{1}{2} = \ln e^{24,360k}$

$\ln \dfrac{1}{2} = 24,360k$

$\dfrac{\ln \frac{1}{2}}{24,360} = k$

$-0.00003 \approx k$

(b) $\quad 1.5 = Ce^{-0.00003(1000)}$

$1.5 = Ce^{-0.03}$

$\dfrac{1.5}{e^{-0.03}} = C$

$1.54$ grams $\approx C$

**66.** (a) $y = Ce^{kt}$

$10 = Ce^{k(0)}$

$10 = C$

(b) $\quad 5 = 10e^{k(24,360)}$

$0.5 = e^{24,360k}$

$\ln 0.5 = \ln e^{24,360k}$

$\dfrac{\ln 0.5}{24,360} = k$

$-0.0000285 = k$

(c) $y = 10e^{-0.0000285(10,000)}$

$y \approx 7.5$ grams

**68.** (a) $\qquad y = Ce^{kt}$

$\dfrac{y}{C} = e^{kt}$

$\dfrac{1}{2} = e^{k(5730)}$

$\ln \dfrac{1}{2} = \ln e^{5730k}$

$\ln \dfrac{1}{2} = 5730k$

$\dfrac{\ln \frac{1}{2}}{5730} = k$

$-0.00012 \approx k$

(b) $\qquad y = Ce^{-0.00012t}$

$\dfrac{y}{C} = e^{-0.00012t}$

$0.15 = e^{-0.00012t}$

$\ln 0.15 = \ln e^{-0.00012t}$

$\ln 0.15 = -0.00012t$

$15,700 \text{ years} \approx t$

**70.** (a) *Keystrokes:*

$\boxed{Y=}$ 32,000 $\boxed{(}$ .8 $\boxed{)}$ $\boxed{\wedge}$ $\boxed{X,T,\theta}$ $\boxed{GRAPH}$

(b) \$25,600   Answers can vary slightly.

(c) 3.1 years   Answers can vary slightly.

**72.** $R = \log_{10} I$

Long Beach: $\quad 6.2 = \log_{10} I$

$10^{6.2} = 10^{\log_{10} I}$

$10^{6.2} = I$

Morocco: $\quad 5.8 = \log_{10} I$

$10^{5.8} = 10^{\log_{10} I}$

$10^{5.8} = I$

Ratio of two intensities:

$\dfrac{I \text{ for Long Beach}}{I \text{ for Morocco}} = \dfrac{10^{6.2}}{10^{5.8}}$

$= 10^{6.2-5.8}$

$= 10^{.4}$

$\approx 2.5$

Earthquake in Long Beach is 2.5 times as great.

**74.** $R = \log_{10} I$

Chile: $8.6 = \log_{10} I$

$10^{8.6} = I$

Armenia: $6.8 = \log_{10} I$

$10^{6.8} = I$

Ratio of 2 intensities:

$\dfrac{I \text{ for Chile}}{I \text{ for Armenia}} = \dfrac{10^{8.6}}{10^{6.8}} = 10^{8.6-6.8} = 10^{1.8} \approx 63$

Earthquake in Chile is 63 times as great.

**76.**
$$pH = -\log_{10}[H^+]$$
$$4.7 = -\log_{10}[H^+]$$
$$-4.7 = -\log_{10}[H^+]$$
$$10^{-4.7} = 10^{\log_{10}[H^+]}$$
$$10^{-4.7} = H^+$$
$$0.00002 = H^+$$
$$2.0 \times 10^{-5} = H^+$$

**78.**
$$pH = -\log_{10}[H^+]$$
$$pH - 1 = -\log_{10}[H^+] - 1$$
$$= -\log_{10}[H^+] - \log_{10} 10$$
$$= -(\log_{10}[H^+] + \log_{10} 10)$$
$$= -\log_{10}[10\, H^+]$$

Thus, a decrease of one unit in the pH of a solution results in the increase of the hydrogen ion concentration by a factor of 10.

**80.** (a) *Keystrokes:*

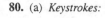 Y= 2000 ÷ ( ( 1 + 4 eˣ ( (−) X,T,θ ÷ 2 ) ) ) GRAPH

(b) 1298 units

(c) 3.2 years

(d) 2000 units

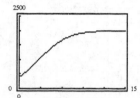

**82.** If the equation $y = Ce^{kt}$ models exponential growth, $k > 0$ because growth is increasing so $k$ must be positive.

**84.** Formulas for periodic and continuous compounding: $A$ is the balance, $P$ is the principal, $r$ is the annual rate, $t$ is the time in years.

**86.** The half-life of a radioactive isotope is the time required for the radioactive material to decay to half of its original amount.

# Review Exercises for Chapter 9

**2.** (a) $g(-2) = 2^{-(-2)} = 2^2 = 4$

  (b) $g(0) = 2^{-0} = 2^0 = 1$

  (c) $g(2) = 2^{-2} = \frac{1}{4}$

**4.** (a) $h(0) = 1 - e^{0.2(0)} = 1 - e^0 = 1 - 1 = 0$

  (b) $h(2) = 1 - e^{0.2(2)} = 1 - e^{0.4} \approx -0.492$

  (c) $h(\sqrt{10}) = 1 - e^{0.2\sqrt{10}} \approx -0.882$

**6.** (d) Basic graph reflected in the $y$-axis.

**8.** (b) Basic graph shifted up 1 unit

**10.** $f(x) = 3^{-x}$

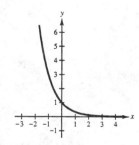

Table of values:

| $x$ | $-1$ | 0 | 1 |
|---|---|---|---|
| $y$ | 3 | 1 | $\frac{1}{3}$ |

**12.** $f(x) = 3^x + 2$

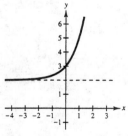

Table of values:

| $x$ | $-1$ | 0 | 1 |
|---|---|---|---|
| $y$ | $\frac{7}{3}$ | 3 | 5 |

**14.** $f(x) = 3^{(x-1)}$

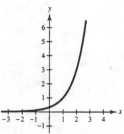

Table of values:

| $x$ | $-1$ | 0 | 1 | 2 |
|---|---|---|---|---|
| $y$ | $\frac{1}{9}$ | $\frac{1}{3}$ | 1 | 3 |

**16.** $f(x) = 3^{-x/2}$

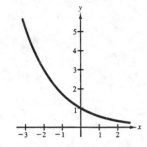

Table of values:

| $x$ | $-2$ | $0$ | $2$ |
|---|---|---|---|
| $y$ | $3$ | $1$ | $\frac{1}{3}$ |

**18.** $f(x) = 3^{x/2} + 3$

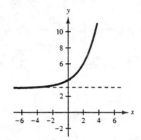

Table of values:

| $x$ | $-2$ | $0$ | $2$ |
|---|---|---|---|
| $y$ | $\frac{10}{3}$ | $4$ | $6$ |

**20.** $y = 6 - e^{x/2}$

*Keystrokes:*

Y= 6 − eˣ ( X,T,θ ÷ 2 ) GRAPH

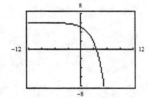

**22.** $h(t) = \dfrac{8}{1 + e^{-t/5}}$

*Keystrokes:*

Y= 8 ÷ ( ( 1 + eˣ ( (−) X,T,θ ÷ 5 ) ) ) GRAPH

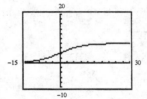

**24.** $f(x) = \sqrt[3]{x},\ g(x) = x + 2$

　(a) $(f \circ g)(x) = \sqrt[3]{x + 2}$

　　$(f \circ g)(6) = \sqrt[3]{8} = 2$

　(b) $(g \circ f)(x) = \sqrt[3]{x} + 2$

　　$(g \circ f)(64) = \sqrt[3]{64} + 2 = 6$

**26.** $f(x) = \dfrac{1}{x - 5},\ g(x) = \dfrac{5x + 1}{x}$

　(a) $(f \circ g)(x) = f(g(x)) = f\left(\dfrac{5x + 1}{x}\right) = \dfrac{1}{\left(\dfrac{5x + 1}{x}\right) - 5} = \dfrac{1}{\dfrac{5x + 1 - 5x}{x}} = \dfrac{1}{\dfrac{1}{x}}$

　　　　　　　　　　　　　　　　　　$= x$

　　so $(f \circ g)(1) = 1$

**—CONTINUED—**

**26.** —CONTINUED—

(b) $(g \circ f)(x) = g(f(x)) = g\left(\dfrac{1}{x-5}\right) = \dfrac{5\left(\dfrac{1}{x-5}\right)+1}{\dfrac{1}{x-5}} \cdot \dfrac{x-5}{x-5}$

$= \dfrac{5+x-5}{1} = x$

so $(g \circ f)\left(\dfrac{1}{5}\right) = \dfrac{1}{5}$

**28.** (a) $(f \circ g)(x) = \dfrac{2}{x^2-4}$

Domain: $(-\infty, -2) \cup (-2, 2) \cup (2, \infty)$

(b) $(g \circ f)(x) = \left(\dfrac{2}{x-4}\right)^2 = \dfrac{4}{(x-4)^2}$

Domain: $(-\infty, 4) \cup (4, \infty)$

**30.** $f(x) = \frac{1}{4}x^3$

*Keystrokes:*

Y=  X,T,θ  ^  3  ÷  4  GRAPH

Yes, $f(x)$ does have an inverse.

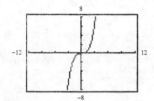

**32.** $g(x) = \sqrt{9 - x^2}$

*Keystrokes:*

Y=  √  (  9  −  X,T,θ  x²  )  GRAPH

No, $g(x)$ does not have an inverse, $g$ is not one-to-one.

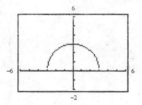

**34.**   $y = 2x - 3$

$x = 2y - 3$

$x + 3 = 2y$

$\dfrac{x+3}{2} = y$

$f^{-1}(x) = \dfrac{x+3}{2}$ or $\dfrac{1}{2}(x+3)$

**36.**   $y = x^2 + 2, x \geq 0$

$x = y^2 + 2$

$x - 2 = y^2$

$\sqrt{x-2} = y$

$g^{-1}(x) = \sqrt{x-2}$

**38.**   $h(t) = \sqrt[3]{t-1}$

$y = \sqrt[3]{t-1}$

$t = \sqrt[3]{y-1}$

$t^3 = y - 1$

$t^3 + 1 = y$

$t^3 + 1 = h^{-1}(t)$

**40.**   $25^{3/2} = 125$

$\log_{25} 125 = \frac{3}{2}$

**42.** $\log_3 \frac{1}{9} = -2$

$3^{-2} = \frac{1}{9}$

**44.** $\log_9 3 = \frac{1}{2}$ because $9^{1/2} = 3$

**46.** $\log_4 \dfrac{1}{16} = -2$ because $4^{-2} = \dfrac{1}{16}$

**48.** $\log_a \dfrac{1}{a} = -1$ because $a^{-1} = \dfrac{1}{a}$

**50.** $\ln e^{-3} = -3$ because $e^{-3} = e^{-3}$

**52.** (a) $g(0.01) = \log_{10}(0.01) = -2$

(b) $g(0.1) = \log_{10}(0.1) = -1$

(c) $g(30) = \log_{10}(30) \approx 1.477$

**54.** (a) $h(e^2) = \ln e^2 = 2$

(b) $h\left(\frac{5}{4}\right) = \ln \frac{5}{4} \approx 0.223$

(c) $h(1200) = \ln 1200 \approx 7.090$

**56.** (a) $f(4) = \log_2\sqrt{4} = \log_2 2 = 1$

(b) $f(64) = \log_2\sqrt{64} = \log_2 8 = 3$

(c) $f(5.2) = \log_2\sqrt{5.2} \approx 1.189$

**58.** $f(x) = -\log_3 x$

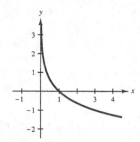

Table of values:

| $x$ | 1 | 3 |
|---|---|---|
| $y$ | 0 | $-1$ |

**60.** $f(x) = 2 + \log_3 x$

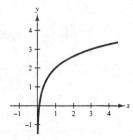

Table of values:

| $x$ | 1 | 3 |
|---|---|---|
| $y$ | 2 | 3 |

**62.** $y = \log_4(x + 1)$

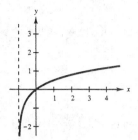

Table of values:

| $x$ | 0 | 3 |
|---|---|---|
| $y$ | 0 | 1 |

**64.** $y = -\ln(x + 2)$

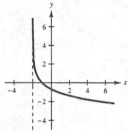

Table of values:

| $x$ | $-1$ | 0 |
|---|---|---|
| $y$ | 0 | $-.7$ |

**66.** $y = 3 + \ln x$

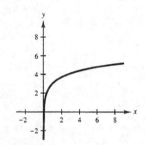

Table of values:

| $x$ | 1 | $e$ |
|---|---|---|
| $y$ | 3 | 4 |

**68.** $\log_{1/2} 5 = \dfrac{\log 5}{\log \frac{1}{2}} \approx -2.322$

**70.** $\log_3 0.28 = \dfrac{\log 0.28}{\log 3} \approx -1.159$

**72.** $\log_5 \sqrt{6} = \log_5 (6)^{1/2}$

$\qquad = \dfrac{1}{2} \log_5 6$

$\qquad = \dfrac{1}{2} \log_5 (2 \cdot 3)$

$\qquad = \dfrac{1}{2} [\log_5 2 + \log_5 3]$

$\qquad \approx \dfrac{1}{2} [0.43068 + 0.6826]$

$\qquad \approx \dfrac{1}{2} (1.11328)$

$\qquad \approx 0.55664$

**74.** $\log_5 \dfrac{2}{3} = \log_5 2 - \log_5 3$

$\qquad \approx 0.43068 - 0.6826$

$\qquad \approx -0.25192$

76. $\log_5(5^2 \cdot 6) = \log_5 5^2 + \log_5 6$

$\qquad = 2\log_5 5 + \log_5(2 \cdot 3)$

$\qquad = 2(1) + \log_5 2 + \log_5 3$

$\qquad = 2 + 0.43068 + 0.6826$

$\qquad = 3.11328$

78. $\log_{10} 2x^{-3} = \log_{10} 2 + \log_{10} x^{-3} = \log_{10} 2 - 3\log_{10} x$

80. $\ln_3 \sqrt{\frac{1}{5}x} = \frac{1}{3}\ln\left(\frac{1}{5}x\right) = \frac{1}{3}[\ln x - \ln 5] = \frac{1}{3}\ln x - \frac{1}{3}\ln 5$

82. $\ln x(x-3)^2 = \ln x + \ln(x-3)^2 = \ln x + 2\ln(x-3)$

84. $\log_3 \frac{a^2\sqrt{b}}{cd^5} = \log_3 a^2 + \log_3 \sqrt{b} - \log_3 c - \log_3 d^5 = 2\log_3 a + \frac{1}{2}\log_3 b - \log_3 c - 5\log_3 d$

86. $5\log_2 y = \log_2 y^5$

88. $\log_4 6x - \log_4 10 = \log_4\left(\frac{6x}{10}\right) = \log_4\left(\frac{3x}{5}\right)$

90. $4(1 + \ln x + \ln x) = 4 + 4\ln x + 4\ln x$

$\qquad = 4 + \ln x^4 + \ln x^4$

$\qquad = 4 + \ln x^8, \ x > 0$

92. $\frac{1}{3}(\log_8 a + 2\log_8 b) = \frac{1}{3}(\log_8 a + \log_8 b^2)$

$\qquad = \frac{1}{3}(\log_8 ab^2)$

$\qquad = \log_8 \sqrt[3]{ab^2}, \ b > 0$

94. $\ln(x+4) - 3\ln x - \ln y = \ln(x+4) - \ln x^3 - \ln y$

$\qquad = \ln\frac{x+4}{x^3 y}, \ x > 0, y > 0$

96. $\frac{\ln 5x}{\ln 10x} = \ln\frac{1}{2}$   False, because $\ln\frac{5x}{10x} = \ln\frac{1}{2}$.

98. $e^{\ln t} = t$   True, Inverse property

100. $6\ln x + 6\ln y = \ln(xy)^6$,   True, because

$\qquad 6\ln x + 6\ln y = 6(\ln x + \ln y)$

$\qquad = 6(\ln xy)$

$\qquad = \ln(xy)^6$

102. $5^x = 25$

$5^x = 5^2$

$x = 2$

104. $3^{x-2} = 81$

$3^{x-2} = 3^4$

$x - 2 = 4$

$x = 6$

106. $\log_4 x = 3$

$4^{\log_4 x} = 4^3$

$x = 64$

108. $\ln(x+4) = \ln 7$

$x + 4 = 7$

$x = 3$

110. $\log_5(x-10) = 2$

$5^{\log_5(x-10)} = 5^2$

$x - 10 = 25$

$x = 35$

112. $8^x = 1000$

$\log_8(8^x) = \log_8 1000$

$x = \log_8 1000$

$x \approx 3.32$

114. $\ln x = -0.5$

$e^{\ln x} = e^{-0.5}$

$x = e^{-0.5}$

$x \approx 0.61$

**116.** $100e^{-0.6x} = 20$

$$e^{-0.6x} = \frac{20}{100}$$

$$e^{-0.6x} = \frac{1}{5}$$

$$\ln e^{-0.6x} = \ln \frac{1}{5}$$

$$-0.6x = \ln \frac{1}{5}$$

$$x = \frac{\ln \frac{1}{5}}{-0.6}$$

$$x \approx 2.68$$

**118.** $25(1 - e^t) = 12$

$$1 - e^t = \frac{12}{25}$$

$$-e^t = -1 + \frac{12}{25}$$

$$-e^t = -\frac{13}{25}$$

$$e^t = \frac{13}{25}$$

$$\ln e^t = \ln \frac{13}{25}$$

$$t = \ln \frac{13}{25}$$

$$t \approx -0.65$$

**120.** $\log_2 2x = -0.65$

$$2^{\log_2 2x} = 2^{-0.65}$$

$$2x = 2^{-0.65}$$

$$x = \frac{2^{-0.65}}{2}$$

$$x \approx 0.32$$

**122.** $4 \log_5(x + 1) = 4.8$

$$\log_5(x + 1) = 1.2$$

$$5^{\log_5(x+1)} = 5^{1.2}$$

$$x + 1 = 5^{1.2}$$

$$x = 5^{1.2} - 1$$

$$x \approx 5.90$$

**124.** $2 \log_4 x - \log_4(x - 1) = 1$

$$\log_4 x^2 - \log_4(x - 1) = 1$$

$$\log_4 \frac{x^2}{x - 1} = 1$$

$$4^{\log_4 x^2/x - 1} = 4^1$$

$$\frac{x^2}{x - 1} = 4$$

$$x^2 = 4(x - 1)$$

$$x^2 = 4x - 4$$

$$x^2 - 4x + 4 = 0$$

$$(x - 2)^2 = 0$$

$$x - 2 = 0$$

$$x = 2.00$$

**126.**

$$A = P\left(1 + \frac{r}{n}\right)^{nt}$$

$$1348.85 = 1000\left(1 + \frac{r}{12}\right)^{12(5)}$$

$$1.34885 = \left(1 + \frac{r}{12}\right)^{60}$$

$$(1.34885)^{1/60} = 1 + \frac{r}{12}$$

$$1.0049 = 1 + \frac{r}{12}$$

$$.0049 = \frac{r}{12}$$

$$.059 = r$$

$$6\% \approx r$$

**128.**

$$A = P\left(1 + \frac{r}{n}\right)^{nt}$$

$$35,236.45 = 10,000\left(1 + \frac{r}{1}\right)^{1(20)}$$

$$3.523645 = (1 + r)^{20}$$

$$(3.523645)^{1/20} = 1 + r$$

$$1.0649 = 1 + r$$

$$.0649 = r$$

$$6.5\% \approx r$$

**130.**
$$A = Pe^{rt}$$
$$15{,}877.50 = 7500e^{r(15)}$$
$$2.117 = e^{15r}$$
$$\ln 2.117 = \ln e^{15r}$$
$$\ln 2.117 = 15r$$
$$\frac{\ln 2.117}{15} = r$$
$$0.050 = r$$
$$5\% \approx r$$

**132.**
$$A = P\left(1 + \frac{r}{n}\right)^{nt}$$
$$A = 1000\left(1 + \frac{0.06}{12}\right)^{12(1)}$$
$$A = 1000(1.005)^{12}$$
$$A = \$1061.68$$
$$\text{Effective} \atop \text{yield} = \frac{61.68}{1000} = 0.06168 \approx 6.17\%$$

**134.**
$$A = P\left(1 + \frac{r}{n}\right)^{nt}$$
$$A = 1000\left(1 + \frac{0.08}{1}\right)^{1(1)}$$
$$A = \$1080$$
$$\text{Effective yield} = \frac{80}{1000} = 0.08 \approx 8\%$$

**136.**
$$A = Pe^{rt}$$
$$A = 1000e^{0.04(1)}$$
$$A = \$1040.81$$
$$\text{Effective yield} = \frac{40.81}{1000} = .0408 \approx 4.08\%$$

**138.** (a)
$$y = Ce^{kt}$$
$$\frac{1}{2}C = Ce^{k(1620)}$$
$$\frac{1}{2} = e^{1620k}$$
$$\ln\frac{1}{2} = \ln e^{1620k}$$
$$\ln\frac{1}{2} = 1620k$$
$$\frac{\ln\frac{1}{2}}{1620} = k$$
$$-0.00043 \approx k$$

(b)
$$0.5 = Ce^{-0.00043(1000)}$$
$$0.5 = Ce^{-0.43}$$
$$\frac{0.5}{e^{-0.43}} = C$$
$$0.769 \text{ grams} \approx C$$

(Answers may vary with rounding.)

**140.** (a)
$$5 = 10e^{k(5730)}$$
$$0.5 = e^{5730k}$$
$$\ln 0.5 = \ln e^{5730k}$$
$$\ln 0.5 = 5730k$$
$$\frac{\ln 0.5}{5730} = k$$
$$-0.00012 \approx k$$

(b) $y = 10e^{-0.00012(1000)}$

$y \approx 8.861$ grams

(Answers may vary with rounding.)

**142.** (a)
$$y = Ce^{kt}$$

$$\frac{1}{2}C = Ce^{k(24,360)}$$

$$\frac{1}{2} = e^{24,360k}$$

$$\ln\frac{1}{2} = \ln e^{24,360k}$$

$$\ln\frac{1}{2} = 24,360k$$

$$\frac{\ln\frac{1}{2}}{24,360} = k$$

$$-0.000028 \approx k$$

(b)
$$2.5 = Ce^{-0.000028(1000)}$$

$$2.5 = Ce^{-0.028}$$

$$\frac{2.5}{e^{-0.028}} = C$$

$$2.571 \text{ grams} \approx C$$

(Answers may vary with rounding.)

**144.**
$$2000 = 1000\left(1 + \frac{0.08}{12}\right)^{12t}$$

$$2 = \left(1 + \frac{0.08}{12}\right)^{12t}$$

$$2 = (1.0066666667)^{12t}$$

$$\log_{1.0066666667} 2 = \log_{1.0066666667}(1.0066666667)^{12t}$$

$$104.3182664 \approx 12t$$

$$8.7 \text{ years} \approx t$$

**146.**
$$p = 25 - 0.4e^{0.02x}$$

$$16.97 = 25 - 0.4e^{0.02x}$$

$$-8.03 = -0.43^{0.02x}$$

$$20.075 = e^{0.02x}$$

$$\ln 20.075 = \ln e^{0.02x}$$

$$\ln 20.075 = 0.02x$$

$$\frac{\ln 20.075}{0.02} = x$$

$$150 \text{ units} \approx x$$

**148.**
$$B = 10 \log_{10}\left(\frac{I}{10^{-16}}\right)$$

$$150 = 10 \log_{10}\left(\frac{I}{10^{-16}}\right)$$

$$15 = \log_{10}\left(\frac{I}{10^{-16}}\right)$$

$$10^{15} = 10^{\log_{10}\left(\frac{I}{10^{-16}}\right)}$$

$$10^{15} = \frac{I}{10^{-16}}$$

$$10^{15} \cdot 10^{-16} = I$$

$$10^{15+(-16)} = I$$

$$10^{-1} = I$$

$$I = 0.1 \text{ watt per square centimeter}$$

**150. (a)** $P = \dfrac{500}{1 + 4e^{-0.36(5)}}$

$P = \dfrac{500}{1 + 4e^{-1.8}}$

$P \approx 300.99$

After 5 years, the deer population
is approximately 301.

**(b)** $250 = \dfrac{500}{1 + 4e^{-0.36t}}$

$250(1 + 4e^{-0.36t}) = 500$

$1 + 4e^{-0.36t} = 2$

$4e^{-0.36t} = 1$

$e^{-0.36t} = \dfrac{1}{4}$

$\ln e^{-0.36t} = \ln \dfrac{1}{4}$

$-0.36t = \ln \dfrac{1}{4}$

$t = \dfrac{\ln\left(\frac{1}{4}\right)}{-0.36} \approx 3.85$ years

**152. (a) (1)** Linear: $A = 156.5t - 8.3$

*Keystrokes:*

$\boxed{Y=}$ 156.5 $\boxed{X,T,\theta}$ $\boxed{-}$ 8.3 $\boxed{\text{GRAPH}}$

**(2)** Quadratic: $A = 50.6t^2 - 349.9t + 1156.5$

*Keystrokes:*

$\boxed{Y=}$ 50.6 $\boxed{X,T,\theta}$ $\boxed{x^2}$ $\boxed{-}$ 349.9 $\boxed{X,T,\theta}$ $\boxed{+}$ 1156.5 $\boxed{\text{GRAPH}}$

**(3)** Exponential: $A = 282.4e^{0.193t}$

*Keystrokes:*

$\boxed{Y=}$ 282.4 $\boxed{e^x}$ .193 $\boxed{X,T,\theta}$ $\boxed{\text{GRAPH}}$

**(4)** Logarithmic: $A = 1133.3 + 620.8t - 2210.9 \ln t$

*Keystrokes:*

$\boxed{Y=}$ 1133.3 $\boxed{+}$ 620.8 $\boxed{X,T,\theta}$ $\boxed{-}$ 2210.9 $\boxed{\text{LN}}$ $\boxed{X,T,\theta}$ $\boxed{\text{GRAPH}}$

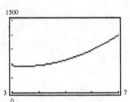

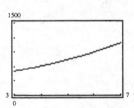

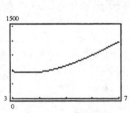

**(b)** The quadratic and logarithmic models "best fit" the data.

# CHAPTER 10
## Sequences, Series, and Probability

# CHAPTER 10
## Sequences, Series, and Probability

### Section 10.1    Sequences and Series

Solutions to Even-Numbered Exercises

**2.** $a_1 = 3(1) = 3$

$a_2 = 3(2) = 6$

$a_3 = 3(3) = 9$

$a_4 = 3(4) = 12$

$a_5 = 3(5) = 15$

$3, 6, 9, 12, 15, \ldots, 3n, \ldots$

**4.** $a_1 = (-1)^{1+1}3(1) = (-1)^2(3)(1) = (1)(3)(1) = 3$

$a_2 = (-1)^{2+1}3(2) = (-1)^3(3)(2) = (-1)(3)(2) = -6$

$a_3 = (-1)^{3+1}3(3) = (-1)^4(3)(3) = (1)(3)(3) = 9$

$a_4 = (-1)^{4+1}3(4) = (-1)^5(3)(4) = (-1)(3)(4) = -12$

$a_5 = (-1)^{5+1}3(5) = (-1)^6(3)(5) = (1)(3)(5) = 15$

$3, -6, 9, -12, 15, \ldots, (-1)^{n+1}3n, \ldots$

**6.** $a_1 = \left(\frac{1}{3}\right)^1 = \frac{1}{3}$

$a_2 = \left(\frac{1}{3}\right)^2 = \frac{1}{9}$

$a_3 = \left(\frac{1}{3}\right)^3 = \frac{1}{27}$

$a_4 = \left(\frac{1}{3}\right)^4 = \frac{1}{81}$

$a_5 = \left(\frac{1}{3}\right)^5 = \frac{1}{243}$

$\frac{1}{3}, \frac{1}{9}, \frac{1}{27}, \frac{1}{81}, \frac{1}{243}, \ldots, \left(\frac{1}{3}\right)^n, \ldots$

**8.** $a_1 = \left(\frac{2}{3}\right)^{1-1} = \left(\frac{2}{3}\right)^0 = 1$

$a_2 = \left(\frac{2}{3}\right)^{2-1} = \left(\frac{2}{3}\right)^1 = \frac{2}{3}$

$a_3 = \left(\frac{2}{3}\right)^{3-1} = \left(\frac{2}{3}\right)^2 = \frac{4}{9}$

$a_4 = \left(\frac{2}{3}\right)^{4-1} = \left(\frac{2}{3}\right)^3 = \frac{8}{27}$

$a_5 = \left(\frac{2}{3}\right)^{5-1} = \left(\frac{2}{3}\right)^4 = \frac{16}{81}$

$1, \frac{2}{3}, \frac{4}{9}, \frac{8}{27}, \frac{16}{81}, \ldots, \left(\frac{2}{3}\right)^{n-1}, \ldots$

**10.** $a_1 = \left(-\frac{2}{3}\right)^{1-1} = \left(-\frac{2}{3}\right)^0 = 1$

$a_2 = \left(-\frac{2}{3}\right)^{2-1} = \left(-\frac{2}{3}\right)^1 = -\frac{2}{3}$

$a_3 = \left(-\frac{2}{3}\right)^{3-1} = \left(-\frac{2}{3}\right)^2 = \frac{4}{9}$

$a_4 = \left(-\frac{2}{3}\right)^{4-1} = \left(-\frac{2}{3}\right)^3 = -\frac{8}{27}$

$a_5 = \left(-\frac{2}{3}\right)^{5-1} = \left(-\frac{2}{3}\right)^4 = \frac{16}{81}$

$1, -\frac{2}{3}, \frac{4}{9}, -\frac{8}{27}, \frac{16}{81}, \ldots, \left(-\frac{2}{3}\right)^{7-1}, \ldots$

**12.** $a_1 = \dfrac{3}{2(1)+1} = \dfrac{3}{2+1} = \dfrac{3}{3} = 1$

$a_2 = \dfrac{3}{2(2)+1} = \dfrac{3}{4+1} = \dfrac{3}{5}$

$a_3 = \dfrac{3}{2(3)+1} = \dfrac{3}{6+1} = \dfrac{3}{7}$

$a_4 = \dfrac{3}{2(4)+1} = \dfrac{3}{8+1} = \dfrac{3}{9} = \dfrac{1}{3}$

$a_5 = \dfrac{3}{2(5)+1} = \dfrac{3}{10+1} = \dfrac{3}{11}$

$1, \frac{3}{5}, \frac{3}{7}, \frac{1}{3}, \frac{3}{11}, \ldots, \dfrac{3}{2n+1}, \ldots$

**14.** $a_1 = \dfrac{5(1)}{4(1)+3} = \dfrac{5}{4+3} = \dfrac{5}{7}$

$a_2 = \dfrac{5(2)}{4(2)+3} = \dfrac{10}{8+3} = \dfrac{10}{11}$

$a_3 = \dfrac{5(3)}{4(3)+3} = \dfrac{15}{12+3} = \dfrac{15}{15} = 1$

$a_4 = \dfrac{5(4)}{4(4)+3} = \dfrac{20}{16+3} = \dfrac{20}{19}$

$a_5 = \dfrac{5(5)}{4(5)+3} = \dfrac{25}{20+3} = \dfrac{25}{23}$

$\frac{5}{7}, \frac{10}{11}, 1, \frac{20}{19}, \frac{25}{23}, \ldots, \dfrac{5n}{4n+3}, \ldots$

**16.** $a_1 = \dfrac{1}{\sqrt{1}} = \dfrac{1}{1} = 1$

$a_2 = \dfrac{1}{\sqrt{2}}$

$a_3 = \dfrac{1}{\sqrt{3}}$

$a_4 = \dfrac{1}{\sqrt{4}} = \dfrac{1}{2}$

$a_5 = \dfrac{1}{\sqrt{5}}$

$1, \dfrac{1}{\sqrt{2}}, \dfrac{1}{\sqrt{3}}, \dfrac{1}{2}, \dfrac{1}{\sqrt{5}}, \ldots, \dfrac{1}{\sqrt{n}}, \ldots$

**64.** $1 + \dfrac{1}{2}, 1 + \dfrac{3}{4}, 1 + \dfrac{7}{8}, 1 + \dfrac{15}{16}, 1 + \dfrac{31}{32}, \ldots$

| $n$: | 1 | 2 | 3 | 4 | 5 |
|------|---|---|---|---|---|
| Terms: | $1 + \dfrac{1}{2}$ | $1 + \dfrac{3}{4}$ | $1 + \dfrac{7}{8}$ | $1 + \dfrac{15}{16}$ | $1 + \dfrac{31}{32}$ |

*Apparent pattern*: Each term is the sum of one and the fraction with numerator two to the $n^{\text{th}}$ power minus one and with denominator two to the $n^{\text{th}}$ power.

$$a_n = 1 + \frac{2^n - 1}{2^n}$$

**66.** $1, 2, \dfrac{2^2}{2}, \dfrac{2^3}{6}, \dfrac{2^4}{24}, \dfrac{2^5}{120}, \ldots$

| $n$: | 1 | 2 | 3 | 4 | 5 | 6 |
|------|---|---|---|---|---|---|
| Terms: | 1 | 2 | $\dfrac{2^2}{2}$ | $\dfrac{2^3}{6}$ | $\dfrac{2^4}{24}$ | $\dfrac{2^5}{120}$ |

*Apparent pattern*: The numerator is two to the $(n-1)$ power and the denominator is $(n-1)$ factorial.

$$a_n = \frac{2^{n-1}}{(n-1)!}$$

**68.** $\displaystyle\sum_{k=1}^{4} 5k = 5(1) + 5(2) + 5(3) + 5(4)$

$$= 5 + 10 + 15 + 20$$

$$= 50$$

**70.** $\displaystyle\sum_{i=0}^{4} (2i + 3) = [2(0) + 3] + [2(1) + 3] + [2(2) + 3] + [2(3) + 3] + [2(4) + 3]$

$$= 3 + 5 + 7 + 9 + 11$$

$$= 35$$

**72.** $\displaystyle\sum_{i=2}^{7} (4i - 1) = (4(2) - 1) + (4(3) - 1) + (4(4) - 1) + (4(5) - 1) + (4(6) - 1) + (4(7) - 1)$

$$= (8 - 1) + (12 - 1) + (16 - 1) + (20 - 1) + (24 - 1) + (28 - 1)$$

$$= 7 + 11 + 15 + 19 + 23 + 27$$

$$= 102$$

**74.** $\displaystyle\sum_{j=0}^{3} \frac{1}{j^2 + 1} = \frac{1}{0^2 + 1} + \frac{1}{1^2 + 1} + \frac{1}{2^2 + 1} + \frac{1}{3^2 + 1}$

$$= \frac{1}{0 + 1} + \frac{1}{1 + 1} + \frac{1}{4 + 1} + \frac{1}{9 + 1}$$

$$= \frac{1}{1} + \frac{1}{2} + \frac{1}{5} + \frac{1}{10}$$

$$= \frac{10}{10} + \frac{5}{10} + \frac{2}{10} + \frac{1}{10}$$

$$= \frac{18}{10}$$

$$= \frac{9}{5}$$

**76.** $\displaystyle\sum_{k=1}^{5} \frac{10k}{k + 2} = \frac{10(1)}{1 + 2} + \frac{10(2)}{2 + 2} + \frac{10(3)}{3 + 2} + \frac{10(4)}{4 + 2} + \frac{10(5)}{5 + 2}$

$$= \frac{10}{3} + \frac{20}{4} + \frac{30}{5} + \frac{40}{6} + \frac{50}{7}$$

$$= \frac{197}{7}$$

$$\approx 28.143$$

**78.** $\displaystyle\sum_{n=3}^{12} 10 = 10 + 10 + 10 + 10 + 10 + 10 + 10 + 10 + 10 + 10$

$$= 100$$

80. $\displaystyle\sum_{k=1}^{5}\left(\frac{2}{k}-\frac{2}{k+2}\right)=\left(\frac{2}{1}-\frac{2}{1+2}\right)+\left(\frac{2}{2}-\frac{2}{2+2}\right)+\left(\frac{2}{3}-\frac{2}{3+2}\right)+\left(\frac{2}{4}-\frac{2}{4+2}\right)+\left(\frac{2}{5}-\frac{2}{5+2}\right)$

$\displaystyle =\left(2-\frac{2}{3}\right)+\left(1-\frac{2}{4}\right)+\left(\frac{2}{3}-\frac{2}{5}\right)+\left(\frac{2}{4}-\frac{2}{6}\right)+\left(\frac{2}{5}-\frac{2}{7}\right)$

$\displaystyle =2-\frac{2}{3}+1-\frac{1}{2}+\frac{2}{3}-\frac{2}{5}+\frac{1}{2}-\frac{1}{3}+\frac{2}{5}-\frac{2}{7}$

$\displaystyle =2+1-\frac{1}{3}-\frac{2}{7}$

$\displaystyle =\frac{42}{21}+\frac{21}{21}-\frac{7}{21}-\frac{6}{21}$

$\displaystyle =\frac{50}{21}$

82. $\displaystyle\sum_{n=0}^{6}\left(\frac{3}{2}\right)^{n}=\left(\frac{3}{2}\right)^{0}+\left(\frac{3}{2}\right)^{1}+\left(\frac{3}{2}\right)^{2}+\left(\frac{3}{2}\right)^{3}+\left(\frac{3}{2}\right)^{4}+\left(\frac{3}{2}\right)^{5}+\left(\frac{3}{2}\right)^{6}$

$\displaystyle =1+\frac{3}{2}+\frac{9}{4}+\frac{27}{8}+\frac{81}{16}+\frac{243}{32}+\frac{729}{64}$

$\displaystyle =\frac{64}{64}+\frac{96}{64}+\frac{144}{64}+\frac{216}{64}+\frac{324}{64}+\frac{486}{64}+\frac{729}{64}$

$\displaystyle =\frac{2059}{64}$

84. $\displaystyle\sum_{n=0}^{5}2n^{2}=110$

*Keystrokes:* [LIST] [MATH 5] [LIST] [OPS 5] 2 [X,T,$\theta$] [$x^2$] [,] [X,T,$\theta$] [,] 0 [,] 5 [,] 1 [)] [ENTER]

86. $\displaystyle\sum_{i=0}^{4}(i!+4)=54$

*Keystrokes:* [LIST] [MATH 5] [LIST] [OPS 5] [X,T,$\theta$] [MATH] [PRB 4] [+] 4 [,] [X,T,$\theta$] [,] 0 [,] 4 [,] 1 [)] [ENTER]

88. $\displaystyle\sum_{k=1}^{6}\left(\frac{1}{2k}-\frac{1}{2k-1}\right)\approx-0.6532$

*Keystrokes:* [LIST] [MATH 5] [LIST] [OPS 5] 1 [÷] [(] 2 [X,T,$\theta$] [)] [−] 1 [÷] [(] 2 [X,T,$\theta$] [−] 1 [)] [,] [X,T,$\theta$] [,] 1 [,] 6 [,] 1 [)]
[ENTER]

90. $\displaystyle\sum_{k=2}^{4}\frac{k}{\ln k}\approx8.5015$

*Keystrokes:* [LIST] [MATH 5] [LIST] [OPS 5] [X,T,$\theta$] [÷] [LN] [X,T,$\theta$] [,] [X,T,$\theta$] [,] 1 [,] 4 [,] 1 [)] [ENTER]

92. $\displaystyle\sum_{k=1}^{6}(k+7)$    94. $\displaystyle\sum_{k=1}^{4}6(k+3)$    96. $\displaystyle\sum_{k=1}^{50}\frac{3}{1+k}$    98. $\displaystyle\sum_{k=0}^{12}\frac{1}{2^{k}}$

100. $\displaystyle\sum_{k=0}^{20}\left(-\frac{2}{3}\right)^{k}$    102. $\displaystyle\sum_{k=1}^{7}\frac{1}{(2k)^{3}}(-1)^{k+1}$    104. $\displaystyle\sum_{k=0}^{9}\frac{2k+2}{3k+4}$    106. $\displaystyle\sum_{k=1}^{25}\left(2+\frac{1}{k}\right)$

**108.** $\displaystyle\sum_{k=0}^{6} \frac{1}{k!}$

**110.** $\bar{x} = \dfrac{1}{n}\displaystyle\sum_{j=1}^{n} x_j$

$\bar{x} = \dfrac{84 + 69 + 66 + 96}{4} = 78.75$

**112.** $\bar{x} = \dfrac{1}{n}\displaystyle\sum_{j=1}^{n} x_j$

$\bar{x} = \dfrac{-1.0 + 4.2 + 5.4 + -3.2 + 3.6}{5} = 1.8$

**114.** (a) $a_3 = 26{,}000\left(\dfrac{3}{4}\right)^3 = 26{,}000\left(\dfrac{27}{64}\right) \approx 10{,}969$

Thus, after 3 years, the car is worth \$10,969.

(b) $a_6 = 26{,}000\left(\dfrac{3}{4}\right)^6 = 26{,}000\left(\dfrac{729}{4096}\right) \approx 4627$

Thus, after 6 years, the car is worth \$4627. This value is less than half the value after 3 years. The value is decreasing at an increasing rate.

**116.** $d_5 = \dfrac{180(5 - 4)}{5} = \dfrac{180(1)}{5} = \dfrac{180}{5} = 36°$

$d_6 = \dfrac{180(6 - 4)}{6} = \dfrac{180(2)}{6} = \dfrac{360}{6} = 60°$

$d_7 = \dfrac{180(7 - 4)}{7} = \dfrac{180(3)}{7} = \dfrac{540}{7} \approx 77.1°$

$d_8 = \dfrac{180(8 - 4)}{8} = \dfrac{180(4)}{8} = \dfrac{720}{8} = 90°$

$d_9 = \dfrac{180(9 - 4)}{9} - \dfrac{180(5)}{9} = \dfrac{900}{9} = 100°$

$d_{10} = \dfrac{180(10 - 4)}{10} = \dfrac{180(6)}{10} = \dfrac{1080}{10} = 108°$

**118.** $a_n = 11{,}791 + 436n, \quad n = 1, 2, \ldots, 5$

Sequence:

$a_1 = 11{,}791 + 436(1) = 12{,}227$

$a_2 = 11{,}791 + 436(2) = 12{,}663$

$a_3 = 11{,}791 + 436(3) = 13{,}099$

$a_4 = 11{,}791 + 436(4) = 13{,}535$

$a_5 = 11{,}791 + 436(5) = 13{,}971$

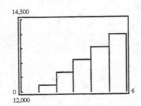

**120.** If $n$ is positive integer, then
$n! = 1 \cdot 2 \cdot 3 \cdot 4 \cdot \ldots (n - 1)n$.
Zero factorial is defined to be
$0! = 1$.

**122.** $\displaystyle\sum_{i=1}^{4} (i^2 + 2i) = \sum_{i=1}^{4} i^2 + \sum_{i=1}^{4} 2i$

True

**124.** $\displaystyle\sum_{j=1}^{4} 2^j = \sum_{j=3}^{6} 2^{j-2}$

True

# Section 10.2   Arithmetic Sequences

**2.** $-8, 0, 8, 16, \ldots$

$0 - -8 = 8$

$8 - 0 = 8$

$16 - 8 = 8$

$d = 8$

**4.** $3200, 2800, 2400, 2000, \ldots$

$2800 - 3200 = -400$

$2400 - 2800 = -400$

$2000 - 2400 = -400$

$d = -400$

**6.** $4, \frac{9}{2}, 5, \frac{11}{2}, 6, \ldots$

$\frac{9}{2} - 4 = \frac{1}{2}$

$5 - \frac{9}{2} = \frac{1}{2}$

$\frac{11}{2} - 5 = \frac{1}{2}$

$6 - \frac{11}{2} = \frac{1}{2}$

$d = \frac{1}{2}$

8. $\frac{11}{2}, \frac{11}{4}, \ldots$

$\frac{5}{4} - \frac{1}{2} = \frac{3}{4}$

$1 - \frac{5}{4} = \frac{3}{4}$

$\frac{11}{2} - 2 = \frac{3}{4}$

$d = \frac{3}{4}$

10. $\frac{5}{2}, \frac{11}{6}, \frac{7}{6}, \frac{1}{2}, -\frac{1}{6}, \ldots$

$\frac{11}{6} - \frac{5}{2} = -\frac{2}{3}$

$\frac{7}{6} - \frac{11}{6} = -\frac{2}{3}$

$\frac{1}{2} - \frac{7}{6} = -\frac{2}{3}$

$-\frac{1}{6} - \frac{1}{2} = -\frac{2}{3}$

$d = -\frac{2}{3}$

12. 1, 2, 4, 8, 16

$2 - 1 = 1$

$4 - 2 = 2$

The sequence is not arithmetic.
The difference is NOT the same.

14. 2, 6, 10, 14, . . .

$6 - 2 = 4$

$10 - 6 = 4$

$14 - 10 = 4$

The sequence is arithmetic.

$d = 4$

16. 32, 16, 8, 4, . . .

$16 - 32 = -16$

$8 - 16 = -8$

The sequence is not arithmetic.

The difference is NOT the same.

18. 8, 4, 2, 1, 0.5, 0.25, . . .

$4 - 8 = -4$

$2 - 4 = -2$

The sequence is not arithmetic.

The difference is NOT the same.

20. $3, \frac{5}{2}, 2, \frac{3}{2}, 1, \ldots$

$\frac{5}{2} - 3 = -\frac{1}{2}$

$2 - \frac{5}{2} = -\frac{1}{2}$

$\frac{3}{2} - 2 = -\frac{1}{2}$

$1 - \frac{3}{2} = -\frac{1}{2}$

The sequence is arithmetic.

$d = -\frac{1}{2}$

22. $\frac{9}{4}, 2, \frac{7}{4}, \frac{3}{2}, \frac{5}{4}, \ldots$

$2 - \frac{9}{4} = -\frac{1}{4}$

$\frac{7}{4} - 2 = -\frac{1}{4}$

$\frac{3}{2} - \frac{7}{4} = -\frac{1}{4}$

$\frac{5}{4} - \frac{3}{2} = -\frac{1}{4}$

The sequence is arithmetic.

$d = -\frac{1}{4}$

24. 1, 4, 9, 16, 25

$4 - 1 = 3$

$9 - 4 = 5$

The sequence is not arithmetic.

The difference is NOT the same.

26. $e, e^2, e^3, e^4, \ldots$

$e^2 - e \approx 4.67$

$e^3 - e^2 \approx 12.70$

The sequence is not arithmetic. The difference is NOT the same.

28. $a_1 = 5(1) - 4 = 5 - 4 = 1$

$a_2 = 5(2) - 4 = 10 - 4 = 6$

$a_3 = 5(3) - 4 = 15 - 4 = 11$

$a_4 = 5(4) - 4 = 20 - 4 = 16$

$a_5 = 5(5) - 4 = 25 - 4 = 21$

30. $a_1 = -10(1) + 100 = -10 + 100 = 90$

$a_2 = -10(2) + 100 = -20 + 100 = 80$

$a_3 = -10(3) + 100 = -30 + 100 = 70$

$a_4 = -10(4) + 100 = -40 + 100 = 60$

$a_5 = -10(5) + 100 = -50 + 100 = 50$

32. $a_1 = \frac{2}{3}(1) + 2 = \frac{2}{3} + \frac{6}{3} = \frac{8}{3}$

$a_2 = \frac{2}{3}(2) + 2 = \frac{4}{3} + \frac{6}{3} = \frac{10}{3}$

$a_3 = \frac{2}{3}(3) + 2 = 2 + 2 = 4$

$a_4 = \frac{2}{3}(4) + 2 = \frac{8}{3} + \frac{6}{3} = \frac{14}{3}$

$a_5 = \frac{2}{3}(5) + 2 = \frac{10}{3} + \frac{6}{3} = \frac{16}{3}$

34. $a_1 = \frac{3}{4}(1) - 2 = -\frac{5}{4}$

$a_2 = \frac{3}{4}(2) - 2 = -\frac{1}{2}$

$a_3 = \frac{3}{4}(3) - 2 = \frac{1}{4}$

$a_4 = \frac{3}{4}(4) - 2 = 1$

$a_5 = \frac{3}{4}(5) - 2 = \frac{7}{4}$

36. $a_1 = 4(1 + 2) + 24 = 4(3) + 24 = 12 + 24 = 36$

$a_2 = 4(2 + 2) + 24 = 4(4) + 24 = 16 + 24 = 40$

$a_3 = 4(3 + 2) + 24 = 4(5) + 24 = 20 + 24 = 44$

$a_4 = 4(4 + 2) + 24 = 4(6) + 24 = 24 + 24 = 48$

$a_5 = 4(5 + 2) + 24 = 4(7) + 24 = 28 + 24 = 52$

**38.** $a_1 = -1, d = 1.2$

$a_n = a_1 + (n - 1)d$

$a_n = -1 + (n - 1)1.2$

$a_n = -1 + 1.2n - 1.2$

$a_n = 1.2n - 2.2$

**40.** $a_1 = 64, d = -8$

$a_n = a_1 + (n - 1)d$

$a_n = 64 + (n - 1)(-8)$

$a_n = 64 - 8n + 8$

$a_n = -8n + 72$

**42.** $a_1 = 12, d = -3$

$a_n = a_1 + (n - 1)d$

$a_n = 12 + (n - 1)(-3)$

$a_n = 12 - 3n + 3$

$a_n = -3n + 15$

**44.** $a_6 = 5, d = \dfrac{3}{2}$

$a_n = a_1 + (n - 1)d$

$5 = a_1 + (6 - 1)\left(\dfrac{3}{2}\right)$

$5 = a_1 + 5\left(\dfrac{3}{2}\right)$

$5 = a_1 + \dfrac{15}{2}$

$\dfrac{10}{2} - \dfrac{15}{2} = a_1$

$-\dfrac{5}{2} = a_1$

So, $a_n = -\dfrac{5}{2} + (n - 1)\left(\dfrac{3}{2}\right)$

$a_n = -\dfrac{5}{2} + \dfrac{3}{2}n - \dfrac{3}{2}$

$a_n = \dfrac{3}{2}n - 4.$

**46.** $a_2 = 93, a_6 = 65$

$d = \dfrac{65 - 93}{4} = \dfrac{-28}{4} = -7$

$a_n = a_1 + (n - 1)d$

$93 = a_1 + (2 - 1)(-7)$

$93 = a_1 - 7$

$100 = a_1$

So, $a_n = 100 + (n - 1)(-7)$

$a_n = 100 - 7n + 7$

$a_n = -7n + 107.$

**48.** $a_5 = 30, a_4 = 25$

$a_n = a_1 + (n - 1)d$

$30 = a_1 + (5 - 1)(5)$

$30 = a_1 + 20$

$10 = a_1$

$a_n = 10 + (n - 1)(5)$

$a_n = 10 + 5n - 5$

$a_n = 5n + 5$

**50.** $a_{10} = 32, a_{12} = 48$

$d = \dfrac{48 - 32}{2} = \dfrac{16}{2} = 8$

$a_n = a_1 + (n - 1)d$

$32 = a_1 + (10 - 1)8$

$32 = a_1 + 72$

$-40 = a_1$

$a_n = -40 + (n - 1)8$

$a_n = -40 + 8n - 8$

$a_n = 8n - 48$

**52.** $a_7 = 8, a_{13} = 6$

$d = \dfrac{6 - 8}{6} = -\dfrac{1}{3}$

$a_n = a_1 + (n - 1)d$

$8 = a_1 + (7 - 1)\left(-\dfrac{1}{3}\right)$

$8 = a_1 - 2$

$10 = a_1$

$a_n = 10 + (n - 1)\left(-\dfrac{1}{3}\right)$

$a_n = 10 - \dfrac{1}{3}n + \dfrac{1}{3}$

$a_n = -\dfrac{1}{3}n + \dfrac{31}{3}$

**54.** $a_1 = 0.08, a_2 = 0.082$

$d = \dfrac{0.082 - 0.08}{1} = 0.002$

$a_n = a_1 + (n - 1)d$

$a_n = 0.08 + (n - 1)0.002$

$a_n = 0.08 + 0.002n - 0.002$

$a_n = 0.002n + 0.078$

**56.** $a_1 = 12, a_{k+1} = a_k - 6$, so $d = -6$.

$a_2 = a_{1+1} = a_1 - 6 = 12 - 6 = 6$

$a_3 = a_{2+1} = a_2 - 6 = 6 - 6 = 0$

$a_4 = a_{3+1} = a_3 - 6 = 0 - 6 = -6$

$a_5 = a_{4+1} = a_4 - 6 = -6 - 6 = -12$

**58.** $a_1 = 8, a_{k+1} = a_k + 7$

$a_n = a_1 + (n-1)d$

$a_1 = 8$ and $d = 7$

$a_2 = a_{1+1} = a_1 + 7 = 8 + 7 = 15$

$a_3 = a_{2+1} = a_2 + 7 = 15 + 7 = 22$

$a_4 = a_{3+1} = a_3 + 7 = 22 + 7 = 29$

$a_5 = a_{4+1} = a_4 + 7 = 29 + 7 = 36$

**60.** $a_1 = -20, a_{k+1} = a_k - 4$

$a_n = a_1 + (n-1)d$

$a_1 = -20$ and $d = -4$

$a_2 = a_{1+1} = a_1 - 4 = -20 - 4 = -24$

$a_3 = a_{2+1} = a_2 - 4 = -13 - 4 = -28$

$a_4 = a_{3+1} = a_3 - 4 = -6 - 4 = -32$

$a_5 = a_{4+1} = a_4 - 4 = 1 - 4 = -36$

**62.** $a_1 = 4.2, a_{k+1} = a_k + 0.4$

$a_1 = 4.2$ and $d = 0.4$

$a_2 = a_{1+1} = a_1 + 0.4 = 4.2 + 0.4 = 4.6$

$a_3 = a_{2+1} = a_2 + 0.4 = 4.6 + 0.4 = 5.0$

$a_4 = a_{3+1} = a_3 + 0.4 = 5.0 + 0.4 = 5.4$

$a_5 = a_{4+1} = a_4 + 0.4 = 5.4 + 0.4 = 5.8$

**64.** $\displaystyle\sum_{k=1}^{30} 4k = 30\left(\frac{4 + 120}{2}\right) = 1860$

**66.** $\displaystyle\sum_{n=1}^{30} (n + 2) = 30\left(\frac{3 + 32}{2}\right) = 525$

**68.** $\displaystyle\sum_{k=1}^{100} (4k - 1) = 100\left(\frac{3 + 399}{2}\right) = 20,100$

**70.** $\displaystyle\sum_{n=1}^{600} \frac{2n}{3} = 600\left(\frac{\frac{2}{3} + 400}{2}\right)$

$= 300\left(\frac{2}{3} + \frac{1200}{3}\right) = 300\left(\frac{1202}{3}\right) = 120,200$

**72.** $\displaystyle\sum_{n=1}^{75} (0.3n + 5) = 75\left(\frac{5.3 + 27.5}{2}\right) = 1230$

**74.** $\displaystyle\sum_{n=1}^{20} (10n - 8) = 20\left(\frac{2 + 192}{2}\right) = 1940$

**76.** $\displaystyle\sum_{n=1}^{20} (520 - 20n) = 20\left(\frac{500 + 120}{2}\right) = 6200$

**78.** $\displaystyle\sum_{n=1}^{25} (815 - 15n) = 25\left(\frac{800 + 440}{2}\right) = 15,500$

**80.** $\displaystyle\sum_{n=1}^{30} (8n - 24) = 30\left(\frac{-16 + 216}{2}\right) = 3000$

**82.** $\displaystyle\sum_{n=1}^{12} (0.6n + 1.6) = 12\left(\frac{2.2 + 8.8}{2}\right) = 66$

**84.** $100\left(\frac{15 + 307}{2}\right) = 16,100$

**86.** (f)

**88.** (a)

**90.** (d)

**92.** *Keystrokes* (calculator in sequence and dot mode):

$\boxed{Y=}$ $\boxed{(-)}$ 25 $\boxed{n}$ $\boxed{+}$ 500 $\boxed{TRACE}$

**94.** *Keystrokes* (calculator in sequence and dot mode):

$\boxed{Y=}$ 3 $\boxed{n}$ $\boxed{\div}$ 2 $\boxed{+}$ 1 $\boxed{TRACE}$

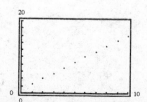

**96.** *Keystrokes* (calculator in sequence and dot mode):

$\boxed{\text{Y=}}$ 6.2 $\boxed{\text{n}}$ $\boxed{+}$ 3 $\boxed{\text{TRACE}}$

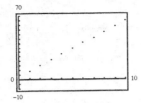

**98.** $\displaystyle\sum_{n=1}^{40} (1000 - 25n)$

*Keystrokes:*

$\boxed{\text{LIST}}$ $\boxed{\text{MATH 5}}$ $\boxed{\text{LIST}}$ $\boxed{\text{OPS 5}}$ 1000 $\boxed{-}$ 25 $\boxed{\text{X,T,}\theta}$ $\boxed{,}$ 1 $\boxed{,}$ 40 $\boxed{,}$ 1 $\boxed{)}$ $\boxed{\text{ENTER}}$

$\displaystyle\sum_{n=1}^{40} (1000 - 25n) = 19,500$

**100.** $\displaystyle\sum_{n=1}^{40} \left(500 - \frac{1}{10}n\right)$

*Keystrokes:*

$\boxed{\text{LIST}}$ $\boxed{\text{MATH 5}}$ $\boxed{\text{LIST}}$ $\boxed{\text{OPS 5}}$ 500 $\boxed{-}$ 0.1 $\boxed{\text{X,T,}\theta}$ $\boxed{,}$ $\boxed{\text{X,T,}\theta}$ $\boxed{,}$ 1 $\boxed{,}$ 20 $\boxed{,}$ 1 $\boxed{)}$ $\boxed{\text{ENTER}}$

$\displaystyle\sum_{n=1}^{20} \left(500 - \frac{1}{10}n\right) = 9979$

**102.** $\displaystyle\sum_{n=1}^{60} (200 - 3.4n)$

*Keystrokes:*

$\boxed{\text{LIST}}$ $\boxed{\text{MATH 5}}$ $\boxed{\text{LIST}}$ $\boxed{\text{OPS 5}}$ 200 $\boxed{-}$ 3.4 $\boxed{\text{X,T,}\theta}$ $\boxed{,}$ $\boxed{\text{X,T,}\theta}$ $\boxed{,}$ 1 $\boxed{,}$ 60 $\boxed{,}$ 1 $\boxed{)}$ $\boxed{\text{ENTER}}$

$\displaystyle\sum_{n=1}^{60} (200 - 3.4n) = 5778$

**104.** $\displaystyle\sum_{n=1}^{66} (n + 34) = 66\left(\frac{35 + 100}{2}\right) = 4455$

**106.** $\displaystyle\sum_{n=1}^{100} (2n - 1) = 100\left(\frac{1 + 199}{2}\right) = 10,000$

**108.** Sequence: 25, 50, 75, . . .

$\displaystyle\sum_{n=1}^{30} 25n = 30\left(\frac{25 + 750}{2}\right) = 11625$ cents $= \$116.25$

**110.** $a_1 = 15, a_7 = 21$

$a_n = a_1 + (n - 1)d$

$a_7 = 15 + (7 - 1)d$

$21 = 15 + 6d$

$6 = 6d$

$1 = d$

$a_n = 15 + (n - 1)(1)$

$a_n = 15 + n - 1$

$a_n = n + 14$

$\displaystyle\sum_{n=1}^{7} (n + 14) = 7\left(\frac{15 + 21}{2}\right)$

$= 7\left(\frac{36}{2}\right) = 7(18) = 126$ logs

**112.** Sequence: 64, 60, 56, 52, . . .

$a_n = 64 + (n - 1)(-4)$

$\quad = 64 - 4n + 4$

$\quad = -4n + 68$

$\displaystyle\sum_{n=1}^{6} (-4n + 68) = 6\left(\frac{64 + 44}{2}\right) = 324$ bales

**114.** Sequence: 1, 2, 3, 4, . . .

$a_n = 1 + (n - 1)(1)$

$a_n = 1 + n - 1$

$a_n = n$

$\displaystyle\sum_{n=1}^{12} n = 12\left(\frac{1 + 12}{2}\right) = 78$ chimes

1 chime each half hour × 12 hours = 12 chimes

Total chimes = 78 + 12 = 90 chimes

**116.** Sequence: 4.9, 14.7, 24.5, . . .

$a_n = 4.9 + (n - 1)(9.8)$

$a_n = 4.9 + 9.8n - 9.8$

$a_n = 9.8n - 4.9$

$\displaystyle\sum_{n=1}^{5} (9.8n - 4.9) = 5\left(\frac{4.9 + 44.1}{2}\right) = 122.5$ feet

**118.** A sequence is arithmetic if the differences between consecutive terms are the same.

**120.** The first two terms of an arithmetic sequence can be used to find the $n$th term by finding the difference between the two terms and substituting into the formula $a_n = a_1 + (n - 1)d$ for $a_1$ and $d$.

**122.** The $n$th partial sum is the sum of the first $n$ terms of a sequence.

**124.** Let $b_n = ca_n$.

$a_n = a_1 + (n - 1)d$

$ca_n = c[a_1 + (n - 1)d]$

$ca_n = ca_1 + (n - 1)cd$

$b_n = b_1 + (n - 1)D$ where $D = cd$.

Yes, $b_n$ is an arithmetic sequence. The common difference changes by a factor of $c$.

# Section 10.3   Geometric Sequences and Series

**2.** 5, -10, 20, -40, . . .

$r = -2$ since $\dfrac{-10}{5} = -2, \dfrac{20}{-10} = -2, \dfrac{-40}{20} = -2$.

**4.** 54, 18, 6, 2, . . .

$r = \dfrac{1}{3}$ since $\dfrac{18}{54} = \dfrac{1}{3}, \dfrac{6}{18} = \dfrac{1}{3}, \dfrac{2}{6} = \dfrac{1}{3}$.

**6.** $9, 6, 4, \dfrac{8}{3}, . . .$

$r = \dfrac{2}{3}$ since $\dfrac{6}{9} = \dfrac{2}{3}, \dfrac{4}{6} = \dfrac{2}{3}, \dfrac{\frac{8}{3}}{4} = \dfrac{2}{3}$.

**8.** $5, -\dfrac{5}{2}, \dfrac{5}{4}, -\dfrac{5}{8}, . . .$

$r = -\dfrac{1}{2}$ since $\dfrac{-\frac{5}{2}}{5} = -\dfrac{1}{2}, \dfrac{\frac{5}{4}}{-\frac{5}{2}} = -\dfrac{1}{2}, \dfrac{-\frac{5}{8}}{\frac{5}{4}} = -\dfrac{1}{2}$.

**10.** $e, e^2, e^3, e^4, . . .$

$r = e$ since $\dfrac{e^2}{e} = e, \dfrac{e^3}{e^2} = e, \dfrac{e^4}{e^3} = e$.

**12.** $1.1, (1.1)^2, (1.1)^3, (1.1)^4, . . .$

$r = 1.1$ since $\dfrac{(1.1)^2}{1.1} = 1.1, \dfrac{(1.1)^3}{(1.1)^2} = 1.1, \dfrac{(1.1)^4}{(1.1)^3} = 1.1$.

**14.** 64, 32, 0, -32, . . .

The sequence is not geometric because $\dfrac{32}{64} = \dfrac{1}{2}$ and $\dfrac{0}{32} = 0$.

**16.** 10, 20, 40, 80

The sequence is geometric: $r = 2$ since $\dfrac{20}{10} = 2, \dfrac{40}{20} = 2, \dfrac{80}{40} = 2$.

**18.** $54, -18, 6, -2, \ldots$

The sequence is geometric:

$$r = -\frac{1}{3} \text{ since } \frac{-18}{54} = -\frac{1}{3}, \frac{6}{-18} = -\frac{1}{3}, \frac{-2}{6} = -\frac{1}{3}.$$

**20.** $12, 7, 2, -3, -8, \ldots$

The sequence is not geometric because $\frac{7}{12} = \frac{7}{12}$ and $\frac{2}{7} = \frac{2}{7}$.

**22.** $\frac{1}{3}, -\frac{2}{3}, \frac{4}{3}, -\frac{8}{3}, \ldots$

The sequence is geometric: $r = -2$ since

$$\frac{-\frac{2}{3}}{\frac{1}{3}} = -2, \frac{\frac{4}{3}}{-\frac{2}{3}} = -2, \frac{-\frac{8}{3}}{\frac{4}{3}} = -2.$$

**24.** $1, 0.2, 0.04, 0.008, \ldots$

The sequence is geometric: $r = 0.2$ since

$$\frac{0.2}{1} = 0.2, \frac{0.04}{0.2} = 0.2, \frac{0.008}{0.04} = 0.2$$

**26.** $a_n = a_1 r^{n-1}$

$a_n = 3(4)^{n-1}$

$a_1 = 3(4)^{1-1} = 3(4)^0 = 3(1) = 3$

$a_2 = 3(4)^{2-1} = 3(4)^1 = 12$

$a_3 = 3(4)^{3-1} = 3(4)^2 = 3(16) = 48$

$a_4 = 3(4)^{4-1} = 3(4)^3 = 3(64) = 192$

$a_5 = 3(4)^{5-1} = 3(4)^4 = 3(256) = 768$

**28.** $a_n = a_1 r^{n-1}$

$a_n = 4\left(\frac{1}{2}\right)^{n-1}$

$a_1 = 4\left(\frac{1}{2}\right)^{1-1} = 4\left(\frac{1}{2}\right)^0 = 4(1) = 4$

$a_2 = 4\left(\frac{1}{2}\right)^{2-1} = 4\left(\frac{1}{2}\right)^1 = 2$

$a_3 = 4\left(\frac{1}{2}\right)^{3-1} = 4\left(\frac{1}{2}\right)^2 = 4\left(\frac{1}{4}\right) = 1$

$a_4 = 4\left(\frac{1}{2}\right)^{4-1} = 4\left(\frac{1}{2}\right)^3 = 4\left(\frac{1}{8}\right) = \frac{1}{2}$

$a_5 = 4\left(\frac{1}{2}\right)^{5-1} = 4\left(\frac{1}{2}\right)^4 = 4\left(\frac{1}{16}\right) = \frac{1}{4}$

**30.** $a_n = a_1 r^{n-1}$

$a_n = 32\left(-\frac{3}{4}\right)^{n-1}$

$a_1 = 32\left(-\frac{3}{4}\right)^{1-1} = 32\left(\frac{3}{4}\right)^0 = 32(1) = 32$

$a_2 = 32\left(-\frac{3}{4}\right)^{2-1} = 32\left(-\frac{3}{4}\right)^1 = -24$

$a_3 = 32\left(-\frac{3}{4}\right)^{3-1} = 32\left(-\frac{3}{4}\right)^2 = 32\left(\frac{9}{16}\right) = 18$

$a_4 = 32\left(-\frac{3}{4}\right)^{4-1} = 32\left(-\frac{3}{4}\right)^3 = 32\left(-\frac{27}{64}\right) = -\frac{27}{2}$

$a_5 = 32\left(-\frac{3}{4}\right)^{5-1} = 32\left(-\frac{3}{4}\right)^4 = 32\left(\frac{81}{256}\right) = \frac{81}{8}$

**32.** $a_n = a_1 r^{n-1}$

$a_n = 4\left(\frac{3}{2}\right)^{n-1}$

$a_1 = 4\left(\frac{3}{2}\right)^{1-1} = 4\left(\frac{3}{2}\right)^0 = 4(1) = 4$

$a_2 = 4\left(\frac{3}{2}\right)^{2-1} = 4\left(\frac{3}{2}\right)^1 = 6$

$a_3 = 4\left(\frac{3}{2}\right)^{3-1} = 4\left(\frac{3}{2}\right)^2 = 4\left(\frac{9}{4}\right) = 9$

$a_4 = 4\left(\frac{3}{2}\right)^{4-1} = 4\left(\frac{3}{2}\right)^3 = 4\left(\frac{27}{8}\right) = \frac{27}{2}$

$a_5 = 4\left(\frac{3}{2}\right)^{5-1} = 4\left(\frac{3}{2}\right)^4 = 4\left(\frac{81}{16}\right) = \frac{81}{4}$

**34.** $a_n = a_1 r^{n-1}$

$a_n = 200(1.07)^{n-1}$

$a_1 = 200(1.07)^{1-1} = 200$

$a_2 = 200(1.07)^{2-1} = 214$

$a_3 = 200(1.07)^{3-1} = 228.98$

$a_4 = 200(1.07)^{4-1} \approx 245.01$

$a_5 = 200(1.07)^{5-1} \approx 262.16$

**36.** $a_n = a_1 r^{n-1}$

$a_n = 1000\left(\frac{1}{1.05}\right)^{n-1}$

$a_1 = 1000\left(\frac{1}{1.05}\right)^{1-1} = 1000\left(\frac{1}{1.05}\right)^0 = 100(1) = 1000$

$a_2 = 1000\left(\frac{1}{1.05}\right)^{2-1} = 1000\left(\frac{1}{1.05}\right)^1 \approx 952.38$

$a_3 = 1000\left(\frac{1}{1.05}\right)^{3-1} = 1000\left(\frac{1}{1.05}\right)^2 \approx 907.03$

$a_4 = 1000\left(\frac{1}{1.05}\right)^{4-1} = 1000\left(\frac{1}{1.05}\right)^3 \approx 863.84$

$a_5 = 1000\left(\frac{1}{1.05}\right)^{5-1} = 1000\left(\frac{1}{1.05}\right)^4 \approx 822.70$

**38.** $a_n = a_1 r^{n-1}$

$a_n = 36\left(\frac{2}{3}\right)^{n-1}$

$a_1 = 36\left(\frac{2}{3}\right)^{1-1} = 36\left(\frac{2}{3}\right)^0 = 36(1) = 36$

$a_2 = 36\left(\frac{2}{3}\right)^{2-1} = 36\left(\frac{2}{3}\right)^1 = 12(2) = 24$

$a_3 = 36\left(\frac{2}{3}\right)^{3-1} = 36\left(\frac{2}{3}\right)^2 = 36\left(\frac{4}{9}\right) = 16$

$a_4 = 36\left(\frac{2}{3}\right)^{4-1} = 36\left(\frac{2}{3}\right)^3 = 36\left(\frac{8}{27}\right) = \frac{32}{3}$

$a_5 = 36\left(\frac{2}{3}\right)^{5-1} = 36\left(\frac{2}{3}\right)^4 = 36\left(\frac{16}{81}\right) = \frac{64}{9}$

**40.** $a_1 = 8, r = \frac{3}{4}, a_8 = $ ▨

$a_n = a_1 r^{n-1}$

$a_8 = 8\left(\frac{3}{4}\right)^{8-1}$

$= \frac{2187}{2048}$

$\approx 1.068$

**42.** $a_1 = 5, r = \sqrt{3}, a_9 = $ ▨

$a_n = a_1 r^{n-1}$

$a_9 = 5\left(\sqrt{3}\right)^{9-1} = 405$

**44.** $a_1 = 500, r = 1.06, a_{40} = $ ▨

$a_n = a_1 r^{n-1}$

$a_{40} = 500(1.06)^{40-1} \approx 4851.754$

**46.** $a_1 = 240, r = -\frac{1}{4}, a_{13} = $ ▨

$a_n = a_1 r^{n-1}$

$a_{13} = 240\left(-\frac{1}{4}\right)^{13-1} = 240\left(-\frac{1}{4}\right)^{12} \approx 0.00001$

**48.** $a_1 = 1, a_2 = 9, a_7 = $ ▨

$a_n = a_1 r^{n-1}$

$a_7 = 1(9)^{7-1} = 531,441$

**50.** $a_3 = 6, a_5 = \frac{8}{3}, a_6 = $ ▨

$r = \frac{a_4}{6} = \frac{\frac{8}{3}}{a_4}$

$a_4^2 = 16$

$a_4 = \pm 4$

So, $r = \pm\frac{4}{6} = \pm\frac{2}{3}$.

$a_3 = a_1\left(\pm\frac{2}{3}\right)^{3-1}$  $\quad a_n = a_1 r^{n-1}$

$6 = a_1\left(\frac{4}{9}\right)$  $\quad a_6 = \frac{27}{2}\left(\pm\frac{2}{3}\right)^{6-1}$

$\frac{27}{2} = a_1$  $\quad a_6 = \frac{27}{2}\left(\pm\frac{32}{243}\right) = \pm\frac{16}{9}$

**52.** $a_4 = 100, a_5 = -25, a_n = $ ▨

$r = \frac{a_5}{a_4} = -\frac{25}{100} = -\frac{1}{4}$

$a_n = a_1 r^{n-1}$

$a_4 = a_1\left(-\frac{1}{4}\right)^{4-1}$

$100 = a_1\left(-\frac{1}{64}\right)$

$-6400 = a_1$

$a_7 = -6400\left(-\frac{1}{4}\right)^{7-1}$

$a_7 = -6400\left(-\frac{1}{4}\right)^6 = -6400\left(\frac{1}{4096}\right)$

$a_7 = -\frac{25}{16}$

**54.** $a_n = a_1 r^{n-1}$

$a_n = 5(4)^{n-1}$

**56.** $a_n = a_1 r^{n-1}$

$a_n = 25(4)^{n-1}$

**58.** $a_n = a_1 r^n - 1$

$a_n = 12\left(-\frac{4}{3}\right)^{n-1}$

**60.** $a_n = a_1 r^{n-1}$

$a_n = 9\left(\frac{2}{3}\right)^{n-1}$

**62.** $a_n = a_1 r^{n-1}$

$a_n = 18\left(\frac{4}{9}\right)^{n-1}$

**64.** $a_n = a_1 r^{n-1}$

$r = \frac{a_2}{a_1} = \frac{\frac{27}{2}}{36} = \frac{3}{8}$

$a_n = 36\left(\frac{3}{8}\right)^{n-1}$

**66.** $a_n = a_1 r^{n-1}$

$a_n = 1\left(\frac{3}{2}\right)^{n-1}$

**68.** (d)

**70.** (c)

**72.** $\displaystyle\sum_{i=1}^{6} 3^{i-1} = 1\left(\frac{3^6 - 1}{3 - 1}\right) = \frac{728}{2} = 364$

**128.** The terms of a geometric sequence decrease when $a_1 > 0$ and $0 < r < 1$ because raising a real number between 0 and 1 to higher powers yields smaller numbers.

**130.** The $n$th partial sum of a sequence is the sum of the first $n$ terms of a sequence.

# Section 10.4    The Binomial Theorem

**2.** $_7C_3 = \dfrac{7 \cdot 6 \cdot 5}{3 \cdot 2 \cdot 1} = 35$

**4.** $_{12}C_9 = {_{12}C_3} = \dfrac{12 \cdot 11 \cdot 10}{3 \cdot 2 \cdot 1} = 220$

**6.** $_{15}C_0 = \dfrac{15!}{15! \cdot 0!} = \dfrac{15!}{15!} = 1$

**8.** $_{200}C_1 = \dfrac{200}{1} = 200$

**10.** $_{75}C_1 = \dfrac{75}{1} = 75$

**12.** $_{18}C_5 = \dfrac{18 \cdot 17 \cdot 16 \cdot 15 \cdot 14}{5 \cdot 4 \cdot 3 \cdot 2 \cdot 1} = 8568$

**14.** *Keystrokes:*

25 [MATH] [PRB 3] 10 [ENTER]    $_{25}C_{10} = 3{,}268{,}760$

**16.** *Keystrokes:*

40 [MATH] [PRB 3] 5 [ENTER]    $_{40}C_5 = 658{,}008$

**18.** *Keystrokes:*

100 [MATH] [PRB 3] 6 [ENTER]    $_{100}C_6 = 1{,}192{,}052{,}400$

**20.** *Keystrokes:*

500 [MATH] [PRB 3] 4 [ENTER]    $_{500}C_4 = 2{,}573{,}031{,}125$

**22.** *Keystrokes:*

1000 [MATH] [PRB 3] 2 [ENTER]  $_{100}C_2 = 499{,}500$

**24.** $_9C_3$

Row 9:  1   9   36   84   126   126   84   36   9   1
                          ↑
                       entry 3

$_9C_3 = 84$

**26.** $_9C_5$

Row 9:  1   9   36   84   126   126   84   36   9   1
                                ↑
                             entry 5

$_9C_5 = 126$

**28.** $_{10}C_6$

Row 10:  1   10   45   120   210   252   210   120   45   10   1
                                ↑
                             entry 6

$_{10}C_6 = 210$

**30.** $(x + 3)^5 = 1x^5 + 5x^4(3) + 10x^3(3)^2 + 10x^2(3)^3 + 5x(3)^4 + 3^5$

$= x^5 + 15x^4 + 90x^3 + 270x^2 + 405x + 243$

**32.** $(r - s)^7 = 1 \cdot r^7 + 7r^6(-s) + 21r^5(-s)^2 + 35r^4(-s)^3 + 35r^3(-s)^4 + 21r^2(-s)^5 + 7r(-s)^6 + 1(-s)^7$

$= r^7 - 7r^6s + 21r^5s^2 - 35r^4s^3 + 35r^3s^4 - 21r^2s^5 + 7rs^6 - s^7$

**34.** $(4 - 3y)^3 = 1(4)^3 + 3(4)^2(-3y) + 3(4)(-3y)^2 + (-3y)^3$

$= 64 - 144y + 108y^2 - 27y^3$

**36.** $(2t - s)^5 = 1(2t)^5 + 5(2t)^4(-s) + 10(2t)^3(-s)^2 + 10(2t)^2(-s)^3 + 5(2t)(-s)^4 + (-s)^5$

$= 32t^5 - 80t^4s + 80t^3s^2 - 40t^2s^3 + 10ts^4 - s^5.$

**38.** $(3 - y^4) = 1(3)^5 + 5(3)^4(-y^4) + 10(3)^3(-y^4)^2 + 10(3)^2(-y^4)^3 + 5(3)(-y^4)^4 + 1(-y^4)^5$

$\qquad = 243 - 405y^4 + 270y^8 - 90y^{12} + 15y^{16} - y^{20}$

**40.** $(x - 5)^4 = 1x^4 + 4x^3(-5) + 6x^2(-5)^2 + 4x(-5)^3 + (-5)^4$

$\qquad = x^4 - 20x^3 + 150x^2 - 500x + 625$

**42.** $(x - 8)^4 = 1x^4 + 4x^3(-8) + 6x^2(-8)^2 + 4x(-8)^3 + (-8)^4$

$\qquad = x^4 - 32x^3 + 384x^2 - 2048x + 4096$

**44.** $(u + v)^6 = 1 \cdot u^6 + 6u^5v + 15u^4v^2 + 20u^3v^3 + 15u^2v^4 + 6uv^5 + v^6$

**46.** $(2x + y)^5 = 1(2x)^5 + 5(2x)^4y + 10(2x)^3y^2 + 10(2x)^2y^3 + 5(2x)y^4 + y^5$

$\qquad = 32x^5 + 80x^4y + 80x^3y^2 + 40x^2y^3 + 10xy^4 + y^5$

**48.** $(4u - 3v)^3 = 1(4u)^3 + 3(4u)^2(-3v) + 3(4u)(-3v)^2 + 1(-3v)^3$

$\qquad = 64u^3 - 144u^2v + 108uv^2 - 27v^3$

**50.** $(x - 4y^3)^4 = 1(x^4) + 4(x^3)(-4y^3) + 6(x^2)(-4y^3)^2 + 4x(-4y^3)^3 + 1(-4y^3)^4$

$\qquad = x^4 - 16x^3y^3 + 96x^2y^6 - 256xy^9 + 256y^{12}$

**52.** $_{12}C_3 = \dfrac{12 \cdot 11 \cdot 10}{3 \cdot 2 \cdot 1} = 220$

Coefficient $= (3)^3 {}_{12}C_3 = 27(220) = 5940$

**54.** $_{14}C_{11} = {}_{14}C_3 = \dfrac{14 \cdot 13 \cdot 12}{3 \cdot 2 \cdot 1} = 364$

Coefficient $= (-3)^{11} {}_{14}C_{11}$

$\qquad = -177,147(364) = -64,481,508$

**56.** $_{10}C_7 = {}_{10}C_3 = \dfrac{10 \cdot 9 \cdot 8}{3 \cdot 2 \cdot 1} = 120$

$120x^7y^3$

Coefficient $= 120$

**58.** $_5C_3 = \dfrac{5 \cdot 4 \cdot 3}{3 \cdot 2 \cdot 1} = 10$

$10(3)^2(-y^3)^3 = -90y^9$

Coefficient $= -90$

**60.** $_6C_4 = {}_6C_2 = \dfrac{6 \cdot 5}{2 \cdot 1} = 15$

$15(2)^2 \left( \dfrac{1}{u} \right)^4 = 60 \left( \dfrac{1}{u^4} \right)$ or $60u^{-4}$

Coefficient $= 60$

**62.** $(2.005)^{10} = (2 + 0.005)^{10}$

$\qquad = (2)^{10} + 10(2)^9(0.005) + 45(2)^8(0.005)^2 + 120(2)^7(0.005)^3 + \cdots$

$\qquad \approx 1024 + 25.6 + 0.288 + 0.00192 + \cdots$

$\qquad \approx 1049.890$

**64.** $(1.98)^9 = (2 - 0.02)^9$

$$= (2)^9 + 9(2)^8(-0.02) + 36(2)^7(-0.02)^2 + 84(2)^6(-0.02)^3 + \cdots$$

$$\approx 512 - 46.08 + 1.8432 - 0.043008$$

$$\approx 467.720$$

**66.** $\left(\frac{2}{3} + \frac{1}{3}\right)^4 = 1\left(\frac{2}{3}\right)^4 + 4\left(\frac{2}{3}\right)^3\left(\frac{1}{3}\right) + 6\left(\frac{2}{3}\right)^2\left(\frac{1}{3}\right)^2 + 4\left(\frac{2}{3}\right)\left(\frac{1}{3}\right)^3 + 1\left(\frac{1}{3}\right)^4$

$$= \frac{16}{81} + \frac{32}{81} + \frac{24}{81} + \frac{8}{81} + \frac{1}{81}$$

**68.** $\left(\frac{2}{5} + \frac{3}{5}\right)^3 = 1\left(\frac{2}{5}\right)^3 + 3\left(\frac{2}{5}\right)^2\left(\frac{3}{5}\right) + 3\left(\frac{2}{5}\right)\left(\frac{3}{5}\right)^2 + 1\left(\frac{3}{5}\right)^3$

$$= \frac{8}{125} + \frac{36}{125} + \frac{54}{125} + \frac{27}{125}$$

**70.** Circle 1: $\begin{vmatrix} 1 & 1 \\ 1 & 2 \end{vmatrix} = 2 - 1 = 1$        Circle 4: $\begin{vmatrix} 4 & 1 \\ 10 & 5 \end{vmatrix} = 20 - 10 = 10$

Circle 2: $\begin{vmatrix} 2 & 1 \\ 3 & 3 \end{vmatrix} = 6 - 3 = 3$        Circle 5: $\begin{vmatrix} 5 & 1 \\ 15 & 6 \end{vmatrix} = 30 - 15 = 15$

Circle 3: $\begin{vmatrix} 3 & 1 \\ 6 & 4 \end{vmatrix} = 12 - 6 = 6$

The difference between each determinant increases by 1.

**72.** The exponents with base $x$ in the expansion of $(x + y)^n$ decreases from $n$ to 0.

**74.** $_{11}C_5 = \dfrac{11!}{5!6!} = \dfrac{11 \cdot 10 \cdot 9 \cdot 8 \cdot 7}{5 \cdot 4 \cdot 3 \cdot 2 \cdot 1}$

**76.** Pascal's Triangle is formed by making the first and last numbers in each row 1. Every other number in the row is formed by adding the two numbers immediately above the number.

# Section 10.5    Counting Principles

**2.** $\{2, 3, 5, 7\}$

4 ways

**4.**

| First number | Second number |
|:---:|:---:|
| 0 | 7 |
| 1 | 6 |
| 2 | 5 |
| 3 | 4 |
| 4 | 3 |
| 5 | 2 |
| 6 | 1 |
| 7 | 0 |

8 ways

**6.**

| First number | Second number |
|:---:|:---:|
| 0 | 7 |
| 1 | 6 |
| 2 | 5 |
| 3 | 4 |
| 4 | 3 |
| 5 | 2 |
| 6 | 1 |
| 7 | 0 |

8 ways

**8.** $\{2, 4, 6, 8, 10, 12, 14, 16, 18, 20\}$

10 ways

**10.** $\{13, 14, 15, 16, 17, 18, 19, 20\}$

8 ways

**12.** $\{6, 12, 18\}$

3 ways

**14.**

| First number | Second number |
|:---:|:---:|
| 1 | 14 |
| 2 | 13 |
| 3 | 12 |
| 4 | 11 |
| 5 | 10 |
| 6 | 9 |
| 7 | 8 |
| 8 | 7 |
| 9 | 6 |
| 10 | 5 |
| 11 | 4 |
| 12 | 3 |
| 13 | 2 |
| 14 | 1 |

14 ways

**16.**

| First number | Sum of second and third numbers | Number of ways to obtain the sum of second and third numbers |
|:---:|:---:|:---:|
| 1 | 14 | 13 |
| 2 | 13 | 12 |
| 3 | 12 | 11 |
| 4 | 11 | 10 |
| 5 | 10 | 9 |
| 6 | 9 | 8 |
| 7 | 8 | 7 |
| 8 | 7 | 6 |
| 9 | 6 | 5 |
| 10 | 5 | 4 |
| 11 | 4 | 3 |
| 12 | 3 | 2 |
| 13 | 2 | 1 |

$13 + 12 + 11 + 10 + 9 + 8 + 7 + 6 + 5 + 4 + 3 + 2 + 1 = 91$ ways

**18.** $3 \cdot 2 \cdot 2 = 12$ ways

**20.** label = letter · number · number

$= 26 \cdot 10 \cdot 10$

$= 2600$ labels

**22.** plate = letter · letter · letter · digit · digit · digit

$= 26 \cdot 26 \cdot 26 \cdot 10 \cdot 10 \cdot 10$

$= 17,576,000$ plates

**24.**

| Front | Seat | Seat | Seat | Seat |
|:---:|:---:|:---:|:---:|:---:|
| seat | 2 | 3 | 4 | 5 |
| 2 | · 4 | · 3 | · 2 | · 1 = 48 ways |

**26.** $3 \cdot 4 \cdot 3 \cdot 2 \cdot 1 = 72$ ways

**28.** *XYZ, XZY, YXZ, YZX, ZXY, ZYX*

**30.** *AB, BA, AC, CA, BC, CB*

**32.** $5! = 5 \cdot 4 \cdot 3 \cdot 2 \cdot 1 = 120$ ways

**34.** $4! = 4 \cdot 3 \cdot 2 \cdot 1 = 24$ ways

**36.** $10^5 = 100,000$ codes

**38.** $8 \cdot 7 \cdot 6 \cdot 5 \cdot 4 = 6720$ ways

**40.** (a) $8! = 8 \cdot 7 \cdot 6 \cdot 5 \cdot 4 \cdot 3 \cdot 2 \cdot 1 = 40,320$ orders

(b) $1 \cdot 7! = 1 \cdot 7 \cdot 6 \cdot 5 \cdot 4 \cdot 3 \cdot 2 \cdot 1 = 5040$ orders

**42.** $_6C_3 = \dfrac{6!}{3!3!} = \dfrac{6 \cdot 5 \cdot 4}{3 \cdot 2 \cdot 1} = 20$ subsets

$\{A, B, C\}$  $\{A, C, D\}$  $\{A, D, F\}$  $\{B, C, F\}$  $\{C, D, E\}$

$\{A, B, D\}$  $\{A, C, E\}$  $\{A, E, F\}$  $\{B, D, E\}$  $\{C, D, F\}$

$\{A, B, E\}$  $\{A, C, F\}$  $\{B, C, D\}$  $\{B, D, F\}$  $\{C, E, F\}$

$\{A, B, F\}$  $\{A, D, E\}$  $\{B, C, E\}$  $\{B, E, F\}$  $\{D, E, F\}$

**44.** $_{30}C_5 = \dfrac{30}{25!5!} = \dfrac{30 \cdot 29 \cdot 28 \cdot 27 \cdot 26}{5 \cdot 4 \cdot 3 \cdot 2 \cdot 1} = 142,056$ ways

**46.** $_{12}C_6 = \dfrac{12!}{6!6!} = \dfrac{12 \cdot 11 \cdot 10 \cdot 9 \cdot 8 \cdot 7}{6 \cdot 5 \cdot 4 \cdot 3 \cdot 2 \cdot 1} = 924$ ways

**48.** $_{10}C_3 = \dfrac{10!}{3!7!} = \dfrac{10 \cdot 9 \cdot 8}{3 \cdot 2 \cdot 1} = 120$ ways

**50.** $_6C_2 = \dfrac{6!}{4!2!} = \dfrac{6 \cdot 5}{2 \cdot 1} = 15$

$2 \cdot {_6C_2} = 2(15) = 30$ ways

**52.** (a) $_8C_3 = 56$

(b) $_8C_2 \cdot {_2C_1} = 28 \cdot 2 = 56$

(c) $_8C_1 \cdot {_2C_2} = 8$

**54.** $_8C_3 = \dfrac{8!}{5!3!} = \dfrac{8 \cdot 7 \cdot 6}{3 \cdot 2 \cdot 1} = 56$ triangles

**56.** Diagonals of pentagon $= {_5C_3} - {_5C_1} = 5$

**58.** Diagonals of octagon $= {_8C_6} - {_8C_1} = 20$

**60.** (a) $_3C_2 = \dfrac{3!}{1!2!} = \dfrac{3 \cdot 2}{2 \cdot 1} = 3$   (b) $_4C_2 = \dfrac{4!}{2!2!} = \dfrac{4 \cdot 3}{2 \cdot 1} = 6$

(c) $_6C_2 = \dfrac{6!}{4!2!} = \dfrac{6 \cdot 5}{2 \cdot 1} = 15$   (d) $_8C_2 = \dfrac{8!}{6!2!} = \dfrac{8 \cdot 7}{2 \cdot 1} = 28$

(e) $_{10}C_2 = \dfrac{10!}{8!2!} = \dfrac{10 \cdot 9}{2 \cdot 1} = 45$   (f) $_{12}C_2 = \dfrac{12!}{10!2!} = \dfrac{12 \cdot 11}{2 \cdot 1} = 66$

**62.** When using the Fundamental Counting Principle, you are counting the number of orderings of a sequence of events.

**64.** Because a different ordering of the six elements selected is a different permutation but the same combination, there are more permutations than combinations.

# Section 10.6    Probability

**2.** $\{2, 3, 4, 5, 6, 7, 8, 9, 10, 11, 12\}$

Number of outcomes $= 11$

**4.** $\{YYY, YYN, YNY, NYY, NNY, NYN, YNN, NNN\}$

Number of outcomes $= 8$

**6.** $\{H1, H2, H3, H4, H5, H6, T1, T2, T3, T4, T5, T6\}$

**8.** $\{AB, AC, AD, BC, BD, CD\}$

**10.** $1 - 0.8 = 0.2$

**12.** $P(E) = 1 - p = 1 - 0.13 = 0.87$

**14.** $P(E) = \dfrac{n(E)}{n(S)} = \dfrac{4}{8} = \dfrac{1}{2}$

**16.** $P(E) = \dfrac{n(E)}{n(S)} = \dfrac{7}{8}$

**18.** $P(E) = \dfrac{n(E)}{n(S)} = \dfrac{4}{52} = \dfrac{1}{13}$

**20.** $P(E) = \dfrac{n(E)}{n(S)} = \dfrac{6}{52} = \dfrac{3}{26}$

**22.** $P(E) = \dfrac{n(E)}{n(S)} = \dfrac{0}{6} = 0$

**24.** $P(E) = \dfrac{n(E)}{n(S)} = \dfrac{6}{6} = 1$

**26.** $P(F) = \dfrac{n(F)}{n(S)} = \dfrac{10}{100} = \dfrac{1}{10}$

$\dfrac{1}{10} = 1 - \dfrac{9}{10}$

($F$ is the event that a person does have type B.)

**28.** $P(E) = \dfrac{n(E)}{n(S)} = \dfrac{35.9}{100} = 0.359$

**30.** $P(E) = \dfrac{n(E)}{n(S)}$

$= \dfrac{14.8 + 24.3}{100}$

$= \dfrac{39.1}{100} = 0.391$

**32.** (a) $P(E) = \dfrac{n(E)}{n(S)} = \dfrac{1}{4}$

(b) $P(E) = \dfrac{n(E)}{n(S)} = \dfrac{1}{2}$

(c) $P(E) = \dfrac{n(E)}{n(S)} = 1$

**34.** Jones & Smith: $P(E) = \dfrac{n(E)}{n(S)} = \dfrac{1}{4}$

Thomas: $P(E) = \dfrac{n(E)}{n(S)} = \dfrac{2}{4} = \dfrac{1}{2}$

**36.** $P(E) = \dfrac{n(E)}{n(S)} = \dfrac{{}_3C_2}{{}_4C_2} = \dfrac{\dfrac{3!}{1!2!}}{\dfrac{4!}{2!2!}} = \dfrac{3}{\dfrac{4 \cdot 3}{2 \cdot 1}} = \dfrac{3}{6} = \dfrac{1}{2}$

**38.** (a) $P(E) = \dfrac{n(E)}{n(S)} = \dfrac{4\pi}{24^2} \approx 0.022$

(b) $P(E) = \dfrac{n(E)}{n(S)} = \dfrac{\pi(10)^2 - \pi(8)^2}{24^2} = \dfrac{100\pi - 64\pi}{576} = \dfrac{36\pi}{576} \approx 0.196$

(c) $P(E) = \dfrac{n(E)}{n(S)} = \dfrac{24^2 - 10^2\pi}{24^2} \approx 0.455$

(d) $P(E) = \dfrac{n(E)}{n(S)} = \dfrac{\pi(6)^2 - \pi(4)^2}{24^2} = \dfrac{36\pi - 16\pi}{576} = \dfrac{20\pi}{576} \approx 0.109$

$E$ is the event that the marble does come to rest in the yellow ring.)

$1 - P(E) = 1 - 0.109 \approx 0.891$

**40.** (a) $P(E) = \dfrac{\text{Area of dime}}{\text{Area of square}} = \dfrac{\pi\left(\dfrac{d}{2}\right)^2}{d^2} = \dfrac{\dfrac{d^2\pi}{4}}{d^2} = \dfrac{\pi}{4}$

(b) Results of experiment will vary. If you multiply the number of times the dime covered a vertex by 4, you will get approximately $\pi$.

**42.** (a)

|   | A | o |
|---|---|---|
| B | AB | Bo |
| o | Ao | oo |

The blood types of the parents were A and B.

(b) Probability of blood type A $= \frac{1}{4}$

Probability of blood type B $= \frac{1}{4}$

Probability of blood type AB $= \frac{1}{4}$

Probability of blood type O $= \frac{1}{4}$

**44.** (a) $P(E) = \dfrac{n(E)}{n(S)} = \dfrac{1}{4 \cdot 3 \cdot 2 \cdot 1} = \dfrac{1}{24}$

(b) $P(E) = \dfrac{n(E)}{n(S)} = \dfrac{1}{3 \cdot 2 \cdot 1} = \dfrac{1}{6}$

**46.** $P(E) = \dfrac{n(E)}{n(S)} = \dfrac{1}{5!} = \dfrac{1}{120}$

**48.** $P(E) = \dfrac{n(E)}{n(S)} = \dfrac{1}{{}_8C_3} = \dfrac{1}{\dfrac{8!}{5!3!}} = \dfrac{1}{\dfrac{8 \cdot 7 \cdot 6}{3 \cdot 2 \cdot 1}} = \dfrac{1}{56}$

**50.** (a) $P(E) = \dfrac{n(E)}{n(S)} = \dfrac{{}_{10}C_3}{{}_{12}C_3} = \dfrac{120}{220} = \dfrac{6}{11}$

(b) $P(E) = \dfrac{n(E)}{n(S)} = \dfrac{{}_2C_1 \cdot {}_{10}C_2 + {}_2C_2 \cdot {}_{10}C_1}{{}_{12}C_3} = \dfrac{100}{220} = \dfrac{5}{11}$

$\left(\text{note that } \tfrac{5}{11} = 1 - \tfrac{6}{11}\right)$

**52.** $P(E) = \dfrac{n(E)}{n(S)} = \dfrac{12}{{}_{52}C_5} = \dfrac{{}_4C_4 \cdot {}_{48}C_1}{\dfrac{52!}{47!5!}} = \dfrac{48}{\dfrac{52 \cdot 51 \cdot 50 \cdot 49 \cdot 48}{5 \cdot 4 \cdot 3 \cdot 2 \cdot 1}} = \dfrac{48}{2,598,960} = \dfrac{1}{54,145}$

**54.** $P(E) = \dfrac{n(E)}{n(S)} = \dfrac{{}_4C_2 \cdot {}_4C_3}{{}_{52}C_5} = \dfrac{\dfrac{4!}{2!2!} \cdot \dfrac{4!}{1!3!}}{\dfrac{52!}{47!5!}} = \dfrac{\dfrac{4 \cdot 3}{2 \cdot 1} \cdot \dfrac{4}{1}}{\dfrac{52 \cdot 51 \cdot 50 \cdot 49 \cdot 48}{5 \cdot 4 \cdot 3 \cdot 2 \cdot 1}} = \dfrac{24}{2,598,960} = \dfrac{1}{108,290} \approx 0.000009$

**56.** The probability of an event must be a real number between 0 and 1, inclusive.

**58.** The sum of the probabilities of all the occurrences of outcomes in a sample space equals 1. By definition, one of the elements of the sample space must occur.

# Review Exercises for Chapter 10

**2.** $a_n = \frac{1}{2}n - 4$

$a_1 = \frac{1}{2}(1) - 4 = \frac{1}{2} - \frac{8}{2} = -\frac{7}{2}$

$a_2 = \frac{1}{2}(2) - 4 = -3$

$a_3 = \frac{1}{2}(3) - 4 = \frac{3}{2} - \frac{8}{2} = -\frac{5}{2}$

$a_4 = \frac{1}{2}(4) - 4 = -2$

$a_5 = \frac{1}{2}(5) - 4 = \frac{5}{2} - \frac{8}{2} = -\frac{3}{2}$

**4.** $a_n = (n + 1)!$

$a_1 = (1 + 1)! = 2! = 2 \cdot 1 = 2$

$a_2 = (2 + 1)! = 3! = 3 \cdot 2 \cdot 1 = 6$

$a_3 = (3 + 1)! = 4! = 4 \cdot 3 \cdot 2 \cdot 1 = 24$

$a_4 = (4 + 1)! = 5! = 5 \cdot 4 \cdot 3 \cdot 2 \cdot 1 = 120$

$a_5 = (5 + 1)! = 6! = 6 \cdot 5 \cdot 4 \cdot 3 \cdot 2 \cdot 1 = 720$

**6.** $3, -6, 9, -12, 15, \ldots$

$a_n = (-1)^{n+1}3n$

**8.** $\dfrac{0}{2}, \dfrac{1}{3}, \dfrac{2}{4}, \dfrac{3}{5}, \dfrac{4}{6}, \ldots$

$a_n = \dfrac{n - 1}{n + 1}$

**10.** (f)      **12.** (e)      **14.** (c)

**16.** $\displaystyle\sum_{k=1}^{4} \dfrac{(-1)^k}{k} = \dfrac{(-1)^1}{1} + \dfrac{(-1)^2}{2} + \dfrac{(-1)^3}{3} + \dfrac{(-1)^4}{4}$

$= -1 + \dfrac{1}{2} - \dfrac{1}{3} + \dfrac{1}{4}$

$= \dfrac{-12 + 6 - 4 + 3}{12}$

$= -\dfrac{7}{12}$

**18.** $\displaystyle\sum_{n=1}^{4}\left(\frac{1}{n}-\frac{1}{n+2}\right)=\left(\frac{1}{1}-\frac{1}{1+2}\right)+\left(\frac{1}{2}-\frac{1}{2+2}\right)+\left(\frac{1}{3}-\frac{1}{3+2}\right)+\left(\frac{1}{4}-\frac{1}{4+2}\right)$

$$=1-\frac{1}{3}+\frac{1}{2}-\frac{1}{4}+\frac{1}{3}-\frac{1}{5}+\frac{1}{4}-\frac{1}{6}$$

$$=1+\frac{1}{2}-\frac{1}{5}-\frac{1}{6}$$

$$=\frac{30+15-6-5}{30}$$

$$=\frac{34}{30}$$

$$=\frac{17}{15}$$

**20.** $\displaystyle\sum_{n=1}^{4}(9-10n)$

**22.** $\displaystyle\sum_{n=0}^{4}\left(-\frac{1}{3}\right)^{n}$

**24.** $d=3$

**26.** $a_1=2(1)+3=2+3=5$
$a_2=2(2)+3=4+3=7$
$a_3=2(3)+3=6+3=9$
$a_4=2(4)+3=8+3=11$
$a_5=2(5)+3=10+3=13$

**28.** $a_1=-\frac{3}{5}(1)+1=-\frac{3}{5}+\frac{5}{5}=\frac{2}{5}$
$a_2=-\frac{3}{5}(2)+1=-\frac{6}{5}+\frac{5}{5}=-\frac{1}{5}$
$a_3=-\frac{3}{5}(3)+1=-\frac{9}{5}+\frac{5}{5}=-\frac{4}{5}$
$a_4=-\frac{3}{5}(4)+1=-\frac{12}{5}+\frac{5}{5}=-\frac{7}{5}$
$a_5=-\frac{3}{5}(5)+1=-3+1=-2$

**30.** $a_1=12$
$a_2=12+1.5=13.5$
$a_3=13.5+1.5=15$
$a_4=15+1.5=16.5$
$a_5=16.5+1.5=18$

**32.** $a_1=25$
$a_2=25-6=19$
$a_3=19-6=13$
$a_4=13-6=7$
$a_5=7-6=1$

**34.** $a_1=a_1+(n-1)d$
$a_n=32+(n-1)(-2)$
$a_n=32-2n+2$
$a_n=34-2n$

**36.** $a_n=a_1+(n-1)d$
$a_n=12+(n-1)(8)$
$a_n=12+8n-8$
$a_n=4+8n$

**38.** $\displaystyle\sum_{k=1}^{10}(100-10k)=10\left(\frac{100-10+100-100}{2}\right)=10\left(\frac{90}{2}\right)=450$

**40.** $\displaystyle\sum_{j=1}^{50}\frac{3j}{2}=50\left(\frac{\frac{3}{2}+\frac{150}{2}}{2}\right)=50\left(\frac{153}{2}\cdot\frac{1}{2}\right)=50\left(\frac{153}{4}\right)=\frac{3825}{2}=1912.5$

**42.** $\displaystyle\sum_{i=1}^{100}(5000-3.5i)=482{,}325$

*Keystrokes:* [LIST] [MATH 5] [LIST] [OPS 5] 5000 [−] 3.5 [X,T,θ] [ , ] [X,T,θ] [ , ] 1 [ , ] 100 [ , ] 1 [ ) ] [ENTER]

**44.** $r=-\frac{2}{3}$

**46.** $a_n=a_1r^{n-1}$
$a_n=2(-5)^{n-1}$
$a_1=2(-5)^{1-1}=2(-5)^0=2(1)=2$
$a_2=2(-5)^{2-1}=2(-5)^1=-10$
$a_3=2(-5)^{3-1}=2(-5)^2=2(25)=50$
$a_4=2(-5)^{4-1}=2(-5)^3=2(-125)=-250$
$a_5=2(-5)^{5-1}=2(-5)^4=2(625)=1250$

**48.** $a_n = a_1 r^{n-1}$

$a_n = 12\left(\frac{1}{6}\right)^{n-1}$

$a_1 = 12\left(\frac{1}{6}\right)^{1-1} = 12\left(\frac{1}{6}\right)^0 = 12(1) = 12$

$a_2 = 12\left(\frac{1}{6}\right)^{2-1} = 12\left(\frac{1}{6}\right) = 2$

$a_3 = 12\left(\frac{1}{6}\right)^{3-1} = 12\left(\frac{1}{6}\right)^2 = 12\left(\frac{1}{36}\right) = \frac{1}{3}$

$a_4 = 12\left(\frac{1}{6}\right)^{4-1} = 12\left(\frac{1}{6}\right)^3 = 12\left(\frac{1}{216}\right) = \frac{1}{18}$

$a_5 = 12\left(\frac{1}{6}\right)^{5-1} = 12\left(\frac{1}{6}\right)^4 = 12\left(\frac{1}{1296}\right) = \frac{1}{108}$

**50.** $a_1 = 36$

$a_2 = \frac{1}{2}(36) = 18$

$a_3 = \frac{1}{2}(18) = 9$

$a_4 = \frac{1}{2}(9) = \frac{9}{2}$

$a_5 = \frac{1}{1}\left(\frac{9}{2}\right) = \frac{9}{4}$

**52.** $a_n = a_1 r^{n-1}$

$a_n = 100(1.07)^{n-1}$

**54.** $a_n = a_1 r^{n-1}$

$a_n = 16\left(-\frac{1}{4}\right)^{n-1}$

**56.** $a_n = a_1 r^{n-1}$

$a_n = 3\left(\frac{1}{3}\right)^{n-1}$

**58.** $\displaystyle\sum_{n=1}^{12} (-2)^n = (-2)^1\left(\frac{(-2)^{12} - 1}{-2 - 1}\right)$

$= -2\left(\frac{4096 - 1}{-3}\right)$

$= -2\left(\frac{4095}{-3}\right)$

$= -2(-1365)$

$= 2730$

**60.** $\displaystyle\sum_{k=1}^{10} 4\left(\frac{3}{2}\right)^k = 6\left(\frac{\left(\frac{3}{2}\right)^{10} - 1}{\frac{3}{2} - 1}\right) \approx 679.98$

**62.** $\displaystyle\sum_{i=1}^{8} (-1.25)^{i-1} = \left(\frac{(-1.25)^8 - 1}{-1.25 - 1}\right) \approx -2.205$

**64.** $\displaystyle\sum_{n=1}^{40} 1000(1.1)^n = 1100\left(\frac{1.1^{40} - 1}{1.1 - 1}\right) \approx 486,851.81$

**66.** $\displaystyle\sum_{i=1}^{\infty} \left(\frac{1}{3}\right)^{i-1} = \frac{1}{1 - \frac{1}{3}} = \frac{1}{\frac{2}{3}} = \frac{3}{2}$

**68.** $\displaystyle\sum_{k=1}^{\infty} 1.3\left(\frac{1}{10}\right)^{k-1} = \frac{1.3}{1 - \frac{1}{10}} = \frac{1.3}{1 - \frac{1}{10}} = \frac{13}{9}$

**70.** $\displaystyle\sum_{j=1}^{60} 25(0.9)^{j-1} \approx 249.551$

*Keystrokes:* $\boxed{\text{LIST}}$ $\boxed{\text{MATH 5}}$ $\boxed{\text{LIST}}$ $\boxed{\text{OPS 5}}$ 25 $\boxed{(}$ 0.9 $\boxed{)}$ $\boxed{\wedge}$ $\boxed{(}$ $\boxed{\text{X,T,}\theta}$ $\boxed{-}$ 1 $\boxed{)}$ $\boxed{,}$ $\boxed{\text{X,T,}\theta}$ $\boxed{,}$ 1 $\boxed{,}$ 60 $\boxed{,}$ 1 $\boxed{)}$ $\boxed{\text{ENTER}}$

**72.** ${}_{12}C_2 = \frac{12!}{10!2!} = \frac{12 \cdot 11 \cdot 10!}{10! \cdot 2 \cdot 1} = 66$

**74.** ${}_{100}C_1 = \frac{100!}{99!1!} = \frac{100 \cdot 99!}{99! \cdot 1} = 100$

**76.** *Keystrokes:*

15 $\boxed{\text{MATH}}$ $\boxed{\text{PRB 3}}$ 9 $\boxed{\text{ENTER}}$   ${}_{15}C_9 = 5005$

**78.** *Keystrokes:*

32 $\boxed{\text{MATH}}$ $\boxed{\text{PRB 3}}$ 2 $\boxed{\text{ENTER}}$   ${}_{32}C_2 = 496$

**80.** $(y - 2)^6 = 1y^6 - 6y^5(2) + 15y^4(2)^2 - 20y^3(2)^3 + 15y^2(2)^4 - 6y(2)^5 + 1(2)^6$

$= y^6 - 12y^5 + 60y^4 - 160y^3 + 240y^2 - 192y + 64$

**82.** $(2u + 5v)^4 = 1(2u)^4 + 4(2u)^3(5v) + 6(2u)^2(5v)^2 + 4(2u)(5v)^3 + 1(5v)^4$

$= 16u^4 + 160u^3v + 600u^2v^2 + 1000uv^3 + 625v^4$

**84.** $(x^4 - y^5)^8 = 1(x^4)^8 + 8(x^4)^7(-y^5) + 28(x^4)^6(-y^5)^2 + 56(x^4)^5(-y^5)^3 + 70(x^4)^4(-y^5)^4 + 56(x^4)^3(-y^5)^5$

$+ 28(x^4)^2(-y^5)^6 + 8(x^4)(-y^5)^7 + (-y^5)^8$

$= x^{32} - 8x^{28}y^5 + 28x^{24}y^{10} - 56x^{20}y^{15} + 70x^{16}y^{20} - 56x^{12}y^{25} + 28x^8y^{30} - 8x^4y^{35} + y^{40}$

**86.** $(x + 4)^9$, $x^6$ term

$_nC_r x^{n-r} y^r$ so $n = 9$, $n - r = 6$, $r = 3$, $x = x$, $y = 4$.

$_9C_3 = \dfrac{9 \cdot 8 \cdot 7}{3 \cdot 2 \cdot 1} = 84$

$84(4)^3 = 84 \cdot 64 = 5376$

**88.**    $_5C_2 = \dfrac{5!}{3!2!} = \dfrac{5 \cdot 4 \cdot 3!}{3! \cdot 2 \cdot 1} = 10$

$10(2x)^2(-3y)^3 = -1080x^2y^3$

Coefficient $= -1080$

**90.** $\displaystyle\sum_{n=1}^{76} (n + 224) = 76\left(\dfrac{225 + 300}{2}\right) = 19{,}950$

**92.** (a) $a_n = a_1 r^n$

$a_n = 120{,}000(0.7)^n$

(b) $a_5 = 120{,}000(0.7)^5 = \$20{,}168.40$

**94.**

$a_n = a_1 r^{n-1}$

$a_n = 32{,}000(4.055)^{n-1}$

$\displaystyle\sum_{n=1}^{40} 32{,}000(1.055)^{n-1} = 32{,}000\left(\dfrac{1.055^{40} - 1}{1.0550 - 1}\right) \approx \$4{,}371{,}379.65$

**96.** $(0,8)$, $(8, 0)$, $(1, 7)$, $(7, 1)$, $(2, 6)$, $(6, 2)$, $(3, 5)$, $(5, 3)$

**98.** $7! = 5040$

**100.** $P(E) = \dfrac{n(E)}{n(S)} = \dfrac{15}{2 \cdot 2 \cdot 2 \cdot 2} = \dfrac{15}{16}$

**102.** $P(E) = \dfrac{n(E)}{n(S)} = \dfrac{1}{6}$

($E$ is the event of rolling a 3 with one six-sided die.)

$P(F) = \dfrac{n(F)}{n(S)} = \dfrac{5}{6 \cdot 6} = \dfrac{5}{36}$

($F$ is the event of rolling a total of 6 with two six-sided dice.)

| First roll | Second roll |
|---|---|
| 1 | 5 |
| 2 | 4 |
| 3 | 3 |
| 4 | 2 |
| 5 | 1 |

5 ways

No. Since $\frac{1}{6} > \frac{5}{36}$, rolling a 3 with one six-sided die has the greater probability of occurring.

**104.** (a) $P(E) = \dfrac{n(E)}{n(S)} = \dfrac{\pi(2)^2}{\pi(10)^2} = \dfrac{4\pi}{100\pi} = \dfrac{1}{25}$

($E$ is the event in which the arrow hits the bull's eye.)

(b) $P(F) = \dfrac{n(F)}{n(S)} = \dfrac{\pi(8)^2 - (6)^2}{\pi(10)^2}$

$= \dfrac{64\pi - 36\pi}{100\pi} = \dfrac{28\pi}{100\pi} = \dfrac{7}{25}$

($F$ is the event in which the arrow hits the blue ring.)

# APPENDICES

# Appendix A    Introduction to Graphing Utilities

**Solutions to Even-Numbered Exercises**

**2.** $y = x - 4$

*Keystrokes:*

Y= X,T,θ − 4 GRAPH

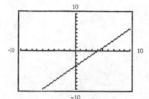

**4.** $y = -3x + 2$

*Keystrokes:*

Y= (−) 3 X,T,θ + 2 GRAPH

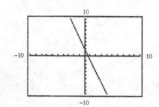

**6.** $y = -\frac{2}{3}x^2$

*Keystrokes:*

Y= (−) ( 2 ÷ 3 ) X,T,θ x² GRAPH

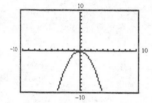

**8.** $y = -0.5x^2 - 2x + 2$

*Keystrokes:*

Y= (−) 0.5 X,T,θ x² − 2 X,T,θ + 2 GRAPH

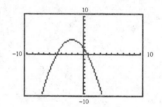

**10.** $y = |x + 4|$

*Keystrokes:*

Y= ABS ( X,T,θ + 4 ) GRAPH

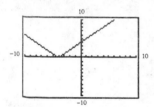

**12.** $y = |x - 2| - 5$

*Keystrokes:*

Y= ABS ( X,T,θ − 2 ) − 5 GRAPH

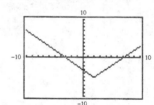

**14.** $y = 50,000 - 6000x$

*Keystrokes:*

Y= 50,000 − 6000 X,T,θ GRAPH

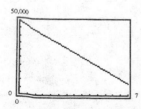

**16.** $y = 100 - 0.5|x|$

*Keystrokes:*

Y= 100 − 0.5 ABS X,T,θ GRAPH

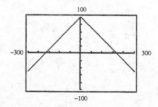

**18.** $y = 15 + (x - 12)^2$

*Keystrokes:*

$\boxed{Y=}$ 15 $\boxed{+}$ $\boxed{(}$ $\boxed{X,T,\theta}$ $\boxed{-}$ 12 $\boxed{)}$ $\boxed{x^2}$ $\boxed{GRAPH}$

```
Xmin= 8
Xmax= 16
Xscl= 1
Ymin= 14
Ymax= 22
Yscl= 1
```

**20.** $y = -15 + (x + 12)^2$

*Keystrokes:*

$\boxed{Y=}$ $\boxed{(-)}$ 15 $\boxed{+}$ $\boxed{(}$ $\boxed{X,T,\theta}$ $\boxed{+}$ 12 $\boxed{)}$ $\boxed{x^2}$ $\boxed{GRAPH}$

```
Xmin= -18
Xmax= -6
Xscl= 1
Ymin= -16
Ymax= 10
Yscl= 1
```

**22.** $y_1 = \frac{1}{2}(3 - 2x)$

$y_2 = \frac{3}{2} - x$

*Keystrokes:*

$y_1$ $\boxed{Y=}$ $\boxed{(}$ 1 $\boxed{\div}$ 2 $\boxed{)}$ $\boxed{(}$ 3 $\boxed{-}$ 2 $\boxed{X,T,\theta}$ $\boxed{)}$ $\boxed{ENTER}$

$y_2$ 3 $\boxed{\div}$ 2 $\boxed{-}$ $\boxed{X,T,\theta}$ $\boxed{GRAPH}$

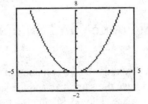

Distributive Property

**24.** $y_1 = x(0.5x)$

$y_2 = (0.5x)x$

*Keystrokes:*

$y_1$ $\boxed{Y=}$ $\boxed{X,T,\theta}$ $\boxed{(}$ 0.5 $\boxed{X,T,\theta}$ $\boxed{)}$ $\boxed{ENTER}$

$y_2$ $\boxed{(}$ 0.5 $\boxed{X,T,\theta}$ $\boxed{)}$ $\boxed{X,T,\theta}$ $\boxed{GRAPH}$

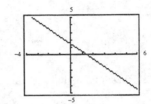

Commutative Property of Multiplication

**26.** $y = 3x^2 - 2x - 5$

*Keystrokes:*

$\boxed{Y=}$ 3 $\boxed{X,T,\theta}$ $\boxed{x^2}$ $\boxed{-}$ 2 $\boxed{X,T,\theta}$ $\boxed{-}$ 5 $\boxed{GRAPH}$

Trace to $x$-intercepts: $(-1, 0)$ and $\left(\frac{5}{3}, 0\right)$

Trace to $y$-intercept: $(0, -5)$

**28.** $y = |x - 2|^2 - 3$

*Keystrokes:*

$\boxed{Y=}$ $\boxed{ABS}$ $\boxed{(}$ $\boxed{X,T,\theta}$ $\boxed{-}$ 2 $\boxed{)}$ $\boxed{x^2}$ $\boxed{-}$ 3 $\boxed{GRAPH}$

Trace to $x$-intercepts: $(0.27, 0)$ and $(3.73, 0)$

Trace to $y$-intercept: $(0, 1)$

**30.** $y = 4 - |x|$

*Keystrokes:*

$\boxed{Y=}$ 4 $\boxed{-}$ $\boxed{ABS}$ $\boxed{X,T,\theta}$ $\boxed{GRAPH}$

Trace to $x$-intercepts: $(-4, 0)$ and $(4, 0)$

Trace to $y$-intercept: $(0, 4)$

**32.** $y = x^3 - 4x$

*Keystrokes:*

$\boxed{Y=}$ $\boxed{X,T,\theta}$ $\boxed{\wedge}$ 3 $\boxed{-}$ 4 $\boxed{X,T,\theta}$ $\boxed{GRAPH}$

Trace to $x$-intercepts: $(0, 0)$, $(-2, 0)$, $(2, 0)$

Trace to $y$-intercept: $(0, 0)$

**34.** $y = |x|$    $y = 5$

*Keystrokes:*

$y_1$ [Y=] [ABS] [X,T,θ] [ENTER]

$y_2$ 5 [GRAPH]

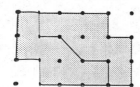

Triangle

**36.** $y = -\frac{1}{2}x + 7,\ y = \frac{8}{3}(x + 5),\ y = \frac{2}{7}(3x - 4)$

*Keystrokes:*

$y_1$ [Y=] [(-)] [(] 1 [÷] 2 [)] [X,T,θ] [+] 7 [ENTER]

$y_2$ [(] 8 [÷] 3 [)] [(] [X,T,θ] [+] 5 [)] [ENTER]

$y_3$ [(] 2 [÷] 7 [)] [(] 3 [X,T,θ] [−] 4 [)] [GRAPH]

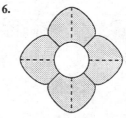

Triangle

**38.** (a) Increasing    (b) Increasing. Revenues from periodicals are not increasing as rapidly as revenues from first-class mail.

# Appendix B    Further Concepts in Geometry

## Appendix B.1    Exploring Congruence and Similarity

**2.** Answers will vary.

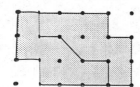

**4.** Two figures are congruent if they have the same size and the same shape. Figures (a) and (c) are congruent.

**6.**

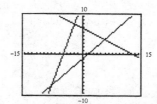

**8.** The grid contains 16 congruent triangles with 1-unit sides.

**10.** The grid contains 3 congruent triangles with 3-unit sides.

**12.** True.
Congruent figures are the same size and the same shape, and any two figures with the same shape are similar.

**14.** False.
A triangle and a square are not the same shape.

**16.** The line through $P$ and $Q$ is matched with notation (c).

**18.** The segment between $P$ and $Q$ is matched with notation (a).

**20.** Statements (a), (b), and (c) are true.

**32.** $m\angle 1 = 80°$ by the Alternating Exterior Angles Theorem.

$m\angle 1 + m\angle 2 = 180°$ because two angles that form a linear pair are supplementary.

$80° + m\angle 2 = 180°$

$m\angle 2 = 100°$

**34.** $a + 100° = 180°$ because two angles that form a linear pair are supplementary.

$a = 80°$

$100° = b - 10°$ because corresponding angles are congruent.

$110° = b$

**36.**    $b = 2b - 30°$ because corresponding angles are congruent.

$b - 2b = -30°$

$-b = -30°$

$b = 30°$

$a + b = 180°$ by the Consecutive Interior Angles Theorem.

$a + 30° = 180°$

$a = 150°$

**38.** $\angle 1$, $\angle 3$, $\angle 4$, $\angle 6$, and $\angle 8$ are the exterior angles of the triangle. (When the sides of the triangles are extended, these are angles which are adjacent to the original three angles of the triangle.

**40.** $m\angle D = 105°$ because corresponding angles of congruent triangles are congruent.

**42.** $m\angle A + m\angle B + m\angle C = 180°$

$105° + 35° + m\angle C = 180°$

$140 + m\angle C = 180°$

$m\angle C = 40°$

**44.** False.

The sum of the measures of the three angles of a triangles is 180°. The measure of the right angle is 90°, so the sum of the measures of the other two angles is $180° - 90° = 90°$. Thus, neither of these other two angles could be an obtuse angle because the measure of an obtuse angle is more than 90°.

**46.** True.

The sum of the measures of the three angles of a triangle is 180°. The measure of the right angle is 90°, so the sum of the measures of the other two angles is $180° - 90° = 90°$. If these two angles are congruent, then both angles must be 45° angles.

**48.** *Step 1:* $m\angle 2 + 125° = 180°$ because two angles that form a linear pair are supplementary.

*Step 2:* $m\angle 1 + 55° + 50° = 180°$ because the sum of the measures of the three angles of a triangle is 180°.

$m\angle 1 + 105° = 180°$

$m\angle 1 = 75°$

*Step 3:* $m\angle 3 = 55°$ because vertical angles are congruent.

*Step 4:* $m\angle 4 + 55° + 85° = 180°$ because the sum of the measures of the three angles of a triangle is 180°.

$m\angle 4 + 140° = 180°$

$m\angle 4 = 40°$

**—CONTINUED—**

**48.** **—CONTINUED—**

*Step 5:* $m\angle 5 + 40° = 180°$ because two angles that form a linear pair are supplementary.

$$m\angle 5 = 140°$$

*Step 6:* $m\angle 6 = 40°$ because vertical angles are congruent.

*Step 7:* $m\angle 8 = 65°$ because vertical angles are congruent.

*Step 8:* $m\angle 7 + 40° + 65° = 180°$ because the sum of the measures of the three angles of a triangle is 180°.

$$m\angle 7 + 105° = 180°$$

$$m\angle 7 = 75°$$

*Step 9:* $m\angle 9 + 65° = 180°$ because two angles that form a linear pair are supplementary.

$$m\angle 9 = 115°$$

**50.**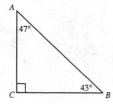

$m\angle A + m\angle B + m\angle C = 180°$ because the sum of the measures of the three interior angles of a triangle is 180°.

$$47° + m\angle B + 90° = 180°$$

$$m\angle B + 137° = 180°$$

$$m\angle B = 43°$$

**52.**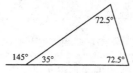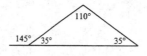

**54.** The sum of the measures of the three interior angles of a triangle is 180°.

$$x + (x + 25°) + (x - 25°) = 180°$$

$$x + x + 25° + x - 25° = 180°$$

$$3x = 180°$$

$$x = 60°$$

$$x + 25° = 60° + 25° = 85°$$

$$x - 25° = 60° - 25° = 35°$$

The measures of the three interior angles are 35°, 60°, and 85°.

**56.** The sum of the measures of the three interior angles of a triangle is 180°.

$$2x + (3x - 10°) + (110° - x) = 180°$$

$$2x + 3x - 10° + 110° - x = 180°$$

$$4x + 100° = 180°$$

$$4x = 80°$$

$$x = 20°$$

$$2x = 2(20°) = 40°$$

$$3x - 10° = 3(20°) - 10° = 50°$$

$$110° - x = 110° - 20° = 90°$$

The measures of the three interior angles are 40°, 50°, and 90°.

# Appendix C   Further Concepts in Statistics

**2.**

| Stems | Leaves |
|-------|--------|
| 0 | 62  65  66  67  80  89  93  98 |
| 1 | 01  09  24  46  90  96 |
| 2 | 40  55  61  92 |
| 3 | 35  68 |
| 4 | 12  38  80  96 |
| 5 | 18  50  66  70  81 |
| 6 | 00  01  54  44 |
| 7 | 00  61  66 |
| 8 | 11  41  57  90 |
| 9 | |
| 10 | |
| 11 | 33  59  60 |
| 12 | 92 |
| 13 | 17  19  37 |
| 27 | 22 |
| 31 | 32 |
| 46 | 80 |
| 65 | 14 |

**4.**

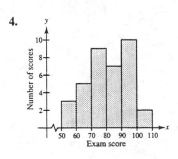

**6.** Organize the data by using a stem-and-leaf plot.

| Stems | Leaves |
|-------|--------|
| 1 | 3.0  3.0  5.7  8.7  8.9  9.1 |
| 2 | 1.1  1.3  1.5  1.6  1.8  1.8  3.3  3.8  5.8  6.2  9.0  9.2 |
| 3 | 0.7  1.0  2.1  2.3  3.6  4.4  8.0  9.8 |
| 4 | 2.1  4.6  7.5  9.0 |

**8.**

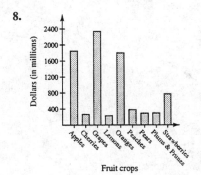

**10.** The amount of incinerated garbage in 1990 is approximately 30 million tons.

**12.** The amount of incinerated garbage decreased from 1960 to 1986.

**14.** Landfill garbage is decreasing because more is being recycled.

**16.**

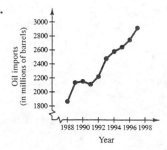

**18.**

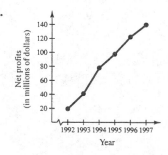

**20.**

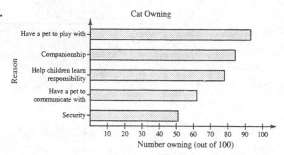

**22.** The scatter plot shows only 28 points because there is some duplication.

**24.** A player can have more runs batted in than hits because one hit can produce up to four runs.

**26.** A student's study time and test scores should have a positive correlation. As study time increases test scores should also increase.

**28.** A student's height and test scores have no correlation.

**30.** *A* and *P* have a negative correlation.

**32.** 28,000 feet is approximately the altitude at which the air pressure is 5.0 pounds per square foot.

**34.** Because *y* tends to increase as *x* increases, the points are positively correlated.

**36.** No, the model is not accurate for large values of *x*.

**38.** Because *v* tends to decrease as *h* increases, the points are negatively correlated.

**40.** Because the model yields $v = 830.3$ feet per second when $h = 70$, we see that the model is inaccurate for large values of *h*.

**42.** $y = 2.2286x + 43.0857$

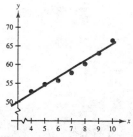

Use graphing utility by entering data in 2 lists.

With ⌈STAT PLOT⌉ graph.

Find regression line with ⌈STAT⌉ ⌈CALC 4⌉.

**44.** $y = -6.0512x + 148.0349$

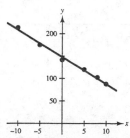

Use graphing utility by entering data in 2 lists.

With ⌈STAT PLOT⌉ graph.

Find regression line with ⌈STAT⌉ ⌈CALC 4⌉.

**46.** (a) $S = 384.1 + 21.2x$; $479.50

(b)

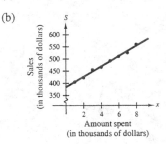

Use graphing utility by entering data in 2 lists.

With $\boxed{\text{STAT PLOT}}$ graph.

Find regression line with $\boxed{\text{STAT}}$ $\boxed{\text{CALC 4}}$.

(c) $r \approx 0.996$

**50.** 20, 37, 32, 39, 33, 34, 32

Mean: $\dfrac{20 + 37 + 32 + 39 + 33 + 34 + 32}{7} = 32.43$

Median: 39  37  34  33  32  32  20 = 33

middle score

Mode: 32 occurs twice

**52.** 410, 260, 320, 320, 460, 150

Mean: $\dfrac{410 + 260 + 320 + 320 + 460 + 150}{6} = 320$

Median: 460  410  320  320  260  150 = 320

average of 320 and 320

Mode: 320 occurs twice

**56.** Mean = 6,   Median = 6,   Mode = 4

Answers may vary. One possibility: $\{4, 4, 6, 7.5, 8.5\}$

Mean: $\dfrac{4 + 4 + 6 + 7.5 + 8.5}{5} = 6$

**58.** $10\frac{1}{2}$, 8, 12, 10, $9\frac{1}{2}$, 11, $10\frac{1}{2}$

Mean: $\dfrac{10\frac{1}{2} + 8 + 12 + 10 + 9\frac{1}{2} + 11 + 10\frac{1}{2}}{7} = \dfrac{7.5}{7} \approx 10.21$

Median: 12  11  $10\frac{1}{2}$  $10\frac{1}{2}$  10  $9\frac{1}{2}$  8 = $10\frac{1}{2}$

middle score

Mode: $10\frac{1}{2}$ occurs twice = $10\frac{1}{2}$

The median and mode give the most representative descriptions.

**48.** 30, 37, 32, 39, 33, 34, 32

Mean: $\dfrac{30 + 37 + 32 + 39 + 33 + 34 + 32}{7} = 33.86$

Median: 39  37  34  33  32  32  30 = 33

middle score

Mode: 32 occurs twice

**54.** (a) Mean or Average number of hits per game =

$\dfrac{0(14) + 1(26) + 2(7) + 3(2) + 4(1)}{50} = \dfrac{50}{50} = 1$

(b) Batting Average = $\dfrac{\text{Number of hits}}{\text{Number at bats}} = \dfrac{50}{200} = 0.250$

# Appendix D    Introduction to Logic

## Appendix D.1    Statements and Truth Tables

**2.** Can I help you?

Nonstatement, because no truth value can be assigned.

**4.** Substitute 4 for $x$.

Nonstatement, because no truth value can be assigned.

**6.** 8 is larger than 4.

Statement, because only one truth value can be assigned.

**8.** $12 + 3 = 14$

Statement, because only one truth value can be assigned.

**10.** One mile is greater than 1 kilometer.

Statement, because only one truth value can be assigned.

**12.** Come to the party.

Nonstatement, because no truth value can be assigned.

**14.** $x^2 - x - 6 = 0$

(a) $x = 2$

(b) $x = -2$

(a) $2^2 - 2 - 6 \overset{?}{=} 0$

$4 - 8 = 0$

$-4 \neq 0$ False

(b) $(-2)^2 - (-2) - 6 \overset{?}{=} 0$

$4 + 2 - 6 = 0$

$0 = 0$ True

**16.** $|x - 3| = 4$

(a) $x = -1$

(b) $x = 7$

(a) $|-1 - 3| \overset{?}{=} 4$

$4 = 4$ True

(b) $|7 - 3| \overset{?}{=} 4$

$4 = 4$ True

**18.** $\sqrt{x^2} = x$

(a) $x = 3$

(b) $x = -3$

(a) $\sqrt{3^2} \overset{?}{=} 3$

$3 = 3$ True

(b) $\sqrt{(-3)^2} \overset{?}{=} -3$

$3 \neq -3$ False

**20.** $\sqrt[3]{x} = -2$

(a) $x = 8$

(b) $x = -8$

(a) $\sqrt[3]{8} \overset{?}{=} -2$

$2 \neq -2$ False

(b) $\sqrt[3]{-8} = -2$

$-2 = -2$ True

**22.** $p$: The car has a radio.    $q$: The car is red.

(a) $\sim p$: The car does not have a radio.

(b) $\sim q$: The car is not red.

(c) $p \wedge q$: The car has a radio and the car is red.

(d) $p \vee q$: The car has a radio or the car is red.

**24.** $p$: Twelve is less than 15.    $q$: Seven is a prime number.

(a) $\sim p$: Twelve is not less than 15.

(b) $\sim q$: Seven is not a prime number.

(c) $p \wedge q$: Twelve is less than 15 and seven is a prime number.

(d) $p \vee q$: Twelve is less than 15 or seven is a prime number.

**26.** $p$: The car has a radio.    $q$: The car is red.

(a) $\sim p \wedge q$: The car does not have a radio and the car is red.

(b) $\sim p \vee q$: The car does not have a radio or the car is red.

(c) $p \wedge \sim q$: The car has a radio and the car is not red.

(d) $p \vee \sim q$: The car has a radio or the car is not red.

**28.** $p$: Twelve is less than 15.    $q$: Seven is a prime number.

(a) $\sim p \wedge q$: Twelve is not less than 15 and seven is a prime number.

(b) $\sim p \vee q$: Twelve is not less than 15 or seven is a prime number.

(c) $p \wedge \sim q$: Twelve is less than 15 and seven is not a prime number.

(d) $p \vee \sim q$: Twelve is less than 15 or seven is not a prime number.

**30.** $p$: It is four o'clock.    $q$: It is time to go home.

It is not four o'clock or it is not time to go home.

$\sim p \vee \sim q$

**32.** $p$: It is four o'clock.    $q$: It is time to go home.

It is four o'clock and it is time to go home.

$p \wedge q$

**34.** *p:* The dog has fleas.   *q:* The dog is scratching.

The dog has fleas and the dog is scratching.

$$p \wedge q$$

**36.** *p:* The dog has fleas.   *q:* The dog is scratching.

The dog has fleas or the dog is not scratching.

$$p \vee \sim q$$

**38.** Statement:  Frank is not 6 feet tall.

Negation:  Frank is 6 feet tall.

**40.** Statement:  *x* is not equal to 4.

Negation:  *x* is equal to 4.

**42.** Statement:  The earth is flat.

Negation:  The earth is not flat.

**44.**

| *p* | *q* | *~p* | *~p* $\vee$ *q* |
|---|---|---|---|
| T | T | F | T |
| T | F | F | F |
| F | T | T | T |
| F | F | T | T |

**46.**

| *p* | *q* | *~p* | *~q* | *~p* $\vee$ *q* |
|---|---|---|---|---|
| T | T | F | F | F |
| T | F | F | T | F |
| F | T | T | F | F |
| F | F | T | T | T |

**48.**

| *p* | *q* | *~q* | *p* $\vee$ *~q* |
|---|---|---|---|
| T | T | F | F |
| T | F | T | T |
| F | T | F | F |
| F | F | T | F |

**50.**

| *p* | *q* | *~q* | *p* $\wedge$ *~q* | *~(p* $\wedge$ *q)* | *~p* | *~p* $\vee$ *q* |
|---|---|---|---|---|---|---|
| T | T | F | F | T | F | T |
| T | F | T | T | F | F | F |
| F | T | F | F | T | T | T |
| F | F | T | F | T | T | T |

identical
logically equivalent

**52.**

| *p* | *q* | *p* $\vee$ *q* | *~(p* $\vee$ *q)* | *~p* | *~q* | *~p* $\vee$ *~q* |
|---|---|---|---|---|---|---|
| T | T | T | F | F | F | F |
| T | F | T | F | F | T | T |
| F | T | T | F | T | F | F |
| F | F | F | T | T | T | T |

not identical
not logically equivalent

**54.**

| *p* | *q* | *~q* | *p* $\wedge$ *~q* | *~p* | *(~p* $\wedge$ *~q)* | *~(~p* $\wedge$ *q)* |
|---|---|---|---|---|---|---|
| T | T | F | F | F | F | T |
| T | F | T | T | F | F | T |
| F | T | F | F | T | T | F |
| F | F | T | F | T | F | T |

not identical
not logically equivalent

**56. (a)** It is not true that the tree is not green.

**(b)** The tree is green.       $(a) = \sim(\sim p)$

let *p* = the tree is green.   $(b) = p$

| *p* | *~p* | *~(~p)* |
|---|---|---|
| T | F | T |
| T | F | T |
| F | T | F |
| F | T | F |

identical
logically equivalent

**58.** (a) I am not 25 years old and I am not applying for this job.

  (b) The statement that I am 25 years old and am applying for this job is not true.

  let $p$ = I am 25 years old    $q$ = I am applying for this job.

  (a) $\sim p \wedge \sim q$    (b) $\sim(p \wedge q)$

| $p$ | $q$ | $\sim q$ | $\sim p$ | $\sim p \wedge \sim q$ | $p \wedge q$ | $\sim(p \wedge q)$ |
|-----|-----|----------|----------|------------------------|--------------|--------------------|
| T | T | F | F | F | T | F |
| T | F | F | T | F | F | T |
| F | T | T | F | F | F | T |
| F | F | T | T | T | F | T |

not identical
not logically equivalent

**60.**

| $p$ | $\sim p$ | $\sim p \vee q$ |
|-----|----------|-----------------|
| T | F | T |
| T | F | T |
| F | T | T |
| F | T | T |

A tautology

**62.**

| $p$ | $\sim p$ | $\sim(\sim p)$ | $\sim(\sim p) \wedge \sim p$ |
|-----|----------|----------------|------------------------------|
| T | F | T | F |
| T | F | T | F |
| F | T | F | F |
| F | T | F | F |

Not a tautology

# Appendix D.2    Implications, Quantifiers, and Venn Diagrams

**2.** *p:* The statement is at school.    *q:* It is nine o'clock.

  (a) $p \longrightarrow q$: If the student is at school, then it is nine o'clock.

  (b) $q \longrightarrow q$: If it is nine o'clock, then the student is at school.

  (c) $\sim q \longrightarrow \sim p$: If it is not nine o'clock, then the student is not at school.

  (d) $p \longrightarrow \sim q$: If the student is at school, then it is not nine o'clock.

**4.** *p:* The person is generous.    *q:* The person is rich.

  (a) $p \longrightarrow q$: If the person is generous, then the person is rich.

  (b) $q \longrightarrow p$: If the person is rich, then the person is generous.

  (c) $\sim q \longrightarrow \sim p$: If the person is not rich, then the person is not generous.

  (d) $p \longrightarrow \sim q$: If the person is generous, then the person is not rich.

**6.** If interest rates are not low, then the economy is not expanding.

  let $p$ = The economy is expanding.

  $q$ = Interest rates are low.

  $\sim q \rightarrow \sim p$

**8.** Low interest are sufficient for an expanding economy.

  let $p$ = The economy is expanding.

  $q$ = Interest rates are low.

  $q \rightarrow p$

**10.** The economy will expand only if interest rates are low.

  let $p$ = The economy is expanding.

  $q$ = Interest rates are low.

  $p \rightarrow q$

**12.** If 4 is even, then 2 is odd.

| Hypothesis | Conclusion | Implication |
|------------|------------|-------------|
| T | F | F |

**14.** If 4 is odd, then 2 is odd.

| Hypothesis | Conclusion | Implication |
|---|---|---|
| F | F | T |

**16.** If $2n + 1$ is even, then $2n + 2$ is odd.

| Hypothesis | Conclusion | Implication |
|---|---|---|
| F | F | T |

**18.** $\frac{1}{6} < \frac{2}{3}$ is necessary for $\frac{1}{2} > 0$.

| Hypothesis | Conclusion | Implication |
|---|---|---|
| T | T | T |

**20.** If $2x = 224$, then $x = 10$.

| Hypothesis | Conclusion | Implication |
|---|---|---|
| T | F | F |

**22.** If the person is near sighted, then he is ineligible for the job.

Converse:

If he is ineligible for the job, then the person is nearsighted.

Inverse:

If the person is not nearsighted, then he is eligible for the job.

Contrapositive:

If he is eligible for the job, then the person is not nearsighted.

**24.** If wages are raised, then the company's profits will decrease.

Converse:

If the company's profits will decrease, then wages are raised.

Inverse:

If wages are not raised, then the company's profits will not decrease.

Contrapositive:

If the company's profit will not decrease, then wages are not raised.

**26.** The number is divisible by 3 only if the sum of its digitis is divisible by 3.

Converse:

The sum of its digits is divisible by 3 only if the number is divisible by 3.

Inverse:

The number is not divisible by 3 only if the sum of its digits is not divisible by 3.

Contrapositive:

The sum of its digits is not divisible by 3 only if the number is not divisible by 3.

**28.** Jack is a senior and he plays varsity basketball.

Negation:  Jack is not a senior or he does not play varsity basketball.

**30.** If the test fails, then the project will be halted.

Negation:  The test will fail and the project will not be halted.

**32.** Completing the pass on this play is necessary if we are going to win the game.

Negation:  We are going to win the game and not complete the pass on this play.

**34.** Some odd integers are not prime numbers.

Negation:  No odd integer is not a prime number.

**36.** All members must pay their dues prior to June 1.

Negation:  Not all members must pay their dues prior to June 1.

**38.** No contestant is over the age of 12.

Negation:  Some contestants are over the age of 12.

**40.** At least one unit is defective.

Negation:  No units are defective.

**42.**

| $p$ | $q$ | $\sim q$ | $p \rightarrow q$ | $\sim q \rightarrow (p \rightarrow q)$ |
|---|---|---|---|---|
| T | T | F | T | T |
| T | F | T | F | F |
| F | T | F | T | T |
| F | F | T | T | T |

**44.**

| $p$ | $q$ | $\sim p$ | $\sim p \vee q$ | $p \rightarrow (\sim p \vee q)$ |
|---|---|---|---|---|
| T | T | F | T | T |
| T | F | F | F | F |
| F | T | T | T | T |
| F | F | T | T | T |

**46.**

| $p$ | $q$ | $p \rightarrow q$ | $\sim q$ | $[(p \rightarrow q) \wedge (\sim q)]$ | $[(p \rightarrow q) \wedge (\sim q)] \rightarrow p$ |
|---|---|---|---|---|---|
| T | T | T | F | F | T |
| T | F | F | T | F | T |
| F | T | T | F | F | T |
| F | F | T | T | T | F |

**48.**

| $p$ | $q$ | $\sim q$ | $(p \vee \sim q)$ | $\sim p$ | $(q \rightarrow \sim p)$ | $(p \vee \sim q) \rightarrow (q \rightarrow \sim p)$ |
|---|---|---|---|---|---|---|
| T | T | F | T | F | F | F |
| T | F | T | T | F | T | T |
| F | T | F | F | T | T | T |
| F | F | T | T | T | T | T |

| $(q \rightarrow \sim p) \rightarrow (p \vee \sim q)$ | $(p \vee \sim q) \leftrightarrow (q \rightarrow \sim p)$ |
|---|---|
| T | F |
| T | T |
| F | F |
| T | T |

**50.**

| $p$ | $q$ | $\sim p$ | $\sim p \rightarrow q$ | $p \vee q$ |
|---|---|---|---|---|
| T | T | F | T | T |
| T | F | F | T | T |
| F | T | T | T | T |
| F | F | T | F | F |

identical

**52.**

| $p$ | $q$ | $p \vee q$ | $(p \vee q) \rightarrow q$ | $p \rightarrow q$ |
|---|---|---|---|---|
| T | T | T | T | T |
| T | F | T | F | F |
| F | T | T | T | T |
| F | F | F | T | T |

identical

**54.**

| $p$ | $q$ | $\sim p$ | $(\sim p \vee q)$ | $q \rightarrow (\sim p \vee q)$ | $\sim q$ | $q \vee \sim q$ |
|---|---|---|---|---|---|---|
| T | T | F | T | T | F | T |
| T | F | F | F | T | T | T |
| F | T | T | T | T | F | T |
| F | F | T | T | T | T | T |

identical

**56.**

| $p$ | $q$ | $(p \wedge q)$ | $\sim(p \wedge q)$ | $\sim q$ | $\sim(p \wedge q) \rightarrow \sim q$ | $p \vee \sim q$ |
|---|---|---|---|---|---|---|
| T | T | T | F | F | T | T |
| T | F | F | T | T | T | T |
| F | T | F | T | F | F | F |
| F | F | F | T | T | T | T |

identical

**58.** It is not truth that Pam is a Conservative and a Democrat.

Let $p$ = Pam is a conservative     $q$ = she is a Democrat

Statement is $\sim(p \wedge q)$

| $p$ | $q$ | $(p \wedge q)$ | $\sim(p \wedge q)$ | $\sim p \vee \sim q$ |
|---|---|---|---|---|
| T | T | T | F | F |
| T | F | F | T | T |
| F | T | F | T | T |
| F | F | F | T | T |

Statement (c) is logically equivalent.

**60.** It is necessary to pay the registration fee to take the course.

(c) is not logically equivalent.

**62.** All happy people are college students.

Let $A$ = people who are happy

  $B$ = college students

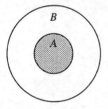

**64.** No happy people are college students.

Let $A$ = people who are happy

  $B$ = college students

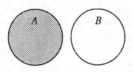

**66.** Some happy people are not college students.

Let $A$ = people who are happy

  $B$ = college students

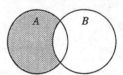

**68.** At least one happy person is not a college students.

Let $A$ = people who are happy

  $B$ = college students

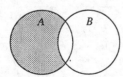

**70.** Each sad person is not a college students.

Let $A$ = people who are happy

  $B$ = college students

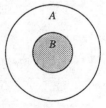

**72.** (a) All men are company presidents.

  Statement follows.

  (b) Some company presidents are women.

  Statement follows.

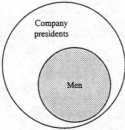

**74.** (a) No football players are more than 6 feet tall.

  Statement does not follow.

  (b) Every football player is more than 6 feet tall.

  Statement does not follow.

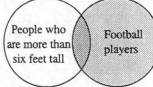

## Appendix D.3    Logical Arguments

**2.** Premise #1: $p \leftrightarrow q$

Premise #2: $p$

Conclusion: $q$

| $p$ | $q$ | $p \leftrightarrow q$ | $(p \leftrightarrow q) \wedge p$ | $[(p \leftrightarrow q) \wedge p] \rightarrow q$ |
|---|---|---|---|---|
| T | T | T | T | T |
| T | F | F | F | T |
| F | T | F | F | T |
| F | F | T | F | T |

**4.** Premise #1: $p \wedge q$

Premise #2: $\sim p$

Conclusion: $q$

| $p$ | $q$ | $\sim p$ | $p \wedge q$ | $(p \wedge q) \wedge \sim p$ | $[(p \wedge q) \wedge \sim p] \rightarrow q$ |
|---|---|---|---|---|---|
| T | T | F | T | F | T |
| T | F | F | F | F | T |
| F | T | T | F | F | T |
| F | F | T | F | F | T |

**6.** Premise #1: $p \rightarrow q$

Premise #2: $\sim p$

Conclusion: $\sim q$

| $p$ | $q$ | $\sim p$ | $\sim q$ | $p \rightarrow q$ | $(p \rightarrow q) \wedge \sim p$ | $[(p \rightarrow q) \wedge \sim p] \rightarrow \sim q$ |
|---|---|---|---|---|---|---|
| T | T | F | F | T | F | T |
| T | F | F | T | F | F | T |
| F | T | T | F | T | T | F |
| F | F | T | T | T | T | T |

**8.** Premise #1: $\sim(p \wedge q)$

Premise #2: $q$

Conclusion: $p$

| $p$ | $q$ | $p \wedge q$ | $\sim(p \wedge q)$ | $\sim(p \wedge q) \wedge q$ | $[\sim(p \wedge q) \wedge q] \rightarrow p$ |
|---|---|---|---|---|---|
| T | T | T | F | F | T |
| T | F | F | T | F | T |
| F | T | F | T | T | F |
| F | F | F | T | F | T |

**10.** Premise #1: If a student does the homework, then a good grade is certain.

Premise #2: Liza does the homework.

Conclusion: So, Liza will receive a good grade for the course.

Let $p$ = student does homework    $q$ = a good grade is certain

Premise #1: $p \rightarrow q$

Premise #2: $p$

Conclusion: $q$

| $p$ | $q$ | $p \rightarrow q$ | $(p \rightarrow q) \wedge p$ | $(p \rightarrow q) \wedge p \rightarrow q$ |
|---|---|---|---|---|
| T | T | T | T | T |
| T | F | F | F | T |
| F | T | T | F | T |
| F | F | T | F | T |

Arguement is valid.

**12.** Premise #1:  If a student does the homework, then a good grade is certain.

Premise #2:  Liza received a good grade for the course.

Conclusion:  Liza did her homework.

Let $p$ = student does homework    $q$ = a good grade is certain

Premise #1:  $p \rightarrow q$

Premise #2:  $q$

Conclusion:  $p$

| $p$ | $q$ | $p \rightarrow q$ | $(p \rightarrow q) \wedge q$ | $(p \rightarrow q) \wedge q \rightarrow p$ |
|---|---|---|---|---|
| T | T | T | T | T |
| T | F | F | F | T |
| F | T | T | T | F |
| F | F | T | F | T |

Argument is invalid.

**14.** Premise #1:  If Jan passes the exam, she is eligible for the position.

Premise #2:  Jan is not eligible for the position.

Conclusion:  So, Jan did not pass the exam.

Let $p$ = Jan passes the exam    $q$ = she is eligible for the position

Premise #1:  $p \rightarrow q$

Premise #2:  $\sim q$

Conclusion:  $\sim p$

| $p$ | $q$ | $p \rightarrow q$ | $\sim p$ | $\sim q$ | $(p \rightarrow q) \wedge \sim q$ | $(p \rightarrow q) \wedge \sim q \rightarrow \sim p$ |
|---|---|---|---|---|---|---|
| T | T | T | F | F | F | T |
| T | F | F | F | T | F | T |
| F | T | T | T | F | F | T |
| F | F | T | T | T | T | T |

Argument is valid.

**16.** Premise #1:  Some cars manufactured by the Ford Motor Company reliable.

Premise #2:  Lincolns are manufactured by Ford.

Conclusion:  So, Lincolns are reliable.

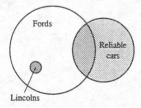

Argument is invalid.

**18.** Premise #1:  All integers divisible by 6, are divisible by 3.

Premise #2:  Eighteen is divisible by 6.

Conclusion:  So, eighteen is divisible by 3.

Let $p$ = all integers divisible by 6    $q$ = integers divisible by 3

Premise #1: $p \rightarrow q$

Premise #2: $p$

Premise #3: $q$

| $p$ | $q$ | $p \rightarrow q$ | $(p \rightarrow q) \wedge p$ | $(p \rightarrow q) \wedge p \rightarrow q$ |
|---|---|---|---|---|
| T | T | T | T | T |
| T | F | F | F | T |
| F | T | T | F | T |
| F | F | T | F | T |

Argument is valid.

**20.** Premise #1:  The book must be returned within 2 weeks or you pay a fine.

Premise #2:  The book was not returned within two weeks.

Conclusion:  So, you must pay a fine.

Let $p$ = book returned within 2 weeks    $q$ = you pay a fine

Premise #1: $p \vee q$

Premise #2: $\sim p$

Premise #3: $q$

| $p$ | $q$ | $\sim p$ | $p \vee q$ | $(p \vee q) \wedge \sim p$ | $(p \vee q) \wedge \sim p \rightarrow q$ |
|---|---|---|---|---|---|
| T | T | F | T | F | T |
| T | F | F | T | F | T |
| F | T | T | T | T | T |
| F | F | T | F | F | T |

Argument is valid.

**22.** Premise #1:  Either I work tonight or I pass the mathematics test.

Premise #2:  I'm going to work tonight.

Conclusion:  So, I will fail the mathematics test.

Let $p$ = I work tonight    $q$ = I pass mathematics test

Premise #1: $p \vee q$

Premise #2: $p$

Premise #3: $\sim q$

| $p$ | $q$ | $\sim p$ | $p \vee q$ | $(p \vee q) \wedge p$ | $(p \vee q) \wedge p \rightarrow \sim q$ |
|---|---|---|---|---|---|
| T | T | F | T | T | F |
| T | F | F | T | T | T |
| F | T | T | T | F | T |
| F | F | T | F | F | T |

Argument is invalid.

**24.** Premise #1: If the fuel is shut off, then the fire will be extinguished.

Premise #2: The fire continues to burn.

Let $p$ = fuel is shut off    $q$ = fire is extinguished

Premise #1: $p \rightarrow q$

Premise #2: $\sim q$

So conclusion must be $\sim p$ which is (a).

**26.** Premise #1: It will snow only if the temperature is below 32° at some level of the atmosphere.

Premise #2: It is snowing.

Let $p$ = it will snow    $q$ = temperature is below 32° at some level of the atmosphere

Premise #1: $p \rightarrow q$

Premise #2: $p$

So conclusion must be $q$ which is (c).

**28.** Premise #1: The library must upgrade its computer system or service will not improve.

Premise #2: Service at the library has improved.

Let $p$ = the library must upgrade its computer system

$q$ = service will not improve

Premise #1: $p \vee q$

Premise #2: $\sim q$

Conclusion must be $q$ which is (a).

**30.** Premise #1: It is necessary to have a ticket and an ID card to get into the arena.

Premise #2: Janice entered the arena.

Let $p$ = Janice entered the arena

$q$ = Janice has an ID card and a ticket

Premise #1: $p \rightarrow q$

Premise #2: $p$

Conclusion must be $q$ which is (b).

**32.** Premise #1: All human beings require adequate rest.

Premise #2: All infants are human beings.

Conclusion: So, all infants require adequate rest.

Valid

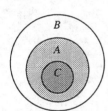

A: All human beings

B: Things that require adequate rest

C: All infants

**34.** Premise #1: Every amateur radio operator has a radio license.

Premise #2: Jackie has a radio license.

Conclusion: So, Jackie is an amateur radio operator.

Invalid

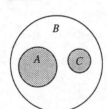

A: Every amateur operator

B: People who have a radio license

C: Jackie

**36.** Premise #1: If Bill is patient, then he will succeed.

Premise #2: Bill will get bonus pay if he succeeds.

Premise #3: Bill did not get bonus pay.

Conclusion: So, Bill is not patient.

Let $p$ represent "Bill is patient." let $q$ represent "He will succeed," and let $r$ represent "Bill will get bonus pay."

First write:

Premise #1: $p \rightarrow q$

Premise #2: $q \rightarrow r$

Premise #3: $\sim r$

Conclusion from Premise #1, Premise #2: $p \rightarrow r$

Conclusion from $p \rightarrow r$, Premise #3: $\sim p$

That is, "Bill is not patient."

**38.** Premise #1:  If it is raining today, Pam will clean her apartment.

Premise #2:  If Pam is cleaning her apartment today, then she is not riding her bike.

Premise #3:  Pam is riding her bike today.

Conclusion:  It is not raining today.

Let $p$ represent "It is raining today," let $q$ represent "Pam is cleaning her apartment," and let $r$ represent "Pam is riding her bike today."

First write:

Premise #1:  $p \rightarrow \sim q$

Premise #2:  $q \rightarrow \sim r$

Premise #3:  $r$

Conclusion from Premise #1, Premise #2:  $p \rightarrow \sim r$

Conclusion from $p \rightarrow \sim r$, Premise #3:  $\sim p$

That is, "It is not raining today."